Engineering Mechanics

Second Vector Edition

SECOND
VECTOR
EDITION

Engineering
Mechanics

VOLUME II: DYNAMICS

ARCHIE HIGDON

Engineering and Technology Education Consultant

WILLIAM B. STILES

Professor of Engineering Mechanics
University of Alabama

ARTHUR W. DAVIS

Professor of Engineering Mechanics
Iowa State University

CHARLES R. EVCES

Associate Professor of Mechanical Engineering
University of Alabama

PRENTICE-HALL, INC., ENGLEWOOD CLIFFS, NEW JERSEY

Library of Congress Cataloging in Publication Data

Main entry under title:

Engineering mechanics.

    CONTENTS: v. 1. Statics. —v. 2. Dynamics.
    1.–Mechanics, Applied. I.–Higdon, Archie–(date) II.–Higdon, Archie–(date) Engineering
mechanics.
TA350.H512–1976   620.1   75–42153
ISBN 0-13-279406-3 (v. 2)

Printed in the United States of America

10 9 8 7 6 5 4 3 2

PRENTICE-HALL INTERNATIONAL, INC., *London*
PRENTICE-HALL OF AUSTRALIA PTY. LIMITED, *Sydney*
PRENTICE-HALL OF CANADA, LTD., *Toronto*
PRENTICE-HALL OF INDIA PRIVATE LIMITED, *New Delhi*
PRENTICE-HALL OF JAPAN, INC., *Tokyo*
PRENTICE-HALL OF SOUTH-EAST ASIA PRIVATE LIMITED, *Singapore*

Contents

8 **KINEMATICS**

9 **KINETICS—**
    FORCE, MASS,
    AND ACCELERATION

10 **KINETICS—**
     WORK AND
     ENERGY

Preface

We have prepared this second vector edition of *Engineering Mechanics* to reflect current and anticipated trends in introductory mechanics courses. Topics included in this revision are essentially the same as in the earlier editions, although many topics have been rearranged and expanded to facilitate adaptation to the numerous methods in which mechanics is integrated into modern engineering curricula. In particular, the material in dynamics is arranged to permit coverage of separate or combined courses in the dynamics of particles and of rigid bodies.

Use of the International System of Units (SI) is becoming more and more common in the United States, particularly among firms operating in several countries. Since it does not appear that there will be a complete and sudden changeover to SI units, between 20 and 25 per cent of the numerical examples and problems in this revision involve metric units. An asterisk by the problem number indicates a problem which uses SI units. When SI units are used they conform essentially to the recommendations of the American National Standards Institute.

The vector method is particularly useful in three-dimensional applications of mechanics, and in all cases it provides a systematic procedure for problem solutions. Physical concepts and visualization of forces and moments are first emphasized in the text without using vector algebra, however, to help develop a "feeling" for these basic concepts. Later it is shown that vector algebra can be used to obtain the same results, often with a considerable reduction in the mathematical manipulation required. Both scalar and vector solutions are presented for some examples to encourage the reader to compare procedures and to check his understanding of the subject matter. Vector methods frequently do not simplify the solution of two-dimensional problems, but such applications are essential to provide practice in using vector notation for the more involved three-dimensional problems.

The principles of mechanics are few and relatively simple; the applications, however, are infinite in their number and variety and frequently appear to be forbiddingly complicated. Since much of engineering is "applied mechanics" (just as mechanics is applied physics and applied mathematics), it is highly desirable that the engineering student grasp mechanics, not as a series of formulas and manipulative operations, but rather as a well-comprehended mechanism for stripping problems of their extraneous confusing aspects. This outlook will enable the student to break problems down into relatively simple, easily analyzed elements.

Emphasis is placed on an understanding of principles employed in the solution of problems rather than reliance on a rote process of substitution in numerous formulas. We believe that every teacher of engineering mechanics must be constantly aware of mechanics as a prelude to engineering design, in which innumerable complicating empirical considerations are superimposed on the relatively simple situations encountered in mechanics. Unless students are well grounded in the principles and well drilled in logical, orderly, step-by-step methods of analysis and procedure, they are likely to experience considerable difficulty as they try to include the design elements of judgment, approximation, and compromise in their solutions.

In an attempt to keep student attention focused on mechanics as a prelude to design, we see no justification for realism carried to the point of introducing time-consuming complexities merely because they are inherent in the problems of engineering. Thus we adhere to the common practice of freely using values for masses, weights, dimensions, forces, and angles that simplify the numerical work involved, in the belief that the time saved can be better employed in solving more problems and gaining a more complete mastery of the principles and their applications. Either a slide rule or a calculator can be used for the solution of problems, but it should be noted that, in general, the data available for engineering problems do not warrant results carried to more than three or four significant figures. In general, problem data are assumed to contain three significant figures regardless of the number of figures shown. Answers, therefore, are given to three significant figures unless the first significant digit is 1, in which case four figures are reported. We have endeavored to maintain a sense of engineering reasonableness by avoiding such absurdities as a wheel having a radius of gyration in excess of its radius, accelerations requiring abnormal coefficients of friction, and weights of bodies that are, for materials normally used, inconsistent with the volume or given dimensions.

Chapters 8 through 11 on kinematics and kinetics are arranged with A and B parts covering the dynamics of particles and of rigid bodies, respectively. Thus a course covering only particle dynamics would use the A part of each of the four chapters. An integrated course on dynamics can be taught by including both parts of each chapter as presented or by studying the A part of each chapter first followed by the B parts.

The selection of specific topics to be covered will vary from school to school. This revision includes those topics common to most courses and a number of additional topics which cannot be included in some instances due to time limitations. The inclusion of such topics in the book provides an opportunity to tailor the course to the needs of each school by the judicious selection of the topics to be covered. The number of problems is intentionally large, not with the idea that problem assignments should be burdensome, but rather to offer selection and afford variation in the problem assignments from term to term.

The closure article at the end of a chapter is intended as a brief review of the important topics covered and to emphasize the relationships of some of the topics after they have been covered individually. Review problems are included with some closures to provide an opportunity to develop judgment in selecting the appropriate methods or theories for solving selected problems.

For numerous suggestions, we are indebted to many users of the earlier editions of *Engineering Mechanics*, including colleagues at Iowa State University, the University of Alabama, the United States Air Force Academy, and the University of Denver. We want to express our particular appreciation to our wives for their patience, their encouragement, and their invaluable help in proofreading, and to Mrs. Afton Earnhart for her help in typing the manuscript.

A. HIGDON
W. B. STILES
A. W. DAVIS
C. R. EVCES

Symbols and Abbreviations*

$\mathbf{a}, \mathbf{a}_G$	linear acceleration, linear acceleration of mass center
A	area, amplitude
Art.	article
avg	average
b	breadth, width
$\mathbf{C}$	couple
C	compression, centroid of area
$C_1, C_2, \ldots$	constants of integration
cu	cubic
d	differential of (as dx), distance
D	diameter
e	coefficient of restitution
$\mathbf{e}_n$	unit vector in normal direction
$\mathbf{e}_t$	unit vector in tangential direction
$\mathbf{e}_r$	unit vector in radial direction
$\mathbf{e}_\theta$	unit vector in transverse direction

---

* The symbols and abbreviations used in this text conform essentially with those approved by the American National Standards Institute.

Eq.	equation
f	frequency, sag of a cable
$\mathbf{F}, \mathbf{F}_1, \mathbf{F}_A, \ldots$	forces or loads
$\mathbf{F}_x, \mathbf{F}_y, \mathbf{F}_n, \ldots$	components of force $\mathbf{F}$
Fig.	figure
fpm	feet per minute
fps	feet per second
fps^2	feet per second per second
ft	foot
$\mathbf{g}$	acceleration of gravity
$\mathbf{G}$	linear momentum
G	center of gravity, mass center
$\mathbf{H}$	angular momentum
h	height, length, depth
hp	horsepower
i	initial when used as subscript, as v_i
$\mathbf{i}, \mathbf{j}, \mathbf{k}$	unit vectors in x, y, z directions, respectively.
$\mathbf{i}', \mathbf{j}', \mathbf{k}'$	unit vectors in X, Y, Z directions, respectively.
$\mathbf{I}_L$	linear impulse
I	moment of inertia of area or mass
I_x, I_0	moment of inertia of area, or mass with respect to axis indicated by subscript
I_{xy}, I_{yz}	product of inertia of area or mass with respect to planes perpendicular to axes shown as subscripts
in.	inch
ips	inches per second
ips^2	inches per second per second
J_0, J_z	polar moment of inertia of area with respect to axis indicated by subscript
J	joule, newton · meter
k	spring constant (load per unit deflection), kip (1000 lb)
$k_x, k_y, \ldots$	radius of gyration with respect to axis indicated by subscript
K	constant
kg	kilogram
kip	kilopound (1000 lb)
l, L	length
lb	pound
ln	logarithm (natural)
log	logarithm (common)
m	mass, W/g, meter
$\mathbf{M}_0, \mathbf{M}_x$	moment of a force with respect to point or line shown as subscript
M_A, M_y	moment of an area or mass with respect to a point or line shown as subscript

m/s	meter per second
m/s^2	meter per second per second
max	maximum
min	minimum
mph	miles per hour
mphps	miles per hour per second
n	normal direction (used only as subscript)
n	unit vector in general direction
N	normal force
N	newton
O	origin of coordinates, instantaneous center of zero velocity
oz	ounce
p	pitch of threads, unit pressure, load per unit area
P	force
P	power
psf	pounds per square foot
psi	pounds per square inch
q	linear displacement
q	load per unit distance, distance from axis of rotation to center of percussion
Q	force
Q	total distance traveled during a time interval
r	position vector
r	radius, polar coordinate
R	resultant force, reaction, position vector
R	radial direction, radius
rad	radian
rpm	revolutions per minute
rps	revolutions per second
s	function indicating position of a particle measured along a path from a fixed point on the path, arc length, length of cable, second
sec	second
sq	square
t	time, thickness, tangential direction
T	torque (moment of a force or couple), tensile force
T	tension, period, transverse direction, kinetic energy
u, v	rectangular coordinates
U	work
v, **v**$_G$	linear velocity, linear velocity of mass center
V	volume, potential function
W	total load, weight
W	watt, newton · meter per second
x, y, z	rectangular coordinates, moving or fixed
X, Y, Z	rectangular coordinates, fixed

x_c, y_c, z_c	rectangular coordinates of centroid
x_G, y_G, z_G	rectangular coordinates of mass center
$\dot{x}, \dot{y}, \dot{z}$	time derivatives of x, y, z
$\ddot{x}, \ddot{y}, \ddot{z}$	second derivatives of x, y, z with respect to time
y_{st}	static deflection
$\boldsymbol{\alpha}$	angular acceleration
α	angle
β	angle of contact for belt friction, angle
γ	specific weight (weight per unit volume)
δ	logarithmic decrement
$\delta x, \delta\theta, \ldots$	virtual displacement (linear or angular)
η	efficiency
θ	function indicating angular position of a line measured from a fixed reference line, a polar coordinate, angle
μ, μ_s, μ_k	coefficient of friction, static or kinetic coefficients
ρ	density (mass per unit volume)
ϕ	phase angle, angular displacement during a time interval
Φ	total angle turned through during a time interval
$\boldsymbol{\omega}$	angular velocity
ω	circular frequency of forced vibration
ω_d	circular frequency of damped vibration
ω_n	circular frequency of free vibration
⟶⟶	vector representing the resultant force
⟹	moment vector

SENSE (OF ROTATION)

counterclockwise looking downward

counterclockwise looking inward

counterclockwise looking to the left

Kinematics

8

PART A—KINEMATICS OF PARTICLES

8-1 INTRODUCTION

Kinematics is that branch of mechanics which deals with the motion of particles, lines, and bodies without consideration of the forces required to produce or maintain the motion. A working knowledge of the relationships between position, time, velocity, acceleration, displacement, and distance traveled for particles, lines, and bodies is essential to the study of the effects of unbalanced force systems on bodies. Kinetics is concerned with the force systems that produce accelerated motion of bodies, the inertial properties of the bodies, and the resulting motion of the bodies.

Kinematic quantities, such as position and velocity, are expressed with respect to a system of reference axes. When axes fixed on the earth are used as a reference, the motion is called *absolute motion*, as distinguished from relative motion, which is measured with respect to moving axes. Strictly speaking, absolute motion is motion measured relative to a coordinate system fixed in space, a Newtonian reference frame, and since the earth is moving, motion measured with respect to axes fixed on the earth is actually relative motion. For elementary mechanics, however, the motion of the earth can be neglected, and in this text *motion measured relative to a coordinate system fixed on the earth is defined as absolute motion.*

The term *particle* is used for any body whose size can be neglected without introducing appreciable errors when studying or describing its motion. In planetary motion, the earth is considered as a particle, whereas the balance wheel of a watch must be treated as a body in an analysis of the motion of the watch mechanism.

The word *linear,* as in linear velocity or linear displacement, is used to describe the motion of particles or points. Angular motion is restricted to lines and bodies because two lines are necessary to define an angle.

Particles are said to have *rectilinear motion* when they move along straight lines and *curvilinear motion* when they travel on curved paths. *Uniform motion of a particle is defined as a motion such that equal distances are traversed during equal intervals of time, regardless of how small the time intervals may be.*

8-2 REVIEW OF VECTOR CALCULUS

Vector calculus is useful in kinematics because many of the quantities studied are vectors. In scalar calculus, only changes in the magnitude of a variable are involved, whereas in vector calculus, changes in both magnitude and direction of variables must be considered.

The following material is a brief review of some aspects of vector calculus. A more complete discussion can be obtained from most calculus texts or from books on vector analysis.

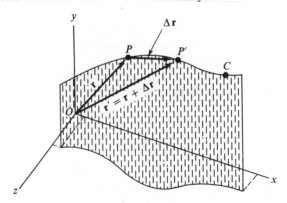

Figure 8-1

Figure 8-1 represents a particle moving along a curved path. Point
O is the origin of the fixed coordinate system, xyz. The position vector $\mathbf{r}$ from O to P, expressed as a function of time, locates the particle at any instant. As the particle moves from point P to point P' the change of $\mathbf{r}$ is

$$\Delta\mathbf{r} = \mathbf{r}' - \mathbf{r}.$$

If $\Delta\mathbf{r}$ occurs during a time interval Δt, the ratio $\Delta\mathbf{r}/\Delta t$ is the average time rate of change of position, and the limit as Δt approaches zero is defined as the derivative of the position vector $\mathbf{r}$ with respect to the time t. Thus

$$\frac{d\mathbf{r}}{dt} = \dot{\mathbf{r}} = \lim_{\Delta t \to 0} \frac{\Delta\mathbf{r}}{\Delta t}. \quad {}^{1}$$

Figure 8-1 shows that $\Delta\mathbf{r}$ is a vector and indicates the change of both magnitude and direction of $\mathbf{r}$. The vector $\mathbf{r}$ is not limited to position vectors but might be any vector which varies with t, and t can be any scalar variable.

To evaluate the derivative of a vector, changes of both magnitude and direction must be considered. If the vector is expressed in terms of unit vectors in fixed directions, the unit vectors can be treated as constants during the differentiation. Consider the function

$$\mathbf{r} = x\mathbf{i} + y\mathbf{j} + z\mathbf{k},$$

where x, y, and z are scalar functions of time and $\mathbf{i}$, $\mathbf{j}$, and $\mathbf{k}$ are unit vectors in the fixed x, y, and z directions. Each term is a product of a variable and a constant, and the derivative with respect to time is

$$\frac{d\mathbf{r}}{dt} = \dot{\mathbf{r}} = \dot{x}\mathbf{i} + \dot{y}\mathbf{j} + \dot{z}\mathbf{k}.$$

If the unit vector or vectors used to express $\mathbf{r}$ change in direction, the differentiation procedure must be modified to account for this change, as explained in Art. 8-8.

The derivative of the sum of two vectors is equal to the sum of the derivatives since vectors obey the distributive law. Thus if $\mathbf{A} = \mathbf{B} + \mathbf{C}$,

$$\frac{d}{dt}\mathbf{A} = \frac{d}{dt}(\mathbf{B} + \mathbf{C}) = \frac{d\mathbf{B}}{dt} + \frac{d\mathbf{C}}{dt}.$$

The following relationships can also be demonstrated by writing the derivative as a limit:

$$\frac{d}{dt}(\mathbf{A} \cdot \mathbf{B}) = \mathbf{A} \cdot \frac{d\mathbf{B}}{dt} + \frac{d\mathbf{A}}{dt} \cdot \mathbf{B} = \mathbf{A} \cdot \dot{\mathbf{B}} + \dot{\mathbf{A}} \cdot \mathbf{B},$$

---

[1] A single dot above a variable indicates the derivative of the variable with respect to time, and two dots, $\ddot{r}$, indicate the second derivative. The dot notation is used only for differentiation with respect to time and was originally adopted by Newton.

and

$$\frac{d}{dt}(\mathbf{A}\times\mathbf{B})=\mathbf{A}\times\frac{d\mathbf{B}}{dt}+\frac{d\mathbf{A}}{dt}\times\mathbf{B}=\mathbf{A}\times\dot{\mathbf{B}}+\dot{\mathbf{A}}\times\mathbf{B}.$$

It is essential that the order of the factors in each term be maintained when differentiating a vector product.

It was shown that the derivative of a vector with respect to a scalar, such as time, is a vector. Sometimes it is desirable to reverse the process, that is, to integrate a vector. Thus, if

$$\mathbf{A}=\frac{d\mathbf{B}}{dt},$$

the inverse expression would be

$$\mathbf{B}=\int\mathbf{A}\,dt+\mathbf{C},$$

where $\mathbf{C}$ is a vector constant of integration, which is constant in both magnitude and direction. One of the most useful methods of carrying out the integration is by expressing the integrand in terms of constant unit vectors, since such constants are not changed by either integration or differentiation. Thus if

$$\mathbf{A}=A_x\mathbf{i}+A_y\mathbf{j}+A_z\mathbf{k},$$

the integral becomes

$$\mathbf{B}=\int\mathbf{A}\,dt=\int(A_x\mathbf{i}+A_y\mathbf{j}+A_z\mathbf{k})\,dt+\mathbf{C}$$

$$=\mathbf{i}\int A_x\,dt+\mathbf{j}\int A_y\,dt+\mathbf{k}\int A_z\,dt+\mathbf{C}.$$

Additional concepts involving the calculus of vectors will be introduced as they are needed.

8-3 BASIC DEFINITIONS

The *instantaneous position* of a particle moving along some path can be specified by the position vector $\mathbf{r}$ from the origin of a fixed coordinate system to the particle. The position of the particle can also be specified by stating the coordinates of the particle or by giving the distance s, measured along the given path, from a fixed point on the path to the particle. It is assumed that all of these quantities are expressed as functions of time.

A consistent and clearly defined sign convention is essential if the algebraic solution is to be given a correct physical interpretation. Positive directions are those of the coordinate axes when given. If axes are not specified, positive directions that are consistent with given data should be selected and shown with problem solutions.

The linear displacement, **q**, of a particle during an interval of time is defined as the change of position of the particle during the time interval. Thus if the particle moves from position P to position P' of Fig. 8-1 during a certain time interval, its displacement is

$$\mathbf{q} = \Delta\mathbf{r} = \mathbf{r}' - \mathbf{r} \qquad (8\text{-}1)$$

and is the vector from P to P'. If the particle travels from P to C and then back to P', its displacement, **q**, is still the vector $\Delta\mathbf{r}$ from P to P' and depends only on the initial and final positions of the particle.

The total distance traveled, represented by Q, is the total accumulated length of path traversed and thus depends on the shape of the path and whether or not the direction of travel is reversed as well as on the initial and final positions of the particle. From this definition it is seen that the total distance traveled will always be equal to or greater than q, the magnitude of the displacement. Only when the path is a straight line and the motion does not reverse will Q be equal to q.

The displacement **q** is a vector quantity and is completely specified only when both the magnitude and direction are given. The total distance Q is a scalar quantity having only magnitude. Linear displacement and total distance traveled represent changes which take place during a time interval, whereas the position, **r**, is a continuous vector function of time with a definite value for each instant of time. All three quantities have fundamental dimensions of length and common units are feet, inches, meters, and similar units.

The linear velocity **v** of a particle is defined as the time rate of change of the position of the particle. If the particle in Fig. 8-1 moves from position P to position P' in time Δt, the average time rate of change of position, the average velocity, is $\Delta\mathbf{r}/\Delta t$, and the instantaneous velocity at position P is

$$\mathbf{v} = \lim_{\Delta t \to 0} \frac{\Delta\mathbf{r}}{\Delta t} = \frac{d\mathbf{r}}{dt} = \dot{\mathbf{r}}. \qquad (8\text{-}2)$$

As the time interval is decreased point P' will approach point P and the displacement $\Delta\mathbf{r}$ will approach the tangent to the curve at P. Therefore the velocity of a point is a vector and is always tangent to the path along which the point moves. The magnitude of the velocity, v, is frequently called the speed. Equation (8-2) indicates that linear velocity has dimensions of length divided by time, LT^{-1}, and common units are feet per second (fps), meters per second (m/s), knots (one nautical mile, 6080 ft, per hour), and similar units.

The linear acceleration **a** of a particle is defined as the time rate of change of the linear velocity of the particle. At point P of Fig. 8-2 the velocity is **v**, and at point P' it is $\mathbf{v} + \Delta\mathbf{v}$. The average time rate of change of linear velocity is $\Delta\mathbf{v}/\Delta t$, and the instantaneous acceleration at position P is

$$\mathbf{a} = \lim_{\Delta t \to 0} \frac{\Delta\mathbf{v}}{\Delta t} = \frac{d\mathbf{v}}{dt} = \frac{d^2\mathbf{r}}{dt^2} = \ddot{\mathbf{r}}. \qquad (8\text{-}3)$$

341

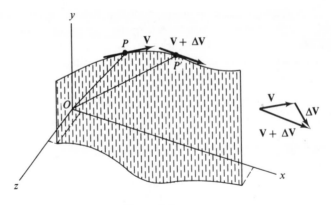

Figure 8-2

Linear acceleration is a vector quantity in the direction of the change of velocity. It should be noted that, although velocity is tangent to the path, the velocity may change in both magnitude and direction, and, in general, the acceleration is not tangent to the path. Equation (8-3) shows that the dimensions of linear acceleration are length divided by time squared, LT^{-2}. Common units are feet per second per second (fps^2), meters per second per second (m/s^2), and similar units. Accelerations are sometimes expressed in gs, where g is the acceleration of gravity (approximately 32.2 fps^2 or 9.81 m/s^2) at the surface of the earth. Thus an acceleration of 5g would be 161 fps^2 or 49.05 m/s^2.

A line has *angular motion* when the angle it makes with a fixed reference line changes. The line may turn about a fixed point on the line or line extended, as in the case of the hand of a clock, or it may be that no point on the line remains fixed, as in the case of a spoke of a rolling wheel. Particles are considered dimensionless, and any angular motion they might have cannot be measured or described; therefore, *angular motion is a property which is restricted to lines and bodies.*

In the following discussion, it is assumed that the line AB in Fig. 8-3 moves only in the xy plane. Noncoplanar motion of a line is discussed in Part B of this chapter. The angular position of line AB at any instant is defined by the angle θ *from* the x axis *to* the line. The angular position function θ, as used here, is a scalar function of time. The net change of angular position of a line during a given time interval is called the angular displacement of the line and is indicated by the symbol **ɸ**. Similarly, the symbol Φ is used to designate the total angle turned through during a time interval and will be more than ɸ whenever the direction of rotation changes.

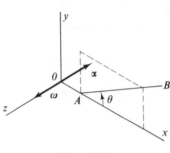

Figure 8-3

The angular velocity, **ω,** *of a line is defined as the time rate of change of the angular position of the line.* Angular velocity is a free vector quantity which can be represented by a vector perpendicular to the plane of motion of the line in a direction determined by the right-hand screw rule. If line AB in Fig. 8-3 is turning so that θ is increasing, the angular velocity vector is in the z direction as shown. When a line remains in one plane the angular velocity vector is always perpendicular to the plane and it is convenient to indicate the direction of the angular velocity with a clockwise or counterclockwise arrow. For coplanar motion of a line, the magnitude of the angular velocity is

$$\omega = \lim_{\Delta t \to 0} \frac{\Delta \theta}{\Delta t} = \frac{d\theta}{dt} = \dot{\theta}. \tag{8-4}$$

Angular velocity has the dimensions of an angle per unit of time, that is, T^{-1}, since an angle is dimensionless. Common units for angular velocity are radians per second (rad per sec), revolutions per minute (rpm), etc.

The angular acceleration **α,** *of a line is defined as the time rate of change of the angular velocity of the line.* Since angular velocity is a vector quantity, it can change in either magnitude or direction or both. Since only coplanar angular motion is considered in this article, the angular velocity can change only in magnitude and sense. The magnitude of the angular acceleration of a line is

$$\alpha = \lim_{\Delta t \to 0} \frac{\Delta \omega}{\Delta t} = \frac{d\omega}{dt} = \frac{d^2\theta}{dt^2} = \ddot{\theta}. \tag{8-5}$$

The angular acceleration of a line can be represented by a free vector, and for coplanar motion of a line, the vector is perpendicular to the plane of motion with the sense of the vector determined by the right-hand screw rule. If line AB of Fig. 8-3 is rotating counterclockwise with decreasing angular velocity, the angular acceleration is as shown on the figure. For coplanar motion it is convenient to use a clockwise or counterclockwise arrow to represent angular acceleration.

Angular acceleration has dimensions of an angle divided by a time squared, T^{-2}. Common units are radians per second per second (rad per sec²), revolutions per second per second (rev per sec²), and similar units.

When angular positions of moving lines are shown on a sketch the direction of the arrow from the fixed reference line is used to indicate the positive direction. Similarly, the coordinate axes are used to indicate positive linear directions. When positive directions are not specified they can frequently be determined from the given data. Where insufficient data permit a choice of positive directions, the assumed directions should be clearly indicated.

343

8-4 RECTILINEAR MOTION

Rectilinear motion of a particle has been defined as motion along a straight line. When the definitions developed in Art. 8-3 are applied to rectilinear motion it is seen that the velocity of the particle, which is always tangent to the path, will always be along the straight line and thus can change only in magnitude and sense. Since the velocity can change only in magnitude, the acceleration (the time rate of change of velocity) must also be along the straight line. If $\mathbf{i}$ is a unit vector along the path, the following equations are obtained from the definitions in Art. 8-3.

The position $\mathbf{r}$ (of the point) is

$$\mathbf{r} = s\mathbf{i},$$

where s is the distance *from* a fixed point on the path *to* the moving point. The velocity is

$$\mathbf{v} = \frac{d\mathbf{r}}{dt} = \frac{ds}{dt}\mathbf{i} = v\mathbf{i}$$

from which

$$v = \frac{ds}{dt}. \tag{a}$$

Similarly, the acceleration of the point is

$$\mathbf{a} = \frac{d\mathbf{v}}{dt} = \frac{dv}{dt}\mathbf{i} = a\mathbf{i};$$

that is,

$$a = \frac{dv}{dt} = \frac{d^2s}{dt^2}. \tag{b}$$

If v is expressed as a function of s, the acceleration can also be expressed (for rectilinear motion) as

$$a = \frac{dv}{ds}\frac{ds}{dt} = v\frac{dv}{ds}. \tag{c}$$

Note that Eqs. (a)–(c) are scalar equations which give the magnitudes of the velocities and accelerations. The directions are indicated by the unit vector along the path and the algebraic sign of the velocity or acceleration. If the magnitude of the velocity is decreasing, the acceleration is sometimes called a *deceleration*.

When some of the kinematic quantities (position, displacement, velocity, acceleration, and time) are known and others are to be determined Eqs. (8-2) and (8-3), or some of their other forms, can be used to obtain the desired quantities by differentiation or integration. The three scalar equations most useful for rectilinear motion are

$$v = \frac{ds}{dt} \quad \text{or} \quad v\,dt = ds,$$

$$a = \frac{dv}{dt} \quad \text{or} \quad a\,dt = dv,$$

$$a = v\frac{dv}{ds} \quad \text{or} \quad a\,ds = v\,dv.$$

The particular equation or form to use will depend on the given data and the desired results.

The similarity of Eqs. (a) and (8-4) and of (b) and (8-5) indicates a direct analogy between the position, velocity, and acceleration for rectilinear motion of a point and for coplanar angular motion of a line. Algebraic expressions for angular quantities can be differentiated and integrated in the same manner as the analogous expressions for linear quantities. It is often useful to relate the linear position of a point to the angular position of a line by means of geometry and obtain expressions relating velocities and accelerations by differentiation. The following examples illustrate some of the principles just discussed.

EXAMPLE 8-1

The velocity of a particle moving along a horizontal straight line is given by the expression

$$\dot{\mathbf{r}} = \mathbf{v} = (6t^2 - 10t)\mathbf{i},$$

where t is time in seconds and r is measured in feet. The particle is moving to the left when t is 2 sec. Determine

(a) The acceleration of the particle when $t = 2$ sec.
(b) The displacement of the particle during the time interval from $t = 1$ sec to $t = 3$ sec.
(c) The total distance traveled by the particle during the interval from $t = 1$ sec to $t = 3$ sec.

SOLUTION

(a) When $t = 2$ sec

$$\mathbf{v} = [6(2)^2 - 10(2)]\mathbf{i} = +4\mathbf{i}\,\text{fps},$$

and the particle is moving to the left; therefore the unit vector $\mathbf{i}$ must be directed to the left. The acceleration of the particle can be obtained by differentiating the expression for the velocity. Thus

$$\mathbf{a} = \dot{\mathbf{v}} = \frac{d}{dt}(6t^2 - 10t)\mathbf{i} = (12t - 10)\mathbf{i},$$

and when $t = 2$ sec

$$\mathbf{a} = [12(2) - 10]\mathbf{i} = 14\mathbf{i}\,\text{fps}^2$$

$$= 14\,\text{fps}^2 \text{ to the left.} \qquad \text{Ans.}$$

(b) Since the velocity is the derivative of the position function, Eq. (8-2), the position function can be determined by integrating the expression for the velocity. Thus

$$\mathbf{v} = \frac{d\mathbf{r}}{dt} \quad \text{and} \quad d\mathbf{r} = \mathbf{v} \, dt$$

from which

$$\mathbf{r} = \int \mathbf{v} \, dt.$$

In this example,

$$\mathbf{r} = \int (6t^2 - 10t)\mathbf{i} \, dt = \mathbf{i} \int (6t^2 - 10t) \, dt = \mathbf{i}(2t^3 - 5t^2) + \mathbf{C},$$

where $\mathbf{C}$ is a constant of integration. The location of the origin is not specified and thus may be selected at any convenient point on the line. If the origin is located at the position of the particle when $t = 0$, the constant $\mathbf{C}$ is zero.

When $t = 1$ sec

$$\mathbf{r}_1 = -3\mathbf{i},$$

and when $t = 3$ sec

$$\mathbf{r}_3 = +9\mathbf{i}.$$

The displacement during the time interval is

$$\mathbf{q} = \mathbf{r}_3 - \mathbf{r}_1 = 9\mathbf{i} - (-3\mathbf{i}) = 12\mathbf{i} \, \text{ft}$$

$$= \underline{12 \text{ ft to the left.}} \qquad \text{Ans.}$$

(c) For rectilinear motion, the magnitude of the displacement and the total distance traveled will be the same unless the particle reverses its velocity. In this example, the velocity is zero when $t = 0$ and when $t = 1.667$ sec. During the interval from $t = 1$ sec to $t = 1.667$ sec, the magnitude of the velocity is negative; that is, the particle is moving to the right, and from $t = 1.667$ sec to $t = 3$ sec, the velocity is to the left. When $t = 1.667$ sec, the position of the particle is

$$\mathbf{r}_{1.667} = \mathbf{i}[2(1.667)^3 - 5(1.667)^2] = -4.63\mathbf{i}.$$

The total distance traveled is the sum of the distances traveled in the two directions; that is,

$$Q = |\mathbf{r}_{1.667} - \mathbf{r}_1| + |\mathbf{r}_3 - \mathbf{r}_{1.667}|$$

$$= |-4.63\mathbf{i} - (-3\mathbf{i})| + |9\mathbf{i} - (-4.63\mathbf{i})|$$

$$= 1.63 + 13.63 = \underline{15.26 \text{ ft.}} \qquad \text{Ans.}$$

The quantity 1.63 ft is the distance traveled to the right during the first two-thirds of a second, and 13.63 ft is the distance traveled to the left after the velocity reverses.

EXAMPLE 8-2

The magnitude of the acceleration of a point moving along a horizontal straight line varies according to the equation $a = 12s^{1/2}$, where a is in feet per second per second and s is the distance of the point from the origin in feet. When the time t is 2 sec, the point is 16 ft to the right of the origin and has a velocity of 32 fps to the right and an acceleration of 48 fps^2 to the right. Determine the velocity and acceleration of the point when the time is 3 sec.

SOLUTION

The positive direction is obtained by observing that s must be positive in order to have the acceleration be real (not imaginary), and the problem specifies that the acceleration is real (48 fps^2) when $s = 16$ ft to the right. Therefore the positive direction is to the right.

When the acceleration is expressed as a function of the position s, it is convenient to write the acceleration in the form

$$a = v\frac{dv}{ds}$$

and

$$v\,dv = a\,ds = 12s^{1/2}\,ds$$

from which

$$\int v\,dv = \int 12s^{1/2}\,ds$$

or

$$\frac{v^2}{2} = 8s^{3/2} + C_1.$$

When $s = +16$ ft, $v = +32$ fps; therefore

$$\frac{32^2}{2} = 8(16)^{3/2} + C_1$$

and

$$C_1 = 0$$

from which

$$v^2 = 16s^{3/2}.$$

The velocity is the derivative of the position function. That is,

$$v = 4s^{3/4} = \frac{ds}{dt}$$

from which

$$4\,dt = s^{-3/4}\,ds$$

and

$$4t = 4s^{1/4} + C_2.$$

When $t = 2$ sec, $s = 16$ ft; therefore

$$4(2) = 4(16)^{1/4} + C_2 = 8 + C_2$$

and $C_2 = 0$. The equation for t thus reduces to

$$t = s^{1/4} \quad\text{or}\quad s = t^4$$

from which

$$v = \frac{ds}{dt} = 4t^3$$

and

$$a = \frac{dv}{dt} = 12t^2.$$

When $t = 3$ sec these expressions give

$$v = 4(3)^3 = 108 \text{ fps}$$
$$\mathbf{v} = \underline{108 \text{ fps to the right.}} \qquad \text{Ans.}$$
$$a = 12(3)^2 = 108 \text{ fps}^2$$
$$\mathbf{a} = \underline{108 \text{ fps}^2 \text{ to the right.}} \qquad \text{Ans.}$$

EXAMPLE 8-3

The acceleration of a particle with rectilinear motion is given by the equation $\mathbf{a} = -k\mathbf{v}$, where a is in meters per second per second, $\mathbf{v}$ is the velocity in meters per second, and k is a constant. When the time t is zero the position and velocity of the particle are $\mathbf{s}_0$ and $\mathbf{v}_0$, respectively.

(a) Derive expressions for the position, velocity, and acceleration of the particle in terms of the time t.
(b) If $\mathbf{s}_0 = 20$ m to the left, $\mathbf{v}_0 = 10$ m/s to the right, and $k = 0.2 \text{ sec}^{-1}$, determine the value of t when $\mathbf{s} = 0$ and the corresponding values of $\mathbf{a}$ and $\mathbf{v}$.

SOLUTION

(a) The positive direction is assumed to be to the right since it is not specified in the problem statement and $\mathbf{i}$ is a unit vector to the right. The magnitude of the acceleration is

$$a = \frac{dv}{dt} = -kv$$

or

$$k\, dt = -\frac{dv}{v}$$

from which

$$kt = -\ln v + C_1.$$

When $t = 0$, $v = v_0$; therefore

$$0 = -\ln v_0 + C_1$$

or

$$C_1 = \ln v_0,$$

and the preceding expression becomes

$$kt = -\ln v + \ln v_0 = \ln \frac{v_0}{v},$$

which can be written as

$$e^{kt} = \frac{v_0}{v}$$

or

$$v = v_0 e^{-kt}.$$

Since $v = ds/dt$, this equation can be written as

$$ds = v_0 e^{-kt}\, dt$$

or

$$s = v_0 \int e^{-kt} \, dt = -\frac{v_0}{k} e^{-kt} + C_2.$$

When $t = 0$, $s = s_0$ and C_2 is found to be

$$C_2 = s_0 + \frac{v_0}{k} e^0 = s_0 + \frac{v_0}{k}.$$

Thus the position function is

$$s = s_0 + \frac{v_0}{k}(1 - e^{-kt}).$$

The acceleration is, from the problem statement,

$$a = -kv = -kv_0 e^{-kt}.$$

The results are

$$\mathbf{s} = \mathbf{s}_0 + \frac{\mathbf{v}_0}{k}(1 - e^{-kt}), \qquad \text{Ans.}$$

$$\mathbf{v} = \mathbf{v}_0 e^{-kt}, \qquad \text{Ans.}$$

and

$$\mathbf{a} = -\mathbf{v}_0 k e^{-kt}. \qquad \text{Ans.}$$

(b) When numerical values for s_0, v_0, and k are substituted into the expression for s it becomes

$$\mathbf{s} = -20\mathbf{i} + \frac{10}{0.2}(1 - e^{-0.2t})\mathbf{i},$$

and when $s = 0$ this equation gives

$$1 - e^{-0.2t} = \frac{20(0.2)}{10} = 0.4$$

$$e^{-0.2t} = 1 - 0.4 = 0.6 = e^{-0.511}$$

and

$$t = \frac{0.511}{0.2} = 2.555 \qquad \text{Ans.}$$

The velocity is

$$\mathbf{v} = 10\mathbf{i}e^{-0.2t} = 6\mathbf{i}\,\text{m/s}, \qquad \text{Ans.}$$

and the acceleration is

$$\mathbf{a} = -k\mathbf{v} = -0.2(6\mathbf{i}) = -1.2\mathbf{i} \,\text{m/s}^2. \qquad \text{Ans.}$$

EXAMPLE 8-4

A spotlight rotates in a horizontal plane with a constant angular velocity ω as shown in Fig. 8-4. The spot of the light, P, moves along the wall as indicated.

(a) Derive expressions for the velocity and acceleration of point P.
(b) The distance b is 100 ft, and the angular velocity of the beam of

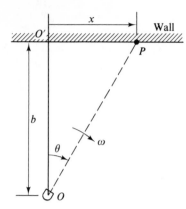

Figure 8-4

light is 5 rpm clockwise. Determine the velocity and acceleration of P when $\theta = 20°$.

SOLUTION

(a) Point P has rectilinear motion, and its position, measured from O', is

$$\mathbf{O'P} = \mathbf{x} = x\mathbf{i} = \mathbf{i}b \tan \theta,$$

where x is positive to the right and θ is positive clockwise. The magnitude of the velocity of P is

$$v_P = \frac{dx}{dt} = b \sec^2 \theta \frac{d\theta}{dt} = b\omega \sec^2 \theta,$$

and the velocity of P is

$$\mathbf{v}_P = \mathbf{i}b\omega \sec^2 \theta. \qquad \text{Ans.}$$

The magnitude of the acceleration of P is

$$a_P = \frac{dv_P}{dt} = b\left[\frac{d\omega}{dt} \sec^2 \theta + \omega 2 \sec \theta (\sec \theta \tan \theta)\frac{d\theta}{dt}\right]$$

$$= b[\alpha \sec^2 \theta + 2\omega^2 \sec^2 \theta \tan \theta].$$

Since ω is constant, α must be zero, and the acceleration is

$$\mathbf{a}_P = \mathbf{i}2b\omega^2 \sec^2 \theta \tan \theta. \qquad \text{Ans.}$$

(b) The angular velocity of the light beam is

$$\omega = 5 \text{ rpm}$$

$$= 5 \frac{\text{rev}}{\text{min}}\left(\frac{1 \text{ min}}{60 \text{ sec}}\right)\left(\frac{2\pi \text{ rad}}{\text{rev}}\right)$$

$$= \frac{\pi}{6} \text{ rad per sec.}$$

When numerical values for b, ω, and θ are substituted into the expressions for the velocity and acceleration of P they become

$$\mathbf{v}_P = \mathbf{i}(100)\left(\frac{\pi}{6}\right)\sec^2 20°$$

$$= 59.3\mathbf{i} \text{ fps}$$
$$= 59.3 \text{ fps to the right.} \qquad \Bigg\}\text{ Ans.}$$

and

$$\mathbf{a}_P = \mathbf{i}(2)(100)\left(\frac{\pi}{6}\right)^2(\sec^2 20°)(\tan 20°)$$

$$= 22.6\mathbf{i} \text{ fps}^2$$
$$= 22.6 \text{ fps}^2 \text{ to the right.} \qquad \Bigg\}\text{ Ans.}$$

PROBLEMS

8-1 The position of a particle moving along the x axis is given by the equation

$$\mathbf{r} = (2t^3 - 15t^2 + 24t - 6)\mathbf{i},$$

where r is measured in feet and t is in seconds. The particle is 5 ft to the left of the origin at $t = 1$ sec. Determine
(a) The velocity at $t = 2$ sec.
(b) The acceleration at $t = 2$ sec.
(c) The total distance traveled during the interval from $t = 0$ to $t = 4$ sec.

8-2 In Problem 8-1 determine
(a) The average velocity during the interval $t = 0$ to $t = 1$ sec.
(b) The average acceleration during the interval $t = 0$ to $t = 1$ sec.
(c) The displacement during the interval $t = 0$ to $t = 4$ sec.

*8-3 The position of a particle with rectilinear motion is given by the equation

$$\mathbf{r} = \mathbf{i}4 \sin^2(0.2t),$$

where r is in meters and t is in seconds. The quantity 0.2 is given in radians per second. The positive direction is to the right. Determine
(a) The velocity when $t = 3$ sec.
(b) The acceleration when $t = 3$ sec.
(c) The displacement during the interval $t = 0$ to $t = 3$ sec.

8-4 A line rotates in a vertical plane according to the equation $\theta = 2t^3 - 12t^2$, where θ gives the angular position of the line in radians and t is time in seconds. The line is turning clockwise when $t = 1$ sec. Determine
(a) The value of t when the line changes direction of rotation.
(b) The angular displacement of the line during the interval $t = 1$ to $t = 5$ sec.
(c) The total angle swept through by the line during the interval $t = 1$ to $t = 5$ sec.

8-5 A line rotates in a horizontal plane in accordance with the equation $\theta = t^3 - 2t^2 - 2$, where θ gives the angular position of the line in radians when t is the time in seconds; the line is turning clockwise when $t = 2$ sec. Determine

(a) The angular acceleration of the line when $t = 2$ sec.
(b) The total angle turned through by the line during the time interval $t = 0$ to $t = 3$ sec.

8-6 The velocity of a particle is given by the equation

$$\mathbf{v} = (5t^2 - 8t)\mathbf{i},$$

where v is in feet per second and t is in seconds. The particle is located at $\mathbf{r} = 3\mathbf{i}$ and moving to the left when $t = 1$ sec. Determine
(a) The displacement of the particle during the interval $t = 0$ to $t = 3$ sec.
(b) The total distance traveled by the particle during the interval $t = 0$ to $t = 3$ sec.
(c) The acceleration of the particle when its velocity is zero.

8-7 A particle starts from rest 2 ft to the left of the origin and moves along a horizontal straight line. The magnitude of the velocity is proportional to the square of the time after starting. During the period from $t = 1$ to $t = 3$ sec the particle has a displacement of 39 ft to the left. When $t = 2$ sec determine
(a) The position of the particle.
(b) The velocity of the particle.
(c) The acceleration of the particle.

*8-8 The magnitude of the velocity of a particle moving along a horizontal straight line is

$$v = kt^2 + 12,$$

where v is in centimeters per second, t is in seconds, and k is a constant (positive to the right). When $t = 3$ s the acceleration is 12 cm/s² to the left. Determine the total distance traveled during the time interval $t = 1$ to $t = 6$ s.

8-9 The velocity of a particle with rectilinear motion is given by the equation

$$\mathbf{v} = (3t^2 - kt)\mathbf{i},$$

where v is in feet per second when the time t is in seconds. The position vector of the particle from the origin is $\mathbf{x} = -8\mathbf{i}$ ft when $t = 0$ and $-16\mathbf{i}$ ft when $t = 4$ sec. Determine
(a) The acceleration of the particle when $t = 2$ sec.
(b) The total distance traveled by the particle during the time interval $t = 1$ to $t = 5$ sec.

8-10 The angular velocity of a line rotating in a vertical plane varies according to the equation $\omega = 3t^2 - 5t$, where ω is in radians per second and t is the time in seconds. The line is rotating counterclockwise when $t = 2$ sec. Determine
(a) The angular acceleration of the line when $t = 2$ sec.
(b) The angular displacement of the line during the interval $t = 1$ to $t = 3$ sec.

8-11 The angular velocity of a line varies according to the equation $\omega = Kt^2 + 12$, where ω is the magnitude of the angular velocity in radians per second, t is time in seconds, and K is a constant. When $t = 0$ the angular position of the line is 4 rad clockwise and the angular velocity is 12 rad per sec counterclockwise. When $t = 3$ sec the angular position is 14 rad counterclockwise. Determine
(a) The angular acceleration of the line when $t = 4$ sec.
(b) The total angle turned through by the line during the interval $t = 1$ to $t = 5$ sec.

*8-12 The magnitude of the acceleration of a particle with rectilinear motion is

$$a = 6t - 10,$$

where a is in meters per second per second and t is in seconds. When $t = 0$ the acceleration is 10 m/s^2 to the right and the velocity is 5 m/s to the left. When $t = 2$ s the particle is 10 m to the right of the origin. Locate the particle when $t = 4$ s.

8-13 The magnitude of the acceleration of a particle with rectilinear motion is

$$a = 12t - 8,$$

where a is in feet per second per second and t is in seconds. When $t = 0$ the acceleration of the particle is 8 fps^2 to the left. During the period from $t = 1$ to $t = 4$ sec the displacement of the particle is 30 ft to the right. Determine the velocity of the particle at the beginning and at the end of the displacement period.

8-14 The acceleration of a particle moving along the x axis is

$$\mathbf{a} = \mathbf{i}Ae^{pt},$$

where a is in inches per second per second, t is

in seconds, $A = 2 \text{ ips}^2$, and $p = 0.5(\text{sec}^{-1})$. When $t = 0$ the particle is at the origin and has a velocity of $-2\mathbf{i}$ ips (to the left). Determine the position, velocity, and acceleration of the particle when $t = 2$ sec.

8-15 In Problem 8-14 determine the total distance the particle travels during the interval $t = 1$ to $t = 4$ sec.

8-16 The angular acceleration of a line rotating in a horizontal plane varies according to the expression $\alpha = 6t - 12$, where α is measured in radians per second per second and t is in seconds. The line is turning clockwise when $t = 1$ sec. The angular velocity is zero, and $\theta = 2$ rad counterclockwise when $t = 0$. Determine the total angle turned through during the time interval $t = 0$ to $t = 5$ sec.

8-17 The angular acceleration of a line rotating in a plane is given by the equation $\alpha = Kt^2 - 6$, where α is in radians per second per second when t is in seconds and where K is a constant. When $t = 0$ the angular acceleration is 6 rad per sec^2 counterclockwise, and the angular velocity is 10 rad per sec clockwise. When $t = 1$ sec the angular velocity is 8 rad per sec clockwise, and the angular position is 5 rad clockwise. Determine the angular position and acceleration of the line when $t = 2$ sec.

8-18 The magnitude of the velocity of a particle with rectilinear motion is given by the equation $v = ks^2$, where v is in feet per second, s is the distance (in feet) of the particle from an origin on the path, and $k = -2 \text{ ft}^{-1} \text{ sec}^{-1}$. The particle is 10 ft to the right of the origin at $t = 0$. When $t = 2$ sec determine
(a) The position of the particle.
(b) The velocity of the particle.
(c) The acceleration of the particle.

*8-19 The velocity of a particle moving along the x axis is given by the equation

$$\mathbf{v} = \mathbf{i}v_0 \cos\left(\frac{x}{b}\right),$$

where $v_0 = 10$ m/s and $b = 5$ m. Determine the velocity and acceleration when $x = 5$ m.

*8-20 The particle in Problem 8-19 is at the origin ($x = 0$) when $t = 0$. Determine the value of t when $\mathbf{x} = 5\mathbf{i}$ m.

*8-21 A French artillerist, Captain A. Le Duc, developed an empirical formula for the speed of a projectile in terms of the distance traveled by the projectile in the bore. For a 155-mm gun, this equation is $v = 1010s/(1.78 + s)$, where s is the distance traveled in the bore in meters and v is the speed of the projectile in meters per second. The projectile travels 4.70 m in the bore. Determine
(a) The muzzle velocity of the projectile.
(b) The magnitude of the acceleration of the projectile when $s = 1.3$ m.

8-22 The acceleration of a particle moving along the x axis is given by the equation $\mathbf{a} = 3x^{1/2}\mathbf{i}$, where a is in feet per second per second and x is the distance (in feet) of the particle from the origin. The particle starts from rest at the origin at $t = 0$. Determine the position, velocity, and acceleration of the particle when $t = 3$ sec.

8-23 The acceleration of a particle with rectilinear motion is given by the equation $a\mathbf{i} = -(k/x)\mathbf{i}$, where a is acceleration in inches per second per second, x is the position of the particle in inches, and k is a constant. The velocity is zero when $x = x_0$.
(a) Derive an expression for the velocity in terms of the position x.
(b) Determine the velocity when $x = x_0/2$ and $k = 18$ in²/sec².

8-24 In Problem 8-23 assume that the velocity is $v_0\mathbf{i}$ at position $x_0\mathbf{i}$. Determine the value of x when the particle comes to rest and the corresponding acceleration.

8-25 The magnitude of the angular acceleration of a line moving in a plane varies in accordance with the expression $\alpha = -K\theta^2$, where α is in radians per second per second, θ is the angular position of the line in radians (clockwise is positive), and K is a constant. The angular velocity of the line changes from 6 rad per sec clockwise to 3 rad per sec clockwise while θ changes from 0 to 3 rad clockwise.
(a) Determine the maximum (positive) value of θ.
(b) Determine the corresponding value of α.

8-26 The acceleration of a particle with rectilinear motion is $\mathbf{a} = -0.05v^2\mathbf{i}$, where a is in

feet per second per second and v is the velocity in feet per second. The particle starts at the origin with a velocity of $50\mathbf{i}$ fps at $t = 0$. Locate the particle when $t = 5$ sec and calculate the corresponding velocity and acceleration.

8-27 In Problem 8-26 determine the time interval for the particle to travel
(a) The first 20 ft.
(b) The second 20 ft.

*8-28 The acceleration of a particle with rectilinear motion is given by the equation $\mathbf{a} = (k/v)\mathbf{i}$, where a is in meters per second per second, v is velocity in meters per second, and k is a constant. The particle is at the origin with a velocity of $v_0\mathbf{i}$ at $t = 0$.
(a) Derive an equation for the position of the particle as a function of t.
(b) What are the fundamental dimensions of k?

8-29 The angular acceleration of a line moving in a plane is given by the equation $\alpha = -k\omega^2$, where α is in radians per second per second, ω is the angular velocity of the line in radians per second, and k is a constant. Clockwise is positive. When $t = 0$, $\omega = +5$ rad per sec, and when $t = 3$ sec, $\omega = +2$ rad per sec. Determine the angular displacement of the line during the interval $t = 0$ to $t = 3$ sec.

8-30 The spotlight in Example 8-4 is used to follow a point P moving along the wall with a constant linear velocity of 50 fps to the right. The light is 200 ft from the wall. Determine the angular velocity and acceleration of the light beam when $\theta = 30°$.

8-31 Point C of the mechanism in Fig. P8-31

Figure P8-31

moves along the horizontal line through AC. Set up an equation relating s and θ and differentiate this expression to relate the linear velocity and acceleration of C to the angular velocity and acceleration of line AB.

8-32 Members AB and BC of Problem 8-31 each have a length of 10 in. When $\theta = 60°$ point C has a velocity of 30 ips and an acceleration of 20 ips², both to the right. Determine the angular velocity and acceleration of member AB for this position.

8-33 Two small vessels are traveling along perpendicular courses at constant speeds as shown in Fig. P8-33. Determine the angular velocity of the line of sight between the ships when OA is 500 ft, OB is 1200 ft, and the speed of each ship is 20 fps.

8-34 A mechanism causes the spotlight of Example 8-4 to oscillate in accordance with the equation

$$\theta = \theta_0 \sin \omega_0 t,$$

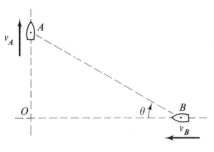

Figure P8-33

where $\theta_0 = \pi/6 \text{ rad} = 30°$ and $\omega_0 = 10$ rad per sec. The distance b is 500 ft. Determine the magnitudes of the velocity and acceleration of point P when
(a) $\omega_0 t = 0$.
(b) $\omega_0 t = 90°$.

8-35 Solve Problem 8-34 when $\omega_0 t = 45°$.

8-5 MOTION CURVES

For many problems a semigraphical or pictorial procedure utilizing motion curves provides an excellent method of solution and leads to ready visualization of the motion. Motion curves are diagrams showing the relationships between various kinematical quantities. For rectilinear motion the position, velocity, or acceleration may be plotted as ordinates against the corresponding time as the abscissa. In some cases the velocity may be plotted against the position (this is known as a phase-plane plot), or the acceleration may be plotted against the position or velocity. One of the most useful diagrams for many problems is the v-t (velocity-time) diagram. An example of a v-t diagram is shown in Fig. 8-5.

The slope of the curve corresponding to any time t represents the change in the magnitude of the velocity dv divided by time interval dt, which, from Eq. (8-3), is the magnitude of the acceleration.

The area under the curve between any two ordinates at t_1 and t_2 of Fig. 8-5 is the definite inte-

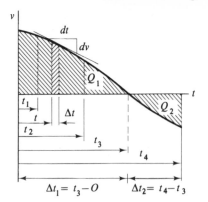

Figure 8-5

gral of $v\,dt$. Since $v = ds/dt$, the expression $v\,dt$ is equal to ds, and the integral is

$$\int_{t_1}^{t_2} v\,dt = \int_{s_1}^{s_2} ds = s_2 - s_1 = \Delta s = q,$$

where s_1 and s_2 are the values of s corresponding to t_1 and t_2, respectively. Thus the area under the v-t curve between the ordinates at t_1 and t_2 is the change of s (the magnitude of the displacement) during the corresponding time interval. A reversal of the velocity occurs at $t = t_3$ when the curve crosses the time axis. The area Q_1 represents the distance traveled during the interval Δt_1 from $t = 0$ to $t = t_3$ (prior to the reversal of the velocity). The area Q_2 represents the distance traveled in the opposite direction during the time interval Δt_2 from $t = t_3$ to $t = t_4$. The sum of the two areas represents the total distance traveled from $t = 0$ to $t = t_4$, and the difference is the net distance (the magnitude of the displacement) during the interval.

EXAMPLE 8-5

Draw a-t, v-t, and s-t curves for Example 8-1.

SOLUTION

The three diagrams are shown in Fig. 8-6. The equations for the curves were obtained from Example 8-1.

Recall that a positive sign indicates a vector quantity to the left. The a-t diagram is a straight line, the v-t diagram is a parabola, and the s-t curve is a cubic.

Since the slope of the v-t curve is the acceleration, the time at which the a-t curve crosses the t axis ($t = 0.833$ sec) should correspond to a zero slope on the v-t curve. Also the times at which the v-t curve crosses the t axis ($t = 0,\ 1.667$ sec) correspond to points of zero slope of the s-t curve.

The area A under the v-t curve represents the distance the particle travels to the right during the interval from $t = 1.000$ to $t = 1.667$ sec. Thus $A = 4.63 - 3.00 = 1.63$ ft. Similarly, the area B represents the distance the particle travels to the left during the period from $t = 1.667$ to $t = 3.00$ sec; that is, $B = 9.00 + 4.63 = 13.63$ ft. The sum of areas A and B is the total distance traveled during the interval, and the difference is the corresponding displacement.

The v-t diagram is probably the most useful diagram in helping to visualize the motion. Its slope gives the acceleration, its area gives the displacement and total distance traveled, and its ordinate gives the velocity.

The v-t diagram is particularly useful when the acceleration is constant for a period of time since the diagram is a straight line and the slope and areas can be computed directly from the diagram. This procedure is illustrated in Examples 8-6 and 8-7.

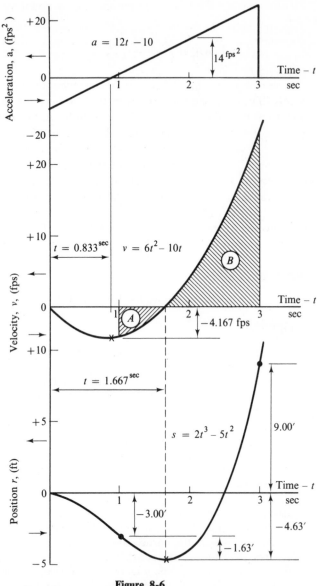

Figure 8-6

EXAMPLE 8-6

A particle moves along a straight line with a constant acceleration. The velocity of the particle changes from 10 fps to the right to 25 fps to the left during a time interval of 7.0 sec. Determine

(a) The acceleration.
(b) The total distance traveled.
(c) The displacement of the particle during the 7-sec interval.

SOLUTION

Since the acceleration is known to be constant, the slope of the v-t diagram must also be constant, and the v-t diagram is a straight line. Initial and final values of v are plotted, and the diagram is drawn as shown in Fig. 8-7.

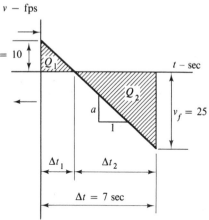

(a) Since the velocity changes from right to left, the acceleration (constant) must be to the left. The magnitude of the acceleration is given by the slope of the v-t curve and is

$$a = \frac{v_f - v_i}{\Delta t} = \frac{-25 - (+10)}{7.0} = -5$$

and

$$\mathbf{a} = \underline{5 \text{ fps}^2 \text{ to the left.}} \qquad \text{Ans.}$$

(b) The time Δt_1 that has elapsed when the motion reverses direction is, from the slope of the line,

$$\Delta t_1 = \frac{v_i}{a} = \frac{10}{5} = 2 \text{ sec.}$$

Figure 8-7

The area Q_1 which represents the distance moved to the right during Δt_1 is

$$Q_1 = \tfrac{1}{2}v_i(\Delta t_1) = \tfrac{1}{2}(10)(2) = 10 \text{ ft.}$$

The time interval Δt_2 is

$$\Delta t_2 = \Delta t - \Delta t_1 = 7 - 2 = 5 \text{ sec,}$$

and the area Q_2, which represents the distance moved to the left during Δt_2, is

$$Q_2 = \tfrac{1}{2}v_f(\Delta t_2) = \tfrac{1}{2}(25)(5) = 62.5 \text{ ft.}$$

The total distance traveled is

$$Q = Q_1 + Q_2 = 10 + 62.5 = \underline{72.5 \text{ ft.}} \qquad \text{Ans.}$$

(c) Since the area Q_2 is greater than Q_1, the displacement during the 7-sec interval must be to the left. Its magnitude is

$$q = Q_2 - Q_1 = 62.5 - 10 = 52.5 \text{ ft,}$$

and the displacement is

$$\mathbf{q} = \underline{52.5 \text{ ft to the left.}} \qquad \text{Ans.}$$

Alternate solution

(a) The following analytical solution is included for comparison. Assume that $t = 0$ and that the particle is at the origin at the beginning of the 7-sec time interval and that the unit vector $\mathbf{i}$ is to the right. Since the acceleration is constant, it can be written as

$$\mathbf{a} = \frac{d\mathbf{v}}{dt} = a\mathbf{i}$$

from which

$$d\mathbf{v} = a\mathbf{i}\, dt$$

and

$$\mathbf{v} = at\mathbf{i} + \mathbf{C}_1.$$

When $t = 0$, $\mathbf{v} = 10\mathbf{i}$ and $\mathbf{C}_1 = 10\mathbf{i}$. Also when $t = 7$ sec, $\mathbf{v} = -25\mathbf{i}$, and when these values are substituted into the equation for $\mathbf{v}$ it becomes

$$-25\mathbf{i} = a7\mathbf{i} + 10\mathbf{i}$$

from which

$$a = -5$$

and

$$\mathbf{a} = -5\mathbf{i}\,\text{fps}^2 = 5\,\text{fps}^2 \text{ to the left.} \qquad \text{Ans.}$$

(c) The equation for $\mathbf{v}$ is

$$\mathbf{v} = -5t\mathbf{i} + 10\mathbf{i} = \frac{d\mathbf{r}}{dt}$$

from which

$$\mathbf{r} = \frac{-5t^2}{2}\mathbf{i} + 10t\mathbf{i} + \mathbf{C}_2.$$

Since $\mathbf{r} = 0$ when $t = 0$, $\mathbf{C}_2 = 0$, and

$$\mathbf{r} = (-2.5t^2 + 10t)\mathbf{i}.$$

When $t = 7$ sec, the preceding equation gives $\mathbf{r} = -52.5\mathbf{i}$ ft, and the displacement from $t = 0$ to $t = 7$ sec is

$$\mathbf{q} = \mathbf{r}_7 - \mathbf{r}_0 = -52.5\mathbf{i} - 0$$
$$= -52.5\mathbf{i}\,\text{ft} = 52.5 \text{ ft to the left.} \qquad \text{Ans.}$$

(b) The time or times when $\mathbf{v} = 0$ must be obtained to determine the total distance traveled. Since $\mathbf{v} = (-5t + 10)\mathbf{i}$, the velocity will be zero at $t = 2$ sec, and the velocity will be to the right from $t = 0$ to $t = 2$ sec and the left after $t = 2$ sec. When $t = 2$ sec

$$\mathbf{r}_2 = [-2.5(2)^2 + 10(2)]\mathbf{i} = 10\mathbf{i},$$

and the total distance traveled from $t = 0$ to $t = 7$ sec is

$$Q = |\mathbf{r}_2 - \mathbf{r}_0| + |\mathbf{r}_7 - \mathbf{r}_2|$$
$$= |10\mathbf{i} - 0| + |-52.5\mathbf{i} - 10\mathbf{i}|$$
$$= 10 + 62.5 = 72.5 \text{ ft.} \qquad \text{Ans.}$$

EXAMPLE 8-7

A particle starts from the origin with a velocity of 2 m/s to the right and moves along a straight line with an acceleration of 4 m/s² to the right for 2.5 s. The acceleration then changes to 6 m/s² to the left for 3 s, at which time the acceleration becomes 1 m/s² to the right for 4 s. Determine

(a) The total distance traveled during the 9.5-s interval.
(b) The displacement of the particle during the first 5.5-s interval.

SOLUTION

Since each of the accelerations is constant during the specified time periods, the v-t diagram provides a convenient solution; see Fig. 8-8. When $t = 0$ the velocity is 2 m/s to the right and the acceleration is 4 m/s² to the right, which indicates an increase in the magnitude of the velocity. The magnitude of the velocity at $t = 2.5$ s can be obtained from the slope of the line; thus

$$a_1 = \text{slope} = \frac{v_{2.5} - v_0}{\Delta t_1} = 4$$

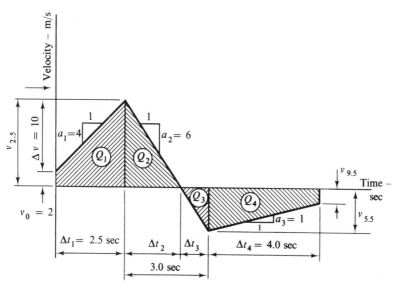

Figure 8-8

from which

$$v_{2.5} = 2 + 4(2.5) = 12 \text{ m/s,}$$

and it is represented by the ordinate which is above the t axis and thus the velocity is to the right. During the time interval $\Delta t_2 + \Delta t_3 = 3$ sec the acceleration is 6 m/s² to the left; hence the velocity is decreasing. The time Δt_2 is obtained as follows:

$$a_2 = \frac{v_{2.5}}{\Delta t_2}$$

from which

$$\Delta t_2 = \frac{v_{2.5}}{a_2} = \frac{12}{6} = 2 \text{ s,}$$

and, from the diagram,

$$\Delta t_3 = 3 - \Delta t_2 = 3 - 2 = 1 \text{ s.}$$

The velocity at the end of 5.5 s can be obtained from the diagram as

$$v_{5.5} = a_2(\Delta t_3) = 6(1) = 6 \text{ m/s}$$

and is represented by an ordinate below the t axis, which indicates a velocity to the left. During the last time interval ($\Delta t_4 = 4$ s) the acceleration is to the right, and the magnitude of the velocity at the end of this interval is

$$v_{9.5} = v_{5.5} - a_3(\Delta t_4) = 6 - 1(4)$$
$$= 2 \text{ m/s}.$$

The ordinate for $v_{9.5}$ is still below the t axis and thus represents a velocity to the left.

(a) The total distance traveled is, from the areas under the diagram,

$$Q = Q_1 + Q_2 + Q_3 + Q_4$$
$$= \frac{(2+12)}{2}(2.5) + \frac{12}{2}(2) + \frac{6}{2}(1) + \frac{(6+2)}{2}(4)$$
$$= 17.5 + 12 + 3 + 16$$
$$= \underline{48.5 \text{ m.}} \qquad\qquad\qquad\qquad \text{Ans.}$$

(b) The magnitude of the displacement during the first 5.5 s is

$$q = Q_1 + Q_2 - Q_3$$
$$= 17.5 + 12 - 3 = 26.5 \text{ m,}$$

and the displacement is

$$\mathbf{q} = \underline{26.5 \text{ m to the right.}} \qquad\qquad \text{Ans.}$$

PROBLEMS

8-36 Using a v-t diagram for rectilinear motion with constant acceleration, prove that

$$v_f = v_i + a(\Delta t)$$

and that

$$q = v_i(\Delta t) + \frac{a(\Delta t)^2}{2}.$$

8-37 Using a v-t diagram, prove that

$$v_f^2 = v_i^2 + 2aq$$

for rectilinear motion with constant acceleration.

8-38 A particle with rectilinear motion has a constant acceleration, and the velocity reverses direction during a certain time interval. Derive an equation relating the magnitudes of the initial and final velocities, the acceleration, and the total distance traveled by the particle. Why is this equation valid *only* when the velocity reverses?

8-39 A particle has a constant acceleration of 5 fps^2 to the left. During a 7-sec interval the particle has a displacement to the right, and the total distance traveled is twice the magnitude of the displacement. Determine the initial and final velocities of the particle.

*8-40 A particle in a mechanism is to travel along a straight line with a constant acceleration. The particle is to have a displacement of 8.50 cm to the left during an interval of 0.50 s and have a velocity of 30 cm/s to the left at the end of the period. Determine the acceleration of the particle.

8-41 A particle which has a constant acceleration of 6 ips^2 downward is given an initial velocity of 30 ips upward. Determine the time required for the particle to travel a total distance of 90 in. and the velocity of the particle at the end of the period.

8-42 Solve Problem 8-41 if the constant acceleration is upward. All other values are unchanged.

8-43 The acceleration of a particle moving along a horizontal path is constant and equal to 4 fps^2 to the left. During a certain time interval, the particle has a displacement of 54 ft to the right and travels a total distance of 90 ft. Determine the initial and final velocities of the particle and the time interval.

8-44 A particle has a rectilinear motion with constant acceleration to the right. During a 5-sec interval the particle has a displacement of 25 ft to the right while traveling a total distance of 65 ft. Determine the velocities at the beginning and end of the time interval and the acceleration during the interval.

8-45 A particle moves along a horizontal straight line with a constant acceleration. The velocity of the particle is 100 fps to the right at the start of a 6-sec interval. During the interval the particle travels a total distance of 260 ft. Determine the acceleration and the final velocity of the particle.

8-46 Solve Problem 8-45 if the displacement (instead of the total distance traveled) is
(a) 260 ft to the right.
(b) 260 ft to the left.

8-47 Determine the acceleration and final velocity of the particle in Problem 8-45 which will result in the minimum total distance traveled in 6 sec.

8-48 A point having a vertical rectilinear motion starts from rest and is given a constant acceleration of 40 ips^2 upward for 0.30 sec, after which it continues to move for 0.20 sec at constant velocity.
(a) What constant downward acceleration must now be given to the point so that it will reach a maximum height of 6.4 in. above the initial position?
(b) Determine the distance traveled by the point during the first second if the final acceleration is held constant until the end of the first second.

8-49 A skier on a mountain slope travels 60 ft during a 2-sec interval and another 60 ft during the next second. The acceleration was constant during the 3-sec period. Determine the velocity at the end of the 3-sec period.

8-50 A particle starts from rest at the origin and moves to the right with a constant acceleration for 4 sec. The acceleration then becomes 12 fps^2 to the left for a second time interval. The particle travels a total distance of 276 ft and is 24 ft to the left of the origin at the end of the total time interval. Determine
(a) The acceleration during the first 4-sec interval.
(b) The distance traveled during the first 4-sec interval.
(c) The total time interval.

8-51 A car on a straight road can accelerate at 20 fps^2 and decelerate at 30 fps^2. Determine the minimum time required for the car to travel 450 ft from rest if it comes to rest at the end of the 450-ft course. What maximum speed is attained?

8-52 A particle starts from rest and moves along a straight line for Δt_1 sec with a constant acceleration of 4 fps^2 to the right. The acceleration is changed to 10 fps^2 to the left for the next 3 sec, after which the velocity becomes constant for a third time interval. During the total time interval the particle has a displacement of 25 ft to the right and travels a total distance of 115 ft. Determine the total time of travel.

8-53 A particle with rectilinear motion with an initial velocity to the right moves 144 ft to the right during the first 4-sec interval and 56 ft to the right during the following second. Determine the constant acceleration required for these conditions.

8-54 The velocity of a particle moving along a straight line changes uniformly from 0 to 32 fps to the right while traveling 64 ft. The acceleration changes to a different value (constant), and the particle travels 130 ft during the next 5 sec. After the 5-sec interval the acceleration is changed to 10 fps^2 to the left until the particle has traveled a total distance of 300 ft. Determine the total traveling time.

8-55 A particle moves along a horizontal line; it starts from point A with a velocity of 4 fps to the right and an acceleration of 3 fps^2 to the right and travels for 4 sec. Then the acceleration changes to 8 fps^2 to the left for 2.5 sec. At the end of this time, the particle travels for 5 sec with a constant velocity.

(a) Determine the position of the particle with reference to point A at the end of the last 5-sec interval.

(b) Determine the total distance traveled.

8-56 A man jumped from a stationary balloon at an elevation of 5000 ft above the ground. He waited 10 sec and then pulled his parachute cord. The parachute decelerated him at an average rate of 20 fps² until a velocity of 22 fps was reached. He then descended to the ground at a constant velocity of 22 fps. How long did it take the man to descend to the ground? Neglect the effect of air friction during the first 10 sec and assume the acceleration of gravity to be 32.2 fps².

8-57 A particle moving along a straight line has a constant acceleration of 8 fps² to the right for 5 sec and zero acceleration for the next second. During the 6-sec interval the particle traveled a total distance of 70 ft. Determine the initial velocity of the particle and the resulting displacement during the 6-sec interval.

8-58 Starting with a velocity of 18 fps to the right, a particle moves along a straight line with an acceleration of 3 fps² to the left for 8 sec. The acceleration then becomes zero for Δt sec, after which the velocity changes uniformly until it is 8 fps to the right. The total distance traveled is 109 ft, and the linear displacement is 31 ft to the right. Determine the time interval during which the particle traveled with constant speed.

8-59 A particle starts from rest with a constant acceleration of 10 fps² to the right for a certain time interval. The acceleration then becomes 20 fps² to the left for a second time interval. The sum of the two intervals is 30 sec, and the particle is at the starting point at the end of the second interval. Determine

(a) The total distance traveled.

(b) The maximum speed of the particle.

(c) The average velocity for the 30-sec period.

(d) The average speed for the 30-sec period.

8-6 RELATIVE MOTION—MOTION OF SEVERAL PARTICLES

When two or more particles are in motion simultaneously their motions may be entirely independent. Consider, for example, two trains moving along parallel tracks. The position, velocity, and acceleration of one train does not affect the position, velocity, or acceleration of the second train, and the motions are independent. If the particles are connected by cords, rods, springs, or other constraints, however, the motion of one of the bodies will usually affect the motion of each of the other bodies. Several examples of such motion are illustrated in Fig. 8-9 in which each particle or body has rectilinear motion and the particles are connected by flexible inextensible cables. In Fig. 8-9(a) if body A moves downward a distance Δs_A, body B must move the same distance to the right, and the position coordinates of the two bodies are related by the length of the cord connecting them. The positions of bodies C and D in Fig. 8-9(b) are also related by the length of the cord but the relationship is a little less direct. Figures 8-9(c) and 8-9(d) are other examples in which the position of one of the bodies is related to the positions of the other bodies by the connecting cord or cable.

Relationships between velocities and accelerations of the connected bodies can be developed by writing one or more equations

(a) (b)

(c) (d)

Figure 8-9

relating the positions of the bodies and differentiating with respect to time. A well-defined sign convention is essential if results are to be interpreted correctly. The coordinate giving the position of each of the bodies in Fig. 8-9 is shown on each sketch with an arrow to show the positive direction. In these examples the position of each particle is measured from a fixed reference rather than from a moving origin. Thus the position of A in Fig. 8-9(d) is measured from the fixed pulleys rather than from the moving pulley D.

In Art. 8-1 reference was made to both absolute and relative motion. When the motion is specified with reference to a system of axes fixed in space the motion is called *absolute motion*. When the position, velocity, and acceleration are measured with respect to a moving origin, the motion is called *relative motion*.

Figure 8-10 represents two particles moving in the fixed xy plane. The absolute positions of A and B are shown as $\mathbf{r}_A$ and $\mathbf{r}_B$, respectively. The position of B can also be measured from the $x'y'$ axes which move with point A. The vector $\mathbf{r}_{B/A}$ is defined as the position of B with respect to a system of axes moving with point A. When x' and y' remain parallel to the fixed axes it is convenient to call the position of B relative to

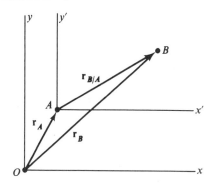

Figure 8-10

363

the $x'y'$ axes the position of B relative to A. Thus the motion of B measured with respect to the $x'y'$ system is the motion of B relative to A.

A vector equation for the position of B relative to A can be written from Fig. 8-10. That is,

$$\mathbf{r}_B = \mathbf{r}_A + \mathbf{r}_{B/A} \quad \text{or} \quad \mathbf{r}_{B/A} = \mathbf{r}_B - \mathbf{r}_A.$$

This equation can be differentiated to give

$$\mathbf{v}_B = \mathbf{v}_A + \mathbf{v}_{B/A} \quad \text{or} \quad \mathbf{v}_{B/A} = \mathbf{v}_B - \mathbf{v}_A$$

and

$$\mathbf{a}_B = \mathbf{a}_A + \mathbf{a}_{B/A} \quad \text{or} \quad \mathbf{a}_{B/A} = \mathbf{a}_B - \mathbf{a}_A.$$

Note that $\mathbf{r}_B$, $\mathbf{r}_A$, and the other quantities are vectors, and when they are differentiated changes in both magnitude and direction must be included as indicated in Art. 8-2. Such differentiation will be discussed in more detail in the following articles.

The following examples illustrate the kinematic relationships for two or more bodies. The procedure of writing one or more equations relating either absolute or relative position functions and differentiating to obtain velocity and acceleration relationships provides a fundamental means of obtaining such relationships and often gives an independent check of other solutions.

EXAMPLE 8-8

Derive equations relating velocities and accelerations of bodies C and D in Fig. 8-9(b).

SOLUTION

The length of the cable connecting bodies C and D is

$$L = 3s_C + s_D + \sum L_p, \tag{a}$$

where L_p represents the length of the rope wrapped around each of the pulleys. Note that $\sum L_p$ is constant; therefore Eq. (a) can be written

$$3s_C + s_D = L - \sum L_p = L', \tag{b}$$

where L' is a constant. When Eq. (b) is differentiated with respect to time it becomes

$$3\frac{ds_C}{dt} + \frac{ds_D}{dt} = 0$$

or

$$\underline{3v_C + v_D = 0.} \qquad \text{Ans.}$$

This equation indicates that if v_C is positive (downward), v_D will be negative, that is, upward, and have a magnitude equal to $3v_C$.

A second differentiation gives

$$3\frac{dv_C}{dt} + \frac{dv_D}{dt} = 0$$

or

$$\underline{3a_C + a_D = 0.} \qquad \text{Ans.}$$

EXAMPLE 8-9

Particles D and E in Fig. 8-9(c) are connected by a 5-ft inextensible cord. When s_E is 3 ft (upward), particle E has a velocity of 8 fps upward and an acceleration of 4 fps² downward. Determine the corresponding velocity and acceleration of D.

SOLUTION

The positive directions for the position functions are selected as shown, and the equation relating the positions is

$$s_D^2 + s_E^2 = 5^2,$$

and when s_E is 3 ft, s_D is 4 ft.

The velocity equation can be obtained by differentiation as

$$2s_D \frac{ds_D}{dt} + 2s_E \frac{ds_E}{dt} = 0$$

or

$$s_D v_D + s_E v_E = 0. \tag{a}$$

The velocity of D can be obtained by substituting numerical values for s_D, s_E, and v_E; thus

$$4v_D + 3(8) = 0$$

or

$$v_D = -6$$

and

$$\mathbf{v}_D = 6 \text{ fps to the right.} \qquad \text{Ans.}$$

Equation (a) can be differentiated to give the following result:

$$\frac{ds_D}{dt}(v_D) + s_D\left(\frac{dv_D}{dt}\right) + \frac{ds_E}{dt}(v_E) + s_E\left(\frac{dv_E}{dt}\right) = 0$$

or

$$v_D^2 + s_D a_D + v_E^2 + s_E a_E = 0.$$

The acceleration of D is obtained by direct substitution. Thus

$$(-6)^2 + 4a_D + (8)^2 + 3(-4) = 0$$

from which

$$4a_D = 12 - 36 - 64 = -88,$$

$$a_D = -22,$$

and

$$\mathbf{a}_D = 22 \text{ fps}^2 \text{ to the right.} \qquad \text{Ans.}$$

EXAMPLE 8-10

Three bodies A, B, and C are connected by cords and pulleys as shown in Fig. 8-9(d). The three bodies are released from rest at the same elevation. Body A has a constant acceleration of 7 ips² upward, and C has a constant acceleration of 2 ips² downward. Determine the time required for the vertical distance between A and B to become 20 in.

SOLUTION

The length of cord wrapped around the pulleys and the horizontal distance between the upper pulleys are all constant and can be neglected when determining velocities and accelerations. The length of the cord between A and B is (downward is positive)

$$L_1 = (s_A - s_D) + (s_B - s_D)$$
$$= s_A + s_B - 2s_D.$$

The length of the other cord is

$$L_2 = s_C + s_D.$$

The position s_D can be eliminated from these two equations to give

$$s_A + s_B - 2(L_2 - s_C) = L_1$$

or

$$s_A + s_B + 2s_C = L_1 + 2L_2.$$

When the preceding expression is differentiated with respect to time it becomes

$$\frac{ds_A}{dt} + \frac{ds_B}{dt} + 2\frac{ds_C}{dt} = 0$$

or

$$v_A + v_B + 2v_C = 0.$$

A second differentiation gives

$$a_A + a_B + 2a_C = 0,$$

and when the given data are substituted into this equation it becomes

$$-7 + a_B + 2(+2) = 0$$

or

$$a_B = 7 - 4 = 3 \text{ fps}^2 \text{ (downward)}.$$

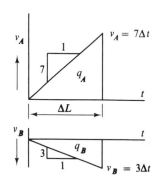

Velocity-time diagrams for the three bodies are drawn in Fig. 8-11 with the accelerations and velocities indicated. Since the displacement of A is upward and that of B is downward, the distance between A and B is equal to the sum of the magnitudes of the displacements:

$$|q_A| + |q_B| = \frac{7(\Delta t)^2}{2} + \frac{3(\Delta t)^2}{2} = 20,$$

and the time interval is found to be

$$(\Delta t)^2 = 4$$

or

$$\Delta t = \underline{2 \text{ sec.}} \qquad \text{Ans.}$$

Figure 8-11

EXAMPLE 8-11

A ball is projected vertically upward from the surface of the earth with an initial velocity of 100 fps. Two seconds later a second ball is thrown upward from the same point with a velocity of 80 fps. How far above the earth will the two balls meet? Neglect air resistance and assume that the acceleration of gravity is 32.2 fps².

SOLUTION

The acceleration of ball 1 is constant, and an equation for the height of ball 1 above the starting point can be obtained as follows (upward is assumed to be positive):

$$a_1 = \frac{dv_1}{dt} = -32.2$$

$$dv_1 = -32.2 \; dt$$

$$v_1 = \int -32.2 \; dt = -32.2t + C_1.$$

Since $v_1 = +100$ fps when $t = 0$ the value of C_1 is 100 and the velocity is

$$v_1 = \frac{dy_1}{dt} = -32.2t + 100,$$

where y_1 is the height of the ball. A second integration gives the position, y_1; that is

$$y_1 = \int (-32.2t + 100) \; dt = -16.1t^2 + 100t + C_2.$$

When $t = 0$, $y_1 = 0$; therefore $C_2 = 0$, and the position of ball 1 is

$$y_1 = -16.1t^2 + 100t. \tag{a}$$

Ball 2 also has a constant acceleration of 32.2 fps² downward, and when $t = 2$ sec it is at the origin $(y_2 = 0)$ with a velocity of 80 fps upward. The equations describing the acceleration and velocity of ball 2 are

$$a_2 = \frac{dv_2}{dt} = -32.2$$

$$v_2 = -32.2t + C_3.$$

Since $v_2 = +80$ when $t = 2$, $C_3 = 80 + 2(32.2) = 144.4$, and v_2 is

$$v_2 = -32.2t + 144.4 = \frac{dy_2}{dt}.$$

The position of ball 2 is

$$y_2 = \int (-32.2t + 144.4) \; dt$$

$$= -16.1t^2 + 144.4t + C_4.$$

Since $y_2 = 0$ when $t = 2$, the value of C_4 is

$$C_4 = 16.1(2)^2 - 144.4(2) = -224.4$$

or

$$y_2 = 16.1t^2 + 144.4t - 224.4 \tag{b}$$

The two balls will meet when $y_1 = y_2$, that is, when

$$-16.1t^2 + 100t = -16.1t^2 + 144.4t - 224.4,$$

which gives

$$t = 5.05 \text{ sec}$$

The value of y when the balls meet, from either Eq. (a) or (b), is

$$y = -16.1(5.05)^2 + 100(5.05)$$
$$= \underline{94.2 \text{ ft.}} \qquad\qquad \text{Ans.}$$

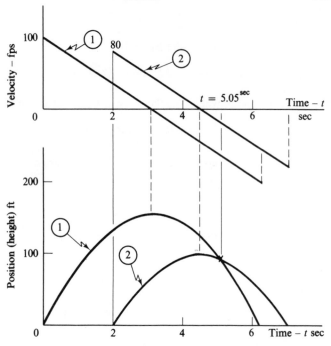

Figure 8-12

Figure 8-12 shows the v-t and s-t diagrams for the two motions. Note that both balls are moving downward when they pass.

EXAMPLE 8-12

Figure 8-13 represents a 1-mile-wide river which has an average velocity of 4 mph south. A motor boat is used to go from A to C. The speed of the boat in still water is 10 mph. In what direction should the boat head in order to go directly to point C, and how much time will be required for the trip?

SOLUTION

Axes are selected as shown on the figure, and angles are measured from the x axis. If the boat is to travel directly from A to C, its absolute velocity must be in the direction from A to C; that is, the angle of the absolute velocity will be $\tan^{-1} 0.5$ from the x axis. The velocity of the boat relative to the water will be 10 mph in a direction to be determined. Note that the speed with which the boat passes a floating leaf or stick, the relative velocity, will be the same whether the water is moving or not. The relative velocity equation for the boat and water is

$$\mathbf{v_B} = \mathbf{v_W} + \mathbf{v_{B/W}}.$$

This equation is illustrated in Fig. 8-14. The angle θ_2 indicates the direction of the velocity of the boat relative to the water, the direction the boat should head, and v_B is the magnitude of the velocity of the boat. These two quantities can be obtained either from the oblique triangle or by writing the equation in terms of the unit vectors $\mathbf{i}$ and $\mathbf{j}$. When the second approach is used the equation becomes

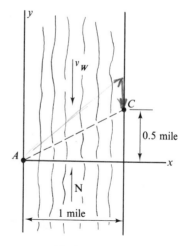

Figure 8-13

$$v_B\left(\frac{2\mathbf{i}+\mathbf{j}}{\sqrt{5}}\right) = -4\mathbf{j} + 10(\mathbf{i}\cos\theta_2 + \mathbf{j}\sin\theta_2)$$

from which

$$\mathbf{i}: \quad v_B\frac{2}{\sqrt{5}} = 10\cos\theta_2, \qquad (a)$$

$$\mathbf{j}: \quad v_B\frac{1}{\sqrt{5}} = -4 + 10\sin\theta_2. \qquad (b)$$

When Eq. (b) is divided by Eq. (a) the result is

$$\frac{1}{2} = \frac{-4 + 10\sin\theta_2}{10\cos\theta_2}$$

or

$$10\cos\theta_2 = -8 + 20\sin\theta_2 = 10\sqrt{1 - \sin^2\theta_2},$$

which gives

$$100(1 - \sin^2\theta_2)$$
$$= 64 - 320\sin\theta_2 + 400\sin^2\theta_2.$$

This equation gives two values of $\sin\theta_2$ as

$$\sin\theta_2 = 0.738 \quad \text{or} \quad -0.0976$$

and

$$\theta_2 = 47.5° \quad \text{or} \quad -5.60°.$$

The first value of θ_2 gives the direction the boat should head to go from A to C, and the second gives the direction of the relative velocity if the boat is to travel from C to A (the heading would be south of due west). The magnitude of the absolute velocity of the boat, when $\theta_2 = 47.5°$ is

Figure 8-14

substituted into Eq. (a), is

$$v_B = \frac{\sqrt{5}}{2}(10 \cos 47.5°) = 7.55 \text{ mph.}$$

The boat must travel $(1^2 + 0.5^2)^{0.5} = 1.118$ miles from A to C at a speed of 7.55 mph. Therefore the time required is

$$\Delta t = \frac{\Delta s}{v} = \frac{1.118}{7.55} = 0.1481 \text{ hr} = \underline{8.88 \text{ min.}} \qquad \text{Ans.}$$

The boat heading should be

$$\theta_2 = 47.5° \text{ north of due east}$$

$$= \underline{42.5° \text{ east of due north (compass heading).}} \qquad \text{Ans.}$$

EXAMPLE 8-13

The two cars A and B in Fig. 8-15 are approaching an overpass intersection. A has a constant velocity of 45 mph east, and B has a constant velocity of 60 mph north. At a given instant A is 500 ft west of the intersection, and B is 300 ft south. Neglect the vertical distance between the roadways. Determine

(a) The initial position of A relative to B.
(b) The velocity of A relative to B.
(c) The minimum distance between A and B.

Figure 8-15

SOLUTION

(a) The position of A relative to B initially is

$$\mathbf{s}_{A/B} = \mathbf{s}_A - \mathbf{s}_B$$

$$= -500\mathbf{i} - (-300\mathbf{j})$$

$$= -500\mathbf{i} + 300\mathbf{j}$$

$$= \underline{583 \text{ ft}} \quad \angle 30.97° \qquad \text{Ans.}$$

$$= \underline{583(-0.858\mathbf{i} + 0.515\mathbf{j}) \text{ ft.}} \qquad \text{Ans.}$$

(b) The velocity of A relative to B is

$$\mathbf{v}_{A/B} = \mathbf{v}_A - \mathbf{v}_B$$

$$= 45\mathbf{i} - 60\mathbf{j}$$

$$= \underline{75 \text{ mph}} \quad \angle 53.15° \qquad \text{Ans.}$$

$$= \underline{75(0.6\mathbf{i} - 0.8\mathbf{j}) \text{ mph.}} \qquad \text{Ans.}$$

(c) At any time t (in seconds) after the initial positions the position of A relative to B is

$$\mathbf{s}_{A/B} = (\mathbf{s}_{A/B})_0 + \Delta(\mathbf{s}_{A/B})$$

$$= (\mathbf{s}_{A/B})_0 + (\mathbf{v}_{A/B})t$$

since $v_{A/B}$ is constant. The relative velocity should be converted to feet per second since $s_{A/B}$ is in feet:

$$v_{A/B} = \left(\frac{75 \text{ miles}}{\text{hr}}\right)\left(\frac{5280 \text{ ft}}{\text{mile}}\right)\left(\frac{\text{hr}}{3600 \text{ sec}}\right)$$
$$= 110 \text{ fps}.$$

Note that 60 mph is the same as 88 fps. When numerical values are substituted into the relative position equation it becomes

$$s_{A/B} = -500\mathbf{i} + 300\mathbf{j} + 110t(0.6\mathbf{i} - 0.8\mathbf{j})$$
$$= (-500 + 66t)\mathbf{i} + (300 - 88t)\mathbf{j}.$$

The value of t which will make the magnitude of $s_{A/B}$ a minimum can be determined by writing an expression for the magnitude of $s_{A/B}$ as a function of t and setting the derivative with respect to t equal to zero. A more convenient solution can be obtained from a graphical representation of the equation for $s_{A/B}$ as shown in Fig. 8-16. Since $s_{A/B}$ is equal to the vector sum of $(s_{A/B})_0$ and $(v_{A/B})t$, it will be minimum when $s_{A/B}$ is perpendicular to $v_{A/B}$, that is, when $s_{A/B} = s_{A/B}(-0.8\mathbf{i} - 0.6\mathbf{j})$. The vector equation becomes

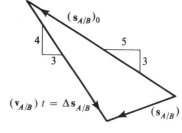

Figure 8-16

$$s_{A/B}(-0.8\mathbf{i} - 0.6\mathbf{j}) = -500\mathbf{i} + 300\mathbf{j} + 66t\mathbf{i} - 88t\mathbf{j},$$

and the component equations are

$$\mathbf{i}: \quad -0.8s_{A/B} = -500 + 66t$$
$$\mathbf{j}: \quad -0.6s_{A/B} = 300 - 88t.$$

The simultaneous solution of these equations gives

$$s_{A/B} = \underline{220 \text{ ft.}} \qquad\qquad \text{Ans.}$$

Note that if either car A or B has an acceleration, the velocity of A relative to B will not be constant in either magnitude or direction and the vector diagram in Fig. 8-16 will not be valid.

PROBLEMS

*8-60 Body A in Fig. P8-60 has a constant velocity of 0.5 m/s to the right. Determine the displacement of body B relative to body A during a 3-s interval.

8-61 Bodies A and B in Fig. P8-61 are connected by a continuous cord as shown. Body A has a velocity of v_A to the right. Determine
(a) The velocity of B relative to A.
(b) The velocity of point C on the cord relative to A.

Figure P8-60

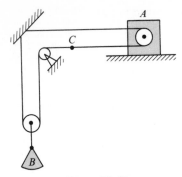

Figure P8-61

8-62 Determine the acceleration of body A in Fig. P8-62 which will result in an acceleration of B relative to A of 12 ips^2 to the right.

Figure P8-62

8-63 Block A in Fig. P8-63 has a velocity of 6 ips to the right when $t = 0$ and a constant acceleration of 5 ips^2 to the left. The cord connecting the bodies remains taut at all times. Determine the total distance traveled by block B during the interval $t = 0$ to $t = 3$ sec.

*8-64 The two bodies in Fig. P8-64 are initially at rest. Determine the constant acceleration of body A which will cause body B to have a displacement of 200 mm to the right during the first 4 s of motion.

8-65 The system of bodies in Fig. P8-65 starts from rest in the position shown. Determine the constant acceleration of C which will cause the right side of body A to slide to the right side of body B in 4 sec.

Figure P8-63

Figure P8-64

8-66 Body A in Fig. P8-66 slides along a fixed horizontal rod, and body B moves in the vertical slot. The cord from body A passes over a *small* peg on B and is tied at C. Derive an equation relating the velocities of A and B. Assume that the cord remains taut.

8-67 Bodies A and B in Fig. P8-67 are connected by a cord which passes over two *small* pulleys.
(a) Write an equation relating s_A and s_B.
(b) Determine the velocity of A in terms of b, s_B, and the velocity of B.

Figure P8-65

Figure P8-66

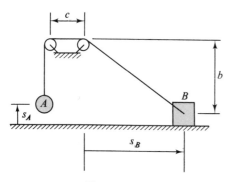

Figure P8-67

8-68 In Problem 8-67 $b = 20$ in. and $c = 10$ in. Body B has a constant velocity of 10 ips to the right. Determine

(a) The velocity of A when $s_B = 0$.
(b) The velocity of A when $s_B = 15$ in.
(c) The acceleration of A when $s_B = 0$.

8-69 Bodies A and B in Fig. P8-69 are con-

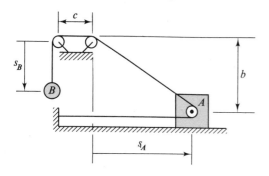

Figure P8-69

nected by a flexible cord which passes over three *small* pulleys.

(a) Write an equation relating s_A and s_B.
(b) Determine the velocity of B in terms of b, s_A, and the velocity of A.

8-70 In Problem 8-69 body A has a constant velocity of 10 ips to the right, $b = 8$ in., and $c = 3$ in. Determine

(a) The velocity of B when $s_A = 0$.
(b) The velocity of B when $s_A = 15$ in.

8-71 Bodies A and B in Fig. P8-71 slide along the horizontal rod and are connected by a cord which passes over the small peg at O. Body A has a constant velocity of 10 ips to the right. When A is directly below O body B is 36 in. to the left of A. Determine

(a) The displacement of B corresponding to a displacement of A of 8 in. to the right from a position below O.
(b) The velocity of B when $s_A = 8$ in. to the right.

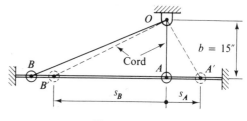

Figure P8-71

8-72 Bodies A and B in Fig. P8-72 slide along fixed rods as shown and are connected by an

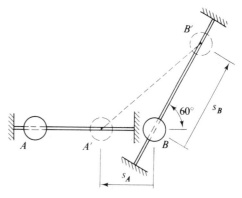

Figure P8-72

inextensible cable. Derive an equation relating the velocities of A and B in terms of their positions.

8-73 The cord connecting A and B in Problem 8-72 is 20 in. long, and B has a constant velocity of 10 ips up and to the right. Determine the acceleration of B when $s_A = s_B$.

8-74 Block C in Fig. P8-74 has a constant acceleration of 1.0 ips^2 and an initial velocity of 2.0 ips, both to the right. Body B has a constant velocity of 3.0 ips to the right. The cord remains taut at all times. Determine the velocity of A when it first hits one of the stops on B.

Figure P8-74

8-75 Bodies A, B, and C in Fig. P8-75 start from rest at a common elevation. Body A has a constant acceleration of 2 ips^2 downward. Body B has a zero acceleration for the first 3 sec, after which its acceleration becomes 3 ips^2 upward. Determine the velocity of C after it has traveled a distance of 12 in.

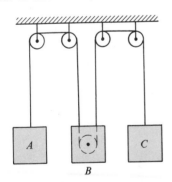

Figure P8-75

*8-76 Blocks A, B, and C in Fig. P8-76 are connected by an inextensible cord as indicated. At a given instant the velocity of C is 90 mm/s to the right, and the velocity of A relative to B is 60 mm/s to the right. Determine the velocity of C relative to A.

Figure P8-76

*8-77 Bodies A, B, and C in Fig. P8-77 move along parallel paths on a horizontal plane. Body C has a velocity of 0.10 m/s to the right. Determine
(a) The velocity of B if A is fixed.
(b) The velocity of A if the velocity of B is 0.1 m/s to the right.

Figure P8-77

8-78 The system of pulleys and cords in Fig. P8-78 support bodies A, B, and C. The system starts from rest in the indicated position. Body A has a constant acceleration of 5 ips^2 downward, and B has a constant acceleration of 5 ips^2

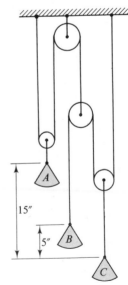

Figure P8-78

upward. Determine the velocity of each body when C first passes either A or B.

8-79 A 64-ft truck and trailer unit has a uniform speed of 60 mph on an interstate highway. The truck is overtaken by a 20-ft passenger car with a uniform speed of 75 mph. The car pulls into the passing lane when it is 100 ft behind the truck and enters the right-hand lane when it is 100 ft ahead of the truck. Determine the total distance the car travels during the passing maneuver.

8-80 Particles A and B move along adjacent parallel paths. Initially, A is 100 ft to the left of B, and each particle has a velocity of 5 fps to the right. The velocity of A remains constant, and the acceleration of B is 2 fps^2 to the left. Determine the total distance traveled by B when the particles pass each other.

8-81 The passenger train A and the freight train B in Fig. P8-81 both start from rest in the position shown and travel to the right. Train A has a constant acceleration of 2 fps^2 to the right until it reaches a maximum speed of 90 mph.

Figure P8-81

The freight has a constant acceleration of 0.5 fps^2 until it attains a maximum speed of 60 mph. Determine
(a) The distance the passenger locomotive travels before overtaking the caboose of the freight.
(b) The time required for the passenger locomotive to pass the freight locomotive.

8-82 A freight elevator is moving upward at a constant rate of 16 fps when it passes a passenger elevator discharging passengers. Two seconds after the freight elevator passes, the passenger elevator starts with a constant acceleration of 12 fps^2 upward. When the velocity of the passenger elevator is 48 fps, the acceleration becomes zero. Determine
(a) The time required by the passenger elevator to overtake the freight elevator.

(b) The distance the passenger elevator travels in overtaking the freight elevator.

8-83 Cars A and B start from rest at a starting line and move along parallel paths. Car A has a constant acceleration of 10 fps^2 to the right for 180 ft, after which it is brought to rest with a constant acceleration of 6 fps^2. Car B has a constant acceleration of 6 fps^2 to the right for 10 sec, after which its velocity remains constant. Determine
(a) The maximum distance A leads B.
(b) The distance A has traveled when the cars have the same velocity.

8-84 Two cars start from rest at the same location and race along a straight track. Car A accelerates at 6.6 fps^2 to a speed of 90 mph and then runs at this constant speed. Car B accelerates at 4.4 fps^2 to a speed of 96 mph and then runs at a constant speed.
(a) Which car will win a 3-mile race and by what distance?
(b) What will be the maximum lead of A over B?
(c) How far will the cars have traveled when B passes A?

8-85 A stone is thrown downward into an abandoned mine shaft with a velocity of 40 fps. Four seconds later the sound of the stone striking water is heard. If sound travels at 1120 fps, determine the distance from the top of the mine shaft to the water surface in the shaft.

8-86 A man is riding in an open hoist (elevator) which is moving upward with a constant velocity of 10 fps. He tosses a stone upward from shoulder height, 5 ft above the floor of the hoist, with a velocity of 20 fps relative to the hoist. Determine
(a) The velocity of the stone when it hits the floor of the hoist.
(b) The maximum height of the stone above the floor of the hoist.
(c) The total vertical distance traveled by the stone before hitting the floor of the hoist.

8-87 Solve Problem 8-86 if the hoist is moving downward with a velocity of 10 fps.

8-88 A stream is 100 ft wide and flows south with an average velocity of 2.0 fps. A man swims

45° north of east and crosses the stream in 30 sec. Determine

(a) The displacement of the man during the 30-sec interval.

(b) The velocity of the man relative to the water.

8-89 An airplane makes a round trip from A to B on each of three days. Point B is 100 miles north of A, and the airspeed of the plane is 120 mph. On the first day there is no wind, on the second day there is a steady 40-mph wind from the north, and on the third day there is a steady 40-mph wind from the west. State which trip will require the least time and which the most.

8-90 A pilot flying a light airplane with an airspeed of 100 mph plans to make a trip to a point due east of his starting point. Upon inquiring at the weather station he learns that he will have a 10-mph head wind at an elevation of 3500 ft and that the wind speed is 20 mph from 20° east of north at an altitude of 5500 ft. At which altitude should he fly to make the best ground speed and what will this ground speed be?

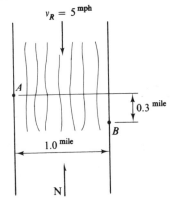

Figure P8-91

8-91 A motor boat is to be used to go from A to B in Fig. P8-91. The speed of the boat in a lake is 10 mph. Determine the least time required for the boat to travel from A to B and the direction the boat should head.

8-92 The boat in Problem 8-91 heads due east for 5 min. What should the heading be for the rest of the trip if the boat is to dock at B?

8-93 Cars A and B in Fig. P8-93 are approaching a grade separation with constant velocities as indicated. A has a speed of 50 mph and is 1.0 mile from the crossover at a given instant. The speed of B is 75 mph, and it is 0.5 mile from the intersection at the same given instant. Determine the location of each car when the distance between them is minimum.

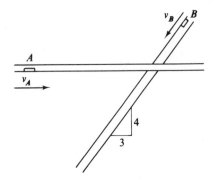

Figure P8-93

8-94 In Example 8-13 at a given instant car A is 500 ft west of the intersection and B is 200 ft south. The velocity of A is 60 fps (constant) to the east. Car B has a constant acceleration of 4 fps^2 north and has a velocity of 30 fps north at the initial instant. Determine the minimum distance between the two cars.

8-7 CURVILINEAR MOTION— RECTANGULAR COMPONENTS

In the preceding articles the particle was assumed to move along a straight line, and the velocity and acceleration were directed along the straight line and could vary only in magnitude (not in direction).

When the particle moves along a curved path these quantities in general vary in both magnitude and direction. It was shown in Art. 8-3 that the velocity is always tangent to the path and thus will change in direction, and perhaps in magnitude, as the particle moves along the path. The acceleration of the particle is not tangent to the path for curvilinear motion. This point will be discussed in more detail in Art. 8-8. Several different coordinate systems can be used to describe the kinematics of curvilinear motion of a particle. The use of rectangular components is discussed in this article.

The position vector from a fixed origin to a point in space can be written in terms of the constant unit vectors $\mathbf{i}$, $\mathbf{j}$, and $\mathbf{k}$ (parallel to the fixed x, y, and z axes, respectively) as

$$\mathbf{r} = x\mathbf{i} + y\mathbf{j} + z\mathbf{k}, \tag{8-6}$$

where the rectangular coordinates (x, y, and z) of the point or particle may vary with time. The displacement of the particle during a time interval is

$$\mathbf{q} = \Delta\mathbf{r} = \mathbf{r}_f - \mathbf{r}_i,$$

where $\mathbf{r}_i$ and $\mathbf{r}_f$ are the initial and final positions of the particle.

The velocity of the particle at any instant was defined in Art. 8-3 as the time rate of change of position of the particle. From Eq. (8-6) the velocity is

$$\mathbf{v} = \frac{d\mathbf{r}}{dt} = \dot{\mathbf{r}} = \dot{x}\mathbf{i} + \dot{y}\mathbf{j} + \dot{z}\mathbf{k}$$

$$= v_x\mathbf{i} + v_y\mathbf{j} + v_z\mathbf{k}, \tag{8-7}$$

where $v_x = \dot{x} = dx/dt$ is the magnitude of the component of the velocity in the x direction with similar expressions for v_y and v_z. Note that the derivative of the product $x\mathbf{i}$ is $\dot{x}\mathbf{i} + x\dot{\mathbf{i}} = \dot{x}\mathbf{i}$ since $\mathbf{i}$ is a constant in both magnitude and direction for fixed axes.

The linear acceleration of the particle at any instant of time is defined as the time rate of change of the velocity of the particle, and from Eq. (8-7) the acceleration is

$$\mathbf{a} = \dot{\mathbf{v}} = \ddot{x}\mathbf{i} + \ddot{y}\mathbf{j} + \ddot{z}\mathbf{k}$$

$$= a_x\mathbf{i} + a_y\mathbf{j} + a_z\mathbf{k} \tag{8-8}$$

in which a_x, a_y, and a_z are the components of the acceleration in the x, y, and z directions, respectively.

Equations (8-6)–(8-8) indicate that curvilinear motion can be considered to be the vector sum of three simultaneous rectilinear motions along the three coordinate axes. This analysis is particularly useful for analyzing motions in which the motion (acceleration and velocity) along each of the coordinate axes is independent.

The motion of a projectile in flight is an example of curvilinear motion. When air resistance is neglected the only force acting on the projectile is its weight[2]; therefore the horizontal component of the acceleration is zero, and the vertical component is the acceleration due to the weight of the body, approximately 32.2 fps² or 9.81 m/s² directed vertically downward. Later in this article it will be shown that a projectile follows a parabolic path when air resistance is neglected. Air resistance is a variable force which changes in both magnitude and direction. The magnitude varies with the velocity and with the density of the air, and the direction is opposite the direction of the velocity. No simple solution is available for such problems, but approximate solutions have been obtained.[3]

Since the horizontal and vertical components of the acceleration of a projectile are known, the resulting motion is the superposition of two rectilinear motions, and rectangular coordinates provide a convenient method of analysis. The accelerations of the component motions are constant (one is zero); therefore the v-t diagrams are straight lines and can often lead to a convenient solution of the resulting motion. The differential equations of the two simultaneous motions can also be used to analyze the motion. Both procedures are discussed in the following examples.

EXAMPLE 8-14

The velocity (in inches per second) of a particle moving in the xy plane is given by

$$\mathbf{v} = 4(t-1)\mathbf{i} + 3(t^2 - 2t + 1)\mathbf{j},$$

where t is time in seconds. The particle is at $\mathbf{r} = (2\mathbf{i} - \mathbf{j})$ in. when $t = 0$. Determine

(a) The equation of the path of the particle.
(b) The position, velocity, and acceleration of the particle when $t = 3$ sec.

SOLUTION

(a) Axes and unit vectors are selected as shown in Fig. 8-17(a). The velocity is

$$\mathbf{v} = \dot{\mathbf{r}} = 4(t-1)\mathbf{i} + 3(t^2 - 2t + 1)\mathbf{j} \qquad \text{(a)}$$

from which

$$\mathbf{r} = \int [4(t-1)\mathbf{i} + 3(t^2 - 2t + 1)\mathbf{j}]\, dt$$

$$= (2t^2 - 4t)\mathbf{i} + (t^3 - 3t^2 + 3t)\mathbf{j} + \mathbf{C},$$

---

[2] A rocket or missile will also be subject to the rocket thrust, and for long-range firings, the curvature of the earth, the rotation of the earth, and sometimes the attraction of the moon or sun may affect the trajectory. Such effects are neglected in this discussion.

[3] D. T. Greenwood, *Principles of Dynamics* (Englewood Cliffs, N.J.: Prentice-Hall, Inc., 1965), pp. 74–76; S. W. Groesberg, *Advanced Mechanics* (New York: John Wiley & Sons, Inc., 1968), pp. 41, 42.

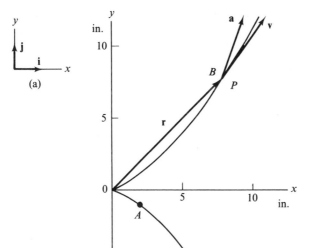

Figure 8-17

where $\mathbf{C}$ is a constant (vector) of integration. When $t = 0$, $\mathbf{r} = 2\mathbf{i} - \mathbf{j}$; therefore

$$\mathbf{C} = 2\mathbf{i} - \mathbf{j},$$

and the position vector is

$$\begin{aligned}\mathbf{r} &= (2t^2 - 4t + 2)\mathbf{i} + (t^3 - 3t^2 + 3t - 1)\mathbf{j} \\ &= 2(t-1)^2\mathbf{i} + (t-1)^3\mathbf{j} \\ &= x\mathbf{i} + y\mathbf{j}. \qquad\qquad\qquad\qquad\qquad\qquad (b)\end{aligned}$$

That is,

$$x = 2(t-1)^2 \quad \text{and} \quad y = (t-1)^3.$$

These are the equations of the path in parametric form. The time t can be eliminated as follows:

$$(t-1) = \left(\frac{x}{2}\right)^{1/2} = y^{1/3},$$

which reduces to

$$\underline{x^3 = 8y^2.} \qquad\qquad \text{Ans.}$$

The curve corresponding to this equation is shown in Fig. 8-17(b).

(b) The acceleration of the particle is obtained by differentiating the velocity expression, Eq. (a), with respect to time, giving

$$\mathbf{a} = \dot{\mathbf{v}} = 4\mathbf{i} + (6t - 6)\mathbf{j}. \qquad\qquad\qquad (c)$$

The position, velocity, and acceleration when $t = 3$ sec are obtained by direct substitution into Eqs. (b), (a), and (c), respectively. Thus

$$\mathbf{r}_{3\,sec} = 2(3-1)^2\mathbf{i} + (3-1)^3\mathbf{j}$$

$$= (8\mathbf{i} + 8\mathbf{j}) \text{ in.} \qquad \text{Ans.}$$

$$\mathbf{v}_{3\,sec} = 4(3-1)\mathbf{i} + 3(9-6+1)\mathbf{j}$$

$$= (8\mathbf{i} + 12\mathbf{j}) \text{ ips,} \qquad \text{Ans.}$$

and

$$\mathbf{a}_{3\,sec} = 4\mathbf{i} + (18-6)\mathbf{j}$$

$$= (4\mathbf{i} + 12\mathbf{j}) \text{ ips}^2. \qquad \text{Ans.}$$

The position of the particle when $t = 3$ sec is shown as point P in Fig. 8-17(b) together with the corresponding velocity and acceleration vectors. Note that the velocity is tangent to the path but that the acceleration is not tangent to the path. As t increases from zero the particle moves along the curved path from A to O and toward B.

EXAMPLE 8-15

A ball is thrown with a velocity of 75 fps with a slope of 4 vertical to 3 horizontal at a building with a flat roof. The building is 45 ft high, and the man throwing the ball is 100 ft from the front wall and is standing in a depression so that the ball leaves his hand at the same elevation as the base of the building. Neglect air resistance.

(a) Will the ball strike the building?
(b) If the ball hits the building, determine where it will hit and with what velocity.

SOLUTION

Figure 8-18(a) shows three possible trajectories of the ball resulting from different magnitudes of the initial velocity. The proper trajectory can be selected by determining the height of the ball when it has traveled 100 ft horizontally. Figure 8-18(b) is drawn on the assumption that the ball will land on the roof, that is, that the ball will be more than 45 ft above the ground when it has traveled 100 ft horizontally. The initial velocity of the ball is

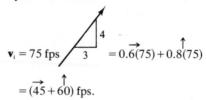

$$= \overset{\rightarrow}{(45} + \overset{\uparrow}{60}) \text{ fps.}$$

Figure 8-18(c) shows the horizontal component of the velocity plotted against time, and Fig. 8-18(d) shows the vertical component of the velocity plotted against time. Although the horizontal distances on the v-t diagrams and on the sketch of the path do not represent the same variable, it is helpful if scales are selected so that points on the v-t diagrams line up vertically with the corresponding points on the sketch

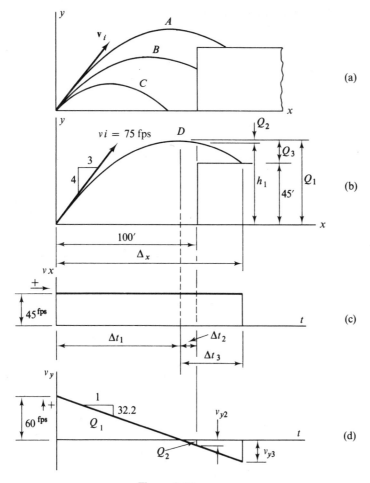

Figure 8-18

of the path. For example, when the ball is at the top of its trajectory, $v_y = 0$ is the corresponding point on the v_y-t diagram. Since a_x is zero, the v_x-t diagram is a horizontal straight line. The vertical component of the acceleration has a constant value of 32.2 fps² downward; thus, the v_y-t diagram in Fig. 8-18(d) is a straight line with a slope of 32.2. The area under the v_x-t diagram represents the horizontal distance traveled during any time interval. Therefore, when the ball has traveled 100 ft horizontally

$$\Delta x = 100 = v_x(\Delta t_1 + \Delta t_2) = 45(\Delta t_1 + \Delta t_2)$$

and

$$\Delta t_1 + \Delta t_2 = \frac{100}{45} = 2.222 \text{ sec.} \qquad (a)$$

The time Δt_1 required for the ball to reach its maximum height at D

381

can be determined from the slope of the v_y-t diagram. Thus

$$\frac{v_{yi}}{\Delta t_1} = a_y = 32.2 = \frac{60}{\Delta t_1}$$

and

$$\Delta t_1 = \frac{60}{32.2} = 1.863 \text{ sec.}$$

From Eq. (a) the value of Δt_2 is

$$\Delta t_2 = 2.222 - \Delta t_1 = 2.222 - 1.863 = 0.359 \text{ sec.}$$

The velocity v_{y2} can be determined from the slope of the v_y-t curve and is

$$v_{y2} = a_y(\Delta t_2) = 32.2(0.359) = 11.56 \text{ fps,}$$

$$\mathbf{v}_{y2} = 11.56 \text{ fps downward.}$$

The difference between areas Q_1 and Q_2 represents the height h_1 of the ball above the ground when it reaches the wall; that is,

$$h_1 = Q_1 - Q_2 = \frac{60(1.863)}{2} - \frac{11.56(0.359)}{2}$$

$$= 55.89 - 2.07 = 53.8 \text{ ft.}$$

Since the value of h_1 is more than the height of the building, the ball will fall on the roof of the building. The vertical distance, Q_3, which the ball must fall from the top of its path before landing on the roof is represented by the corresponding area under the v_y-t diagram. That is,

$$Q_3 = Q_1 - 45 = 55.89 - 45 = 10.89 \text{ ft,}$$

which is equal to the area of the corresponding triangle. Thus

$$\frac{v_{y3}(\Delta t_3)}{2} = 10.89.$$

The slope of the v_y-t diagram gives

$$a_y = \frac{v_{y3}}{\Delta t_3} = 32.2,$$

and the simultaneous solution of these two equations gives

$$\Delta t_3 = 0.822 \text{ sec}$$

and

$$v_{y3} = 26.5 \text{ fps.}$$

The horizontal distance traveled by the ball, Δx, when it is 45 ft above the ground is, from the area under the v_x-t diagram,

$$\Delta x = v_x(\Delta t_1 + \Delta t_3) = 45(1.863 + 0.822)$$

$$= 120.8 \text{ ft.}$$

Therefore the ball will land on the roof of the building 20.8 ft from the left edge.

The velocity of the ball as it strikes the roof is the vector sum of the component velocities and is

$$\mathbf{v}_f = \mathbf{v}_x + \mathbf{v}_y$$

$$= (45\mathbf{i} - 26.5\mathbf{j})\ \text{fps} = 52.2\ \text{fps} \quad \angle 30.6° \qquad \text{Ans.}$$

ALTERNATE SOLUTION

In some instances, it may be desirable to solve the differential equations of motion instead of using v-t diagrams for projectile motion. The position vector is

$$\mathbf{r} = x\mathbf{i} + y\mathbf{j},$$

and the equation for the acceleration is

$$\ddot{\mathbf{r}} = -g\mathbf{j}. \qquad (b)$$

Integration of Eq. (b) gives the velocity as

$$\dot{\mathbf{r}} = -gt\mathbf{j} + \mathbf{C}_1,$$

where $\mathbf{C}_1$ is a constant of integration. If $\mathbf{u}$ is the velocity of the projectile when t is zero, $\mathbf{C}_1$ must be equal to $\mathbf{u}$, and the equation for the velocity becomes

$$\dot{\mathbf{r}} = -gt\mathbf{j} + \mathbf{u} = u_x\mathbf{i} + (u_y - gt)\mathbf{j},$$

where u_x and u_y are the components of the initial velocity. The velocity equation can be integrated to give the position. Thus,

$$\mathbf{r} = u_x t\mathbf{i} + \left(u_y t - \frac{gt^2}{2}\right)\mathbf{j} + \mathbf{C}_2.$$

The constant $\mathbf{C}_2$ gives the position of the projectile when $t = 0$, and if the origin is selected at this position, this constant is zero and the position equation becomes

$$\mathbf{r} = (x\mathbf{i} + y\mathbf{j}) = u_x t\mathbf{i} + \left(u_y t - \frac{gt^2}{2}\right)\mathbf{j}.$$

The parametric equations of the path are

$$x = u_x t \quad \text{and} \quad y = u_y t - \frac{gt^2}{2}. \qquad (c)$$

When t is eliminated from these two expressions, the equation of the path in cartesian coordinates becomes

$$y = \frac{u_y}{u_x} x - \frac{gx^2}{2u_x^2}$$

or $\qquad (d)$

$$2u_x^2 y = 2u_x u_y x - gx^2,$$

which is the equation of a parabola.

When numerical data are substituted into Eq. (d) it becomes

$$y = \frac{60}{45}x - \frac{32.2x^2}{2(45)^2}$$

$$= 1.333x - 0.00795x^2. \tag{e}$$

The height of the ball when it reaches the building is the value of y when $x = 100$ ft, that is,

$$y = h_1 = 1.333(100) - 0.00795(100)^2$$

$$= 53.8 \text{ ft},$$

which shows that the ball passes above the front of the building. When $y = 45$ ft (the height of the building) the resulting quadratic equation gives two values of x, which are

$$x = 46.8 \text{ ft} \quad \text{and} \quad 120.9 \text{ ft}.$$

The second value locates the point where the ball lands on the roof. The value of t when $x = 120.9$ ft can be obtained from Eq. (c) as

$$t = \frac{x}{u_x} = \frac{120.9}{45} = 2.69 \text{ sec}.$$

The velocity of the ball as it strikes the roof is obtained by substituting numerical values in the expression for $\dot{r}$, that is,

$$\dot{r} = u_x \mathbf{i} + (u_y - gt)\mathbf{j}$$

$$= 45\mathbf{i} + [60 - 32.2(2.69)]\mathbf{j} = 45\mathbf{i} - 26.6\mathbf{j}$$

$$= \underline{52.3(0.861\mathbf{i} - 0.509\mathbf{j}) \text{ fps}.} \qquad \text{Ans.}$$

PROBLEMS

Note: Neglect air resistance in all projectile problems unless otherwise stated.

8-95 The position of a point moving in the xy plane is given by

$$\mathbf{r} = 2t^3\mathbf{i} + 5t^2\mathbf{j},$$

where r is in inches when t is time in seconds. Determine
(a) The displacement during the interval $t = 1$ to $t = 3$ sec.
(b) The average velocity during the interval $t = 1$ to $t = 3$ sec.
(c) The velocity at $t = 2$ sec.
(d) The acceleration at $t = 2$ sec.

*8-96 The position of a particle moving in the xy plane is given by

$$\mathbf{r} = Bt^2\mathbf{i} + \left(\frac{C}{t^2}\right)\mathbf{j},$$

where $B = 2$ m/s^2, $C = 4$ m · s^2, and t is time in seconds. Determine

(a) The displacement during the interval $t = 1$ to $t = 2$ s.
(b) The velocity at $t = 2$ s.
(c) The acceleration at $t = 2$ s.

8-97 The position vector for a particle is

$$\mathbf{r} = 2t\mathbf{i} + (2t^2 - 3)\mathbf{j},$$

where r is measured in feet when the time, t, is in seconds. Determine
(a) The displacement of the particle during the time interval from $t = 1$ to $t = 3$ sec.
(b) The velocity of the particle when $t = 1$ sec.

8-98 The position of a particle is given by

$$\mathbf{r} = (t^2 - 2)\mathbf{i} + t^3\mathbf{j} + e^{-t}\mathbf{k},$$

where r is in inches and t is in seconds.
(a) Locate the particle when $t = 1$ sec.
(b) Determine the velocity of the particle when $t = 1$ sec.
(c) Determine the acceleration of the particle when $t = 1$ sec.

8-99 A point moves along the curve $y^2 = 25x$ (x and y in feet) in such a manner that the y component of the position vector at any time t is $5t^2\mathbf{j}$. Determine the acceleration of the point when $t = 2$ sec.

*8-100 The x coordinate of a particle moving along the path $by = x^2$ is given by $x = b^{nt}$. The constants are $b = 4$ m and $n = 0.2$ s^{-1}. Determine the acceleration of the particle when $t = 2$ s.

8-101 The velocity of a particle moving in the xy plane is given by the expression $\mathbf{v} = (4t - 1)\mathbf{i} + 2\mathbf{j}$, where v is in feet per second when t is in seconds. The particle is located at $\mathbf{r} = (3\mathbf{i} + 4\mathbf{j})$ ft when $t = 1$ sec. Determine the equation of the path in terms of x and y.

8-102 The acceleration of a particle moving in the xy plane is given by

$$\mathbf{a} = -\omega^2(\mathbf{i}b\cos\omega t + \mathbf{j}c\sin\omega t),$$

where ω, b, and c are constants. When the time $t = 0$ the position and velocity of the particle are $\mathbf{r} = 2b\mathbf{i}$ and $\dot{\mathbf{r}} = c\omega\mathbf{j}$, respectively.
(a) Determine the equation of the path of the particle.
(b) The values of the constants are $b = 2c = 6$ in. and $\omega = (\pi/2)$ rad per sec. Calculate the position and velocity of the particle at $t = 3$ sec.

8-103 Block A in Fig. P8-103 slides in the vertical slot in member BC as BC moves along the x axis. The curved guide passes through a hole in A and thus causes A to move vertically as BC moves horizontally. The value of b in the

equation is 2 in. Determine the velocity and acceleration of A when $\mathbf{x} = 5\mathbf{i}$ in., $\dot{\mathbf{x}} = 3\mathbf{i}$ ips, and $\ddot{\mathbf{x}} = 0$.

8-104 Determine the acceleration of BC in Problem 8-103 which will result in a zero vertical component of the acceleration of A. Other data for Problem 8-103 are unchanged.

*8-105 Roller A in Fig. P8-105 slides along the fixed curved track as it moves in the vertical slot in member BC. Member BC moves horizontally along the fixed track as shown.
(a) Derive equations for the velocity and acceleration of A in terms of b, x, $\dot{x}$, and $\ddot{x}$.
(b) Determine the velocity and acceleration of A when $b = 10$ cm, $\mathbf{x} = 4\mathbf{i}$ cm, $\dot{\mathbf{x}} = 10\mathbf{i}$ cm/s, and $\ddot{\mathbf{x}} = -8\mathbf{i}$ cm/s^2.

Figure P8-105

8-106 The right triangular plate in Fig. P8-106 moves in a vertical plane, with vertices B and C moving in horizontal and vertical grooves as shown. Determine, in terms of b, θ, $\dot{\theta}$, and $\ddot{\theta}$, the velocity and acceleration of point A.

Figure P8-106

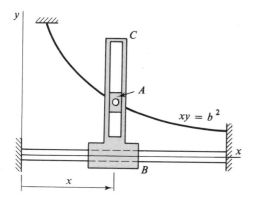

Figure P8-103

8-107 Block B in Fig. P8-107 slides along the fixed curve whose equation is $x^2 = by$ and at the same time slides in a slot in member AC. The value of b is 1 in., and when $x = 2.0$ in. the horizontal component of the velocity of B is 2.0 ips to the right and the horizontal component of its acceleration is zero. Determine the angular velocity and acceleration of AB at this instant.

Figure P8-107

8-108 The triangular block ABC in Fig. P8-108 moves in a vertical plane with its sides AB and BC in contact with fixed points D and E, respectively. The angular velocity of the triangle is 5.0 rad per sec counterclockwise when $\tan \theta = 0.75$. Determine the velocity of point B when in this position.

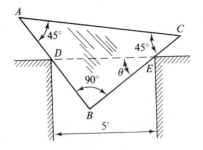

Figure P8-108

8-109 Bar AB in Fig. P8-109 moves with end A in a fixed horizontal slot while the slot in the member slides along the fixed pin C. When $\theta = 60°$, the velocity of A is $\mathbf{v}_A = -16\mathbf{i}$ ips. Determine, for this position,
(a) The angular velocity of the bar.
(b) The velocity of point B.

Figure P8-109

8-110 Determine the acceleration of pin A in Problem 8-109 which will result in zero angular acceleration of AB with the conditions given.

8-111 Determine the angular acceleration of the bar in Problem 8-109 which will result in zero acceleration of pin A.

8-112 In Fig. P8-112, AB is a slot fixed in the xy plane. Member C slides horizontally along the x axis and contains a curved slot whose equation is $bx_1 = y_1^2$, where the x_1y_1 axes move with C. The roller P extends through both slots. The position of O_1 from the xy axes is x_2 as shown. Prove that the velocity of P is given by

$$\dot{\mathbf{r}} = (\mathbf{i} - \mathbf{j})\dot{x}_2 \left[9 - \left(\frac{4x_2}{b} \right) \right]^{-0.5}.$$

8-113 A ball is thrown horizontally from a 240-ft building with an initial velocity of 80 fps.
(a) Determine the distance from the bottom of the building to the point where the ball hits the ground.
(b) What is the height of the ball from the ground when it is 250 ft from the building?

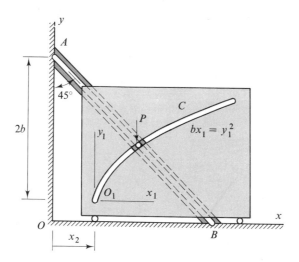

Figure P8-112

8-114 A firing table for a 105-mm howitzer gives a range of 5200 yd for an elevation of 443 mil with an initial velocity of 875 fps. Given that 1600 mil = 90°, determine the reduction in range caused by air resistance.

8-115 A ball is thrown at a building which is 200 ft from the thrower. The initial speed of the ball is 100 fps, and the ball strikes the side of the building 3.0 sec after being thrown. Determine the height (above the initial position) at which the ball strikes the building.

8-116 An airplane at 20,000-ft altitude (see Fig. P8-116) is flying along a course which passes over an antiaircraft gun. The speed of the plane is 800 fps. The gun angle is 60°, and the muzzle velocity of the projectile is 2000 fps. Determine the angle, β, of the line of sight at which the projectile should be fired if it is to hit the plane during the upward flight of the projectile.

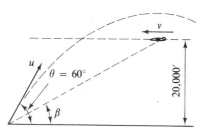

Figure P8-116

8-117 A baseball is hit 3 ft above the ground and just clears the 3-ft center-field fence 400 ft away. The initial speed of the ball is 150 fps. Determine the angle of the initial velocity of the ball (two values) and its maximum elevation corresponding to each angle.

8-118 A batter hit a baseball 4 ft above the ground toward the 10-ft-high center-field fence 455 ft away. The ball left the bat with a velocity of 125 fps with a slope of 3 vertical to 4 horizontal. The center fielder can catch a ball 9 ft above the ground. Did the batter fly out, hit a double against the fence above the fielder's glove, or hit a home run? Show all calculations.

*8-119 A mortar is fired with a muzzle velocity of 250 m/s. A radar range station indicates that the projectile has an elevation of 200 m 23 s after firing. Determine the horizontal distance traveled by the projectile during the 23-s interval and the speed of the projectile at the end of the interval.

*8-120 A low-speed projectile with a muzzle velocity of 100 m/s is fired at a target which is 600 m horizontally and 150 m vertically from the gun. Determine the angle of elevation of the gun necessary for the projectile to hit the target.

8-121 A three-stage rocket fired from Cape Kennedy was known to have reached a speed of 10,000 fps with an unknown slope at an altitude of 150,000 ft when the third stage burned out. The FM telemeter signal faded out exactly 600 sec after burnout as the rocket fell below the eastern horizon. Assume that $g = 30.0$ fps^2 (an average value) and neglect curvature and rotation of the earth. Calculate

(a) The maximum altitude attained by the rocket.

(b) The horizontal distance traveled.

8-122 A golf ball is driven from a tee which is 50 ft above a level fairway. The maximum height of the ball in flight above the tee is 70 ft, and the horizontal distance traveled by the ball before it strikes the fairway is 700 ft. Determine the velocity of the ball as it hits the fairway.

8-123 The stream from a fire hose is directed toward a building as shown in Fig. P8-123 and is to hit point P. Show that the minimum required nozzle velocity will occur when θ is given by the equation $\tan 2\theta = -x/y$.

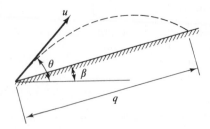

Figure P8-123

8-124 Determine the minimum required nozzle velocity in Problem 8-123 if the stream is to hit point P when $x = 80$ ft and $y = 50$ ft.

8-125 A stream of water falls on a sloping plane as shown in Fig. P8-125. Show that the

Figure P8-125

distance q where the water strikes the plane will be greatest when the angle θ satisfies the equation $\tan 2\theta = -\cot \beta$.

8-126 Determine the minimum speed a jeep must have at A in Fig. P8-126 to clear the stream and land on the opposite bank at B.

Figure P8-126

Figure P8-127

Figure P8-128

Figure P8-129

8-127 A ball is thrown with an initial velocity of 100 fps as shown in Fig. P8-127 and just clears the edges A and B of a building 120 ft away. Determine the height, h, and the width, b, of the building.

8-128 Determine the initial velocity u of a ball thrown down the slope as shown in Fig. P8-128.

8-129 A fire hose is directed at a building from a point x ft away as shown in Fig. P8-129. The nozzle velocity is u fps.
(a) Show that the maximum value of y will result when $\tan \theta = u^2/gx$.
(b) Determine the velocity of the water as it hits the building at the maximum height when $u = 100$ fps and $x = 120$ ft.

8-8 CURVILINEAR MOTION—NORMAL AND TANGENTIAL COMPONENTS

Curvilinear motion was analyzed by means of rectangular components in Art. 8-7, and in certain types of problems this analysis provides the simplest and most direct solution. Rectangular components of velocity and acceleration, however, fail to relate the components directly to the path being traversed, and this may be a disadvantage for some problems. For certain problems in curvilinear motion, particularly circular motion of a particle, tangential and normal components of velocity and acceleration may be more convenient than rectangular components.

The position of a particle which moves along a curved path is completely specified when the path is known and when the distance, measured along the path, from some fixed point on the path to the particle is known as a function of time. The position of particle P in Fig. 8-19(a) as it moves along the plane curve is given by the distance s measured along the curve from the fixed point A.

It is convenient to introduce unit vectors in the tangential and normal directions, as shown in Fig. 8-19(a). The unit vector $\mathbf{e}_n$ is along

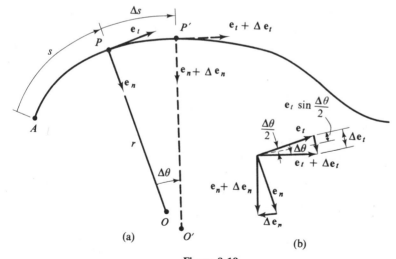

Figure 8-19

the principal radius of curvature[4] at the point, and the positive direction is from the curve toward the center of curvature at O. The unit vector $\mathbf{e}_t$ is tangent to the path at the point, and the positive direction is in the direction of increasing s as shown. *Since these unit vectors will change in direction, they have time derivatives different from zero.*

The velocity of a particle is always tangent to the path as shown in Art. 8-3. The magnitude of the velocity can be obtained by analyzing Fig. 8-19(a). As the particle moves from P to P' it travels a distance Δs in a time interval Δt and has an average speed of $\Delta s/\Delta t$. The magnitude of the instantaneous velocity is the limit of this ratio as Δt approaches zero; that is,

$$v = \lim_{\Delta t \to 0} \frac{\Delta s}{\Delta t} = \frac{ds}{dt} = \dot{s}$$

and

$$\mathbf{v} = \dot{s}\mathbf{e}_t. \tag{8-9}$$

The acceleration of the particle is defined as the time rate of change of the velocity. The velocity from Eq. (8-9) is the product of two factors, and its derivative is

$$\mathbf{a} = \dot{\mathbf{v}} = \ddot{s}\mathbf{e}_t + \dot{s}\dot{\mathbf{e}}_t.$$

The quantity $\dot{\mathbf{e}}_t$ is the time rate of change of $\mathbf{e}_t$ and can be evaluated as indicated in Fig. 8-19(b). Since $\mathbf{e}_t$ has a constant magnitude (unity), it can change only in direction. The magnitude of $\Delta\mathbf{e}_t$ is

$$\Delta e_t = 2\left(e_t \sin\frac{\Delta\theta}{2}\right) = 2(1)\sin\frac{\Delta\theta}{2},$$

and as $\Delta\theta$ becomes very small $\sin(\Delta\theta/2)$ is approximately equal to $\Delta\theta/2$ and Δe_t becomes

$$\Delta e_t = 2\left(\frac{\Delta\theta}{2}\right) = \Delta\theta.$$

The direction of $\Delta\mathbf{e}_t$ is perpendicular to $\mathbf{e}_t$, that is, in the direction of $\mathbf{e}_n$, and the time derivative of $\mathbf{e}_t$ is

$$\dot{\mathbf{e}}_t = \lim_{\Delta t \to 0}\frac{\Delta \mathbf{e}_t}{\Delta t} = \lim_{\Delta t \to 0}\frac{\Delta\theta}{\Delta t}\mathbf{e}_n = \dot{\theta}\mathbf{e}_n. \tag{8-10}$$

Note that $\dot{\theta}$ must be given in radians per unit of time.

Although it is not needed for this analysis, it is useful to note that $\mathbf{e}_n$ can be differentiated in the same manner to give

$$\dot{\mathbf{e}}_n = -\dot{\theta}\mathbf{e}_t \tag{8-10a}$$

since the direction of $\Delta\mathbf{e}_n$ is opposite that of $\mathbf{e}_t$.

---

[4] The following discussion is limited to coplanar curvilinear motion, but the same reasoning applies for noncoplanar motion, provided that the principal radius of curvature is used.

The linear acceleration of the particle thus becomes

$$\mathbf{a} = \dot{\mathbf{v}} = \ddot{s}\mathbf{e}_t + \dot{s}\dot{\theta}\mathbf{e}_n. \qquad (8\text{-}11)$$

It is sometimes helpful to relate angular and linear quantities in the preceding equations. Points O and O' in Fig. 8-19(a) are the centers of curvature of the curve at P and P' respectively, and r is the radius of curvature at P. It can be observed that as $\Delta\theta$ approaches zero, O' will approach O and ds will be equal to $r\,d\theta$,[5] where $d\theta$ is in radians.

When this quantity is divided by dt, it becomes

$$v = \frac{ds}{dt} = r\frac{d\theta}{dt} = r\dot{\theta} = \dot{s}, \qquad (8\text{-}12)$$

and Eq. (8-11) can be written as

$$\mathbf{a} = \ddot{s}\mathbf{e}_t + \frac{\dot{s}^2}{r}\,\mathbf{e}_n = \ddot{s}\mathbf{e}_t + r\dot{\theta}^2\mathbf{e}_n = \mathbf{a}_t + \mathbf{a}_n, \qquad (8\text{-}13)$$

where $\mathbf{a}_t$ and $\mathbf{a}_n$ are the tangential and normal components, respectively, of the acceleration of the particle.

It should be noted that the velocity is always tangent to the path and thus has no normal component. The velocity of the point has a magnitude $\dot{s}$ in the direction $\mathbf{e}_t$—see Eq. (8-9). Note that the magnitude of the tangential component of the acceleration, $\ddot{s}$, results from a change in the magnitude of v. The normal component of the acceleration is due to the change of the direction of the velocity as indicated by $\dot{\mathbf{e}}_t$. The normal component is always directed from the particle toward the center of curvature of the path and is zero only when the velocity is zero or is not changing in direction (rectilinear motion). The tangential component is zero when the speed is constant.

For the special case of motion on a circular path, r is constant and from Eqs. (8-12) and (8-5)

$$\ddot{s} = r\ddot{\theta},$$

which makes it possible to write Eq. (8-13) as

$$\mathbf{a} = r\ddot{\theta}\mathbf{e}_t + r\dot{\theta}^2\mathbf{e}_n.$$

Sometimes it is convenient to change the independent variable in the expression for a_t in Eq. (8-13) from t to s. This can be accomplished by writing

$$a_t = \ddot{s} = \dot{v} = \frac{dv}{dt} = \frac{dv}{ds}\frac{ds}{dt} = v\frac{dv}{ds},$$

---

[5] The radius of curvature is

$$r = \frac{[1 + (dy/dx)^2]^{3/2}}{d^2y/dx^2}$$

as derived in any standard calculus text.

or in vector form

$$\mathbf{a}_t = v\frac{dv}{ds}\,\mathbf{e}_t.$$

The following examples illustrate the use of the normal and tangential components of acceleration.

EXAMPLE 8-16

The position of a particle moving on a circular path with a radius of 32 ft varies according to the law $s = 3t^2 + 4t$, where s is the distance in feet from a fixed point to the particle measured along the path and t is the time in seconds. The line from the center of the path to the particle is turning counterclockwise when $t = 1$ sec, and the particle is at the top of the path when $t = 2$ sec. Determine the acceleration of the particle when $t = 2$ sec.

SOLUTION

The positive direction of motion along the path can be determined from the value of the velocity when $t = 1$ sec. The magnitude of the velocity is

$$v = \frac{ds}{dt} = 6t + 4.$$

Thus v is positive when $t = 1$ sec and P is moving counterclockwise along the path; therefore counterclockwise is the positive direction. The magnitude of the normal component of the acceleration is v^2/r, and when $t = 2$ sec the magnitude of the velocity is

$$v = 6(2) + 4 = 16 \text{ fps.}$$

The normal component of the acceleration is directed from the particle toward the center of the circular path, which is downward when $t = 2$ sec. Therefore

$$\mathbf{a}_n = \frac{v^2}{r}\,\mathbf{e}_n = \frac{16^2}{32}\,\mathbf{e}_n = 8 \text{ fps}^2 \downarrow.$$

The magnitude of the tangential component of the acceleration is

$$a_t = \frac{dv}{dt} = 6 \text{ fps}^2,$$

which is positive for all values of t and shows that the velocity is increasing in magnitude and the direction of the tangential component of acceleration is the same as the direction of the velocity, that is, to the left when $t = 2$ sec. The acceleration when $t = 2$ sec is

$$\mathbf{a} = \mathbf{a}_n + \mathbf{a}_t = \overset{\downarrow}{8} + \overset{\leftarrow}{6} = 10 \text{ fps}^2 \qquad \text{Ans.}$$

EXAMPLE 8-17

Example 8-15 analyzed the motion of a ball which was thrown toward a building. Determine the normal and tangential components of its acceleration and the radius of curvature of the trajectory

(a) When the ball is at the top of the trajectory.

(b) Just before the ball hits the roof.

SOLUTION

(a) At the top of the path the velocity is

$$\mathbf{v} = 45\mathbf{i} \text{ fps},$$

and the acceleration is

$$\mathbf{a} = -32.2\mathbf{j} \text{fps}^2;$$

therefore

$$a_t = \underline{0} \qquad \text{Ans.}$$

and

$$\mathbf{a}_n = \underline{-32.2\mathbf{j} \text{ fps}^2.} \qquad \text{Ans.}$$

The magnitude of the normal component of the acceleration [see Eq. (8-13)] is

$$a_n = \frac{\dot{s}^2}{r} = \frac{v^2}{r} = 32.2$$

from which

$$r = \frac{(45)^2}{32.2} = \underline{62.9 \text{ ft.}} \qquad \text{Ans.}$$

(b) Just before the ball hits the roof its velocity (see Example 8-15) is

$$\mathbf{v} = 45\mathbf{i} - 26.5\mathbf{j}$$

or

$$v\mathbf{e}_t = 52.2(0.862\mathbf{i} - 0.508\mathbf{j}) \text{ fps},$$

and the acceleration is

$$\mathbf{a} = -32.2\mathbf{j} \text{ fps}^2.$$

The acceleration can be resolved into n and t components by various methods. For example,

$$\mathbf{a}_t = (\mathbf{a} \cdot \mathbf{e}_t)\mathbf{e}_t$$

$$= (-32.2\mathbf{j}) \cdot (0.862\mathbf{i} - 0.508\mathbf{j})(0.862\mathbf{i} - 0.508\mathbf{j})$$

$$= \underline{(14.10\mathbf{i} - 8.31\mathbf{j}) \text{ fps}^2}$$

$$= \underline{16.38\mathbf{e}_t \text{ fps}^2.} \qquad \text{Ans.}$$

and

$$\mathbf{a}_n = \mathbf{a} - \mathbf{a}_t = -32.2\mathbf{j} - (14.10\mathbf{i} - 8.31\mathbf{j})$$

$$= \underline{(-14.10\mathbf{i} - 23.89\mathbf{j}) \text{ fps}^2}$$

$$= \underline{27.7\mathbf{e}_n \text{ fps}^2.} \qquad \text{Ans.}$$

The radius curvature of the trajectory just before impact is

$$r = \frac{v^2}{a_n} = \frac{(52.2)^2}{27.7}$$

$$= \underline{98.4 \text{ ft.}}$$ Ans.

PROBLEMS

8-130 A particle travels in a vertical plane along a 10-ft-diameter circular path. The distance along the path from the bottom of the circle to the particle is $s = t^2 - 3t$, where s is in feet and t is in seconds (counterclockwise motion is positive). Determine the velocity and acceleration of the particle when $t = 3$ sec.

8-131 The velocity of a particle P moving along a circular path varies according to the expression $v = 2t^3 - 6$, where v is the magnitude of the velocity in feet per second and t is time in seconds. The particle is moving downward in the position shown in Fig. P8-131 when $t = 2$ sec. Determine the acceleration of the particle at this instant.

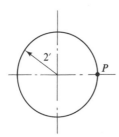

Figure P8-131

*8-132 The particle P in Fig. P8-132 has a tangential acceleration whose magnitude is given by the expression $a_t = 12t - 8$, where a_t is in meters per second per second and t is in seconds. Line OP starts from rest with a clockwise angular acceleration when t is zero. When $t = 2$ s the particle is at point A. Determine the linear acceleration of P when $t = 2$ s.

8-133 The velocity of a point moving along a 40-ft-diameter circular track is given by the equation $v^2 = 16s$, where v is in feet per second and s is the distance in feet from the origin to the point (measured along the track). Determine the magnitude of the acceleration of the point when $s = 4$ ft.

Figure P8-132

8-134 Point A moves on a circle with decreasing magnitude of velocity. At a given instant, the point has the acceleration and position shown in Fig. P8-134. Determine the velocity of point A.

Figure P8-134

8-135 A small cart starts from rest at point A in Fig. P8-135 and moves counterclockwise along the circular track. The speed of the cart increases uniformly at the rate of 25 ips². Determine the velocity of the cart when the magnitude of the acceleration is 65 ips².

*8-136 The magnitude of the velocity of a particle changes uniformly from 14 m/s downward at A to 4 m/s upward at B (see Fig. P8-136) as the particle moves along the path from A to B. Determine the acceleration of the particle when it is at point C.

8-137 The body in Fig. P8-137 rotates about point O. When line OA is in the position shown; point A has a speed of 8.0 fps and an acceleration of 34.0 fps² downward. Determine the angular acceleration of line OA.

Figure P8-135

Figure P8-136

Figure P8-137

8-138 The position of a point moving in the xy plane is given by the equation $\mathbf{r} = 2t^3\mathbf{i} + 5t^2\mathbf{j}$, where r and t are in inches and seconds, respectively. Determine, for $t = 2$ sec,
(a) The tangential component of the acceleration of the point.
(b) The normal component of the acceleration of the point.
(c) The radius of curvature of the path.

*8-139 The position of a particle is given by

$$\mathbf{r} = 2t^2\mathbf{i} + \left(\frac{4}{t^2}\right)\mathbf{j},$$

where r and t are in meters and seconds, respectively. Determine, when $t = 1$ s,
(a) The magnitudes of the normal and tangential components of the acceleration of the particle.
(b) The radius of curvature of the path.

8-140 A ball is thrown from the top of a 240-ft building with a horizontal velocity of 80 fps. Determine the radius of curvature of the trajectory of the ball when the ball is 250 ft from the building.

8-141 A particle moves along the curve $y = 0.2x^2$, where x and y are in feet. When the particle is at the point $x = 2.4$ ft, the acceleration is $-125\mathbf{i}$ fps^2 and the speed of the particle is decreasing. Determine the tangential and normal components of the acceleration of the particle and show them on a sketch.

8-142 Solve Problem 8-141 if the acceleration at $x = 2.4$ ft is $75\mathbf{j}$ fps^2.

*8-143 A particle starts from rest at $x = 0$ and moves to the right along the curve $y = (36/\pi)\cos(\pi x/24)$, where x and y are in meters. The speed of the particle increases uniformly with time. When $x = 4$ m the acceleration of the particle is $-16\mathbf{j}$ m/s^2. Determine the elapsed time.

8-144 In Problem 8-98 the position of a particle was given as
$$\mathbf{r} = (t^2 - 2)\mathbf{i} + t^3\mathbf{j} + e^{-t}\mathbf{k},$$
where r and t are in inches and seconds. Determine, for $t = 1$ sec,
(a) The tangential and normal components of the acceleration of the particle.
(b) The position vector from the origin to the center of curvature of the path.

8-9 CURVILINEAR MOTION—RADIAL AND TRANSVERSE COMPONENTS

When a particle travels on a curved path in a plane, it is sometimes convenient to express its position in polar coordinates. The radial distance r from a fixed point and the angular position θ from a fixed reference line, as shown in Fig. 8-20(a), expressed as functions of time, completely determine the location of the particle. When using

(a)

(b)

Figure 8-20

polar coordinates, the velocity and acceleration are expressed in terms of radial and transverse components. The radial component is along the radius vector to the particle, and the transverse component is perpendicular to the radius vector. Figure 8-20(a) shows unit vectors $\mathbf{e}_r$ and $\mathbf{e}_\theta$ in the positive directions of increasing r and θ.

The position vector for the particle P of Fig. 8-20(a) can be written as

$$\mathbf{r} = r\mathbf{e}_r, \qquad (8\text{-}14)$$

where r is the radial polar coordinate.

The velocity of the particle is defined as the time rate of change of position of the particle. The unit vectors $\mathbf{e}_r$ and $\mathbf{e}_\theta$ are differentiated in the same manner as were $\mathbf{e}_n$ and $\mathbf{e}_t$ in the preceding section. From Fig. 8-20(b), it is seen that

$$\dot{\mathbf{e}}_r = \lim_{\Delta t \to 0} \frac{(\Delta\theta)\mathbf{e}_\theta}{\Delta t} = \dot{\theta}\mathbf{e}_\theta, \qquad (8\text{-}15)$$

and in the same manner

$$\dot{\mathbf{e}}_\theta = -\dot{\theta}\mathbf{e}_r. \qquad (8\text{-}15a)$$

The linear velocity of the particle P, as obtained by differentiation of Eq. (8-14), is

$$\mathbf{v} = \dot{\mathbf{r}} = \dot{r}\mathbf{e}_r + r\dot{\mathbf{e}}_r = \dot{r}\mathbf{e}_r + r\dot{\theta}\mathbf{e}_\theta = \mathbf{v}_R + \mathbf{v}_T, \quad (8\text{-}16)$$

where $\mathbf{v}_R$ and $\mathbf{v}_T$ are the radial and transverse components, respectively, of the velocity.

The linear acceleration of P is the time rate of change of velocity and, from Eq. (8-16), is

$$\mathbf{a} = \ddot{\mathbf{r}} = \ddot{r}\mathbf{e}_r + \dot{r}\dot{\mathbf{e}}_r + \dot{r}\dot{\theta}\mathbf{e}_\theta + r\ddot{\theta}\mathbf{e}_\theta + r\dot{\theta}\dot{\mathbf{e}}_\theta$$

$$= \ddot{r}\mathbf{e}_r + \dot{r}\dot{\theta}\mathbf{e}_\theta + \dot{r}\dot{\theta}\mathbf{e}_\theta + r\ddot{\theta}\mathbf{e}_\theta - r\dot{\theta}^2\mathbf{e}_r$$

$$\left.\begin{aligned} &= (\ddot{r} - r\dot{\theta}^2)\mathbf{e}_r + (r\ddot{\theta} + 2\dot{r}\dot{\theta})\mathbf{e}_\theta \\ &= (\ddot{r} - r\dot{\theta}^2)\mathbf{e}_r + \frac{1}{r}\frac{d}{dt}(r^2\dot{\theta})\mathbf{e}_\theta = \mathbf{a}_R + \mathbf{a}_T, \end{aligned}\right\} \qquad (8\text{-}17)$$

where $\mathbf{a}_R$ and $\mathbf{a}_T$ are the radial and transverse components, respectively, of the acceleration. The equivalence of the two forms for $\mathbf{a}_T$ can be verified by expanding the second form. The second form is often useful when it is necessary to integrate the tangential component of the acceleration.

The following example illustrates the use of polar coordinates for finding the velocity and acceleration of a particle moving on a curved path in a plane.

EXAMPLE 8-18
Pin A in Fig. 8-21 moves along the curved fixed slot and also along the slot in member OB as OB rotates about O. The equation of the curved slot is $r = b + c \cos\theta$.

(a) Derive expressions for the velocity and acceleration of A in terms of the angular position, velocity, and acceleration of OB.

(b) Determine the velocity and acceleration of A when $b = 3$ cm, $c = 2$ cm, $\theta = 90°$, $\dot{\theta} = 5$ rad/s counterclockwise, and $\ddot{\theta} = 20$ rad/s² clockwise.

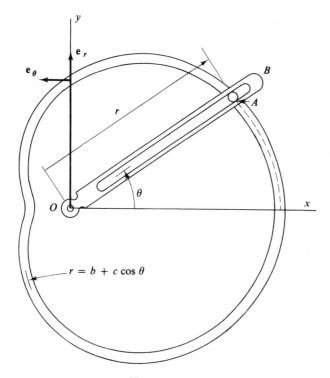

Figure 8-21

SOLUTION

(a) The position of point A, in polar coordinates, is

$$\mathbf{r} = r\mathbf{e}_r = (b + c \cos \theta)\mathbf{e}_r.$$

The velocity and acceleration of A can be obtained by successive differentiations of the position function; thus the velocity is

$$\underline{\mathbf{v} = \dot{\mathbf{r}} = -c \sin \theta \dot{\theta} \mathbf{e}_r + (b + c \cos \theta)\dot{\theta}\mathbf{e}_\theta.} \qquad \text{Ans.}$$

A second differentiation will give the acceleration of A. This acceleration can also be obtained by differentiating the expression for the magnitude of r and substituting into Eq. (8-17) as follows:

$$r = b + c \cos \theta$$

$$\dot{r} = -c \sin \theta \, \dot{\theta}$$

$$\ddot{r} = -c \cos \theta \, \dot{\theta}^2 - c \sin \theta \, \ddot{\theta}$$

397

from which

$$\mathbf{a} = (\ddot{r} - r\dot{\theta}^2)\mathbf{e}_r + (r\ddot{\theta} + 2\dot{r}\dot{\theta})\mathbf{e}_\theta$$
$$= [-c \cos\theta\,\dot{\theta}^2 - c\sin\theta\,\ddot{\theta} - (b + c\cos\theta)\,\dot{\theta}^2]\mathbf{e}_r$$
$$+ [(b + c\cos\theta)\,\ddot{\theta} + 2(-c\sin\theta\,\dot{\theta})\,\dot{\theta}]\mathbf{e}_\theta$$
$$= [-(b + 2c\cos\theta)\,\dot{\theta}^2 - c\sin\theta\,\ddot{\theta}]\mathbf{e}_r$$
$$\underline{+ [(b + c\cos\theta)\,\ddot{\theta} - 2c\sin\theta\,\dot{\theta}^2]\mathbf{e}_\theta} \qquad \text{Ans.}$$

(b) From the given data $b = 3$ cm, $c = 2$ cm, $\theta = 90°$, $\dot{\theta} = 5$ rad/s, and $\ddot{\theta} = -20$ rad/s², which give

$$r = 3 - 2(\cos 90°) = 3 \text{ cm},$$
$$\dot{r} = -2(\sin 90°)(5) = -10 \text{ cm/s},$$

and

$$\ddot{r} = -2(\cos 90°)(5)^2 - 2(\sin 90°)(-20) = +40 \text{ cm/s}^2.$$

The directions of $\mathbf{e}_r$ and $\mathbf{e}_\theta$ for $\theta = 90°$ are shown on Fig. 8-21. The velocity of A can be obtained by substituting numerical data into either Eq. (8-16) or in the result of part (a), either of which gives

$$\mathbf{v} = \dot{r}\mathbf{e}_r + r\dot{\theta}\mathbf{e}_\theta$$
$$= -10\mathbf{e}_r + 3(5)\mathbf{e}_\theta$$
$$= -10\mathbf{e}_r + 15\mathbf{e}_\theta = \underline{18.03 \text{ cm/s}} \quad \angle 33.7° \qquad \text{Ans.}$$

Similarly, the acceleration of A is

$$\mathbf{a} = (\ddot{r} - r\dot{\theta}^2)\mathbf{e}_r + (r\ddot{\theta} + 2\dot{r}\dot{\theta})\mathbf{e}_\theta$$
$$= [40 - 3(5)^2]\mathbf{e}_r + [3(-20) + 2(-10)(5)]\mathbf{e}_\theta$$
$$= -35\mathbf{e}_r - 160\mathbf{e}_\theta = \underline{163.8 \text{ cm/s}^2} \quad \angle 12.34° \qquad \text{Ans.}$$

The velocity and acceleration are shown on Figs. 8-22(a) and 8-22(b), respectively.

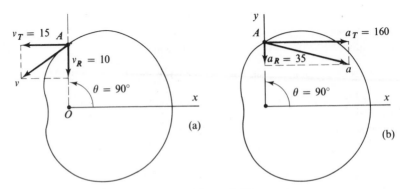

Figure 8-22

PROBLEMS

8-145 Blocks *A* and *B* in Fig. P8-145 are pinned together, and *A* slides along the fixed curved rod, while *B* slides along member *OC*. Rod *OC* rotates about point *O*, and the equation of the curved rod is

$$r = 10 \sin 2\theta \qquad (r \text{ is in inches}).$$

(a) When $\theta = 45°$, $\dot{\theta} = 10$ rad per sec counterclockwise. Determine the velocity of block *A*.
(b) When $\theta = 60°$ the speed of *A* is 100 ips. Determine the angular velocity of *OC*.

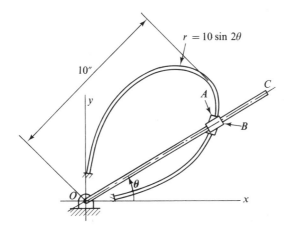

Figure P8-145

8-146 Determine the radial and transverse components of the acceleration of block *A* in Problem 8-145 when $\theta = 30°$, $\dot{\theta} = 10$ rad per sec counterclockwise, and $\ddot{\theta} = 0$.

*8-147 In Example 8-18 let $b = c = 5$ cm. When $\theta = 45°$ the radial component of the velocity of *A* is 10 cm/s from *O* toward *B*. Determine the velocity of *A*.

*8-148 In addition to the data in Problem 8-147 the radial component of the acceleration of *A* is 150 cm/s² from *B* toward *O*. Determine the acceleration of *A*.

8-149 Particle *A* in Fig. P8-149 travels counterclockwise along the circular path with a constant speed *v*.

(a) Determine the radial and transverse components of the acceleration of *A* when $h = 0.8b$.
(b) Show that the result from part (a) is the same as the *normal* component of the acceleration of *A*.

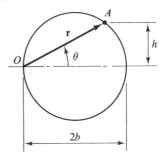

Figure P8-149

8-150 The circular table in Fig. P8-150 rotates in a horizontal plane, while the ball B slides along a slot in the table. At a given instant, the ball is 14 in. from O and is moving outward at the rate of 6 ips, while the plate has an angular velocity of 2 rad per sec and an angular acceleration of 8 rad per sec², both clockwise. Determine

(a) The magnitude of the acceleration of B if $\ddot{r} = 0$.

(b) The value of $\ddot{r}$ which will result in the minimum value of a_B.

Figure P8-150

*8-151 The ring in Fig. P8-151 slides along the rotating rod OB. In the position shown $r = 0.3$ m, $\tan \theta = \frac{4}{3}$, r is increasing at the rate of 0.5 m/s, and $\dot{r}$ is decreasing at 0.6 m/s². Member AB has an angular velocity of 2 rad/s clockwise and an angular acceleration of 5 rad/s² clockwise.

Figure P8-151

(a) Set up an expression for the vertical component of the position of A and determine the vertical component of the acceleration of A by differentiation.

(b) Determine the radial and transverse components of the acceleration of A and verify the answer to part (a).

8-152 The equation of the spiral in Fig. P8-152 is $(r/b)^2 = \theta$, where r and b are in inches and θ is in radians. The angle θ is equal to $6t$, where t is time in seconds. Point P is the intersection of line OA and the spiral. When $b = 2$ in. and $t = \pi/12$ sec, determine

(a) The velocity of P.

(b) The acceleration of P.

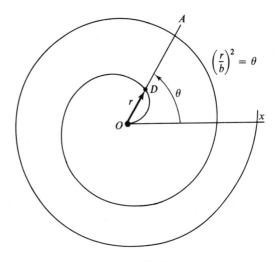

$$\left(\frac{r}{b}\right)^2 = \theta$$

Figure P8-152

8-153 Solve Problem 8-152 when $b = 2$ in. and $t = 5\pi/12$ sec.

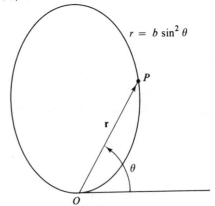

$$r = b \sin^2 \theta$$

Figure P8-156

8-154 A particle travels along the cardioid $r = b(1 - \cos\theta)$, where r and b are in inches. The distance r increases uniformly at 3 ips for $10° < \theta < 170°$. The distance b is 6 in. Determine the value of θ, in the specified range, for which the velocity of the particle will be horizontal and the corresponding velocity.

8-155 Determine the value of θ, in the specified range, in Problem 8-154 for which the velocity of the particle will be vertical and the corresponding acceleration of the particle.

8-156 Point P in Fig. P8-156 moves along the curve whose equation, in polar coordinates, is $r = b\sin^2\theta$, where $b = 2$ in. The radius vector has a constant angular velocity of 5 rad per sec counterclockwise. Determine the velocity and acceleration of P when $\theta = 45°$.

8-10 KINEMATICS OF A PARTICLE USING CYLINDRICAL COORDINATES

The analysis of particle motion using polar coordinates as explained in Art. 8-9 can be extended to three dimensions by the addition of a z coordinate and a unit vector $\mathbf{e}_z$. The added unit vector is constant in direction (parallel to the z axis) as well as in magnitude. The positive directions are in the directions of increasing r, θ, and z, as shown by the unit vectors of Fig. 8-23. It should be noted that $\mathbf{e}_r$, $\mathbf{e}_\theta$, and $\mathbf{e}_z$ define a right-hand system of coordinates.

The position vector from the origin to any particle P is, as shown in Fig. 8-23,

$$\mathbf{r}_1 = r\mathbf{e}_r + z\mathbf{e}_z. \tag{8-18}$$

The velocity of P is defined as the time rate of change of position. The derivative of $\mathbf{e}_r$ is given by Eq. (8-15) of Art. 8-9. Since $\mathbf{e}_z$ is constant in both magnitude and direction, its derivative is zero. The time derivative of Eq. (8-18) is

$$\mathbf{v} = \dot{\mathbf{r}}_1 = \dot{r}\mathbf{e}_r + r\dot{\theta}\mathbf{e}_\theta + \dot{z}\mathbf{e}_z$$
$$= \mathbf{v}_R + \mathbf{v}_T + \mathbf{v}_z, \tag{8-19}$$

where $\mathbf{v}_R$, $\mathbf{v}_T$, and $\mathbf{v}_z$ are the velocity components in the radial, transverse, and z directions, respectively.

In the same manner, differentiation of Eq. (8-19) gives

$$\mathbf{a} = \ddot{\mathbf{r}}_1 = (\ddot{r} - r\dot{\theta}^2)\mathbf{e}_r + (r\ddot{\theta} + 2\dot{\theta}\dot{r})\mathbf{e}_\theta + \ddot{z}\mathbf{e}_z$$
$$= \mathbf{a}_R + \mathbf{a}_T + \mathbf{a}_z, \tag{8-20}$$

where $\mathbf{a}_R$, $\mathbf{a}_T$, and $\mathbf{a}_z$ are the corresponding acceleration components.

The following example illustrates the use of cylindrical coordinates for problems in kinematics.

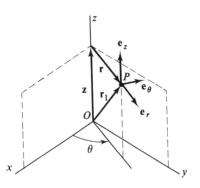

Figure 8-23

EXAMPLE 8-19

A T-29 airplane is cruising at 180 knots east in straight and level flight. Each propeller has a diameter of 13 ft and rotates at 1035 rpm clockwise as viewed from the rear. Determine the velocity and acceleration of the particle at the tip of a propeller when the tip is in the farthest south position. *Note:* A knot is one nautical mile (6080 ft) per hour.

SOLUTION

x

z (East)

y (South)

Figure 8-24

A set of axes is selected as shown in Fig. 8-24 with the z axis to the east (along the line of flight), the x axis vertical, and the y axis to the south. For the specified position of the propeller tip $\mathbf{e}_r = \mathbf{e}_y$ and $\mathbf{e}_\theta = -\mathbf{e}_x$. From the problem statement, the z component of the velocity of the tip is

$$\dot{z} = 180 \text{ knots} = \frac{180(6080)}{3600} = 304 \text{ fps}$$

and

$$\dot{\mathbf{z}} = 304\mathbf{e}_z \text{ fps (east).}$$

The magnitude of the transverse component of the tip velocity is

$$v_T = r\dot{\theta} = \frac{(13/2)(1035)(2\pi)}{60} = 705 \text{ fps}$$

and the radial component is zero, since the radius vector has a constant magnitude. The velocity of the tip of the propeller is

$$\mathbf{v} = \mathbf{v}_T + \mathbf{v}_R + \mathbf{v}_z = 705\mathbf{e}_\theta + 304\mathbf{e}_z,$$

and when the propeller is horizontal the velocity of the south tip is

$$\mathbf{v} = -705\mathbf{e}_x + 304\mathbf{e}_z$$

$$= 768(-0.918\mathbf{e}_x + 0.396\mathbf{e}_z) \text{ fps.} \qquad \text{Ans.}$$

From the statement of the problem, $\ddot{z}$, $\dot{r}$, $\ddot{r}$, and $\ddot{\theta}$ are all zero; therefore, the acceleration of the tip of the propeller is

$$\mathbf{a} = (\ddot{r} - r\dot{\theta}^2)\mathbf{e}_r + (r\ddot{\theta} + 2\dot{r}\dot{\theta})\mathbf{e}_\theta + \ddot{z}\mathbf{e}_z$$

$$= -r\dot{\theta}^2\mathbf{e}_r = -6.5\left[\frac{1035(2\pi)}{60}\right]^2\mathbf{e}_r$$

$$= -76,400\mathbf{e}_r$$

and when the propeller is horizontal the acceleration of the south tip is

$$\mathbf{a} = -76,400\mathbf{e}_y \text{ fps}^2 \text{ (to the north).} \qquad \text{Ans.}$$

PROBLEMS

8-157 A flexible hose is wrapped around the cylinder in Fig. P8-157 to form a helix. The mean radius of the helix is R, and the pitch is p. A small ball P is forced to slide along the helix by the vertical roller AB which rotates about the center of the cylinder. The angular position of the ball in front of the roller is given by the angle θ, where θ is a function of time.

(a) Write an equation for the position of P using cylindrical coordinates.

(b) Derive expressions for the velocity and acceleration of the ball P.

(c) Given $R = 6$ in., $p = 3$ in., and $\theta = ct^2$, where $c = 0.2$ rad per sec^2, determine the velocity and acceleration of P when $t = 5$ sec.

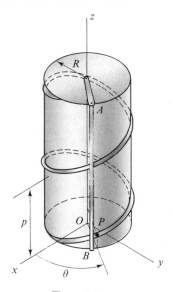

Figure P8-157

8-158 Particle P in Fig. P8-158 is constrained to move along the spiral path on the inside of the cone. The particle makes one revolution about the z axis in 2 sec during which time it moves downward 5 in. The particle starts from point A. Determine the velocity and acceleration of the particle at point B after one complete revolution. (Note that $\dot{\theta}$ and $\dot{r}$ are both constant.)

8-159 Solve Problem 8-158 if $\ddot{\theta} = \pi$ rad per sec^2. The dimensions of the path are the same as

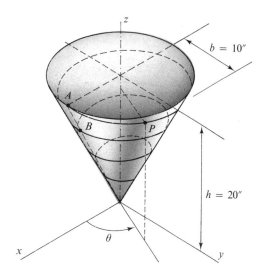

Figure P8-158

in Problem 8-158. The particle starts from rest at point A.

8-160 The equation of the paraboloid of revolution in Fig. P8-160 is

$$h(x^2 + y^2) = hr^2 = b^2 z.$$

The particle moves along a path on the surface

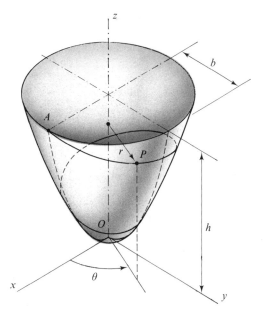

Figure P8-160

403

in accordance with

$$\theta = c_1 t$$

$$r = b - c_2 t,$$

where c_1 and c_2 are constants.

(a) Derive an expression for the velocity of P.

(b) Let $h = 20$ in., $b = 10$ in., $c_1 = 0.2\pi$ rad per

sec, and $c_2 = 0.5$ ips. Determine the velocity of P when $t = 0$ sec and when $t = 10$ sec.

8-161 Determine the acceleration of the point P in Problem 8-160(b) when

(a) $t = 0$ sec.

(b) $t = 10$ sec.

8-11 CLOSURE

The kinematic quantities used to describe the motion of a particle are time, position (including displacement and total distance traveled), velocity, and acceleration. Similar quantities are used to describe the angular motion of a line. The problems of kinematics consist of determining one or more of the preceding quantities from the given data. The various kinematic quantities are related by differential equations. The relationships have been developed using cartesian, normal and tangential, radial and transverse, and cylindrical coordinates or components. Parametric equations are frequently convenient. Motion curves may be used to supplement or replace the analytical solution. Since many of the quantities involved are vectors, it is frequently necessary to differentiate and integrate vector quantities.

REVIEW PROBLEMS

8-162 A particle starts from rest from a point 6 ft to the right of the origin when $t = 0$. The point moves along a horizontal straight line, and the magnitude of its acceleration is directly proportional to the time after starting. The displacement of the particle from $t = 0$ to $t = 2$ sec is 2 ft to the left. Determine the velocity of the particle when it passes the origin.

*__8-163__ A particle starts from rest and moves along a straight line with an acceleration of 5 m/s² to the right for 12 sec. The acceleration is then changed to a different constant value such that the displacement for the éntire period is 180 m to the right and the total distance traveled is 780 m. Determine

(a) The acceleration during the second time interval.

(b) The total time interval.

8-164 Body BC in Fig. P8-164 slides in the fixed horizontal slot as shown. The arm OA has an angular velocity of $\dot{\theta}$ and an angular acceleration of $\ddot{\theta}$. Write an equation for the position of B

and differentiate it to determine its velocity and acceleration in terms of θ, $\dot{\theta}$, and $\ddot{\theta}$.

Figure P8-164

8-165 The 12-ft bar AB in Fig. P8-165 moves in a vertical plane with its ends in the guides as shown. When $\tan \theta = \frac{3}{4}$ the bar has an angular velocity of 3 rad per sec clockwise and an angular acceleration of 2 rad per sec² counterclockwise. Write an equation for the position of the center, C, of the bar at any instant in terms of θ and differentiate to determine the acceleration of C when $\tan \theta = \frac{3}{4}$.

Figure P8-165

8-166 A stone is thrown with an initial velocity of 80 fps in the direction shown in Fig. P8-166 and falls on the inclined plane. Determine
(a) The distance from A to B.
(b) The maximum height z of the stone above the inclined plane.

Figure P8-166

*8-167 A point moves along the circular path in Fig. P8-167. When the point is at position A the angular velocity of line OA is 2 rad/s clockwise. The angular velocity of OA changes uniformly in such a manner than the angular displacement is 6 rad clockwise, while the line turns through a total angle of 10 rad. Determine the acceleration of the point 3 s after leaving position A.

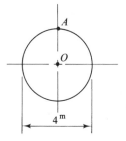

Figure P8-167

*8-168 A particle travels on an elliptical path given by the equation $x^2 + 4y^2 = 16$, where x and y are in centimeters. At the point where $y = 2$ cm, the particle has a velocity of 6 cm/s to the right, and the velocity is increasing at the rate of 2 cm/s². Determine the acceleration of the point when $y = 2$ cm.

8-169 A particle moves along the curve $y = 0.2x^2$, where x and y are in feet. When the particle is at the point $x = 2.4$ ft, the acceleration is $-125\mathbf{i}$ fps² and the speed is decreasing. Determine the values of $\ddot{s}\mathbf{e}_t$ and $\ddot{s}\mathbf{e}_t$ and show them on a sketch of the curve.

8-170 The velocity of a particle moving along a horizontal path is given by the expression $\mathbf{v} = (s^2 + 9)\mathbf{i}$, where v is in feet per second and positive is to the right. When $t = 0$, $s = 3$ ft. Determine s as a function of the time t.

8-171 Rod AB in Fig. P8-171 rotates about an axis at A and is connected to E by a flexible and inextensible cable. As AB turns from the horizontal position through the angle θ, body E moves upward a distance s.
(a) Derive an expression relating s and θ.
(b) Differentiate the equation from part (a) to determine the velocity of E in terms of the angular velocity of AB and the positions of the bodies.
(c) When the angle θ is 60°, the angular velocity of AB is 0.5 rad per sec clockwise and the acceleration of E is 10 fps² upward. Determine the angular acceleration of AB for this position.

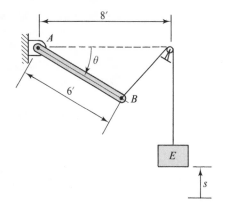

Figure P8-171

PART B—RIGID BODY KINEMATICS

8-12 DEFINITIONS

The motion of points or particles, and of lines connecting them, was analyzed in the preceding articles of this chapter. When the dimensions of a body are small compared with the dimensions of the path of motion of the body, it is usually permissible to treat the body as a particle in kinetics. When the dimensions of the body are not negligible compared to the dimensions of the path it usually is necessary to investigate the relationships between the positions, velocities, and accelerations of various points and lines of the body. A rigid body was defined in Art. 1-2 as one in which the particles remain at fixed distances from each other. Any rigid body motion may be classified as one of the following types.

Translation: When a rigid body moves in such a manner that all lines in the body remain parallel to their original positions, the body has a motion of translation. For such motion, the angular velocity and angular acceleration of all lines in the body must be zero. If the mass center of the body remains in one plane, the motion is defined as *coplanar translation.* Most bodies having a motion of translation will be found to have coplanar translation. Some examples of translation are (1) the piston rod of a locomotive running along a straight level track and also the parallel rod connecting the drivers of a locomotive, (2) the bumper of an automobile traveling in a straight line along a road, and (3) a sliding door whose rollers run on straight tracks.

From the definition of translation, it is apparent that all points of the body must have the same displacement during any time interval and consequently must have identical velocities and identical accelerations at any instant. In other words, *the motion of any particle of a rigid body having a motion of translation determines the motion of every particle of the body.*

Rotation: A rigid body has a motion of rotation if one line in the body or the body extended is fixed and all particles in the body not on the fixed line travel in circular paths with centers on the fixed axis. The fixed line is called the *axis of rotation,* and each particle not on the axis moves in a plane perpendicular to the axis of rotation. Sometimes this motion is called *pure rotation,* to distinguish it from motion in which no line is fixed but in which the body turns or rotates about a moving line as in plane motion (to be defined later).

The rotor of a stationary steam turbine, a water wheel, the hands of a clock, and the crankshaft of an automobile when the car is standing still and the motor is running are examples of rotating rigid bodies. If the car is traveling along a straight road, the crankshaft does not have rotation because the points travel along helical paths instead of circular paths.

Plane motion: When a rigid body moves in such a manner that every particle of the body remains a constant distance from a fixed reference plane, the body has plane motion. Thus, all particles in a rigid body with plane motion move in the same plane or in parallel planes. Rotation is a special case of plane motion, and coplanar translation is also plane motion. The plane in which the mass center moves is called the plane of motion of the body. In general, plane motion may be thought of as a combination of rotation and translation. Two common examples of plane motion are the connecting rod of a stationary engine and the wheel of an automobile running straight along a road.

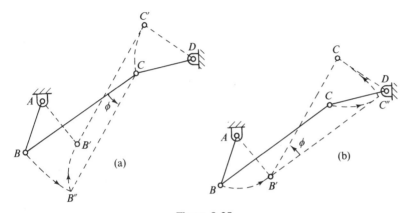

Figure 8-25

Plane motion can be analyzed by treating it as a combination of a rotation of the body about some convenient reference axis in the body, or the body extended, and a translation of the body in which all points have the same motion as the reference point. Member BC of the link mechanism in Fig. 8-25(a) has plane motion. As arm AB rotates to position AB', member DC rotates to DC' and BC moves to $B'C'$. The motion of BC can be visualized as a rotation about an axis at C to position $B''C$ followed by a curvilinear translation to $B'C'$ with all particles of BC moving along curved paths similar to CC' from $B''C$ to position $B'C'$. The motion can also be thought of as a curvilinear translation of BC to position $B'C''$ in Fig. 8-25(b) followed by a rotation about B' to position $B'C'$. *The angular displacement of BC when moving from BC to $B'C'$ is ϕ counterclockwise whether the rotation is considered as occurring about an axis at C or one at B or about any other axis perpendicular to the plane of motion.* Furthermore, it is immaterial whether the rotation or the translation is considered as occurring first.

Motion about a fixed point: When a body moves in such a manner that one point, O, remains fixed in space, it can be shown that the resulting displacement is a rotation about some axis through O with an angular velocity $\boldsymbol{\omega}$.[6] A general infinitesimal rotation associated with the angular velocity $\boldsymbol{\omega}$ can be considered to be successive rotations about a set of x, y, and z axes with angular velocities $\boldsymbol{\omega}_x = \omega_x\mathbf{i}$, $\boldsymbol{\omega}_y = \omega_y\mathbf{j}$, and $\boldsymbol{\omega}_z = \omega_z\mathbf{k}$. Infinitesimal angular rotations can be shown to be vectors; therefore, the angular velocity $\boldsymbol{\omega}$ is the sum of the three component angular velocities.

General motion. Euler's theorem for motion of a rigid body about a fixed point can be extended to describe the general motion of a rigid body. The general theorem, known as *Chasles' theorem,*[7] states that any displacement can be analyzed as a combination of a translation and a rotation.

When a body has plane motion, all lines in the body parallel to the plane of motion have common angular velocities and accelerations,

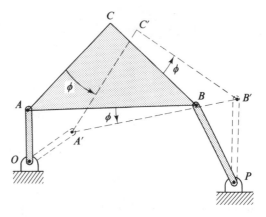

Figure 8-26

and the values are referred to as the angular velocity or angular acceleration of the body. The plate ABC in Fig. 8-26 has plane motion, and as link OA moves to OA' member PB will move to PB' and ABC will move to $A'B'C'$. Line AB on plate ABC will turn through the angle ϕ, and lines BC and AC must also turn through the same angle ϕ since the angles ABC and BAC must remain constant on a rigid body. Thus it is seen that all lines in body ABC will turn through the same angle in the same time interval and thus will have the same angular velocity at any instant. This same reasoning leads to the conclusion that all lines in the body (parallel to the plane of motion) must have a common angular acceleration at any instant.

---

[6] This statement is sometimes called *Euler's theorem*; see Herbert Goldstein, *Classical Mechanics* (Reading, Mass.: Addison-Wesley Publishing Company, Inc., 1950), p. 118.

[7] Goldstein, *Ibid.*, p. 124.

The motion of various points and lines of bodies in a mechanism can be related by deriving expressions relating the linear and angular positions (from geometry) and differentiating these expressions to obtain velocities and accelerations as illustrated in some of the preceding articles. The equations of relative motion, derived in Art. 8-6, can often be applied to rigid body problems in kinematics to obtain more direct solutions. Familiarity with various methods is important if the most direct solution is to be selected.

The notation for absolute and relative motion of positions, velocities, and accelerations of points was discussed in Art. 8-6. The symbol $\mathbf{v}_A$ represents the velocity of point A measured with respect to a fixed set of axes (the earth is assumed to be fixed for most problems). The quantity $\mathbf{v}_{A/B}$ is the velocity of point A measured with respect to a set of translating (nonrotating) axes whose origin moves with point B. Motion with respect to rotating axes is analyzed in Art. 8-21.

Angular motion of a line or body can also be expressed as either an absolute or a relative quantity. The absolute angular motion of a line is defined as the angular motion of the line measured with respect to a set of nonrotating axes. Motion of a line measured with respect to rotating axes is called the *relative angular motion of the line*. In this text, all values of linear and angular quantities are considered absolute unless otherwise specified.

8-13 ROTATION OF A RIGID BODY

The velocity and acceleration of any point on a rotating rigid body can be expressed in terms of the angular motion of the body. Figure 8-27 represents a rigid body which is rotating about the fixed z axis with an angular velocity $\omega = \dot{\theta}$ and an angular acceleration $\alpha = \ddot{\theta}$ as

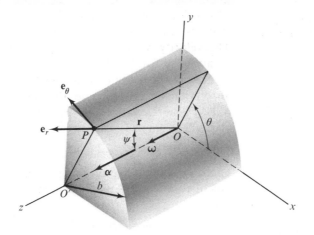

Figure 8-27

409

shown. The point P moves along a circular path of radius b, and the position of P from an arbitrary origin on the axis of rotation is

$$\mathbf{r} = r\mathbf{e}_r.$$

The velocity of P is tangent to the path and from Art. 8-8 is

$$\mathbf{v}_P = \dot{\mathbf{r}} = b\omega\mathbf{e}_\theta = r(\sin\psi)\omega\mathbf{e}_\theta = \boldsymbol{\omega}\times\mathbf{r}. \qquad (8\text{-}21)$$

Note that the order of the terms in the vector product must be as indicated in order to give the correct direction for the velocity. The derivative of the vector $\mathbf{r}$ is given in Eq. (8-21) only when the magnitude of $\mathbf{r}$ is constant as when O and P are two points in a rigid body.

The acceleration of P can be determined by differentiating Eq. (8-21) and is

$$\mathbf{a}_P = \dot{\mathbf{v}}_P = \dot{\boldsymbol{\omega}}\times\mathbf{r} + \boldsymbol{\omega}\times\dot{\mathbf{r}}$$

$$= \boldsymbol{\alpha}\times\mathbf{r} + \boldsymbol{\omega}\times(\boldsymbol{\omega}\times\mathbf{r}). \qquad (a)$$

The first term on the right can be written as

$$\boldsymbol{\alpha}\times\mathbf{r} = \alpha r \sin\psi\mathbf{e}_\theta = \alpha b\mathbf{e}_\theta = \mathbf{a}_t.$$

Since $\boldsymbol{\omega}$ and $\dot{\mathbf{r}}$ are perpendicular, the magnitude of the second term of Eq. (a) is

$$|\boldsymbol{\omega}\times\dot{\mathbf{r}}| = \omega v_P = b\omega^2,$$

and its direction, perpendicular to $\boldsymbol{\omega}$ and $\mathbf{e}_\theta$, is from P toward O' and is thus seen to be the normal acceleration of P. That is,

$$\mathbf{a}_P = \boldsymbol{\alpha}\times\mathbf{r} + \boldsymbol{\omega}\times(\boldsymbol{\omega}\times\mathbf{r}). \qquad (8\text{-}22)$$

Euler's theorem states that rotation of a body about a fixed *point* is also rotation about a line in the body through the fixed point. The angular velocity of the body lies along the instantaneous axis of rotation as shown in Fig. 8-28. The velocity of any point P can be determined by means of Eq. (8-21).

Since the direction of the instantaneous axis of rotation varies with time, the angular velocity will, in general, vary in both magnitude and direction. The angular velocity can be expressed as

$$\boldsymbol{\omega} = \omega\mathbf{e}_1,$$

where $\mathbf{e}_1$ is a unit vector along the instantaneous axis of rotation. The angular acceleration of the body is

$$\dot{\boldsymbol{\omega}} = \dot{\omega}\mathbf{e}_1 + \omega\dot{\mathbf{e}}_1$$

$$= \dot{\omega}\mathbf{e}_1 + \omega(\boldsymbol{\omega}_2\times\mathbf{e}_1),$$

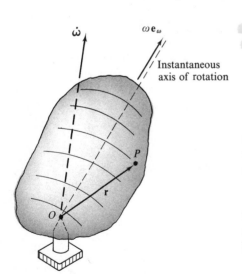

Figure 8-28

where $\boldsymbol{\omega}_2$ is the angular velocity of the axis of rotation. The first term of this expression is the component of the angular acceleration due to a change in magnitude of the angular velocity, and the second term is the component resulting from a change in direction of the angular velocity. These two components are mutually perpendicular, and since the first component is parallel to the angular velocity, the resultant angular acceleration will be parallel to the angular velocity only when the angular velocity does not change direction. Equation (8-22) can be used to calculate the acceleration of point P in Fig. 8-28, but since $\boldsymbol{\alpha}$ and $\boldsymbol{\omega}$ are not necessarily parallel, the quantity $\boldsymbol{\alpha} \times \mathbf{r}$ is not necessarily parallel to $\boldsymbol{\omega} \times \mathbf{r}$, and the acceleration component given by $\boldsymbol{\alpha} \times \mathbf{r}$ may not be tangent to the path of the particle.

8-14 RELATIVE VELOCITY— NONROTATING AXES

The concept of the motion of point B in Fig. 8-9 with respect to a set of nonrotating axes whose origin moved with point A was discussed in Art. 8-6. The equation relating the velocities of points A and B is

$$\mathbf{v}_B = \mathbf{v}_A + \mathbf{v}_{B/A}. \tag{8-23}$$

Equation (8-23) can be rearranged to give

$$\mathbf{v}_{B/A} = \mathbf{v}_B - \mathbf{v}_A.$$

A helpful interpretation of the last equation can be obtained by considering points A and B in Fig. 8-29, which have velocities $\mathbf{v}_A$ and $\mathbf{v}_B$. If a velocity $-\mathbf{v}_A$ is added to the velocities of each of the particles, the resultant velocity of particle A will be zero and the resulting velocity of B will be $\mathbf{v}_{B/A}$. In other words, the velocity of B relative to A is the velocity B would appear to have to an observer moving with A (having the same velocity as A) but not rotating. The concept of thinking of the motion of B with A fixed will often aid in determining correctly the direction of the velocity of B relative to A.

Figure 8-29

When points A and B in Eq. (8-23) are two points on a rigid body the motion of B relative to A must be along a spherical path with its center at A and with a radius equal to the distance from A to B since the two points must remain the same distance apart. In this case the velocity of B relative to A will be given by Eq. (8-21), and the direction of the relative velocity will be perpendicular to the line AB. The velocity of B is thus

$$\mathbf{v}_B = \mathbf{v}_A + \boldsymbol{\omega} \times \mathbf{r}, \tag{8-23a}$$

where **ω** is the angular velocity of the member containing A and B and **r** is the position of B relative to A.

When Eq. (8-23) is applied to two points which are not on the same rigid body the motion of B relative to A will usually not be on a spherical surface, and the relative velocity can be more difficult to analyze. The use of Eq. (8-23) for plane motion and for three-dimensional motion will be discussed in Art. 8-15 and 8-17.

8-15 RELATIVE VELOCITY—PLANE MOTION

Many of the kinematic problems encountered in engineering practice involve bodies which have plane motion. When points A and B in Eq. (8-23) are points on a rigid body with plane motion the motion of B relative to A will be along a circular path with its center at A, and all velocity vectors will lie in a common plane. The vector equation can be resolved into two scalar equations, which in turn can be solved for two unknown quantities. Note that the angular velocity vector is always perpendicular to the plane of motion and thus can change only in magnitude.

An important case of plane motion occurs when a wheel rolls without slipping along a plane surface. The center of the wheel has rectilinear motion parallel to the plane. Figure 8-30 shows a wheel in its initial position and after it has rolled a distance s_O to the right. The line OA on the wheel turns clockwise through an angle θ to the position $O'A'$, and if there is no slipping, the arc $A'B = r\theta$ must be the same as the distance $AB = s_O$; that is,

$$s_O = r\theta. \qquad (a)$$

The magnitude of the velocity of point O is

$$v_O = \frac{d}{dt}(r\theta) = r\omega, \qquad (b)$$

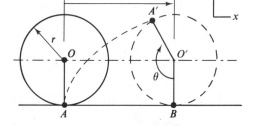

Figure 8-30

where $\omega = d\theta/dt$ is the magnitude of the angular velocity of the wheel. Note that when s_O is to the right θ is clockwise; consequently when v_O is to the right ω must be clockwise.

The following examples illustrate the application of Eq. (8-23) to problems involving plane motion.

EXAMPLE 8-20

The wheel in Fig. 8-30 rolls without slipping along the horizontal plane. The center of the wheel has a constant velocity of v_O to the right. Determine the velocity of point A' on the rim of the wheel when the angle θ is 150°.

SOLUTION

The velocity of the center of the wheel is

$$\mathbf{v}_O = v_O \mathbf{i},$$

and the angular velocity of the wheel, from Eq. (b), is

$$\boldsymbol{\omega} = \frac{v_O}{r} \mathbf{)} = -\frac{v_O}{r} \mathbf{k}.$$

The velocity of A' can be determined by means of the relative velocity equation for points A' and O' since the velocity of O' is known. The equation is

$$\mathbf{v}_{A'} = \mathbf{v}_{O'} + \mathbf{v}_{A'/O'}$$

$$= v_O \mathbf{i} + \boldsymbol{\omega} \times \mathbf{r}_{O'A'}.$$

Note that when θ is 150° point A' will be above and to the left of O' and the vector $r_{O'A'}$ will be $r(-0.500\mathbf{i}+0.866\mathbf{j})$. Thus the equation becomes

$$\mathbf{v}_{A'} = v_O \mathbf{i} - \frac{v_O}{r} \mathbf{k} \times r(-0.500\mathbf{i}+0.866\mathbf{j})$$

$$= v_O \mathbf{i} + 0.500 v_O \mathbf{j} + 0.866 v_O \mathbf{i}$$

$$= v_O (1.866 \mathbf{i} + 0.500 \mathbf{j})$$

$$= 1.932 v_O \qquad \underline{} 15° \qquad\qquad \text{Ans.}$$

EXAMPLE 8-21

The mechanism in Fig. 8-31 moves in a vertical plane, and member CD, when in the position shown, has an angular velocity of 8 rad per sec clockwise. Determine

(a) The angular velocities of members AB and BC.
(b) The velocity of point M, the midpoint of member BC.

SOLUTION

(a) Members AB and CD each have a motion of rotation, and pins B and C move in circular paths about centers at A and D, respectively. Pins B and C are also points on the rigid body BC, which has plane motion, and thus the relative motion of B with respect to C (or of C with respect to B) must be along a circular path whose radius is the distance from B to C. The absolute and relative velocities of points B and C can be expressed in terms of the angular velocities of the members and the radius vectors by means of Eq. (8-21). Unknown angular velocities are assumed to be counterclockwise (in this example), and a negative result will indicate that the assumed direction was incorrect. The angular velocities are

$$\boldsymbol{\omega}_{AB} = \omega_{AB} \mathbf{k}, \qquad \boldsymbol{\omega}_{BC} = \omega_{BC} \mathbf{k},$$

$$\boldsymbol{\omega}_{CD} = -8\mathbf{k} \text{ rad per sec,}$$

and the radius vectors are

$$\mathbf{r}_{AB} = 3\mathbf{j} \text{ in.}, \qquad \mathbf{r}_{CB} = (-12\mathbf{i}+5\mathbf{j}) \text{ in.}, \qquad \mathbf{r}_{DC} = (-3\mathbf{i}-4\mathbf{j}) \text{ in.}$$

Figure 8-31

The relative velocity equation for points B and C is

$$\mathbf{v}_B = \mathbf{v}_C + \mathbf{v}_{B/C}$$

or

$$\boldsymbol{\omega}_{AB} \times \mathbf{r}_{AB} = \boldsymbol{\omega}_{CD} \times \mathbf{r}_{DC} + \boldsymbol{\omega}_{BC} \times \mathbf{r}_{CB}$$

which becomes

$$\omega_{AB}\mathbf{k} \times 3\mathbf{j} = -8\mathbf{k} \times (-3\mathbf{i} - 4\mathbf{j}) + \omega_{BC}\mathbf{k} \times (-12\mathbf{i} + 5\mathbf{j}). \qquad (a)$$

Note that the radius vector is the vector *from* the fixed point (A or D) or the reference point (point C for the velocity of B with respect to C) *to* the moving point. When Eq. (a) is expanded it becomes

$$-3\omega_{AB}\mathbf{i} = +24\mathbf{j} - 32\mathbf{i} - 12\omega_{BC}\mathbf{j} - 5\omega_{BC}\mathbf{i}$$

or

$$-3\omega_{AB}\mathbf{i} = \mathbf{i}(-32 - 5\omega_{BC}) + \mathbf{j}(24 - 12\omega_{BC}).$$

If two vectors are equal, their components (the coefficients of the unit vectors) must be equal. Thus the coplanar vector equation can be reduced to the following two scalar equations:

$$\mathbf{i}: \quad -3\omega_{AB} = -32 - 5\omega_{BC},$$

$$\mathbf{j}: \quad 0 = +24 - 12\omega_{BC}.$$

The simultaneous solution of these equations gives

$$\omega_{BC} = 2.00 \text{ rad per sec}; \qquad \boldsymbol{\omega}_{BC} = \underline{2.00\mathbf{k} \text{ rad per sec, Ans.}}$$

$$\omega_{AB} = 14.00 \text{ rad per sec}; \qquad \boldsymbol{\omega}_{AB} = \underline{14.00\mathbf{k} \text{ rad per sec. Ans.}}$$

(b) The velocity of point M can be obtained by writing a relative velocity equation between M and either B or C. If point B is used for the reference point, the relative velocity equation becomes

$$\mathbf{v}_M = \mathbf{v}_B + \mathbf{v}_{M/B}$$

$$= \boldsymbol{\omega}_{AB} \times \mathbf{r}_{AB} + \boldsymbol{\omega}_{BC} \times \mathbf{r}_{BM}$$

$$= 14.00\mathbf{k} \times 3\mathbf{j} + 2.00\mathbf{k} \times (6\mathbf{i} - 2.5\mathbf{j})$$

$$= -42.0\mathbf{i} + 12.0\mathbf{j} + 5.0\mathbf{i}$$

$$= -37.0\mathbf{i} + 12.0\mathbf{j}$$

$$= \underline{38.9 \text{ ips}} \quad \diagdown \underline{17.97°}. \qquad \text{Ans.}$$

The solution can also be accomplished by a semigraphical procedure. The relative velocity equation for points B and C in algebraic and graphical form is

$$\mathbf{v}_B \quad = \quad \mathbf{v}_C \quad + \quad \mathbf{v}_{B/C}.$$

These sketches show the slope and sense of each of the velocity terms beneath the corresponding terms of the vector equation. Since B is traveling along a circular path about A, its velocity must be tangent to the path, and its sense must agree with the assumed angular velocity of AB. If ω_{AB} is found to have a negative sign, the assumed sense is wrong. In the diagram showing the velocity of B relative to C, point C is not actually fixed, but it is convenient to show it fixed temporarily to indicate the correspondence of the direction of ω_{BC} and $\mathbf{v}_{B/C}$. The vector equation contains two unknown quantities, ω_{AB} and ω_{BC} (or v_B and $v_{B/C}$), and the two unknowns can be obtained by equating the horizontal and vertical components of the various terms. The sum of the horizontal components (assume positive to the right) gives

$$\left(\xrightarrow{+} \right) \quad -3\omega_{AB} = -\tfrac{4}{5}(8)(5) - \tfrac{5}{13}(13\omega_{BC}) = -32 - 5\omega_{BC}.$$

In a similar manner the sum of the vertical components of the velocities gives

$$\left(\uparrow + \right) \quad 0 = \tfrac{3}{5}(8)(5) - \tfrac{12}{13}(13\omega_{BC}) = 24 - 12\omega_{BC}.$$

The solution of these two algebraic equations gives

$$\omega_{BC} = 2 \text{ rad per sec}, \quad \omega_{AB} = 14 \text{ rad per sec}.$$

Both quantities have positive signs, indicating that the assumed directions were correct. Therefore

$$\omega_{BC} = \underline{2.00 \text{ rad per sec counterclockwise}}, \qquad \text{Ans.}$$

$$\omega_{AB} = \underline{14.00 \text{ rad per sec counterclockwise}}. \qquad \text{Ans.}$$

The velocity of M can be obtained by writing an equation for the velocity of M relative to either B or C. The equation relating v_M and v_B is

$$\mathbf{v}_M \quad = \quad \mathbf{v}_B \quad + \quad \mathbf{v}_{M/B}$$

$$v_B = 3(14) = 42 \qquad\qquad v_{M/B} = 6.5(2) = 13$$

$$\omega = 2$$

$$\omega = 14$$

$$= \overset{12}{\underset{5}{\diagup}}$$

$$= \overleftarrow{42} + 13 = \overleftarrow{42} + \overrightarrow{5} + \overset{\uparrow}{12}$$

$$= \overleftarrow{37} + \overset{\uparrow}{12} = 38.9 \text{ ips} \quad \diagdown 19.97° .$$ Ans.

EXAMPLE 8-22

Blocks D and E in Fig. 8-32(a) are pinned together. Block D slides in a slot in the rotating member OA, and E slides in the fixed vertical slot. Determine the velocity of E as a function of θ and $\dot{\theta}$ by

(a) Differentiating the position vector of E.
(b) Using the equation of relative velocity.

SOLUTION

(a) The position vector for block E [see Fig. 8-32(b)] can be written using either polar or rectangular coordinates as

$$\mathbf{r} = r\mathbf{e}_r = b\mathbf{i} + b \tan \theta \mathbf{j}. \tag{a}$$

Since both r and $\mathbf{e}_r$ vary with time, the second (rectangular form of Eq. (a) is more convenient to use. The velocity of E is

$$\mathbf{v}_E = \dot{\mathbf{r}} = b \sec^2 \theta \, \dot{\theta} \mathbf{j}. \qquad \text{Ans.}$$

(b) Point B is on block E at the center of the pin connecting the two blocks, and C is a point on the rotating arm OA, which is coincident with B in the position shown. A relative velocity equation can be written between points B and C, and diagrams representing each of these terms are drawn below each term of the equation. An explanation of these diagrams is given below.

$$\mathbf{v}_B = \mathbf{v}_C + \mathbf{v}_{B/C}$$

$$v_C = r\dot{\theta}$$

The velocity of B is vertical since B is on block E, which is constrained to move in the vertical slot. Point C is on the rotating member and its velocity is perpendicular to OA with a magnitude of $r\dot{\theta}$. The velocity of B relative to C is the velocity B would appear to have to an observer at point C, and this would be along OA,[8] when B and C

(a)

(b)

Figure 8-32

---

[8] It should be noted that, although the relative velocity of B with respect to C is along OA, the relative velocity may be changing in both magnitude and direction, and the relative acceleration of the two points will not, in general, be along the member OA.

are coincident. In these diagrams there are two unknowns, the magnitudes v_B and $v_{B/C}$. Note that $r = b \sec \theta$. The value of v_B can be determined by summing velocity components in the direction perpendicular to member OA:

$$v_B \cos \theta = b \sec \theta\ \dot\theta$$
$$v_B = b \sec^2 \theta\ \dot\theta$$
$$\mathbf{v}_E = \mathbf{v}_B = \underline{b \sec^2 \theta\ \dot\theta\ \uparrow}.$$

Ans.

The velocity of B relative to C ($v_{B/C}$) can be determined by summing velocity components in the direction along OA but is not needed for this problem.

PROBLEMS

8-172 The angular velocity of member AB in Fig. P8-172 is 9 rad-per sec clockwise in the position shown. Determine
(a) The angular velocities of members CE and CD.
(b) The velocity of point E.

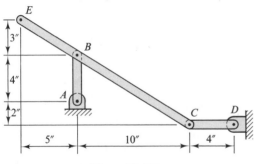

Figure P8-172

8-173 The angular velocity of member AB in Fig. P8-173 is 21 rad per sec counterclockwise in the position shown. Determine
(a) The angular velocities of members BC and CD.
(b) The velocity of point E.

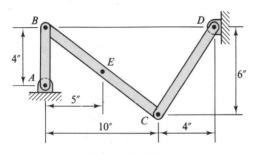

Figure P8-173

8-174 A simplified version of the mechanism for operating the hydraulic dump truck illustrated in Fig. P8-174(a) is shown in Fig. P8-174(b). When the truck body is in the position shown, the angular velocity of link A is 0.035 rad per sec clockwise. Determine the angular velocity of the truck body for this position.

(a)

Courtesy Hercules Steel Products Corp., Galion, Ohio.

(b)

Figure P8-174

8-175 Determine the velocity of the piston relative to the cylinder in Problem 8-174. All data in the original problem apply to this problem.

8-176 The crank, connecting rod, and piston arrangement for a large reciprocating gas engine is shown in Fig. P8-176. The angular velocity of the crank shaft is 300 rpm clockwise. Determine the velocity of the piston for the position indicated.

Figure P8-176

*8-177 Roller C of member CD in Fig. P8-177 has a velocity of 0.30 m/s upward along the slot. Determine

(a) The angular velocities of members AB and CD.

(b) The velocity of point D.

Figure P8-177

*8-178 The angular velocity of member BCD in Fig. P8-178 is 0.50 rad/s clockwise. Determine the velocity of point D.

8-179 The wheel in Fig. P8-179 rolls without slipping along an elevated track with an angular velocity of 4 rad per sec clockwise. Determine the linear velocities of points B, C, D, and E.

Figure P8-178

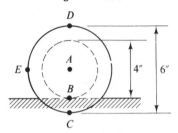

Figure P8-179

8-180 Solve Problem 8-179 if the wheel rolls and slips on the track. The velocity of point A is 6.0 ips to the left, and the angular velocity of the wheel is 4.0 rad per sec clockwise.

8-181 The wheel in Fig. P8-181 rolls without slipping along the curved track. As the center moves from O to O' the radius of the wheel will turn through an angle θ as shown. Derive the relationship between θ and ϕ and show that the speed of the center of the wheel is equal to $r\dot{\theta}$.

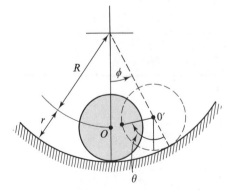

Figure P8-181

8-182 The gear in Fig. P8-182 rotates about a fixed axis at C with an angular velocity of 2.0 rad per sec clockwise in the position shown. The rack D slides along the horizontal plane, and block A slides in a slot in rack D. Determine the velocity of block A relative to the rack.

Figure P8-182

8-183 The crank OA in Fig. P8-183 has a constant clockwise angular velocity of 6.25 rad per sec. When the mechanism is in the position shown, determine
(a) The angular velocity of member AB.
(b) The linear velocity of the point on AB which is coincident with the roller at C.

Figure P8-184

(b) The velocity of A and the angular velocity of AB.
(c) The angular velocities of the wheel and of AB.

Figure P8-183

8-184 Member OA in Fig. P8-184 has an angular velocity of 2 rad per sec clockwise when in the position shown. Determine
(a) The angular velocity of body C.
(b) The velocity of point B.
(c) The velocity of AB relative to C.

8-185 Roller A on member AB in Fig. P8-185 slides along the horizontal slot, while the member causes the wheel to rotate. There is no slipping between AB and the wheel. Derive expressions relating
(a) The velocity of A and the angular velocity of the wheel.

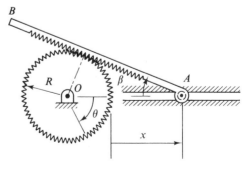

Figure P8-185

419

8-186 Member OP in Fig. P8-186 has an angular velocity of 5.0 rad per sec clockwise in the position shown. Determine the velocity of point Q.

Figure P8-186

8-16 INSTANTANEOUS CENTERS FOR PLANE MOTION

Frequently, it is necessary to determine the linear velocity of some point on a rigid body or the angular velocity of a line in a rigid body when the body has plane motion. These velocities can usually be obtained from Eq. (8-23) for relative velocities. The relative velocity equation is

$$\mathbf{v}_B = \mathbf{v}_A + \mathbf{v}_{B/A},$$

where B and A are any two points on the rigid body. If a point A with zero velocity can be located, the velocity of B [see Eq. (8-21)] becomes

$$\mathbf{v}_B = \mathbf{v}_{B/A} = \boldsymbol{\omega} \times \mathbf{r}, \qquad (8\text{-}24)$$

where $\mathbf{r}$ is the position vector of B from the point of zero velocity and $\boldsymbol{\omega}$ is the angular velocity of the body. When A and B are in any plane parallel to the plane of motion, the two vectors $\boldsymbol{\omega}$ and $\mathbf{r}$ in Eq. (8-24) are perpendicular, and the equation can be written in scalar form as

$$v_B = \omega r. \qquad (8\text{-}25)$$

Either Eq. (8-24) or (8-25) provides a convenient method for determining the velocity of any point B on a rigid body having plane motion provided a point (on the body, or the body extended) whose velocity is zero at the instant in question can be readily located. For kinematics, a body, such as the rod CD in Fig. 8-33, can be considered to be extended or enlarged in any manner, provided the enlarged body remains a rigid body. The concept of a body extended is needed

Figure 8-33

because the point with zero velocity is not necessarily a point on the actual body.

When a rigid body has plane motion, the axis in the body, or the body extended, *whose particles*, at any instant, *have zero velocity is defined as the instantaneous axis of zero velocity* or the instantaneous axis of the body. The instantaneous axis of zero velocity is always perpendicular to the plane of motion of the body. The point of intersection of the instantaneous axis and the plane of motion is defined as the *instantaneous center* of zero velocity. The velocity of any point of the body not on the instantaneous axis is perpendicular to the line from the instantaneous axis of the body to the point. Thus the velocities (but not the accelerations) of points on a body with plane motion can be obtained by considering the body to be rotating about the instantaneous axis of zero velocity at any instant. At the next instant, the instantaneous center is usually another point in the rigid body or body extended. The instantaneous axis is not a fixed axis in the body, nor is it a fixed axis in space.

The point in the plane of motion of the body, or body extended, which at any instant has zero velocity can be readily located if the directions of the velocities of any two points in the plane of motion of the body are known, provided the velocities are not parallel. The rigid body in Fig. 8-34 has plane motion, and the directions of the absolute velocities of two points A and B are as shown. Since at the instant in question the velocity of A is perpendicular to the line from the instantaneous center to A, the instantaneous center must be somewhere on the line ac through A perpendicular to $\mathbf{v}_A$. Likewise, the instantaneous center must be on the line bd through B and perpendicular to $\mathbf{v}_B$. Consequently, the instantaneous center of zero velocity is the intersection of these lines at point O. The magnitude of the angular velocity of the body is

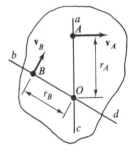

Figure 8-34

$$\omega = \frac{v_A}{r_A} = \frac{v_B}{r_B}. \qquad (a)$$

Thus if the directions of two nonparallel velocities in the plane of motion and the magnitude of one of them are known, the location of the instantaneous center and the angular velocity of the body can be obtained, and consequently the linear velocity of any other point on the body can be determined.

When the velocities of two points in the plane of motion of a rigid body with plane motion are parallel, equal in magnitude, and in the same direction, the instantaneous center is at infinity, the angular velocity of the body is zero, and all points on the body have the same velocity. When the velocities of the two points are parallel and unequal in magnitude, lines drawn through the two points perpendicular to their velocities will be collinear, and the instantaneous center of the body will be on this line. If the sense and magnitudes of the velocities

421

Figure 8-35

are known, the instantaneous center can be located by using Eq. (a). If the velocities of C and A in Fig. 8-35 are to the right and if v_C is less than v_A, the instantaneous center will be a distance r_C below C on ab. The distance r_C can be obtained from Eq. (a) as follows:

$$\omega = \frac{v_C}{r_C} = \frac{v_A}{r_A} = \frac{v_A}{d + r_C},$$

and if v_A, v_C, and d are known, r_C can be determined.

The instantaneous center of zero velocity of a rigid body can also be located if the velocity of one point on the body and the angular velocity of the body are known for the instant in question. If the velocity of point A on the rigid body in Fig. 8-36 has a known value to the right and if the angular velocity of the body has a known value counterclockwise, the distance from A to the instantaneous center is

$$r_A = \frac{v_A}{\omega}$$

from Eq. (a). The instantaneous center must be on the line ab, and if the velocity of A is to be to the right when the angular velocity of the body is counterclockwise, the instantaneous center must be above A.

Once the instantaneous center of zero velocity is located, the magnitude of the velocity of any particle of the body at that instant is given by Eq. (a), where ω is the angular velocity of any line in the body in, or parallel to, the plane of motion and r is the distance from the instantaneous axis to the particle whose velocity is to be found. The value of ω can be obtained from Eq. (a) if it is not given. The direction of the velocity is obtained by inspection, since it is perpendicular to the line from O to the point and its sense is consistent with the sense of ω.

The instantaneous center of zero velocity is, in general, not a point of zero acceleration. No convenient method is available for locating the point of zero acceleration. The principles of relative motion as developed in Art. 8-14 for velocities will be extended in Art. 8-18 to apply to the accelerations of particles of a rigid body. It should be emphasized that *the instantaneous center of zero velocity cannot be used to obtain accelerations because the instantaneous center does not, in general, have zero acceleration.*

The instantaneous center of zero velocity as presented in this article is the instantaneous center of a body or member with respect to the earth. For use in courses in kinematics, this concept can be readily extended to determine the instantaneous center of one moving member of a mechanism relative to another moving member of the mechanism. The instantaneous center of two bodies with plane motion is the point common to the two bodies, or bodies extended, which has the same linear velocity in each body. In the previous discussion no mention was made of the second body, but if the instantaneous center of zero velocity is desired, the second body is understood to be the

Figure 8-36

earth. For example, when a wheel rolls without slipping on a fixed surface, the point on the body which is in contact with the fixed surface (the earth) is the instantaneous center of zero velocity.

EXAMPLE 8-23

The wheel shown in Fig. 8-37 rolls without slipping on a horizontal elevated track with an angular velocity of 5 rad per sec clockwise. Locate the instantaneous center of zero velocity and determine the velocities of points A, D, and E.

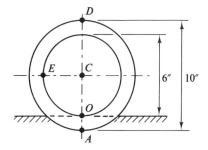

Figure 8-37

SOLUTION

The magnitude of the velocity of C, as determined by Example 8-21, is

$$v_C = r\omega = 3(5) = 15$$

and

$$\mathbf{v}_C = 15 \text{ ips} \rightarrow.$$

The instantaneous center of zero velocity must therefore be on a vertical line through C. Furthermore, from Eq. (8-25),

$$v_C = r_C\omega,$$

where r_C is the distance from the instantaneous center to C. Thus, $r_C = 3$ in., and point O is the instantaneous center of zero velocity. This result verifies the last statement preceding this example.

Again from Eq. (8-25),

$$v_A = r_A\omega = 2(5) = 10 \text{ ips}$$

and

$$\mathbf{v}_A = 10 \text{ ips} \leftarrow. \qquad \text{Ans.}$$

In a similar manner,

$$v_E = r_E\omega = 3\sqrt{2}(5) = 21.2 \text{ ips}$$

and

$$\mathbf{v}_E = 21.2 \text{ ips} \quad \angle 45° \qquad \text{Ans.}$$

Also

$$v_D = r_D\omega = 8(5) = 40 \text{ ips}$$

and

$$\mathbf{v}_D = 40 \text{ ips} \rightarrow. \qquad \text{Ans.}$$

EXAMPLE 8-24

The angular velocity of member AB of the link mechanism in Fig. 8-38(a) is 8 rad per sec clockwise. Determine, for the given position,
(a) The location of the instantaneous center of zero velocity of member BC.
(b) The angular velocity of BC.
(c) The angular velocity of CD.
(d) The velocity of point E, the midpoint of BC.

423

Figure 8-38

SOLUTION

(a) For the position indicated, point B has a horizontal velocity; therefore, the instantaneous center of member BC is on a vertical line through B. Point C has a velocity perpendicular to DC, and the instantaneous center of BC is also along line CD. These lines intersect at O, the instantaneous center of member BC, located as shown in Fig. 8-38(b).

(b) The magnitude of the velocity of point B is

$$v_B = r_{AB}\omega_{AB} = 3(8) = 24 \text{ fps}$$

and

$$\mathbf{v}_B = 24 \text{ fps} \leftarrow.$$

From Eq. (a),

$$\omega_{OB} = \frac{v_B}{r_{OB}} = \frac{24}{6} = 4 \text{ rad per sec}$$

and

$$\boldsymbol{\omega}_{BC} = \boldsymbol{\omega}_{OB} = \boldsymbol{\omega}_{OC} = \underline{4 \text{ rad per sec } \circlearrowright}. \qquad \text{Ans.}$$

(c) From Eq. (8-25),

$$v_C = \omega_{OC} r_{OC} = 4(5) = 20 \text{ fps}$$

and

$$\mathbf{v}_C = 20 \text{ fps}$$

Finally,

$$\omega_{CD} = \frac{v_C}{r_{CD}} = \frac{20}{2.5} = 8 \text{ rad per sec}$$

and

$$\boldsymbol{\omega}_{CD} = \underline{8 \text{ rad per sec } \circlearrowright}. \qquad \text{Ans.}$$

(d) Point E is 2.0 ft to the left of AB and 1.5 ft below AC; therefore,

$$r_{OE} = \sqrt{(2.0)^2 + (3.0 + 1.5)^2} = 4.92 \text{ ft}.$$

The magnitude of the velocity of point E is

$$v_E = r_{OE}\omega_{BC} = 4.92(4.0) = 19.68 \text{ fps},$$

and the velocity of E is

$$\mathbf{v}_E = 19.68 \text{ fps} \quad \underline{23.96°}. \qquad \text{Ans.}$$

PROBLEMS

*8-187 Member AB in Fig. P8-187 has an angular velocity of 12 rad/s clockwise in the position shown. Determine the angular velocities of members BC and CD.

Figure P8-187

8-188 Member AB in Fig. P8-188 has an angular velocity of 6 rad per sec counterclockwise

Figure P8-188

in the position shown. Determine the angular velocities of members BC and CD.

*8-189 Member AB in Fig. P8-189 has a constant angular velocity of 15 rad/s clockwise. Determine

(a) The angular velocity of member CD.

(b) The velocity of E, the midpoint of member BC.

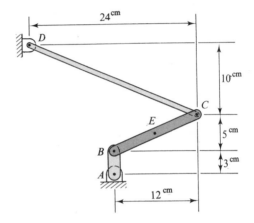

Figure P8-189

8-190 Member AB in Fig. P8-190 has an angular velocity of 4 rad per sec counterclockwise in the position shown. Determine

(a) The velocity of point C.

(b) The velocity of point E.

Figure P8-190

8-191 Determine the angular velocity of member CD in Problem 8-190 for which the speed of E will be 30 ips. (The angular velocity of AB will not be the same as for Problem 8-190.)

8-192 Member AB in Fig. P8-192 has an angular velocity of 3.0 rad per sec clockwise in the position shown. Determine
(a) The velocity of point B.
(b) The angular velocity of the crank OA.

Figure P8-192

8-193 The velocity of point A on member AC in Fig. P8-193 is 75 ips up the slot in the position shown. Determine the velocity of point C and the angular velocity of member BD.

Figure P8-193

8-194 Plate $ABCD$ in Fig. P8-194 is pinned to OA at A, and the roller at C slides in the horizontal slot. In the position shown the veloc-

ity of C is 35 ips to the right. Determine the velocities of points A, B, and D.

Figure P8-194

8-195 The angular velocity of member C in Fig. P8-195 is 0.80 rad per sec counterclockwise in the position shown. Determine the velocities of points A and B.

Figure P8-195

8-196 The angular velocity of member A in Fig. P8-196 is 2.0 rad per sec clockwise. Determine the velocities of points P and Q.

*__8-197__ The wheel in Fig. P8-197 rolls on the hub without slipping along the horizontal plane. The velocity of point C is 2.0 m/s to the right. Determine the velocities of points A and M.

*__8-198__ The wheel in Problem 8-197 rolls and slips along the plane in such a manner that the velocity of C is 1.5 m/s to the right and the

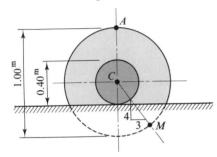

Figure P8-196

Figure P8-197

angular velocity of the wheel is 5 rad/s counterclockwise. Determine the velocities of points A and M.

*8-199 Point A on the wheel in Fig. P8-199 has a velocity of 2.0 m/s in the direction shown on the figure. Determine the angular velocity of the wheel if it remains in contact with the plane.

Figure P8-199

*8-200 The planetary gear in Fig. P8-200 has an outer gear A, an inner gear B, and a spider on which are mounted three small planet gears D. The pitch diameters are 30 cm for A, 20 cm

for B, and 5 cm for D. The angular velocity of A is zero, and the angular velocity of C is 10 rpm clockwise. Determine

(a) The angular velocity of B.
(b) The velocity of point P on D.

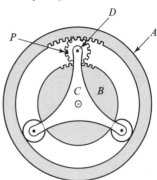

Figure P8-200

*8-201 In Problem 8-200 the inner gear B is stationary, and the spider C has an angular velocity of 10 rpm clockwise. Determine
(a) The angular velocity of A.
(b) The velocity of point P on D.

*8-202 In Problem 8-200 the spider C is stationary, and the inner gear B has an angular velocity of 10 rpm clockwise. Determine
(a) The angular velocity of A.
(b) The velocity of point P on D.

*8-203 The bar AB in Fig. P8-203 moves with its ends in contact with the horizontal and vertical walls. When $\theta = 60°$ the velocity of A is 50 cm/s downward. Determine the velocities of points B and G for this position.

Figure P8-203

427

8-204 Body *ABC* in Fig. P8-204 is a rigid body with plane motion. In the position shown the velocity of *C* is 42 ips downward. Determine the velocities of points *A* and *B*.

8-205 The angular velocity of member *ABC* in Fig. P8-205 is 2.0 rad per sec counterclockwise in the position shown. Determine the velocities of points *A*, *B*, and *C*.

Figure P8-205

Figure P8-204

*8-206 The wheel in Fig. P8-206 rolls without slipping, and point *O* has a constant velocity of 40 cm/s to the left. Determine the linear velocity of *B* for the position shown.

Figure P8-206

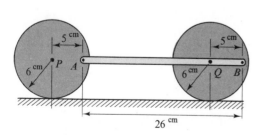

Figure P8-209

*8-207 Solve Problem 8-206 when point *A* is directly above point *O*.

*8-208 Solve Problem 8-206 when point *A* is to the left of point *O*.

*8-209 The two wheels in Fig. P8-209 roll along the horizontal plane without slipping and are connected by the rod *AB*. The velocity of point *P* is 12 cm/s to the right. Determine the angular velocity of *AB* and the linear velocity of *Q* (a) When *A* is to the right of *P* and *B* is to the right of *Q* as shown.

428

(b) When A is to the right of P and B is to the left of Q.

*8-210 Solve Problem 8-209 when the bodies are arranged as shown in Fig. P8-210.

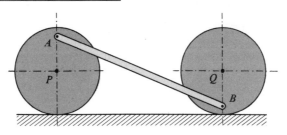

Figure P8-210

8-211 Gear A in Fig. P8-211 rotates with an angular velocity of 2 rad per sec clockwise. Gear B rolls without slipping on gear A. Arm C connects the gears as indicated and turns with an angular velocity of 3 rad per sec clockwise. Determine

(a) The angular velocity of gear B.

(b) The linear velocity of point D on gear B.

Figure P8-211

*8-212 Gears A and B in Fig. P8-212 are connected by smooth pins to the horizontal arm C at E and D. Arm C rotates with a constant angular velocity of 3 rad/s clockwise. Gear B rolls without slipping on either the fixed track or gear A. Determine the velocity of point P on gear A in the position shown.

*8-213 Gears D and E in Fig. P8-213 are pinned to body ABC at B and C. Member ABC rotates about A with an angular velocity of 2 rad/s clockwise. Gear D rolls without slipping on either gear E or the fixed surface. Determine the velocity of point P on gear E.

8-214 The three rigid bodies in Fig. P8-214 all move in the same plane. Member AE has an angular velocity of 5 rad per sec counterclockwise in the indicated position. Determine the linear velocities of points B and F.

Figure P8-212

Figure P8-213

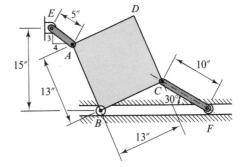

Figure P8-214

429

*8-215 Member *ABC* in Fig. P8-215 is pinned at *C* to bar *CD*. The blocks *A*, *B*, and *D* move in fixed slots. In the position shown *A* has a velocity of 10 cm/s to the right. Determine the velocities of points *B* and *D* for this position.

Figure P8-215

8-17 RELATIVE VELOCITY—
THREE-DIMENSIONAL MOTION

As indicated in Art. 8-14, the equation

$$\mathbf{v}_B = \mathbf{v}_A + \mathbf{v}_{B/A}$$

is valid for any two points *A* and *B* provided nonrotating reference axes are used to describe the three terms. When points *A* and *B* are points on a rigid body with three-dimensional motion, the motion of *B* relative to *A* will be tangent to a spherical surface with its center at *A* and with a radius equal to the distance from *A* to *B*. The velocity of *B* relative to *A* can be expressed as

$$\mathbf{v}_{B/A} = \boldsymbol{\omega} \times \mathbf{r},$$

where **ω** is the absolute angular velocity of the body and **r** is the vector *from A to B* as shown in Art. 8-13. Even though the velocities of points *A* and *B* are known, if the system is not fully constrained, the angular velocity **ω** cannot be uniquely determined. An example of such a case is indicated by Fig. 8-39 in which member *AB* is connected to a rotating wheel and to a sliding collar at *B* by ball and socket joints. The angular velocity, **ω**, of *AB* can be resolved into two components, **ω**_t parallel to *AB* and **ω**_n normal to *AB*. The component **ω**_t represents the *spin* of the member about its axis, and its magnitude will have no effect on the motion of points *A* and *B*. If one of the ball and socket joints is replaced by a clevis as shown in Fig. 8-41 (p. 432) or if the angular motion of AB is constrained by other means, the quantity $\boldsymbol{\omega}_t$, as well and $\boldsymbol{\omega}_n$, will have a definite value and $\boldsymbol{\omega}$ is therefore determinate. The angular velocity can be expressed as

$$\boldsymbol{\omega} = \omega_x \mathbf{i} + \omega_y \mathbf{j} + \omega_z \mathbf{k},$$

and when $\boldsymbol{\omega}$ is indeterminate it is sometimes convenient to assume one of the components to be equal to zero and solve for the remaining components. If the resulting angular velocity is resolved into $\boldsymbol{\omega}_t$ and $\boldsymbol{\omega}_n$ components, the $\boldsymbol{\omega}_n$ component will be unique, but the value of $\boldsymbol{\omega}_t$ will depend on which of the axial components (ω_x, ω_y, or ω_z) was assumed to be zero or some other arbitrary value. The following examples will demonstrate these concepts.

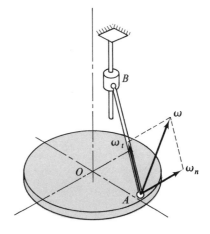

Figure 8-39

EXAMPLE 8-25

Collars A and B in Fig. 8-40 slide along fixed rods. The rod C is connected to A by a clevis and to B by a ball and socket joint. The velocity of A is $v_A \mathbf{i}$. Determine the velocity of B and the angular velocity of member C.

SOLUTION

The system is fully constrained; therefore the angular velocity of C can be uniquely determined. Rod C will have an angular velocity component due to the rotation of the collar and clevis pin and a component due to its rotation about the clevis pin. These components are shown in Fig. 8-41(a), and the angular velocity is

$$\boldsymbol{\omega} = \boldsymbol{\omega}_1 + \boldsymbol{\omega}_2$$

$$= \omega_1 \mathbf{i} + \omega_2 \mathbf{e}_p$$

$$= \omega_1 \mathbf{i} + \omega_2 \frac{d\mathbf{j} + c\mathbf{k}}{\sqrt{c^2 + d^2}}. \qquad (a)$$

The unit vector $\mathbf{e}_p$ (parallel to the pin) can often be visualized more readily by drawing a sketch looking at the end of the collar as in Fig. 8-41(b). Since $\mathbf{e}_p$ is perpendicular to both member C and the rod along which A slides, it can also be calculated as

$$\mathbf{e}_p = \frac{\mathbf{r}_{AB} \times \mathbf{r}_{DE}}{|\mathbf{r}_{AB} \times \mathbf{r}_{DE}|}.$$

The relative velocity equation is

$$\mathbf{v}_B = \mathbf{v}_A + \mathbf{v}_{B/A} = \mathbf{v}_A + \boldsymbol{\omega} \times \mathbf{r}_{B/A},$$

Figure 8-40

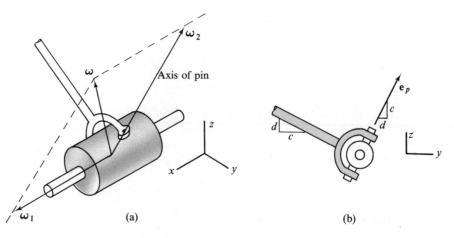

(a)

(b)

Figure 8-41

where $\mathbf{r}_{B/A} = -b\mathbf{i} - c\mathbf{j} + d\mathbf{k}$ is the position of B relative to A. The velocity equation can be expanded as

$$v_B\mathbf{k} = v_A\mathbf{i} + \begin{vmatrix} \mathbf{i} & \mathbf{j} & \mathbf{k} \\ \omega_1 & \dfrac{d\omega_2}{\sqrt{c^2+d^2}} & \dfrac{c\omega_2}{\sqrt{c^2+d^2}} \\ -b & -c & d \end{vmatrix}$$

$$= \mathbf{i}\left(v_A + \frac{d^2+c^2}{\sqrt{c^2+d^2}}\,\omega_2\right) + \mathbf{j}\left(\frac{-bc\omega_2}{\sqrt{c^2+d^2}} - d\omega_1\right)$$

$$+ \mathbf{k}\left(-c\omega_1 + \frac{bd\omega_2}{\sqrt{c^2+d^2}}\right).$$

The component equations are

$$\sqrt{c^2+d^2}\,\omega_2 = -v_A, \tag{b}$$

$$d\omega_1 + \frac{bc}{\sqrt{c^2+d^2}}\,\omega_2 = 0, \tag{c}$$

$$-c\omega_1 + \frac{bd}{\sqrt{c^2+d^2}}\,\omega_2 = v_B. \tag{d}$$

Equations (b) and (c) give

$$\omega_2 = \frac{-v_A}{\sqrt{c^2+d^2}},$$

$$\omega_1 = \frac{bcv_A}{d(c^2+d^2)},$$

and when these values are substituted into Eq. (a) the angular velocity of C is

$$\boldsymbol{\omega} = \frac{v_A}{d(c^2+d^2)}\,(bc\mathbf{i} - d^2\mathbf{j} - cd\mathbf{k}). \qquad \text{Ans.}$$

The velocity of B is obtained by substituting values of ω_1 and ω_2 into Eq. (d), thus

$$v_B = \frac{-bc^2v_A}{d(c^2+d^2)} + \frac{bd}{\sqrt{c^2+d^2}}\left(\frac{-v_A}{\sqrt{c^2+d^2}}\right),$$

which reduces to

$$v_B = -\frac{b}{d}\,v_A$$

and

$$\mathbf{v}_B = -\frac{b}{d}\,v_A\mathbf{k}. \qquad \text{Ans.}$$

When numerical data are given, numbers can be substituted for symbols and thus simplify the calculations, but this makes it impossible to use dimensional homogeneity to check the algebraic manipulations.

EXAMPLE 8-26

Solve Example 8-25 if the clevis at A is replaced a ball and socket joint.

SOLUTION

When the clevis at A is replaced by a ball and socket the system is no longer completely constrained and the angular velocity of C cannot be uniquely determined. The velocity of B, however, can be uniquely determined, and an angular velocity of C can be obtained by introducing an arbitrary constraint on the system. The problem will be solved twice, first by introducing the constraint $\omega_x = 0$ and second by using the constraint ω_t (the spin velocity) $= 0$.

The relative velocity equation for points A and B is

$$\mathbf{v_B} = \mathbf{v_A} + \mathbf{v_{B/A}},$$

$$v_B\mathbf{k} = v_A\mathbf{i} + \boldsymbol{\omega} \times \mathbf{r_{B/A}},$$

where

$$\boldsymbol{\omega} = \omega_x\mathbf{i} + \omega_y\mathbf{j} + \omega_z\mathbf{k}.$$

Thus the velocity equation becomes

$$v_B\mathbf{k} = v_A\mathbf{i} + \begin{vmatrix} \mathbf{i} & \mathbf{j} & \mathbf{k} \\ \omega_x & \omega_y & \omega_z \\ -b & -c & d \end{vmatrix}$$

$$= v_A\mathbf{i} + \mathbf{i}(d\omega_y + c\omega_z) - \mathbf{j}(d\omega_x + b\omega_z) + \mathbf{k}(-c\omega_x + b\omega_y),$$

which is equivalent to the following three algebraic equations:

$$\mathbf{i}: \qquad +d\omega_y + c\omega_z = -v_A, \qquad (a)$$

$$\mathbf{j}: \quad -d\omega_x \quad -b\omega_z = 0 \qquad (b)$$

$$\mathbf{k}: \quad -c\omega_x + b\omega_y \qquad = v_B. \qquad (c)$$

Since there are four unknown quantities, ω_x, ω_y, ω_z, and v_B, and only three equations, not all the unknowns can be determined. When $\boldsymbol{\omega}$ is indeterminate a possible solution can be obtained by assuming a value for one of the components and solving for the other two. From the assumption that $\omega_x = 0$

$$\omega_z = -\frac{d}{b}\omega_x = 0,$$

$$\omega_y = \frac{-v_A}{d},$$

and

$$\boldsymbol{\omega} = \underline{\frac{-v_A}{d}\mathbf{j}.} \qquad \text{Ans.}$$

From Eq. (c)

$$v_B = b\omega_y = -\frac{b}{d}\,v_A$$

and

$$\mathbf{v}_B = -\frac{b}{d}\,v_A\mathbf{k}. \qquad \text{Ans.}$$

It is seen that an initial assumption of *either* $\omega_x = 0$ or $\omega_y = 0$ yields the same axial components for $\boldsymbol{\omega}$. Assuming $\omega_y = 0$ will yield a different set of axial components for $\boldsymbol{\omega}$ and will also increase the complexity of the calculations for $\boldsymbol{\omega}$. All three of these assumptions, however, result in the same answer for $\mathbf{v}_B$.

The angular velocity of C can be resolved into normal and tangential components as follows. The component parallel to C is

$$\boldsymbol{\omega}_t = (\boldsymbol{\omega} \cdot \mathbf{e}_t)\mathbf{e}_t$$

$$= \frac{-v_A}{d}\,\mathbf{j} \cdot \left(\frac{-b\mathbf{i} - c\mathbf{j} + d\mathbf{k}}{\sqrt{b^2 + c^2 + d^2}}\right)\left(\frac{-b\mathbf{i} - c\mathbf{j} + d\mathbf{k}}{\sqrt{b^2 + c^2 + d^2}}\right)$$

$$= \frac{v_A c}{d(b^2 + c^2 + d^2)}\,(-b\mathbf{i} - c\mathbf{j} + d\mathbf{k}).$$

Note that $\mathbf{e}_t$ is a unit vector parallel to AB. The component of the angular velocity normal to AB is

$$\boldsymbol{\omega}_n = \boldsymbol{\omega} - \boldsymbol{\omega}_t$$

$$= -\frac{v_A}{d}\,\mathbf{j} - \frac{v_A c}{d(b^2 + c^2 + d^2)}\,(-b\mathbf{i} - c\mathbf{j} + d\mathbf{k})$$

$$= \frac{v_A}{d(b^2 + c^2 + d^2)}\,[bc\mathbf{i} - (b^2 + d^2)\mathbf{j} - cd\mathbf{k}].$$

The magnitude of the tangential component of the angular velocity (the spin velocity) is $\boldsymbol{\omega} \cdot \mathbf{e}_t$, and when this quantity is set equal to zero it provides a fourth equation,

$$\omega_t = (\omega_x\mathbf{i} + \omega_y\mathbf{j} + \omega_z\mathbf{k}) \cdot \left(\frac{-b\mathbf{i} - c\mathbf{j} + d\mathbf{k}}{\sqrt{b^2 + c^2 + d^2}}\right) = 0$$

or

$$-b\omega_x - c\omega_y + d\omega_z = 0. \qquad (d)$$

There are now four equations (a), (b), (c), and (d) which can be solved simultaneously for the four unknowns ω_x, ω_y, ω_z, and v_B.

From Eq. (a),

$$\omega_y = -\frac{v_A}{d} - \frac{c\omega_z}{d}, \qquad (e)$$

and from Eq. (b),

$$\omega_x = -\frac{b\omega_z}{d}.$$ (f)

Equations (e) and (f) can be substituted into Eq. (d) to give

$$\frac{b^2\omega_z}{d} + \frac{cv_A}{d} + \frac{c^2\omega_z}{d} + d\omega_z = 0$$

from which

$$\omega_z = -\frac{cv_A}{b^2 + c^2 + d^2}.$$

The values for ω_x and ω_y can be determined by substituting into Eq. (e) and (f), and the final result is

$$\boldsymbol{\omega} = \frac{v_A}{d(b^2+c^2+d^2)}[bc\mathbf{i} - (b^2+d^2)\mathbf{j} - cd\mathbf{k}]. \qquad \text{Ans.}$$

This result is the same as the value of $\boldsymbol{\omega}_n$ obtained in the first solution, which serves to confirm the fact that $\boldsymbol{\omega}_n$ is independent of the assumptions made in obtaining a value for the indeterminate angular velocity of the bar.

The velocity of B is obtained by substituting into Eq. (c); that is

$$v_B = -c\omega_x + b\omega_y$$

$$= \frac{v_A}{d(b^2+c^2+d^2)}(-bc^2 - b^3 - bd^2)$$

$$= -\frac{b}{d}v_A$$

and

$$\mathbf{v}_B = -\frac{b}{d}v_A\mathbf{k}. \qquad \text{Ans.}$$

The procedures used in the preceding examples for solving the relative velocity equation

$$\mathbf{v}_B = \mathbf{v}_A + \mathbf{v}_{B/A}$$

can be summarized as follows:

1. If the system is fully constrained, the relative velocity equation together with the equations of constraint give the same number of algebraic equations as the number of unknowns, and $\boldsymbol{\omega}$ can be uniquely determined (as in Example 8-25).

2. If the system is not fully constrained, the number of algebraic equations is less than the number of unknowns. Additional arbitrary constraints (which do not violate any existing constraints) must be introduced to solve for $\boldsymbol{\omega}$, which will then not be unique (as in Example 8-26).

8-216 The bent rod *CDE* is welded to rod *AB* in Fig. P8-216, and point *C* slides on the horizontal *xy* plane. Rod *AB* is connected to the two collars by means of ball and socket joints. Determine the velocity of *C* and the angular velocity of member *ABC* in terms of v_A.

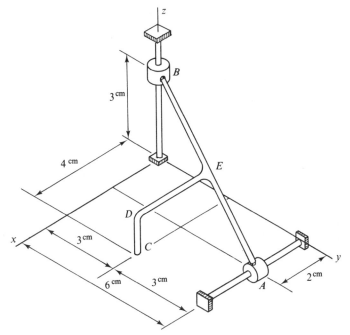

Figure P8-216

8-217 The crank *OA* in Fig. P8-217 rotates in the *yz* plane about an axis through *O* with a constant angular velocity of $5\mathbf{i}$ rad per sec. Link *AB* has a ball and socket joint at each end.

(a) Determine the velocity of the collar at *B*.
(b) Determine the component of the angular velocity of *AB* which is perpendicular to *AB*.

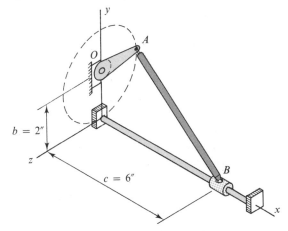

Figure P8-217

8-218 Determine the angular velocity of member AB in Problem 8-217 if it is connected to the collar at B by a clevis as shown in Fig. P8-218.

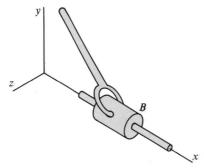

Figure P8-218

8-219 Figure P8-219 represents a pair of bevel gears A and B. Gear A rotates in a horizontal plane about a vertical axis at O. Gear B rotates around the shaft OC, which in turn rotates about the vertical axis at O. The angular velocity of the shaft OC is $10\mathbf{j}$ rad per sec, and gear A is stationary. Determine
(a) The angular velocity of gear B.
(b) The velocity of point D at the top of the gear.

Figure P8-219

8-220 Solve Problem 8-219 if gear A ha an angular velocity of $4\mathbf{j}$ rad per sec. All other data are the same as for Problem 8-219.

8-221 Disk D in Fig. P8-221 rotates about an axis at O with an angular velocity of $\omega_1\mathbf{k}$. The collar B slides along the inclined rod which is parallel to the yz plane. Member AB is

Figure P8-221

Figure P8-222

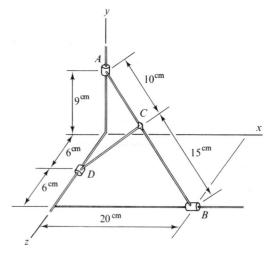

Figure P8-223

connected to the disk and to the collar by ball and socket joints. Determine

(a) The velocity of collar *B*.
(b) The component of the angular velocity of *AB* which is normal to *AB*.

8-222 The ball and socket at *A* in Problem 8-221 is replaced by the joint shown in Fig. P8-222 in which the vertical member rotates in a hole in the disk and is connected to *AB* by a horizontal pin. Determine the angular velocity of member *AB*.

*8-223 Cylinders *A*, *B*, and *D* in Fig. P8-223 slide along smooth rods as shown. The rigid member *AB* is connected to *A* and *B* with ball and socket joints. Member *CD* is connected to point *C* on member *AB* and to the cylinder *D* by ball and socket joints. The velocity of *A* is 12 cm/s downward in the position shown. Determine the velocities of points *B* and *D*.

8-18 RELATIVE ACCELERATION— NONROTATING AXES

In Art 8-6 equations were developed relating the absolute motion of a moving point to its motion measured relative to a set of nonrotating axes at a second moving point and the absolute motion of A. The equation relating the accelerations is

$$\mathbf{a}_B = \mathbf{a}_A + \mathbf{a}_{B/A}, \qquad (8\text{-}26)$$

where $\mathbf{a}_B$ and $\mathbf{a}_A$ are the absolute acceleration of points B and A, respectively, and $\mathbf{a}_{B/A}$ is the acceleration of B relative to a set of nonrotating axes attached to A. The acceleration of B relative to A is the acceleration B would appear to have to an observer moving with point A (but not rotating).

When A and B are two points on a rigid body, the motion of B relative to A must be along a path on a spherical surface with its center at A and with a radius equal to the distance from A to B since the two points must remain the same distance apart. The acceleration of B relative to A is given by Eq. (8-22) as

$$\mathbf{a}_{B/A} = \boldsymbol{\alpha} \times \mathbf{r} + \boldsymbol{\omega} \times (\boldsymbol{\omega} \times \mathbf{r})$$

$$= \boldsymbol{\alpha} \times \mathbf{r} + \boldsymbol{\omega} \times \mathbf{v}_{B/A},$$

where $\boldsymbol{\alpha}$ and $\boldsymbol{\omega}$ are the absolute angular acceleration and velocity of the rigid body and $\mathbf{r}$ is the position vector *from A to B*. When the body has plane motion the angular acceleration and angular velocity vectors are parallel. Thus for plane motion $\boldsymbol{\alpha} \times \mathbf{r}$ is parallel to $\boldsymbol{\omega} \times \mathbf{r}$ and is tangent to the path of B relative to A. The term $\boldsymbol{\omega} \times \mathbf{v}_{B/A}$ is normal to the relative velocity and thus is normal to the path. When the body containing A and B has three-dimensional motion the vectors representing $\boldsymbol{\alpha}$ and $\boldsymbol{\omega}$ are usually not parallel, and the quantity $\boldsymbol{\alpha} \times \mathbf{r}$ will not, in general, be parallel to $\boldsymbol{\omega} \times \mathbf{r}$ and thus will not be the tangential component of the acceleration of B relative to A.

If Eq. (8-26) is applied to two points not on the same rigid body, the motion of B relative to A will not, in general, be along a circular path, and the relative acceleration is usually more difficult to analyze. Application of Eq. (8-26) to problems involving plane motion and three-dimensional motion are discussed in the following two articles.

8-19 RELATIVE ACCELERATION— PLANE MOTION

In Art. 8-18 it was pointed out that when a rigid body has plane motion the angular velocity and angular acceleration vectors are parallel and can change in magnitude but not in direction. When the position vector $\mathbf{r}$ from A to B is parallel to the plane of motion it is normal to $\boldsymbol{\alpha}$ and $\boldsymbol{\omega}$ and Eq. (8-26) can be written as

$$\mathbf{a}_{B/A} = \boldsymbol{\alpha} \times \mathbf{r} + \boldsymbol{\omega} \times (\boldsymbol{\omega} \times \mathbf{r})$$

$$= \boldsymbol{\alpha} \times \mathbf{r} - \omega^2 \mathbf{r}$$

$$= \alpha r \mathbf{e}_t + \omega^2 r \mathbf{e}_n,$$

where $\mathbf{e}_t$ is a unit vector tangent to the path of B relative to A and $\mathbf{e}_n$ is a unit vector normal to the path and directed from B to A. Note that $\boldsymbol{\alpha}$ and $\boldsymbol{\omega}$ are absolute angular quantities (measured with respect to nonrotating axes) and that $\mathbf{r}$ is the position vector of B relative to A.

In Art. 8-15 it was noted that a wheel which rolls without slipping along a fixed plane surface is an important case of plane motion of a rigid body. It was shown that the center O of the wheel has rectilinear motion and that the magnitudes of the position and velocity of point O are related to the magnitudes of the angular position and velocity of the wheel by the expressions [see Fig. 8-30]

$$s_O = r\theta \tag{a}$$

$$v_O = r\dot{\theta} = r\omega \tag{b}$$

in which s_O and v_O are to the right when θ and ω are clockwise. The acceleration of point O can be obtained by differentiating Eq. (b) and is

$$a_O = r\dot{\omega} = r\alpha. \tag{c}$$

Note that the acceleration of O will be to the right when $\dot{\omega} = \alpha$ is clockwise and that r is the distance from the fixed plane to the center of the wheel. These three expressions are valid when the wheel rolls without slipping along a fixed plane. If the wheel rolls without slipping along a curved surface, a similar proof shows that the quantity $r\alpha$ is the magnitude of the tangential component of the acceleration of the center of the wheel. The normal component of the acceleration of the center of the wheel will depend on the velocity of the center and the radius of curvature of the path. If the wheel rolls without slipping along a moving surface, for example, a drum rolling on the bed of a moving truck, the preceding expressions give the velocity and acceleration of the center of the wheel *relative* to the moving surface. None of the preceding expressions are valid if the wheel rolls and slips on a fixed or moving surface.

The following examples illustrate the use of Eq. (8-26) for problems involving plane motion.

EXAMPLE 8-27

The large unbalanced wheel in Fig. 8-42 rolls without slipping along the horizontal plane. In the position shown the center O has a velocity of 4.0 m/s to the right and an acceleration of 10.0 m/s^2 to the left. Determine the velocity and acceleration of the mass center, G.

SOLUTION

The wheel rolls without slipping along a fixed horizontal plane; consequently the angular velocity and acceleration of the wheel are

$$\omega = \frac{v_O}{r} = \frac{4.0}{0.8} = 5.0 \text{ rad/s}\circlearrowleft,$$

$$\alpha = \frac{a_O}{r} = \frac{10.0}{0.8} = 12.5 \text{ rad/} \circlearrowright \text{s}^2.$$

The instantaneous center of zero velocity of the wheel is at point P, and the velocity of G is

$$\mathbf{v}_G = r_{PG}\omega\mathbf{e}_t$$
$$= [(0.75)^2 + (0.0866)^2]^{1/2}(5.0)\mathbf{e}_t$$
$$= 3.77 \text{ m/s} \quad 6.59° \text{ Ans.}$$

The acceleration of G is related to the acceleration of O by

$$\mathbf{a}_G = \mathbf{a}_O + \mathbf{a}_{G/O}$$
$$= \overleftarrow{10} + \quad$$

$$= -10\mathbf{i} + 1.25(\mathbf{i}\sin 30° + \mathbf{j}\cos 30°) + 2.5(-\mathbf{i}\cos 30° + \mathbf{j}\sin 30°)$$
$$= -10\mathbf{i} + 0.625\mathbf{i} + 1.082\mathbf{j} - 2.165\mathbf{i} + 1.25\mathbf{j} = -11.54\mathbf{i} + 2.33\mathbf{j}$$
$$= 11.77 \text{ m/s}^2 \quad 11.40° \qquad \text{Ans.}$$

Figure 8-42

EXAMPLE 8-28

The links of the mechanism in Fig. 8-43 move in a vertical plane. When they are in the position shown, CD has an angular velocity of 8 rad per sec clockwise and AB has an angular acceleration of 30 rad per sec^2 clockwise, see Example 8-21. Determine the acceleration of point C for the indicated position.

441

Figure 8-43

SOLUTION

The angular velocities of members AB and BC were found in Example 8-21 to be

$$\omega_{AB} = 14.00 \text{ rad per sec clockwise}$$

and

$$\omega_{BC} = 2.00 \text{ rad per sec counterclockwise.}$$

The angular velocities can also be determined by use of instantaneous centers. In any case they must be computed before the relative acceleration equation can be solved for the angular accelerations. A relative acceleration equation can be written for any two points. Whenever possible, two points on a rigid body should be selected since the path of the relative motion is a circle. In addition, the points selected should be points which move along known paths and whose accelerations can be expressed in terms of known or desired values. For this example points B and C are chosen since they are points on member BC and since they move along circular paths. The relative acceleration equation is

The diagrams below the equation represent the normal and tangential components of the accelerations and provide a graphical representation of the vector equation. All directions are known and there are two unknown magnitudes, α_{CD} and α_{BC}. This coplanar vector equation is equivalent to two scalar equations representing the horizontal and vertical components of the various terms. Thus

$$\left(\underset{\longrightarrow}{+} \right) \quad \frac{4}{5}(5)\alpha_{CD} + \frac{3}{5}(5)(8^2) = 3(30) + \frac{5}{13}(13)\alpha_{BC} - \frac{12}{13}(13)(2^2)$$

$$\left(\underset{+}{\uparrow} \right) \quad \frac{-3}{5}(5)\alpha_{CD} + \frac{4}{5}(5)(8^2) = -3(14^2) + \frac{12}{13}(13)\alpha_{BC} + \frac{5}{13}(13)(2^2),$$

which become

$$4\alpha_{CD} + 192 = 90 + 5\alpha_{BC} - 48 - 3\alpha_{CD} + 256 = -588 + 12\alpha_{BC} + 20$$

or

$$5\alpha_{BC} - 4\alpha_{CD} = 150$$

$$12\alpha_{BC} + 3\alpha_{CD} = 824.$$

The solution of these simultaneous equations gives

$$\alpha_{BC} = 59.5 \text{ rad per sec}^2$$

and

$$\alpha_{CD} = 36.8 \text{ rad per sec}^2.$$

Both of these results are positive, which indicates that the assumed directions (both counterclockwise) were correct. The acceleration of C is

$$\mathbf{a}_C = 5(64) \quad + \quad 5(36.8)$$

$$= 18\overrightarrow{2}.0 \quad + \quad 256.0 + 14\overrightarrow{7}.3 + 110.5$$

$$= 32\overrightarrow{9}.3 \quad + \quad 145.5$$

$$= 360 \text{ ips}^2 \qquad 23.85° \qquad \text{Ans.}$$

ALTERNATE SOLUTION

The solution can also be obtained using vector notation. When the axes are selected as shown on the figure the angular velocities and accelerations are

$$\boldsymbol{\omega}_{AB} = +14\mathbf{k}, \qquad \boldsymbol{\omega}_{BC} = 2\mathbf{k}, \qquad \boldsymbol{\omega}_{CD} = -8\mathbf{k},$$

$$\boldsymbol{\alpha}_{AB} = -30\mathbf{k}, \qquad \boldsymbol{\alpha}_{BC} = \alpha_{BC}\mathbf{k}, \qquad \boldsymbol{\alpha}_{CD} = \alpha_{CD}\mathbf{k}.$$

Note that $\boldsymbol{\omega}_{AB}$ and $\boldsymbol{\omega}_{BC}$ were obtained from Example 8-21. The radius vectors are

$$\mathbf{r}_{AB} = 3\mathbf{j}, \qquad \mathbf{r}_{BC} = 12\mathbf{i} - 5\mathbf{j}, \qquad \mathbf{r}_{CD} = -3\mathbf{i} - 4\mathbf{j}.$$

In each case the radius vector is the vector from the reference (or fixed) point to the corresponding moving point. The acceleration equation is

$$\mathbf{a}_C = \mathbf{a}_B + \mathbf{a}_{C/B}. \tag{a}$$

The acceleration of point C is

$$\mathbf{a}_C = \boldsymbol{\alpha}_{CD} \times \mathbf{r}_{CD} + \boldsymbol{\omega}_{CD} \times (\boldsymbol{\omega}_{CD} \times \mathbf{r}_{CD})$$

$$= \boldsymbol{\alpha}_{CD} \times \mathbf{r}_{CD} - \omega_{CD}^2 \mathbf{r}_{CD}$$

$$= \alpha_{CD}\mathbf{k} \times (-3\mathbf{i} - 4\mathbf{j}) - (8)^2(-3\mathbf{i} - 4\mathbf{j})$$

$$= -3\alpha_{CD}\mathbf{j} + 4\alpha_{CD}\mathbf{i} + 192\mathbf{i} + 256\mathbf{j}$$

$$= (4\alpha_{CD} + 192)\mathbf{i} + (-3\alpha_{CD} + 256)\mathbf{j}.$$

Similarly,

$$\mathbf{a}_B = \boldsymbol{\alpha}_{AB} \times \mathbf{r}_{AB} - \omega_{AB}^2 \mathbf{r}_{AB}$$

$$= -30\mathbf{k} \times 3\mathbf{j} - (14)^2 3\mathbf{j}$$

$$= +90\mathbf{i} - 588\mathbf{j}$$

and

$$\mathbf{a}_{C/B} = \boldsymbol{\alpha}_{BC} \times \mathbf{r}_{BC} - (\omega_{BC})^2 \mathbf{r}_{BC}$$

$$= \alpha_{BC}\mathbf{k} \times (12\mathbf{i} - 5\mathbf{j}) - (2)^2(12\mathbf{i} - 5\mathbf{j})$$

$$= 12\alpha_{BC}\mathbf{j} + 5\alpha_{BC}\mathbf{i} - 48\mathbf{i} + 20\mathbf{j}$$

$$= (5\alpha_{BC} - 48)\mathbf{i} + (12\alpha_{BC} + 20)\mathbf{j}.$$

When these expressions are substituted into Eq. (a) it becomes

$$(4\alpha_{CD} + 192)\mathbf{i} + (-3\alpha_{CD} + 256)\mathbf{j} = 90\mathbf{i} - 588\mathbf{j} + (5\alpha_{BC} - 48)\mathbf{i} + (12\alpha_{BC} + 20)\mathbf{j}.$$

When the coefficients of the unit vectors on the left-hand side of the equation are equated to the corresponding coefficients of the right-hand side the vector equation gives

$$\mathbf{i}: \quad 4\alpha_{CD} + 192 = 90 + 5\alpha_{BC} - 48,$$

$$\mathbf{j}: \quad -3\alpha_{CD} + 256 = -588 + 12\alpha_{BC} + 20,$$

which, of course, is the same as the result obtained by means of the "pictorial" analysis.

EXAMPLE 8-29

The bottom of the 25-ft ladder in Fig. 8-44 slides along the floor with a velocity of 6 fps to the left and an acceleration of 8 fps^2 to the left when $\theta = \tan^{-1}\frac{24}{7}$ (as shown). The top of the ladder remains in contact with the vertical wall. Determine, for this position, the velocity and acceleration of the mass center G.

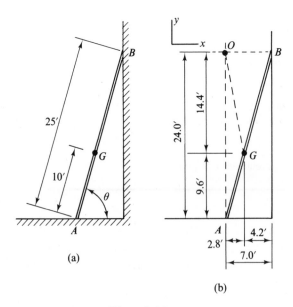

(a)

(b)

Figure 8-44

SOLUTION

The velocity of G can be obtained by applying the equations of relative velocity, as discussed in Art. 8-15, or by means of instantaneous centers. The instantaneous center is at point O of Fig. 8-44(b), and the magnitude of the angular velocity is

$$\omega = \frac{V_A}{OA} = \frac{6}{24} = 0.25 \text{ rad per sec}$$

and the angular velocity is

$$\boldsymbol{\omega} = 0.25 \text{ rad per sec clockwise.}$$

The velocity of G has a magnitude of

$$v_G = OG(\omega) = [(14.4)^2 + (2.8)^2]^{1/2}(0.25) = 3.67 \text{ fps,}$$

and its direction is perpendicular to OG; thus

$$\mathbf{v}_G = \underline{3.67 \text{ fps}} \qquad \text{Ans.}$$

Since the acceleration of point A is given, points A and G are selected for the relative acceleration equation. That is,

$$\mathbf{a}_G = \mathbf{a}_A \quad + \quad \mathbf{a}_{G/A}$$

and the terms are represented by the sketches just below the equation. Both the magnitude and direction of $\mathbf{a}_G$ are unknown, and the magnitude of α_{AB} (assumed to be clockwise) is also unknown. The angular velocity of the ladder ($\omega = 0.25$ rad per sec) was determined by means of instantaneous centers. The acceleration equation can be written in terms of $\mathbf{i}$ and $\mathbf{j}$ as

$$\mathbf{a}_G = a_{Gx}\mathbf{i} + a_{Gy}\mathbf{j}$$

$$= -8\mathbf{i} + 10(0.25)^2\left(\frac{-7\mathbf{i} - 24\mathbf{j}}{25}\right) + 10\alpha\left(\frac{24\mathbf{i} - 7\mathbf{j}}{25}\right).$$

This vector equation contains three unknowns and is equivalent to two scalar equations; therefore it cannot be solved without more information. Since point B moves vertically, the direction of its acceleration must be vertical, and a relative acceleration equation for points A and B will contain only two unknown magnitudes. The equation is

$$\mathbf{a}_B = \mathbf{a}_A \quad + \quad \mathbf{a}_{B/A},$$

445

which can be written in terms of $\mathbf{i}$ and $\mathbf{j}$ as

$$-a_B\mathbf{j}=-8\mathbf{i}+25(0.25)^2\left[\frac{-7\mathbf{i}-24\mathbf{j}}{25}\right]+25\alpha\left[\frac{24\mathbf{i}-7\mathbf{j}}{25}\right].$$

This equation contains two unknown magnitudes, a_B and α, which can be determined by solving the two scalar equations representing the coefficients of $\mathbf{i}$ and $\mathbf{j}$. The coefficients of $\mathbf{i}$ give

$$0=-8+25(0.25)^2\left(\frac{-7}{25}\right)+25\alpha\left(\frac{24}{25}\right)$$

from which

$$\alpha=0.352 \text{ rad per sec}^2,$$

and the positive sign indicates that the assumed direction of α is correct; therefore

$$\boldsymbol{\alpha}=0.352 \text{ rad per sec}^2 \text{ clockwise}$$

$$=-0.352\mathbf{k} \text{ rad per sec}^2.$$

When this value of α is substituted into Eq. (a) it becomes

$$\mathbf{a}_G=-8\mathbf{i}+10(0.25)^2\left(\frac{-7\mathbf{i}-24\mathbf{j}}{25}\right)+10(0.352)\left(\frac{24\mathbf{i}-7\mathbf{j}}{25}\right)$$

$$=-8\mathbf{i}-0.175\mathbf{i}-0.600\mathbf{j}+3.375\mathbf{i}-0.984\mathbf{j}$$

$$=-4.800\mathbf{i}-1.584\mathbf{j}$$

$$=5.05 \text{ fps}^2 \qquad \boxed{18.27°}$$

Ans.

PROBLEMS

8-224 The wheel in Fig. P8-224 rolls without slipping along the fixed surface. The velocity and acceleration of the center O are 20 ips to the left and 30 ips^2 to the right for the position shown. Determine the velocity and acceleration of point A for this position.

*8-225 Disk A in Fig. P8-225 rotates about a fixed axis through O. Disk B is connected to A by a smooth pin. When in the position shown, A has an angular velocity of 4 rad per sec counterclockwise and an angular acceleration of 6 rad per sec^2 clockwise, and B has an angular velocity of 12 rad per sec and an angular acceleration of 20 rad per sec^2, both counterclockwise. Determine the acceleration of point Q on disk B.

Figure P8-224

Figure P8-225

8-226 When a wheel rolls along a circular surface, the center of the wheel also moves along a circular path as indicated in Fig. P8-226. Determine the relationships between the angular velocity and acceleration of the wheel and the linear velocity and acceleration of the center O when the wheel rolls without slipping. *Hint:* Write an expression for the position of O in terms of the angular position of the wheel and differentiate to determine the magnitude of the velocity and of the tangential acceleration of O.]

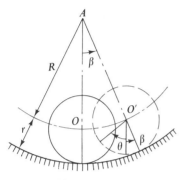

Figure P8-226

8-227 The wheel in Fig. P8-227 rolls without slipping with decreasing angular velocity along the horizontal plane. When in the position shown, point B has an acceleration of 5 ips² to the right. Determine the velocity and acceleration of point O.

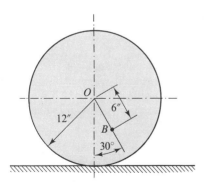

Figure P8-227

*8-228 Wheel A in Fig. P8-228 rolls along the horizontal plane without· slipping and is connected to body B by a flexible cord. In the

position shown, point C has a velocity of 8 cm/s and an acceleration of 20 cm/s², both to the left. Determine the acceleration of body B with respect to point D.

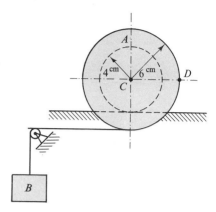

Figure P8-228

*8-229 Crank A in Fig. P8-229 is connected to member C by the connecting rod B. Crank A has a constant angular velocity of 10 rad/s clockwise. Determine the angular accelerations of members B and C when in the position shown.

Figure P8-229

8-230 In the position shown in Fig. P8-230 member CD has an angular velocity of 2 rad per sec counterclockwise and an angular acceleration of 2 rad per sec² clockwise. Determine the acceleration of point B.

8-231 Member AB in Fig. P8-231 has an angular velocity of 4 rad per sec clockwise and an angular acceleration of 6 rad per sec² counterclockwise in the indicated position. Determine the acceleration of point C when in this position.

Figure P8-230

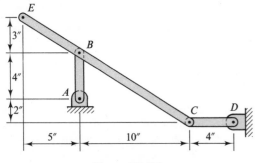

Figure P8-231

8-232 Member *AB* in Fig. P8-232 has an angular velocity of 3 rad per sec and an angular acceleration of 10 rad per sec², both counterclockwise, when in the indicated position. De-

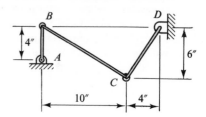

Figure P8-232

termine the angular acceleration of member *BC*.

8-233 Bar *CD* in Fig. P8-233 has an angular velocity of 1.0 rad per sec and an angular acceleration of 2.0 rad per sec², both clockwise, at the instant shown. Determine the angular acceleration of bar *BC* for this position.

Figure P8-233

8-234 Arm *AB* in Fig. P8-234 has an angular velocity of 2 rad per sec clockwise and an angular acceleration of 10 rad per sec² counterclockwise when in the position shown. Determine the velocity and acceleration of block *C*.

Figure P8-234

8-235 Determine the angular acceleration of member *BC* in Fig. P8-235 which will result in a zero acceleration of point *C* when *C* has a velocity of 28 ips upward.

8-236 Roller *C* of the mechanism in Fig. P8-236 is to have a velocity of 42 ips to the right and be slowing down at the rate of 100 ips² in the position shown. Determine the required angular velocity and acceleration of member *OA*.

8-237 The block in Fig. P8-237 has a clockwise angular velocity of 5 rad per sec and a counterclockwise angular acceleration of 20 rad per

Figure P8-235

Figure P8-238

mine the angular velocity and acceleration of member *AB* for this position.

Figure P8-236

sec² in the indicated position. Determine the acceleration of point *D*.

Figure P8-239

8-240 The disk in Fig. P8-240 rolls without slipping along the circular track. The disk has a constant clockwise angular velocity of 30 rpm.
(a) Determine the time required for the center of the disk to make one complete revolution about the point *B*.
(b) Determine the acceleration of point *A* when the disk is in the position shown.
(c) If, in addition to the given angular velocity, the disk has a counterclockwise angular acceleration of 4 rad per sec², determine the acceleration of point *A*.

Figure P8-237

8-238 The velocity of point *A* in Fig. P8-238 is 10.5 ips down the slot and its acceleration is 100 ips² up the slot when in the position shown. Determine the acceleration of point *C*.

*8-239 Roller *A* in Fig. P8-239 has a velocity of 0.70 m/s downward and an acceleration of 1.50 m/s² upward in the position shown. Deter-

Figure P8-240

8-241 In the planetary gear assembly in Fig. P8-241 the inner gear A has an angular velocity of 2 rad per sec and an angular acceleration of 4 rad per sec², both counterclockwise, and the outer gear is clamped. The planet gears B, C, and D roll between the inner and outer gears. Determine the velocity and acceleration of point P on gear B.

8-243 The wheel in Fig. P8-243 rolls without slipping along the curved surface, and in the position shown it has an angular velocity of 3 rad per sec and an angular acceleration of 5 rad per sec², both clockwise. Determine
(a) The angular acceleration of member AB.
(b) The acceleration of slider B.

Figure P8-243

Figure P8-241

8-242 The wheel in Fig. P8-242 rolls without slipping, and when in the indicated position it has a clockwise angular velocity of 2 rad per sec and a counterclockwise angular acceleration of 6 rad per sec². Determine the angular accelerations of AB and BC.

*8-244 Block D in Fig. P8-244 has a velocity of 16 cm/s to the left and an acceleration of 30 cm/s² to the right in the indicated position. Determine the velocity and acceleration of block A for this position.

Figure P8-242

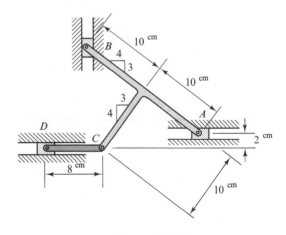

Figure P8-244

450

8-20 RELATIVE ACCELERATION—
THREE-DIMENSIONAL MOTION

Equation (8-26) of Art. 8-18 was applied to rigid bodies with plane motion in Art. 8-19. The equation is also valid for rigid bodies which have three-dimensional motion provided nonrotating reference axes are used. The equation is

$$\mathbf{a}_B = \mathbf{a}_A + \mathbf{a}_{B/A},$$

and when A and B are points on a rigid body the relative acceleration term is

$$\mathbf{a}_{B/A} = \dot{\boldsymbol{\omega}} \times \mathbf{r} + \boldsymbol{\omega} \times (\boldsymbol{\omega} \times \mathbf{r}), \qquad \text{(a)}$$

where $\dot{\boldsymbol{\omega}}$ is the absolute angular acceleration of the body, $\boldsymbol{\omega}$ is the absolute angular velocity of the body, and $\mathbf{r}$ is the vector *from A to B* (the position of B relative to A). As noted in Art. 8-18, $\dot{\boldsymbol{\omega}}$ or $\boldsymbol{\alpha}$ is usually not parallel to $\boldsymbol{\omega}$ for three-dimensional motion.

Figure 8-45 represents a member AB connected by ball and socket joints to a rotating disk and to a sliding collar. The angular velocity is shown resolved into two components, one tangent to AB and one normal to AB. Since the angular velocity can change in both magnitude and direction, the angular acceleration will, in general, not be parallel to the angular velocity. The angular acceleration can also be resolved into two components tangent and normal to member AB, that is,

$$\dot{\boldsymbol{\omega}} = \dot{\boldsymbol{\omega}}_t + \dot{\boldsymbol{\omega}}_n,$$

and the quantity $\dot{\boldsymbol{\omega}} \times \mathbf{r}$ becomes

$$\dot{\boldsymbol{\omega}} \times \mathbf{r} = (\dot{\boldsymbol{\omega}}_t + \dot{\boldsymbol{\omega}}_n) \times \mathbf{r} = \dot{\boldsymbol{\omega}}_n \times \mathbf{r}$$

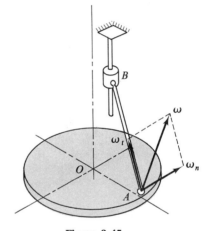

Figure 8-45

since $\dot{\boldsymbol{\omega}}_t$ is parallel to $\mathbf{r}$. This result indicates that $\dot{\boldsymbol{\omega}} \times \mathbf{r}$ is independent of $\dot{\boldsymbol{\omega}}_t$; that is $\dot{\boldsymbol{\omega}}_t$ can have any value without affecting Eq. (a).

It was shown in Art. 8-17 that the value of $\boldsymbol{\omega}_t$ was indeterminate for this system and could have any assigned magnitude. When $\boldsymbol{\omega}_t$ is made equal to zero the value of $\boldsymbol{\omega}$ becomes $\boldsymbol{\omega}_n$ and is perpendicular to $\mathbf{r}$. In this case, the right-hand term in Eq. (a) becomes

$$\boldsymbol{\omega} \times (\boldsymbol{\omega} \times \mathbf{r}) = \boldsymbol{\omega}_n \times (\boldsymbol{\omega}_n \times \mathbf{r}) = -\omega_n^2 \mathbf{r}.$$

Note that this result is true only when $\boldsymbol{\omega}$ and $\mathbf{r}$ are normal to each other.

If the ball and socket joint at B in Fig. 8-45 is replaced by the clevis in Fig. 8-46, the angular velocity and acceleration will not be indeterminate. The angular velocity can be expressed as the vector sum

Figure 8-46

of the angular velocity of the collar and the angular velocity of member AB around the pin. Thus

$$\boldsymbol{\omega} = \omega_1\mathbf{k} + \omega_2\mathbf{e}_p,$$

where ω_1 is the magnitude of the angular velocity of the collar, ω_2 is the magnitude of the angular velocity of AB around the pin, and $\mathbf{e}_p$ is a unit vector parallel to the pin. The pin must be perpendicular to the shaft on which the collar slides and to the member AB; therefore

$$\mathbf{e}_p = \frac{\mathbf{k}\times\mathbf{r}_{AB}}{|\mathbf{k}\times\mathbf{r}_{AB}|}.$$

The angular acceleration of AB is

$$\boldsymbol{\alpha} = \dot{\boldsymbol{\omega}} = \dot{\omega}_1\mathbf{k} + \omega_1\dot{\mathbf{k}} + \dot{\omega}_2\mathbf{e}_p + \omega_2\dot{\mathbf{e}}_p.$$

The vector $\mathbf{k}$ is constant; therefore $\dot{\mathbf{k}} = 0$. Since $\mathbf{e}_p$ is parallel to the pin, it must rotate with the pin so that

$$\dot{\mathbf{e}}_p = \boldsymbol{\omega}_{pin}\times\mathbf{e}_p = \omega_1\mathbf{k}\times\mathbf{e}_p$$

where

$$\boldsymbol{\omega}_{pin} = \boldsymbol{\omega}_{collar} = \omega_1\mathbf{k}.$$

Thus

$$\boldsymbol{\alpha} = \dot{\omega}_1\mathbf{k} + \dot{\omega}_2\mathbf{e}_p + \omega_2(\omega_1\mathbf{k}\times\mathbf{e}_p). \qquad (b)$$

The first term in Eq. (b) represents the rate of change of magnitude of $\boldsymbol{\omega}_1$, the second term is the rate of change of magnitude of $\boldsymbol{\omega}_2$, and the third term is the change of $\boldsymbol{\omega}_2$ resulting from the change of direction of the pin (or of the collar). The following examples illustrate the application of Eq. (8-26).

EXAMPLE 8-30

Collars A and B of the mechanism in Fig. 8-47 slide along fixed rods and are connected to member C by a clevis at A and by a ball and socket at B. The clevis at A is illustrated in Fig. 8-41 (which is repeated here for convenience). In the position shown $c = d = 2b$ and collar A has a velocity of $v_A\mathbf{i}$ and an acceleration of $a_A\mathbf{i}$. Determine the acceleration of B and the angular acceleration of member C.

SOLUTION

The angular velocity of C is the sum of the angular velocity, $\boldsymbol{\omega}_1$, of the collar and the angular velocity, $\boldsymbol{\omega}_2$, of C about the clevis pin. The unit

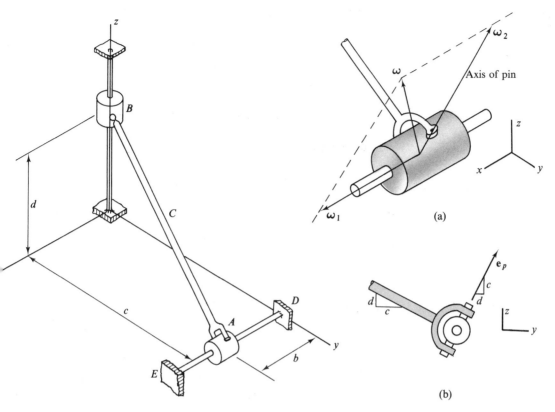

Figure 8-47

Figure 8-41

vector along the pin is

$$\mathbf{e}_p = \frac{\mathbf{r}_{AB} \times \mathbf{i}}{|\mathbf{r}_{AB} \times \mathbf{i}|} = \frac{(-b\mathbf{i} - 2b\mathbf{j} + 2b\mathbf{k}) \times \mathbf{i}}{|(-b\mathbf{i} - 2b\mathbf{j} + 2b\mathbf{k}) \times \mathbf{i}|}$$

$$= \frac{\mathbf{j} + \mathbf{k}}{\sqrt{2}},$$

and the angular velocity of C is

$$\left.\begin{array}{c} \boldsymbol{\omega} = \omega_1 \mathbf{i} + \omega_2 \mathbf{e}_p \\[2mm] = \omega_1 \mathbf{i} + \dfrac{\omega_2}{\sqrt{2}} (\mathbf{j} + \mathbf{k}). \end{array}\right\} \qquad (a)$$

The relative velocity equation of B with respect to A is

$$v_B \mathbf{k} = v_A \mathbf{i} + \boldsymbol{\omega} \times \mathbf{r}$$

$$= v_A \mathbf{i} + b \begin{vmatrix} \mathbf{i} & \mathbf{j} & \mathbf{k} \\ \omega_1 & \dfrac{\omega_2}{\sqrt{2}} & \dfrac{\omega_2}{\sqrt{2}} \\ -1 & -2 & +2 \end{vmatrix}$$

$$= v_A \mathbf{i} + b \left[\mathbf{i}(2)\sqrt{2}\,\omega_2 + \mathbf{j}\left(-\frac{\sqrt{2}}{2}\,\omega_2 - 2\omega_1 \right) + \mathbf{k}\left(-2\omega_1 + \frac{\sqrt{2}}{2}\,\omega_2 \right) \right]$$

from which

$$\mathbf{i:} \quad 2\sqrt{2}\,\omega_2 = -\frac{v_A}{b},$$

$$\mathbf{j:} \quad -2\omega_1 - \frac{\sqrt{2}}{2}\,\omega_2 = 0,$$

$$\mathbf{k:} \quad -2\omega_1 + \frac{\sqrt{2}}{2}\,\omega_2 = \frac{v_B}{b},$$

which in turn gives

$$\omega_2 = -\frac{\sqrt{2}}{4}\frac{v_A}{b},$$

$$\omega_1 = \frac{1}{8}\frac{v_A}{b}.$$

The angular velocity of AB is

$$\boldsymbol{\omega} = \frac{v_A}{8b}(\mathbf{i} - 2\mathbf{j} - 2\mathbf{k}).$$

The acceleration of B is

$$\mathbf{a}_B = \mathbf{a}_A + \mathbf{a}_{B/A}$$
$$= \mathbf{a}_A + \boldsymbol{\alpha} \times \mathbf{r} + \boldsymbol{\omega} \times (\boldsymbol{\omega} \times \mathbf{r}). \tag{b}$$

The angular velocity as given in Eq. (a) is

$$\boldsymbol{\omega} = \omega_1 \mathbf{i} + \omega_2 \mathbf{e}_p,$$

and the angular acceleration is

$$\boldsymbol{\alpha} = \dot{\boldsymbol{\omega}} = \dot{\omega}_1 \mathbf{i} + \omega_1 \dot{\mathbf{i}} + \dot{\omega}_2 \mathbf{e}_p + \omega_2 \dot{\mathbf{e}}_p.$$

Since $\mathbf{i}$ is constant, $\dot{\mathbf{i}}$ is zero. The quantity $\mathbf{e}_p$ is a unit vector which can change in direction but not in magnitude. The angular velocity of $\mathbf{e}_p$ is the angular velocity of the collar, that is, $\omega_1 \mathbf{i}$; therefore

$$\dot{\mathbf{e}}_p = \omega_1 \mathbf{i} \times \mathbf{e}_p$$

$$= \omega_1 \mathbf{i} \times \left(\frac{\mathbf{j} + \mathbf{k}}{\sqrt{2}}\right)$$

$$= \frac{\omega_1}{\sqrt{2}}(-\mathbf{j} + \mathbf{k}).$$

The angular acceleration thus becomes

$$\boldsymbol{\alpha} = \dot{\omega}_1 \mathbf{i} + \frac{\dot{\omega}_2}{\sqrt{2}}(\mathbf{j} + \mathbf{k}) + \frac{\omega_2 \omega_1}{\sqrt{2}}(-\mathbf{j} + \mathbf{k})$$

$$= \dot{\omega}_1 \mathbf{i} + \frac{\dot{\omega}_2}{\sqrt{2}}(\mathbf{j} + \mathbf{k}) - \frac{\sqrt{2}}{32}\left(\frac{v_A}{b}\right)^2 \frac{1}{\sqrt{2}}(-\mathbf{j} + \mathbf{k})$$

$$= \dot{\omega}_1 \mathbf{i} + \left[\frac{\dot{\omega}_2}{\sqrt{2}} + \frac{1}{32}\left(\frac{v_A}{b}\right)^2\right]\mathbf{j} + \left[\frac{\dot{\omega}_2}{\sqrt{2}} - \frac{1}{32}\left(\frac{v_A}{b}\right)^2\right]\mathbf{k}, \tag{c}$$

and the quantity $\boldsymbol{\alpha} \times \mathbf{r}$ is

$$\boldsymbol{\alpha} \times \mathbf{r} = b \begin{vmatrix} \mathbf{i} & \mathbf{j} & \mathbf{k} \\ \dot{\omega}_1 & \dfrac{\dot{\omega}_2}{\sqrt{2}} + \dfrac{1}{32}\left(\dfrac{v_A}{b}\right)^2 & \dfrac{\dot{\omega}_2}{\sqrt{2}} - \dfrac{1}{32}\left(\dfrac{v_A}{b}\right)^2 \\ -1 & -2 & 2 \end{vmatrix}$$

$$= b\left\{ 2\sqrt{2}\,\dot{\omega}_2\mathbf{i} + \left[-2\dot{\omega}_1 - \dfrac{\sqrt{2}}{2}\,\dot{\omega}_2 + \dfrac{1}{32}\left(\dfrac{v_A}{b}\right)^2 \right]\mathbf{j} \right.$$

$$\left. + \left[-2\dot{\omega}_1 + \dfrac{\sqrt{2}}{2}\,\dot{\omega}_2 + \dfrac{1}{32}\left(\dfrac{v_A}{b}\right)^2 \right]\mathbf{k} \right\}.$$

The quantity $\boldsymbol{\omega} \times \mathbf{r}$ is

$$\boldsymbol{\omega} \times \mathbf{r} = \dfrac{v_A b}{8b} \begin{vmatrix} \mathbf{i} & \mathbf{j} & \mathbf{k} \\ 1 & -2 & -2 \\ -1 & -2 & 2 \end{vmatrix}$$

$$= \dfrac{v_A}{8}\left[\mathbf{i}(-4-4) + \mathbf{j}(2-2) + \mathbf{k}(-2-2) \right]$$

$$= \dfrac{v_A}{2}(-2\mathbf{i} - \mathbf{k}),$$

and the vector triple product becomes

$$\boldsymbol{\omega} \times (\boldsymbol{\omega} \times \mathbf{r}) = \dfrac{v_A}{8b}\dfrac{v_A}{2} \begin{vmatrix} \mathbf{i} & \mathbf{j} & \mathbf{k} \\ 1 & -2 & -2 \\ -2 & 0 & -1 \end{vmatrix}$$

$$= \dfrac{v_A^2}{16b}(2\mathbf{i} + 5\mathbf{j} - 4\mathbf{k}).$$

The preceding values can be substituted in Eq. (b) to give

$$a_B\mathbf{k} = a_A\mathbf{i} + b\left\{ 2\sqrt{2}\,\dot{\omega}_2\mathbf{i} + \left[-2\dot{\omega}_1 - \dfrac{\sqrt{2}}{2}\,\dot{\omega}_2 + \dfrac{1}{32}\left(\dfrac{v_A}{b}\right)^2 \right]\mathbf{j} \right.$$

$$\left. + \left[-2\dot{\omega}_1 + \dfrac{\sqrt{2}}{2}\,\dot{\omega}_2 + \dfrac{1}{32}\left(\dfrac{v_A}{b}\right)^2 \right]\mathbf{k} \right\} + \dfrac{(v_A)^2}{16b}(2\mathbf{i} + 5\mathbf{j} - 4\mathbf{k}),$$

which yields the following three scalar equations:

$$\mathbf{i}: \quad 2\sqrt{2}\,\dot{\omega}_2 = -\dfrac{1}{8}\left(\dfrac{v_A}{b}\right)^2 - \dfrac{a_A}{b},$$

$$\mathbf{j}: \quad -2\dot{\omega}_1 - \dfrac{\sqrt{2}}{2}\,\dot{\omega}_2 = -\dfrac{11}{32}\left(\dfrac{v_A}{b}\right)^2,$$

$$\mathbf{k}: \quad -2\dot{\omega}_1 + \dfrac{\sqrt{2}}{2}\,\dot{\omega}_2 = \dfrac{7}{32}\left(\dfrac{v_A}{b}\right)^2 + \dfrac{a_B}{b}.$$

These three equations contain three unknown quantities, $\dot{\omega}_1$, $\dot{\omega}_2$, and

a_B. The simultaneous solution of these equations gives

$$\dot{\omega}_1 = \frac{1}{8}\frac{a_A}{b} + \frac{3}{16}\left(\frac{v_A}{b}\right)^2,$$

$$\dot{\omega}_2 = -\frac{\sqrt{2}}{4}\frac{a_A}{b} - \frac{\sqrt{2}}{32}\left(\frac{v_A}{b}\right)^2,$$

and

$$a_B = -\frac{a_A}{2} - \frac{5}{8}\frac{v_A^2}{b}.$$

The angular acceleration of C is obtained by substituting these values into Eq. (c); that is

$$\boldsymbol{\alpha} = \frac{a_A}{8b}(\mathbf{i} - 2\mathbf{j} - 2\mathbf{k}) + \frac{1}{16}\left(\frac{v_A}{b}\right)^2(3\mathbf{i} - \mathbf{k}). \qquad \text{Ans.}$$

The acceleration of B is

$$\mathbf{a}_B = a_B\mathbf{k} = -\left(\frac{a_A}{2} + \frac{5v_A^2}{8b}\right)\mathbf{k}. \qquad \text{Ans.}$$

EXAMPLE 8-31

Solve Example 8-30 if the clevis at A is replaced by a ball and socket joint.

SOLUTION

The component of the angular velocity of the bar parallel to the bar, $\boldsymbol{\omega}_t$, is indeterminate when both ends are connected to ball and socket joints as noted in Art. 8-17. The same reasoning indicates that the component of the angular acceleration parallel to the bar is indeterminate. Since the direction of $\boldsymbol{\omega}_t$ is changing, the value assumed for the magnitude of $\boldsymbol{\omega}_t$ will be a factor in determining the value of $\boldsymbol{\alpha}_n$ which is normal to the bar. For convenience it is assumed that $\boldsymbol{\omega}_t$ is zero. A different assumed value for $\boldsymbol{\omega}_t$ would result in a different value for $\boldsymbol{\alpha}_n$, but it would not affect the acceleration of B. The relative velocity equation for points A and B is

$$\mathbf{v}_B = \mathbf{v}_A + \mathbf{v}_{B/A} = \mathbf{v}_A + \boldsymbol{\omega} \times \mathbf{r},$$

$$v_B\mathbf{k} = v_A\mathbf{i} + (\omega_x\mathbf{i} + \omega_y\mathbf{j} + \omega_z\mathbf{k}) \times (-b\mathbf{i} - 2b\mathbf{j} + 2b\mathbf{k}),$$

where $\boldsymbol{\omega} = \omega_x\mathbf{i} + \omega_y\mathbf{j} + \omega_z\mathbf{k}$ is the angular velocity of AB. When this equation is expanded it becomes

$$v_B\mathbf{k} = v_A\mathbf{i} + b\begin{vmatrix} \mathbf{i} & \mathbf{j} & \mathbf{k} \\ \omega_x & \omega_y & \omega_z \\ -1 & -2 & 2 \end{vmatrix}$$

$$= v_A\mathbf{i} + b[\mathbf{i}(2\omega_y + 2\omega_z) + \mathbf{j}(-\omega_z - 2\omega_x) + \mathbf{k}(-2\omega_x + \omega_y)],$$

and when the coefficients of $\mathbf{i}$, $\mathbf{j}$, and $\mathbf{k}$ are equated the equation gives

$$\mathbf{i}: \quad 2\omega_y + 2\omega_z = -\frac{v_A}{b} \qquad\qquad (a)$$

$$\mathbf{j}: \quad -2\omega_x - \omega_z = 0 \qquad\qquad (b)$$

$$\mathbf{k}: \quad -2\omega_x + \omega_y = \frac{v_B}{b} \qquad\qquad (c)$$

The requirement that $\omega_t = \boldsymbol{\omega}\cdot(\mathbf{r}/r) = 0$ gives

$$-\omega_x - 2\omega_y + 2\omega_z = 0. \qquad\qquad (d)$$

When Eq. (b) is multiplied by $-\frac{1}{2}$ and added to the sum of Eq. (a) and (d) the result is

$$2\omega_z + \frac{1}{2}\,\omega_z + 2\omega_z = -\frac{v_A}{b}$$

from which

$$\omega_z = -\frac{2v_A}{9b}.$$

From Eq. (b),

$$\omega_x = -\frac{1}{2}\,\omega_z = \frac{v_A}{9b},$$

and from Eq. (a) or (d),

$$\omega_y = -\frac{5v_A}{18b}.$$

The angular velocity is

$$\boldsymbol{\omega} = \boldsymbol{\omega}_t + \boldsymbol{\omega}_n = 0 + \boldsymbol{\omega}_n$$

$$= \frac{v_A}{18b}\,(2\mathbf{i} - 5\mathbf{j} - 4\mathbf{k}) \text{ rad per sec.}$$

This result can also be obtained from Example 8-25 by substituting $c = d = 2b$.

The relative acceleration equation for points A and B is

$$\mathbf{a}_B = \mathbf{a}_A + \mathbf{a}_{B/A}$$

$$= \mathbf{a}_A + \boldsymbol{\alpha}\times\mathbf{r} + \boldsymbol{\omega}\times(\boldsymbol{\omega}\times\mathbf{r}),$$

where $\mathbf{r}$ is the position of B relative to A and $\boldsymbol{\alpha} = \alpha_x\mathbf{i} + \alpha_y\mathbf{j} + \alpha_z\mathbf{k}$ is the angular acceleration of AB. Since $\boldsymbol{\omega}_t$ was set equal to zero, the quantity $\boldsymbol{\omega}\times(\boldsymbol{\omega}\times\mathbf{r})$ is equal to $-\omega^2\mathbf{r}$. The magnitude of $\boldsymbol{\omega}$ is

$$\omega_n = \frac{v_A}{18b}\sqrt{2^2 + 5^2 + 4^2} = \frac{\sqrt{45}\,v_A}{18b},$$

and the vector triple product becomes

$$\boldsymbol{\omega}\times(\boldsymbol{\omega}\times\mathbf{r}) = -\frac{45}{(18)^2}\left(\frac{v_A}{b}\right)^2 b(-\mathbf{i} - 2\mathbf{j} + 2\mathbf{k})$$

$$= -\frac{5v_A^2}{36b}(-\mathbf{i} - 2\mathbf{j} + 2\mathbf{k}).$$

457

The relative acceleration equation, in terms of $\mathbf{i}$, $\mathbf{j}$, and $\mathbf{k}$, is

$$a_B\mathbf{k} = a_A\mathbf{i} + b\begin{vmatrix} \mathbf{i} & \mathbf{j} & \mathbf{k} \\ \alpha_x & \alpha_y & \alpha_z \\ -1 & -2 & 2 \end{vmatrix} - \frac{5v_A^2}{36b}(-\mathbf{i} - 2\mathbf{j} + 2\mathbf{k})$$

$$= a_A\mathbf{i} + b[\mathbf{i}(2\alpha_y + 2\alpha_z) + \mathbf{j}(-\alpha_z - 2\alpha_x) + \mathbf{k}(-2\alpha_x + \alpha_y)]$$

$$+ \frac{5v_A^2}{36b}(+\mathbf{i} + 2\mathbf{j} - 2\mathbf{k}),$$

which reduces to the following three equations:

$$\mathbf{i}: \quad +2\alpha_y + 2\alpha_z = -\frac{5}{36}\left(\frac{v_A}{b}\right)^2 - \frac{a_A}{b} \tag{e}$$

$$\mathbf{j}: \quad -2\alpha_x - \alpha_z = -\frac{10}{36}\left(\frac{v_A}{b}\right)^2 \tag{f}$$

$$\mathbf{k}: \quad -2\alpha_x + \alpha_y = +\frac{10}{36}\left(\frac{v_A}{b}\right)^2 + \frac{a_B}{b}. \tag{g}$$

The requirement that α_t be zero is satisfied by $\boldsymbol{\alpha} \cdot \mathbf{r} = 0$; that is

$$b(-\alpha_x - 2\alpha_y + 2\alpha_z) = 0. \tag{h}$$

The components α_x, α_y, and α_z can be eliminated by adding one-half of Eq. (e) to Eq. (f) and subtracting Eq. (g), which gives

$$0 = -\frac{5}{72}\left(\frac{v_A}{b}\right)^2 - \frac{a_A}{2b} - \frac{10}{36}\left(\frac{v_A}{b}\right)^2 - \frac{10}{36}\left(\frac{v_A}{b}\right)^2 - \frac{a_B}{b}$$

from which

$$\frac{a_B}{b} = -\frac{a_A}{2b} - \frac{5}{8}\left(\frac{v_A}{b}\right)^2,$$

and the acceleration of B is

$$\mathbf{a}_B = -\left(\frac{a_A}{2} + \frac{5v_A^2}{8b}\right)\mathbf{k}. \qquad \text{Ans.}$$

The values of α_x, α_y, and α_z, obtained by solving Eq. (e), (f), and (h) simultaneously, are

$$\alpha_x = \frac{5}{36}\left(\frac{v_A}{b}\right)^2 + \frac{1}{9}\left(\frac{a_A}{b}\right),$$

$$\alpha_y = -\frac{5}{72}\left(\frac{v_A}{b}\right)^2 - \frac{5}{18}\left(\frac{a_A}{b}\right),$$

$$\alpha_z = -\frac{2}{9}\left(\frac{a_A}{b}\right),$$

and the component of the angular acceleration normal to AB is

$$\boldsymbol{\alpha}_n = \frac{5}{72}\left(\frac{v_A}{b}\right)^2(2\mathbf{i} - \mathbf{j}) + \frac{1}{18}\left(\frac{a_A}{b}\right)(2\mathbf{i} - 5\mathbf{j} - 4\mathbf{k}). \qquad \text{Ans.}$$

PROBLEMS

8-245 Crank OA in Fig. P8-245 rotates in the yz plane with a constant angular velocity of 6**i** rad per sec. Member AB has a ball and socket joint at each end. Determine the acceleration of collar B.

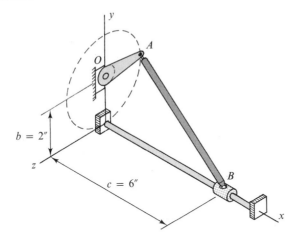

Figure P8-245

8-246 Assume the component of the angular velocity of AB in Problem 8-245 parallel to AB to be zero and calculate the component of the angular acceleration of AB normal to AB.

8-247 The ball and socket at B in Problem 8-245 is replaced by a clevis; see Fig. P8-218. Determine the angular acceleration of member AB.

8-248 Figure P8-248 represents a pair of bevel gears A and B. Gear A rotates in a horizontal plane about a vertical axis at O. Gear B rotates around the shaft OC, which in turn rotates about the vertical axis at O (see Problem 8-219). The shaft OC has a constant angular velocity of $\omega_2\mathbf{j}$, and gear A is stationary. Determine, for the position shown,
(a) The angular acceleration of gear B.
(b) The linear acceleration of point D on gear B.

Figure P8-248

8-249 Determine the angular acceleration of gear B in Problem 8-248 when gear A has an angular velocity of 30**j** rad per sec and an angu-lar acceleration of $-20\mathbf{j}$ rad per sec^2 and member OC has an angular velocity of 10**j** rad per sec and an angular acceleration of 25**j** rad per sec^2.

8-250 Disk D in Fig. P8-250 rotates about an axis at O with a constant angular velocity of $\omega_1 \mathbf{k}$ rad per sec. The collar B slides along the inclined rod which is parallel to the yz plane. The link AB is connected to the disk and collar by ball and socket joints. Determine the acceleration of collar B.

8-251 The ball and socket joint at A in Problem 8-250 is replaced by the joint shown in Fig. P8-251 in which the vertical member rotates in a hole in the disk D and is connected to AB by a horizontal pin at A. Determine the angular acceleration of member AB.

Figure P8-251

*8-252** Rod CDE in Fig. P8-252 is welded to AB, and point C slides on the horizontal xy plane. Rod AB is connected to two collars by means of ball and socket joints. In the position shown the velocity of A is $27\mathbf{i}$ cm/s and its acceleration is zero. Determine
(a) The angular acceleration of ABC.
(b) The linear acceleration of point C.

Figure P8-250

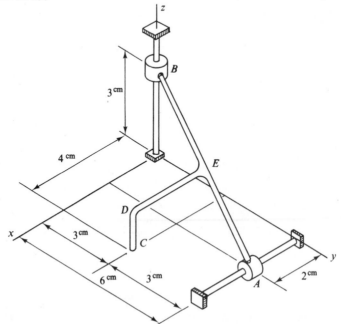

Figure P8-252

8-21 RELATIVE MOTION WITH RESPECT TO ROTATING AXES

Equations relating absolute and relative positions, velocities, and accelerations of two points were developed in Art. 8-6. Application of these equations to problems in which the moving system of coordinates did not rotate was discussed in some detail in Art. 8-14 through 8-20. The equations were particularly useful when the two points were on a rigid body. In some instances it is convenient to use axes that rotate with the body. Moments and products of inertia are usually functions of time unless the moving axes are fixed in the body. Rotating axes may be advantageous when the two points are not on the same rigid body as when a point on one body slides along a path on the second body.

Consider the XYZ system of coordinates in Fig. 8-48 to be fixed in space so that the unit vectors $\mathbf{i'}$, $\mathbf{j'}$, and $\mathbf{k'}$ are constant in direction.

Figure 8-48

The origin of the xyz coordinates moves with point O, and at the same time the axes have an angular velocity $\boldsymbol{\omega}$ and an angular acceleration $\boldsymbol{\alpha} = \dot{\boldsymbol{\omega}}$. The quantities $\boldsymbol{\omega}$ and $\boldsymbol{\alpha}$ are measured with respect to the fixed XYZ axes and are thus absolute quantities. In general, $\boldsymbol{\omega}$ and $\boldsymbol{\alpha}$ will not be collinear unless the xyz axes have plane motion. Note that the unit vectors $\mathbf{i}$, $\mathbf{j}$, and $\mathbf{k}$ are not constant since they rotate with the xyz axes. The time rate of change of these unit vectors will be needed in the following development and can be determined from the principles developed in Art. 8-13.

Consider **i** to be the position vector for the tip of the unit vector along the x axis. The velocity of the tip relative to a set of nonrotating axes at O (not shown) is

$$\dot{\mathbf{i}} = \boldsymbol{\omega} \times \mathbf{i},$$

Similarly,

$$\dot{\mathbf{j}} = \boldsymbol{\omega} \times \mathbf{j} \quad \text{and} \quad \dot{\mathbf{k}} = \boldsymbol{\omega} \times \mathbf{k}.$$

The position of point P in Fig. 8-48 is

$$\mathbf{r}_1 = \mathbf{R} + \mathbf{r},$$

where

$$\mathbf{r} = x\mathbf{i} + y\mathbf{j} + z\mathbf{k} \tag{a}$$

is the position of P relative to the moving axes (as seen by an observer moving with the xyz reference system). The velocity of P is

$$\mathbf{v}_P = \dot{\mathbf{r}}_1 = \dot{\mathbf{R}} + \dot{\mathbf{r}},$$

and the value of $\dot{\mathbf{r}}$, from Eq. (a), is

$$\dot{\mathbf{r}} = (\dot{x}\mathbf{i} + \dot{y}\mathbf{j} + \dot{z}\mathbf{k}) + (x\dot{\mathbf{i}} + y\dot{\mathbf{j}} + z\dot{\mathbf{k}})$$

$$= (\dot{x}\mathbf{i} + \dot{y}\mathbf{j} + \dot{z}\mathbf{k}) + (x\boldsymbol{\omega} \times \mathbf{i} + y\boldsymbol{\omega} \times \mathbf{j} + z\boldsymbol{\omega} \times \mathbf{k})$$

$$= (\dot{x}\mathbf{i} + \dot{y}\mathbf{j} + \dot{z}\mathbf{k}) + \boldsymbol{\omega} \times (x\mathbf{i} + y\mathbf{j} + z\mathbf{k})$$

$$= \mathbf{v}_{P/xyz} + \boldsymbol{\omega} \times \mathbf{r} \tag{b}$$

in which $\mathbf{v}_{P/xyz} = (\dot{x}\mathbf{i} + \dot{y}\mathbf{j} + \dot{z}\mathbf{k})$ is the velocity of P measured relative to the moving references axes. Thus the velocity of P is

$$\mathbf{v}_P = \dot{\mathbf{R}} + \mathbf{v}_{P/xyz} + \boldsymbol{\omega} \times \mathbf{r}. \tag{8-27}$$

If P' is a point fixed to the rotating axes which is coincident with P at the instant under consideration, the quantity $\boldsymbol{\omega} \times \mathbf{r}$ is the velocity of P' relative to a set of nonrotating axes at O (the velocity of P' relative to O). The velocity of P is seen to consist of three terms:

$\dot{\mathbf{R}}$: the velocity of the moving origin.
$\mathbf{v}_{P/xyz} = (\dot{x}\mathbf{i} + \dot{y}\mathbf{j} + \dot{x}\mathbf{k})$: the velocity of P relative to the moving axes.
$\boldsymbol{\omega} \times \mathbf{r}$: the velocity of P' relative to O.

The sum of the second and third terms on the right-hand side of Eq. (8-27) is the velocity of P relative to a set of nonrotating axes at O, and in Art. 8-14 it was designated as $\mathbf{v}_{P/O}$; that is

$$\mathbf{v}_{P/O} = \mathbf{v}_{P/xyz} + \boldsymbol{\omega} \times \mathbf{r}$$

and

$$\mathbf{v}_P = \mathbf{v}_O + \mathbf{v}_{P/O}.$$

The following example illustrates the use of Eq. (8-27) with various sets of moving axes.

EXAMPLE 8-32

Section 8-21
Relative motion with
respect to rotating
axes

The small disk D in Fig. 8-49 is pinned to the larger disk C at point A. Disk C rotates about a vertical axis at O' with an absolute angular velocity of $5\mathbf{k}$ rad per sec, and D has an absolute angular velocity of $8\mathbf{k}$ rad per sec. Determine the velocity of point P on disk D.

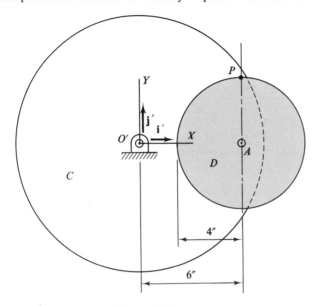

Figure 8-49

SOLUTION

Four different choices of moving axes xyz are used to demonstrate the various terms in Eq. (8-27). The x and y axes are chosen parallel to the X and Y axes for each application; consequently $\mathbf{i} = \mathbf{i}'$, $\mathbf{j} = \mathbf{j}'$, and $\mathbf{k} = \mathbf{k}'$.

(a) *Nonrotating axes, origin at A.* The velocity equation is

$$\mathbf{v}_P = \mathbf{v}_O + \mathbf{v}_{P/xyz} + \boldsymbol{\omega} \times \mathbf{r},$$

where

$$\mathbf{v}_O = \mathbf{v}_A = \boldsymbol{\omega}_C \times \mathbf{O'A} = 5\mathbf{k} \times 6\mathbf{i} = 30\mathbf{j} \text{ ips}$$

and

$$\mathbf{v}_{P/xyz} = \boldsymbol{\omega}_D \times \mathbf{AP} = 8\mathbf{k} \times 4\mathbf{j} = -32\mathbf{i} \text{ ips}.$$

Since the xy axes are nonrotating, $\boldsymbol{\omega} = \boldsymbol{\omega}_{xyz} = 0$, and the third term in the equation is zero. Therefore

$$\mathbf{v}_P = (30\mathbf{j} - 32\mathbf{i}) \text{ ips} = 43.9 \text{ ips} \qquad \text{Ans.}$$

$43.2°$

(b) *Axes rotate with body D, origin at A.* The angular velocity of the xyz axes is $8\mathbf{k}$ rad per sec. Since P is a point fixed on the rotating axes, $\mathbf{v}_{P/xyz}$ must be zero.

The velocity equation becomes

$$\mathbf{v}_P = \mathbf{v}_O + \mathbf{v}_{P/xyz} + \boldsymbol{\omega} \times \mathbf{r}$$

$$= \mathbf{v}_A + 0 + 8\mathbf{k} \times 4\mathbf{j}$$

$$= \underline{(30\mathbf{j} - 32\mathbf{i})\ \text{ips as before.}} \qquad \text{Ans.}$$

Note that when P is fixed on the moving axes the velocity of P relative to the axes is zero.

(c) *Axes rotate with body C, origin at P.* The velocity of the origin is the velocity of the point on C which is coincident with P at the instant under consideration; thus

$$\mathbf{v}_O = \boldsymbol{\omega}_C \times \mathbf{r} = 5\mathbf{k} \times (6\mathbf{i} + 4\mathbf{j})$$

$$= (-20\mathbf{i} + 30\mathbf{j})\ \text{ips.}$$

The velocity of P relative to the rotating axes is the velocity it would *appear* to have to an observer who was moving (rotating) with the axes. The angular velocity of disk D relative to the xyz axes is

$$\boldsymbol{\omega}_{D/xyz} = \boldsymbol{\omega}_D - \boldsymbol{\omega}_{xyz} = 8\mathbf{k} - 5\mathbf{k} = 3\mathbf{k}\ \text{rad per sec,}$$

and the relative velocity of P is

$$\mathbf{v}_{P/xyz} = \boldsymbol{\omega}_{D/xyz} \times \mathbf{r}_{AP} = 3\mathbf{k} \times 4\mathbf{j} = -12\mathbf{i}\ \text{ips.}$$

Note that the path of P relative to the rotating axes is a circular path which moves with the axes. The third term in the velocity equation is zero since P is at the origin and the value of $\mathbf{r}$ (the position of P from the origin) is zero. Thus the velocity of P is

$$\mathbf{v}_P = \mathbf{v}_O + \mathbf{v}_{P/xyz} + \boldsymbol{\omega} \times \mathbf{r}$$

$$= (-20\mathbf{i} + 30\mathbf{j}) + (-12\mathbf{i}) + 5\mathbf{k} \times 0$$

$$= \underline{(-32\mathbf{i} + 30\mathbf{j})\ \text{ips as before.}} \qquad \text{Ans.}$$

(d) *Axes rotate with body C, origin at A.* The angular velocity of the rotating axes is $5\mathbf{k}$ rad per sec, and the angular velocity of body D relative to the rotating axes is $3\mathbf{k}$ rad per sec. The velocity of P is

$$\mathbf{v}_P = \mathbf{v}_O + \mathbf{v}_{P/xyz} + \boldsymbol{\omega} \times \mathbf{r}$$

$$= 5\mathbf{k} \times 6\mathbf{i} + 3\mathbf{k} \times 4\mathbf{j} + 5\mathbf{k} \times 4\mathbf{j}$$

$$= 30\mathbf{j} - 12\mathbf{i} - 20\mathbf{i}$$

$$= \underline{(-32\mathbf{i} + 30\mathbf{j})\ \text{ips.}} \qquad \text{Ans.}$$

The result must, of course, be the same for any choice of rotating or nonrotating axes and for any origin. In this particular example each set of axes gave results with about the same effort. It is important, however, that the effects of the different sets be completely understood.

The acceleration of P can be obtained by differentiating Eq. (8-27) with respect to time. The acceleration is

$$\mathbf{a}_P = \dot{\mathbf{v}}_P = \ddot{\mathbf{R}} + \dot{\mathbf{v}}_{P/xyz} + \dot{\boldsymbol{\omega}} \times \mathbf{r} + \boldsymbol{\omega} \times \dot{\mathbf{r}}. \qquad (c)$$

The quantity $\mathbf{v}_{P/xyz} = (\dot{x}\mathbf{i} + \dot{y}\mathbf{j} + \dot{z}\mathbf{k})$ can be differentiated to give

Section 8-21
*Relative motion with
respect to rotating
axes*

$$\dot{\mathbf{v}}_{P/xyz} = (\ddot{x}\mathbf{i} + \ddot{y}\mathbf{j} + \ddot{z}\mathbf{k}) + (\dot{x}\dot{\mathbf{i}} + \dot{y}\dot{\mathbf{j}} + \dot{z}\dot{\mathbf{k}})$$
$$= (\ddot{x}\mathbf{i} + \ddot{y}\mathbf{j} + \ddot{z}\mathbf{k}) + (\dot{x}\boldsymbol{\omega}\times\mathbf{i} + \dot{y}\boldsymbol{\omega}\times\mathbf{j} + \dot{z}\boldsymbol{\omega}\times\mathbf{k})$$
$$= \mathbf{a}_{P/xyz} + \boldsymbol{\omega}\times\mathbf{v}_{P/xyz},$$

where $\mathbf{a}_{P/xyz}$ is the acceleration of P relative to the moving xyz coordinate system. It is the acceleration P would *appear* to have to an observer who was rotating with the xyz axes.

The quantity $\dot{\mathbf{r}}$ from Eq. (b) is substituted into the last term of Eq. (c) to give

$$\boldsymbol{\omega}\times\dot{\mathbf{r}} = \boldsymbol{\omega}\times(\mathbf{v}_{P/xyz} + \boldsymbol{\omega}\times\mathbf{r})$$
$$= \boldsymbol{\omega}\times\mathbf{v}_{P/xyz} + \boldsymbol{\omega}\times(\boldsymbol{\omega}\times\mathbf{r}).$$

The acceleration of P thus becomes

$$\mathbf{a}_P = \ddot{\mathbf{R}} + \mathbf{a}_{P/xyz} + \boldsymbol{\omega}\times\mathbf{v}_{P/xyz} + \dot{\boldsymbol{\omega}}\times\mathbf{r} + \boldsymbol{\omega}\times\mathbf{v}_{P/xyz} + \boldsymbol{\omega}\times(\boldsymbol{\omega}\times\mathbf{r})$$
$$= \ddot{\mathbf{R}} + \dot{\boldsymbol{\omega}}\times\mathbf{r} + \boldsymbol{\omega}\times(\boldsymbol{\omega}\times\mathbf{r}) + \mathbf{a}_{P/xyz} + 2\boldsymbol{\omega}\times\mathbf{v}_{P/xyz}. \qquad (8\text{-}28)$$

The sum of the first three terms on the right-hand side of Eq. (8-28) gives the acceleration of a point P', fixed in the xyz coordinate system, which is coincident with particle P at the instant under consideration. Thus the acceleration of P can be written as

$$\mathbf{a}_P = \mathbf{a}_{P'} + \mathbf{a}_{P/xyz} + 2\boldsymbol{\omega}\times\mathbf{v}_{P/xyz}. \qquad (8\text{-}28a)$$

The quantity $\mathbf{a}_{P/xyz}$ has already been defined as the acceleration of P measured relative to the moving (xyz) coordinate system. Any angular velocity or acceleration used to compute this quantity must be measured relative to the moving axes.

The last term of Eq. (8-28) is the Coriolis[9] or supplementary acceleration. This component is normal to the plane containing the vectors $\boldsymbol{\omega}$ and $\mathbf{v}_{P/xyz}$ as determined by the right-hand rule for vector products. The sense of the Coriolis component can be visualized by noting the motion of the tip of the vector $\mathbf{v}_{P/xyz}$ as it is rotated in the direction indicated by $\boldsymbol{\omega}$.

If a rigid body is attached to the moving axes and if points O and P are points on the rigid body (P and P' will be identical), the last two terms of Eq. (8-28) will be zero since both $\mathbf{a}_{P/xyz}$ and $\mathbf{v}_{P/xyz}$ are zero. The second and third terms of Eq. (8-28) are the acceleration of P' relative to nonrotating axes at O.

It is sometimes convenient to express the angular velocity of one rigid body relative to a second rigid body as in the discussion of the

---

[9] Named after G. C. Coriolis (1792–1843), a French scientist, who was the first to publish a discussion of this acceleration.

angular velocity of a member connected to a rotating collar by a clevis in Art. 8-17. As another example, assume that the wheel B in Fig. 8-50(a) rotates about a horizontal axis along line OP on the disk A as A rotates about the z axis. The absolute angular velocity of A is $\boldsymbol{\omega}_A$, and wheel B has an angular velocity relative to disk A of $\boldsymbol{\omega}_{B/A}$ along the axis OP. The absolute angular velocity of B is

$$\boldsymbol{\omega}_B = \boldsymbol{\omega}_A + \boldsymbol{\omega}_{B/A}$$

$$= \boldsymbol{\omega}_A + \omega_{B/A}\,\mathbf{e}_{OP},$$

where $\mathbf{e}_{OP}$ is a unit vector parallel to line OP, the axis of the relative angular velocity. The angular velocity of B is shown in Fig. 8-50(b).

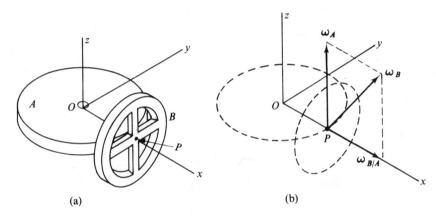

(a)   (b)

Figure 8-50

The angular acceleration of B is obtained by differentiating the angular velocity; that is,

$$\boldsymbol{\alpha}_B = \dot{\boldsymbol{\omega}}_B = \dot{\boldsymbol{\omega}}_A + \dot{\omega}_{B/A}\mathbf{e}_{OP} + \omega_{B/A}\dot{\mathbf{e}}_{OP}$$

$$= \dot{\boldsymbol{\omega}}_A + \dot{\omega}_{B/A}\mathbf{e}_{OP} + \omega_{B/A}\boldsymbol{\omega}_A \times \mathbf{e}_{OP}$$

$$= \boldsymbol{\alpha}_B + \boldsymbol{\alpha}_{B/A} + \boldsymbol{\omega}_A \times \boldsymbol{\omega}_{B/A}.$$

The third term in the preceding equation results from the change in direction of the relative angular velocity vector. This term will be zero only when $\boldsymbol{\omega}_A$ or $\boldsymbol{\omega}_{B/A}$ is zero or when $\boldsymbol{\omega}_A$ and $\boldsymbol{\omega}_{B/A}$ are parallel.

    The determination of accelerations using rotating axes, using nonrotating axes, and by differentiation is discussed in the following examples. Link mechanisms with no sliding connections can usually be solved most directly by using nonrotating axes as in Art. 8-19 and 8-20. Mechanisms with sliding connections can often be solved more readily using rotating axes and Eq. (8-28); see Example 8-34. Many problems, such as Example 8-33, can be solved most directly by differentiating the position function to determine the velocity and acceleration. Most problems can be solved by more than one method, and personal preference dictates the choice. Solution by two different methods provides an excellent check on the work.

EXAMPLE 8-33

Blocks D and E in Fig. 8-51 are pinned together, and D slides along the slot in member OA, while E slides in the vertical slot. The angular velocity of OA is $\omega\mathbf{k}$, and its angular acceleration is $\dot{\omega}\mathbf{k}$ when the angle θ is as shown. Determine the acceleration of block E for this position.

Figure 8-51

SOLUTION

Since block D slides along member OA, a set of rotating, xyz, axes are selected as shown. The fixed, XYZ, axes and the rotating xyz axes both have their origins at O in this example. The angular velocity and acceleration of the rotating axes are the same as for member OA. The unit vectors $\mathbf{i}'$ and $\mathbf{j}'$ are parallel to X and Y and $\mathbf{i}$ and $\mathbf{j}$ are parallel to x and y, respectively. Equation (8-28) is

$$\mathbf{a}_E = \ddot{\mathbf{R}} + \dot{\boldsymbol{\omega}}\times\mathbf{r} + \boldsymbol{\omega}\times(\boldsymbol{\omega}\times\mathbf{r}) + \mathbf{a}_{P/xyz} + 2\boldsymbol{\omega}\times\mathbf{v}_{P/xyz}.$$

The following values are available from the given data and the selection of axes:

$$\mathbf{R} = \dot{\mathbf{R}} = \ddot{\mathbf{R}} = 0, \qquad \mathbf{r} = r\mathbf{i}, \qquad \dot{\boldsymbol{\omega}} = \dot{\omega}\mathbf{k}, \qquad \mathbf{v}_{P/xyz} = \dot{r}\mathbf{i},$$

and

$$\mathbf{a}_{P/xyz} = \ddot{r}\mathbf{i}.$$

Note that r is the distance from O to D. Since E moves along the vertical slot, its acceleration is $\mathbf{a}_E = a_E\mathbf{j}'$. When these values are substituted into the acceleration equation it becomes

$$a_E\mathbf{j}' = 0 + \dot{\omega}\mathbf{k}\times r\mathbf{i} + \omega\mathbf{k}\times(\omega\mathbf{k}\times r\mathbf{i}) + \ddot{r}\mathbf{i} + 2\omega\mathbf{k}\times\dot{r}\mathbf{i}$$

$$= \dot{\omega}r\mathbf{j} - \omega^2 r\mathbf{i} + \ddot{r}\mathbf{i} + 2\omega\dot{r}\mathbf{j}. \tag{a}$$

The quantity $r = b \sec\theta$ can be determined for any angle θ, but the equation still contains three unknown magnitudes: a_E, $\dot{r}$, and $\ddot{r}$. The value of $\dot{r}$ can be determined from Eq. (8-27); thus

$$\mathbf{v}_E = \dot{\mathbf{R}} + \mathbf{v}_{E/xyz} + \boldsymbol{\omega} \times \mathbf{r}$$

or

$$v_E \mathbf{j}' = 0 + \dot{r}\mathbf{i} + \omega r \mathbf{j}.$$

Since $\mathbf{v}_E$ is vertical, the horizontal component of $\mathbf{v}_E$ must be zero; that is,

$$\mathbf{v}_E \cdot \mathbf{i}' = 0. \tag{b}$$

Note that

$$\mathbf{i}' = \mathbf{i} \cos \theta - \mathbf{j} \sin \theta;$$

therefore Eq. (b) becomes

$$(\dot{r}\mathbf{i} + \omega r \mathbf{j}) \cdot (\mathbf{i} \cos \theta - \mathbf{j} \sin \theta) = \dot{r} \cos \theta - \omega r \sin \theta = 0$$

from which

$$\dot{r} = \omega r \tan \theta.$$

When this quantity is substituted in Eq. (a) it becomes

$$a_E \mathbf{j}' = \dot{\omega} r \mathbf{j} - \omega^2 r \mathbf{i} + \ddot{r}\mathbf{i} + 2\omega^2 r \tan \theta \mathbf{j}$$
$$= (\ddot{r} - r\omega^2)\mathbf{i} + r(\dot{\omega} + 2\omega^2 \tan \theta)\mathbf{j} \tag{c}$$

in which the two unknowns are a_E and $\ddot{r}$. The first term on the right in this equation can be eliminated by replacing $\mathbf{j}'$ by its equivalent $(\mathbf{i} \sin \theta + \mathbf{j} \cos \theta)$ and taking the scalar or dot product of the equation with $\mathbf{j}$. That is,

$$a_E (\mathbf{i} \sin \theta + \mathbf{j} \cos \theta) \cdot \mathbf{j} = (\ddot{r} - r\omega^2)\mathbf{i} \cdot \mathbf{j} + r(\dot{\omega} + 2\omega^2 \tan \theta)\mathbf{j} \cdot \mathbf{j}$$

or

$$a_E \cos \theta = 0 + r(\dot{\omega} + 2\omega^2 \tan \theta)$$

from which

$$\mathbf{a}_E = r \sec \theta (\dot{\omega} + 2\omega^2 \tan \theta)\mathbf{j}'$$
$$= \underline{b \sec^2 \theta (\dot{\omega} + 2\omega^2 \tan \theta)\mathbf{j}'.} \qquad \text{Ans.}$$

As indicated in the paragraph preceding this example, this particular problem could be solved more readily by deriving an expression relating the position of D to the angular position of OA and differentiating twice. The purpose of this example is to illustrate the application of relative acceleration with respect to rotating axes.

EXAMPLE 8-34

Block D in Fig. 8-52 is pinned to member CD and slides along member OAB. The mechanism moves in a common plane and in the position shown $\boldsymbol{\omega}_{CD} = -4\mathbf{k}$ rad per sec, and $\dot{\boldsymbol{\omega}}_{CD} = -10\mathbf{k}$ rad per sec^2. Determine the angular acceleration of member OAB.

SOLUTION

Since block D slides along member AB, the motion of D relative to axes which rotate with OAB will be rectilinear and should provide a

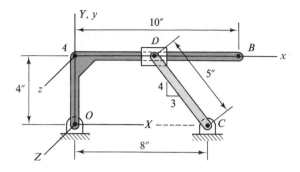

Section 8-21
Relative motion with
respect to rotating
axes

Figure 8-52

direct solution. The origin of the moving axes can be selected at any convenient point, such as O, A, B, or the point on AB which is coincident with D. Point A will be selected as the origin for this example in which case $\mathbf{R}$ and $\mathbf{r}$ in Eq. (8-28) are

$$\mathbf{R} = \mathbf{R}_{OA} = 4\mathbf{j} \quad \text{and} \quad \mathbf{r} = \mathbf{r}_{AD} = 5\mathbf{i}.$$

The relative velocity $\mathbf{v}_{D/xyz}$ and the angular velocity of the xyz axes are needed for the relative acceleration equation and can be obtained from the relative velocity equation; thus

$$\mathbf{v}_D = \dot{\mathbf{R}} + \boldsymbol{\omega} \times \mathbf{r} + \mathbf{v}_{D/xyz}$$

in which $\boldsymbol{\omega}$ is the angular velocity of the xyz axes. The velocity of D is

$$\mathbf{v}_D = \boldsymbol{\omega}_{CD} \times \mathbf{r}_{CD} = -4\mathbf{k} \times (-3\mathbf{i} + 4\mathbf{j}) = 16\mathbf{i} + 12\mathbf{j},$$

the quantity $\dot{\mathbf{R}}$ is

$$\dot{\mathbf{R}} = \boldsymbol{\omega} \times \mathbf{R} = \omega \mathbf{k} \times (4\mathbf{j}) = -4\omega \mathbf{i},$$

the vector $\boldsymbol{\omega} \times \mathbf{r}$ is

$$\boldsymbol{\omega} \times \mathbf{r} = \omega \mathbf{k} \times (5\mathbf{i}) = 5\omega \mathbf{j},$$

and

$$\mathbf{v}_{D/xyz} = v_{D/xyz}\mathbf{i}$$

from which

$$16\mathbf{i} + 12\mathbf{j} = -4\omega\mathbf{i} + 5\omega\mathbf{j} + v_{D/xyz}\mathbf{i}$$

This vector equation is equivalent to

$$16 = -4\omega + v_{D/xyz}$$

and

$$12 = 5\omega$$

from which

$$\boldsymbol{\omega} = 2.4\mathbf{k} \quad \text{and} \quad \mathbf{v}_{D/xyz} = 25.6\mathbf{i}.$$

The acceleration of D is

$$\begin{aligned}
\mathbf{a}_D &= \dot{\boldsymbol{\omega}}_{CD} \times \mathbf{r}_{CD} + \boldsymbol{\omega}_{CD} \times (\boldsymbol{\omega}_{CD} \times \mathbf{r}_{CD}) \\
&= -10\mathbf{k} \times (-3\mathbf{i} + 4\mathbf{j}) + (-4\mathbf{k}) \times [(-4\mathbf{k}) \times (-3\mathbf{i} + 4\mathbf{j})] \\
&= 88\mathbf{i} - 34\mathbf{j}.
\end{aligned}$$

469

The acceleration of D can also be expressed as

$$\mathbf{a}_D = \ddot{\mathbf{R}} + \dot{\boldsymbol{\omega}} \times \mathbf{r} + \boldsymbol{\omega} \times (\boldsymbol{\omega} \times \mathbf{r}) + \mathbf{a}_{D/xyz} + 2\boldsymbol{\omega} \times \mathbf{v}_{D/xyz},$$

where $\boldsymbol{\omega}$ and $\dot{\boldsymbol{\omega}}$ refer to the rotating axes. The vector $\ddot{\mathbf{R}}$ is

$$\ddot{\mathbf{R}} = \dot{\omega}\mathbf{k} \times (4\mathbf{j}) + 2.4\mathbf{k} \times (2.4\mathbf{k} \times 4\mathbf{j}) = -4\dot{\omega}\mathbf{i} - 23.04\mathbf{j}.$$

When numerical data are substituted for $\boldsymbol{\omega}$, $\mathbf{r}$, and $\mathbf{v}_{D/xyz}$ the acceleration equation becomes

$$88\mathbf{i} - 34\mathbf{j} = (-4\dot{\omega}\mathbf{i} - 23.04\mathbf{j}) + \dot{\omega}\mathbf{k} \times 5\mathbf{i} + 2.4\mathbf{k} \times (2.4\mathbf{k} \times 5\mathbf{i})$$
$$+ a_{D/xyz}\mathbf{i} + 2(2.4\mathbf{k} \times 25.6\mathbf{i})$$
$$= -4\dot{\omega}\mathbf{i} - 23.04\mathbf{j} + 5\dot{\omega}\mathbf{j} - 28.8\mathbf{i} + a_{D/xyz}\mathbf{i} + 122.9\mathbf{j}.$$

This vector equation is equivalent to

$$88\mathbf{i} = (-4\dot{\omega} - 28.8 + a_{D/xyz})\mathbf{i}$$

and

$$-34\mathbf{j} = (-23.04 + 5\dot{\omega} + 122.9)\mathbf{j}$$

from which

$$\dot{\omega} = -26.8 \text{ rad per sec}^2$$

or

$$\dot{\boldsymbol{\omega}} = \boldsymbol{\alpha}_{OAB} = -26.8\mathbf{k} \text{ rad per sec}^2 \text{ (clockwise).} \qquad \text{Ans.}$$

The first equation, in terms of the $\mathbf{i}$ components, is not used for this problem.

EXAMPLE 8-35

The helicopter in Fig. 8-53 is hovering above a point on the ground so that point O' is stationary. The body A has an angular velocity of

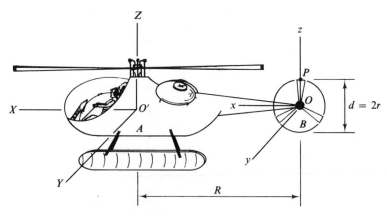

Figure 8-53

$\boldsymbol{\omega}_A = \omega_A \mathbf{k}'$ which is slowing down with an angular acceleration $\dot{\boldsymbol{\omega}}_A = \dot{\omega}_A \mathbf{k}'$. Note that the numerical value of $\dot{\omega}_A$ will be negative. The tail rotor B has a constant angular velocity relative to A and in the position shown $\boldsymbol{\omega}_{B/A} = \omega_{B/A} \mathbf{j}$. Determine

Section 8-21
Relative motion with
respect to rotating
axes

(a) The angular acceleration of the tail rotor.

(b) The acceleration of the point P at the top of the tail rotor.

SOLUTION

(a) The XYZ axes ($\mathbf{i}'$, $\mathbf{j}'$, $\mathbf{k}'$ unit vectors) are fixed, and the moving xyz axes ($\mathbf{i}$, $\mathbf{j}$, $\mathbf{k}$ unit vectors) at O move with the body A. The angular velocity of the xyz axes therefore is $\boldsymbol{\omega}_A$. The angular velocity of B is

$$\boldsymbol{\omega}_B = \boldsymbol{\omega}_A + \boldsymbol{\omega}_{B/A}$$

$$= \omega_A \mathbf{k}' + \omega_{B/A} \mathbf{j},$$

and the angular acceleration of B is

$$\boldsymbol{\alpha}_B = \dot{\boldsymbol{\omega}}_B = \dot{\omega}_A \mathbf{k}' + \omega_A \dot{\mathbf{k}}' + \dot{\omega}_{B/A} \mathbf{j} + \omega_{B/A} \dot{\mathbf{j}}.$$

The factor $\dot{\mathbf{k}}'$ is zero since $\mathbf{k}'$ does not change direction. The quantity $\dot{\omega}_{B/A}$ is equal to zero, and $\dot{\mathbf{j}} = \boldsymbol{\omega}_A \times \mathbf{j}$; therefore

$$\boldsymbol{\alpha}_B = \dot{\omega}_A \mathbf{k}' + 0 + 0 + \omega_{B/A} \mathbf{k}' \times \mathbf{j}$$

$$= \underline{\dot{\omega}_A \mathbf{k}' - \omega_{B/A} \omega_A \mathbf{i}'}. \qquad \text{Ans.}$$

Note that $\mathbf{i} = \mathbf{i}'$ at the instant under consideration.

(b) The acceleration of P, from Eq. (8-28), is

$$\mathbf{a}_P = \ddot{\mathbf{R}} + \dot{\boldsymbol{\omega}}_A \times \mathbf{r} + \boldsymbol{\omega}_A \times (\boldsymbol{\omega}_A \times \mathbf{r}) + \mathbf{a}_{P/xyz} + 2\boldsymbol{\omega}_A \times \mathbf{v}_{P/xyz}.$$

The same axes are used as for part (a), and the following quantities can be written for this choice of axes. The vector $\ddot{\mathbf{R}}$ is the acceleration of O which moves in a horizontal circle around O'; that is

$$\ddot{\mathbf{R}} = \dot{\omega}_A \mathbf{k} \times (-R\mathbf{i}) + \omega_A \mathbf{k} \times [\omega_A \mathbf{k} \times (-R\mathbf{i})]$$

$$= -R\dot{\omega}_A \mathbf{j} + R\omega_A^2 \mathbf{i}.$$

The vector $\mathbf{r}$ is $r\mathbf{k}$; therefore

$$\dot{\boldsymbol{\omega}}_A \times \mathbf{r} + \boldsymbol{\omega}_A \times (\boldsymbol{\omega}_A \times \mathbf{r}) = 0$$

since $\mathbf{r}$ is parallel to both $\boldsymbol{\omega}_A$ and $\dot{\boldsymbol{\omega}}_A$. Point A moves on a circle of radius r relative to the xyz axes; therefore

$$\mathbf{v}_{P/xyz} = \boldsymbol{\omega}_{B/A} \times \mathbf{r} = r\omega_{B/A} \mathbf{i}$$

and

$$\mathbf{a}_{P/xyz} = \boldsymbol{\alpha}_{B/A} \times \mathbf{r} + \boldsymbol{\omega}_{B/A} \times (\boldsymbol{\omega}_{B/A} \times \mathbf{r})$$

$$= 0 - r(\omega_{B/A})^2 \mathbf{k}.$$

The acceleration of P thus becomes

$$\mathbf{a}_P = -R\dot{\omega}_A \mathbf{j} + R\omega_A^2 \mathbf{i} + 0 + 0 - r(\omega_{B/A})^2 \mathbf{k} + 2\omega_A \mathbf{k} \times r\omega_{B/A} \mathbf{i}$$

$$= -R\dot{\omega}_A \mathbf{j} + R\omega_A^2 \mathbf{i} - r(\omega_{B/A})^2 \mathbf{k} + 2r\omega_A \omega_{B/A} \mathbf{j}$$

$$= \underline{R\omega_A^2 \mathbf{i} + (2r\omega_A \omega_{B/A} - R\dot{\omega}_A)\mathbf{j} - r(\omega_{B/A})^2 \mathbf{k}} \qquad \text{Ans.}$$

A check on this solution can be made by selecting xyz axes at O which do not rotate or by letting the xyz axes rotate with the rotor B.

PROBLEMS

8-253 Rods AB and CD in Fig. P8-253 rotate in vertical planes and are connected by a slider which is pinned to CD and slides along AB. In the position shown, member CD has an angular velocity of 3 rad per sec clockwise and an angular acceleration of 5 rad per sec² counterclockwise. Determine the angular velocity and acceleration of member AB.

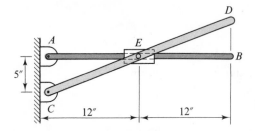

Figure P8-253

8-254 Solve Problem 8-253 when the slider is pinned to AB and slides along CD as in Fig. P8-254. All other data are the same as in Problem 8-253.

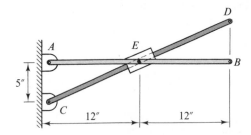

Figure P8-254

*__8-255__ Bar AB in Fig. P8-255 is pinned to block A, which slides along the horizontal slot, while the bar slides past the fixed pin C. Block A has a velocity of 26 cm/s to the right and an acceleration of 65 cm/s² to the left when in the position shown. Determine the angular velocity and acceleration of the bar. Use axes which rotate with AB and whose origin is at the moving point A.

*__8-256__ Solve Problem 8-255 using axes which rotate with AB and whose origin is at C.

8-257 Member AB in Fig. P8-257 is pinned to crank OA and slides in block C. Block C is free

Figure P8-255

to rotate. Crank OA has an angular velocity of 4 rad per sec counterclockwise and an angular acceleration of 10 rad per sec² clockwise in the indicated position. Determine the angular velocity and acceleration of member AB using axes which rotate with AB and whose origin is at
(a) Point A (moving).
(b) Point C (fixed).

Figure P8-257

8-258 Pin P on crank B in Fig. P8-258 slides in the slot in body A. In the position shown, B has an angular velocity of 8 rad per sec clockwise and an angular acceleration of 10 rad per sec² counterclockwise. Determine the angular acceleration of body A and the acceleration of P relative to body A.

8-259 A 20-ft ladder leans against a wall as shown in Fig. P8-259, and a cat is on the ladder 6 ft from end B. For the given position, the velocity of end B is 6 fps downward and the acceleration of B is 7 fps² downward. The cat has a velocity relative to the ladder of 9 fps along the ladder toward B and an acceleration relative to the ladder of 2 fps² along the ladder

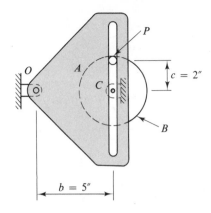

Figure P8-258

toward *A*. Determine the acceleration of the cat. Both ends of the ladder remain in contact with the surfaces.

Figure P8-259

8-260 Particle *P* in Fig. P8-260 travels along a straight rod, *AB*, which is attached to disk *C* that rotates at a constant angular velocity of $2\mathbf{k}$ rad per sec. The *xyz* axes rotate with the disk. At the instant the particle crosses the *y* axes, its velocity and acceleration relative to the rod are observed to be 25 fps toward *B* and 50 fps² toward *A*. Determine the velocity and acceleration of particle *P*.

8-261 In the mechanism in Fig. P8-261, the

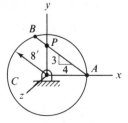

Figure P8-260

angular velocity and acceleration of rod *AB* are 6 rad per sec clockwise and 8 rad per sec² counterclockwise. Determine the angular velocity and acceleration of rod *BD*.

Figure P8-261

8-262 Member *OA* of Fig. P8-262 has an angular velocity of 2 rad per sec clockwise and an angular acceleration of 8 rad per sec² counterclockwise in the position shown. Determine (a) The angular acceleration of member *C*. (b) The linear acceleration of point *B*.

Figure P8-262

8-263 Pin *E* on member *AB* in Fig. P8-263 slides in the slot in member *CD*. In the position shown, block *A* has a velocity of 16 ips to the left and an acceleration of 56 ips² to the right. Determine the angular acceleration of member *CD*.

Figure P8-263

8-264 Block A in Fig. P8-264 has a velocity of 16 ips to the left and an acceleration of 56 ips² to the right in the position shown. Determine the angular acceleration of member CD.

Figure P8-264

8-265 Wheel C in Fig. P8-265 rolls without slipping along a horizontal plane. Crank B rotates about a fixed axis at O, and pin P on the crank slides along the slot in the wheel. The crank has a constant angular velocity of ω_B rad per sec counterclockwise. Determine the angular velocity and acceleration of C and the acceleration of P relative to C for the position shown.

*8-266 Wheel A in Fig. P8-266 rolls without slipping along the horizontal plane. The wheel is driven by the pin P on the crank OP. The pin slides along the slot in the wheel. In the position shown, OP has an angular velocity of 8 rad/s clockwise and an angular acceleration of 10 rad/s² counterclockwise. Determine the velocity and acceleration of the center of the wheel.

Figure P8-265

Figure P8-266

Figure P8-267

Figure P8-268

Figure P8-269

8-267 The mechanism in Fig. P8-267, consisting of rigid members *AB*, *BC*, *DE*, and *EF*, is pinned at *A*, *B*, *D*, *E*, and *F* and moves in the vertical plane. End *C* of member *BC* is pinned to a block which slides on member *DE*. For the given position $\omega_{AB} = 2$ rad per sec clockwise and $\alpha_{AB} = 3$ rad per sec^2 counterclockwise. Determine the acceleration of the block at *C* for this position.

8-268 Figure P8-268 represents the bull gear and rocker arm for the quick return mechanism of a shaper. The rocker arm *AD* is connected to the ram *F* by the link *DE*. The bull gear *H* operates with a constant angular velocity of 45 rpm counterclockwise. When the rocker arm, *AD*, is in the position shown, determine
the position shown, determine
(a) The velocity of the ram, *F*.
(b) The angular acceleration of *AD*.
(c) The acceleration of the ram, *F*.

8-269 Figure P8-269 shows a "back-hoe" used by the construction industry. When in the indicated position, the entire machine is rotating about the *Y* axis with a constant angular velocity of 1.2**j** rad per sec. Determine the velocity and acceleration of point *C* when θ_1 is increasing at the constant rate of 0.8 rad per sec and θ_2 is constant.

8-270 Solve Problem 8-269 if θ_1 is constant and θ_2 is increasing at the constant rate of 1.5 rad per sec.

8-271 A tube is formed into a circular ring and mounted on a vertical shaft *AB* as shown in Fig. P8-271. The ring rotates about *AB* with a constant angular velocity ω_0. A small ball *P* slides inside the tube from *C* to *D* with a constant speed v relative to the tube. Determine the velocity and acceleration of *P* when it reaches point *D* and the ring is in the *yz* plane.

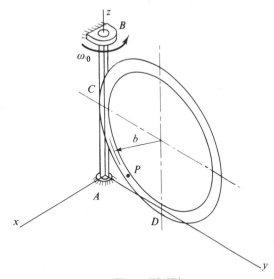

Figure P8-271

8-272 An amusement park ride called the "Rolla-Plane" is illustrated in Fig. P8-272. Member *AB* rotates about a horizontal axle at *O*. The cars *C* and *D* are mounted in bearings at the ends of *A* and *B* and can rotate with an angular velocity of ω_1 relative to member AB. Assume that the axle at O is fixed in the horizontal direction, while AB has a constant angular velocity of 1.0 rad per sec counterclockwise and car C has a constant relative angular velocity of 3.0 rad per sec clockwise looking from G to H. Determine the velocity and acceleration of the passenger's head P when the angle θ is zero and the passenger is vertical as shown.

Figure P8-273

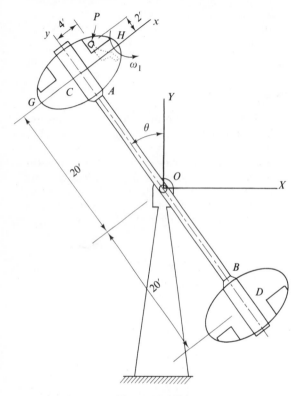

Figure P8-272

8-273 Disk A of Fig. P8-273 rotates about the Y axis, and rod BC is connected to A by a horizontal shaft which rotates with A. In the position shown, disk A has an angular acceleration of $3\mathbf{j}'$ rad per sec^2 and an angular velocity of $-5\mathbf{j}'$ rad per sec, and the rod has an angular acceleration of $-4\mathbf{i}$ rad per sec^2 and an angular velocity of $6\mathbf{i}$ rad per sec, both relative to the disk A. Determine
(a) The linear velocity of point B.
(b) The linear acceleration of point B.
(c) The angular acceleration of BC.

8-274 Disk W in Fig. P8-274 slides along the rod OD and also rotates around the rod as OD rotates in the xy plane around the z axis. When in the position shown, the angular velocity of OD is $2\mathbf{k}$ rad per sec and its angular acceleration is $3\mathbf{k}$ rad per sec^2; the angular velocity and acceleration of W relative of OD are $4\mathbf{j}$ rad per sec and $5\mathbf{j}$ rad per sec^2, respectively; and the disk is sliding along the rod from O to D with a constant speed of 6 ips. Determine
(a) The angular acceleration of W.
(b) The velocity of point A on the disk.
(c) The acceleration of point A.

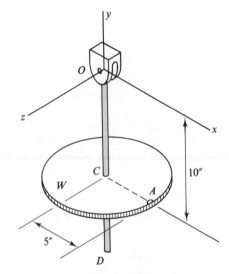

Figure P8-274

8-275 The wheel in Fig. P8-275 rotates about the shaft AB as member OAB rotates in the xy plane about the z axis. The ring R slides along a spoke on the wheel. When the system is in the position shown OAB has an angular velocity of $2\mathbf{k}$ rad per sec and an angular acceleration of $5\mathbf{k}$ rad per sec^2, and the wheel has an angular velocity of $6\mathbf{i}$ rad per sec and an angular acceleration of $8\mathbf{i}$ rad per sec^2, both relative to member OAB. At the same instant the ring R has a velocity of $10\mathbf{j}$ ips and an acceleration of $-7\mathbf{j}$ ips^2, both relative to the wheel. Determine
(a) The angular acceleration of the wheel.
(b) The acceleration of R.

position shown, A has an angular velocity of $4\mathbf{j}$ rad per sec and an angular acceleration of $3\mathbf{j}$ rad per sec^2; B has an angular velocity of $2\mathbf{k}$ rad per sec and an angular acceleration of $6\mathbf{k}$ rad per sec^2, both with respect to body A; and C has a constant angular velocity of $-2\mathbf{k}$ rad per sec with respect to B. Determine
(a) The angular acceleration of body B.
(b) The acceleration of point D on body C.

Figure P8-275

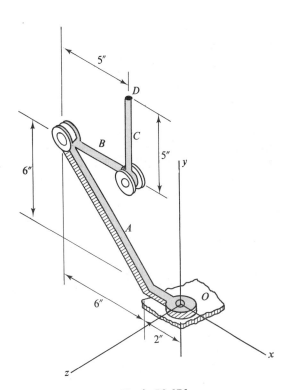

Figure P8-276

8-276 Body A in Fig. P8-276 rotates about a fixed vertical axis at O. Body B is connected to bodies A and C by horizontal pins. When in the

8-277 The turntable T in Fig. P8-277 rotates about the y axis. The rod OA is connected to T by a horizontal pin at O, and R is a ring which slides along OA. In the position shown, T has an angular velocity of $2\mathbf{j}$ rad per sec and an angular acceleration of $4\mathbf{j}$ rad per sec²; OA has an angular velocity of $3\mathbf{k}$ rad per sec and an angular acceleration of $-5\mathbf{k}$ rad per sec², both relative to the turntable; and the ring has a relative velocity along the rod of 10 ips toward A and a relative acceleration of 15 ips² toward O. Determine the acceleration of the ring in the indicated position.

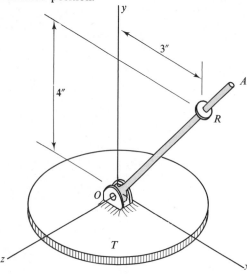

Figure P8-277

8-278 Member AB of Fig. P8-278 rotates in a horizontal plane around the z axis with a con-

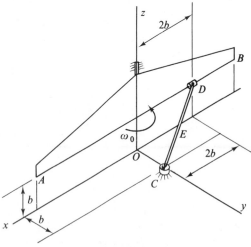

Figure P8-278

stant angular velocity of ω_0. The collar D slides along AB. Member E is connected to point C and to the collar by ball and socket joints. Determine the velocity and acceleration of the collar D when in the indicated position.

8-279 Disk A in Fig. P8-279 rotates about the z axis with a constant angular velocity of ω_0 rad per sec as shown. The rod BC extends through a small hole in A and is pivoted about a ball and socket joint at B. Point P is the point on BC which is passing through the plate in the position shown. Determine the velocity and acceleration of point P.

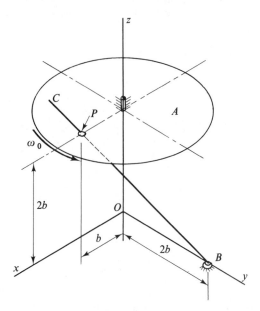

Figure P8-279

8-280 Crank OA in Fig. P8-280 (p. 479) rotates about the z axis (in a horizontal plane) with a constant angular velocity of ω_0 rad per sec. The slender rod AB is connected to the crank by a ball and socket joint and passes through a small hole in the fixed plate Q. Point P on AB is coincident with the hole in the plate in the position shown. Determine the velocity and acceleration of P for this position.

8-281 The bent rod $OABC$ in Fig. P8-281 oscillates about the z axis, and when in the position shown it has an angular velocity of 3 rad per sec and no angular acceleration.

Member D is connected to O' and to the sliding collar P by ball and socket joints. Determine the velocity and acceleration of P for the given position.

Figure P8-280

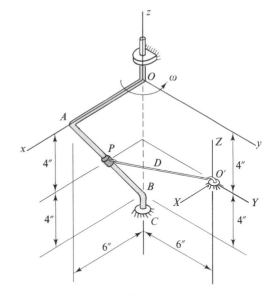

Figure P8-281

8-22 CLOSURE

The position, velocity, and acceleration of a particle can be expressed in terms of the motion of the particle relative to a set of moving axes and the motion of the axes. When the relative motion is expressed with respect to a set of translating (nonrotating) axes, the relative velocity and acceleration are given by Eq. (8-21) and (8-22). When the relative motion is expressed with respect to a set of rotating axes, Eq. (8-27) and (8-28) give the velocity and acceleration of the moving point. Note that Eq. (8-21) and (8-22) apply to two points on a *rigid body*, while Eq. (8-27) and (8-28) apply to any two points and are *not* restricted to rigid bodies.

The *velocities*, but not the accelerations, of points and lines on rigid bodies which have plane motion can be obtained by means of instantaneous centers.

The kinematic relationships between positions, velocities, and accelerations of points and lines can also be obtained by writing an expression for the position of the point or line and differentiating, as was discussed in Part A of this chapter. This mathematical approach to the problem provides an independent check on the solution by relative motion and sometimes leads to a more direct solution to the problem.

PROBLEMS

8-282 In the linkage of Fig. P8-282 the angular velocity of AB is 3 rad per sec clockwise. Determine the velocities of points C and P.

Figure P8-282

*8-283 Link AB in Fig. P8-283 has an angular velocity of 6 rad/s clockwise. Determine, for this position,

Figure P8-283

(a) The location of the instantaneous center of zero velocity of link BC.
(b) The angular velocity of link BC.

8-284 The two meshed gears A and B in Fig. P8-284 are rotating as indicated. When body A is in the given position, its angular velocity is 40 rpm clockwise and its angular acceleration is 8 rad per sec² counterclockwise. Determine the acceleration of point D with respect to point C.

Figure P8-284

8-285 The six rods in Fig. P8-285 are welded to form a rigid structure which rotates about AC with a constant angular velocity. At the instant shown the x component of the velocity of D is 80 ips to the right. Determine
(a) The angular velocity of the structure.
(b) The acceleration of point D.

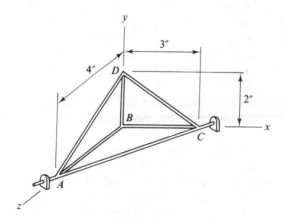

Figure P8-285

480

8-286 The block in Fig. P8-286 rotates about an axle along *AB*. In the position shown the block has an angular velocity of 5 rad per sec and an angular acceleration of 10 rad per sec², both clockwise when looking from *B* to *A*. Determine the velocity and acceleration of point *C*.

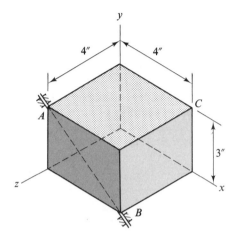

Figure P8-286

8-287 Block *A* in Fig. P8-287 slides along the horizontal slot as member *AB* slides past the fixed pin at *C*. Write an equation relating the position of *A* to the angular position of *AB* and, by differentiation, derive equations for the velocity and acceleration of *A* in terms of θ and its derivatives.

Figure P8-287

8-288 The crank in Fig. P8-288 rotates with a constant angular velocity of 10 rad per sec clockwise. Determine the velocity and acceleration of point *C* when in the position shown.

Figure P8-288

8-289 The angular velocity of *BC* of Fig. P8-289 is 2 rad per sec counterclockwise, and the angular acceleration of *BC* is 4 rad per sec² clockwise when in the indicated position. Determine the acceleration of block *C*.

Figure P8-289

481

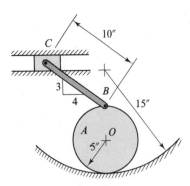

Figure P8-290

8-290 Block C of Fig. P8-290 has a velocity of 20 ips to the right and an acceleration of 30 ips² to the left when in the indicated position. The wheel rolls without slipping. Determine the angular acceleration of the wheel.

*8-291 The wheel in Fig. P8-291 rolls along the horizontal plane without slipping and is connected to B by a flexible cord which is wrapped around a drum on the wheel. When the bodies are in the position shown the acceleration of point A relative to point B is 50 cm/s² to the right and the speed of B is decreasing.
(a) Determine the velocity and acceleration of point C.
(b) During a certain time interval C is displaced 3 cm to the left. Determine the displacement of B during this time interval.

Figure P8-291

8-292 Rod BC in Fig. P8-292 is pinned to the wheel at B in such a manner that the rim of the wheel does not strike the rod. Block C slides in the horizontal slot, and the wheel rolls without slipping. When the assembly is in the position shown the acceleration of point O is 20 ips² to the right and the angular velocity of the wheel is 3 rad per sec clockwise. Determine

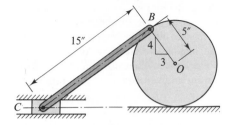

Figure P8-292

(a) The velocity of C.
(b) The acceleration of C.

8-293 Wheel D of Fig. P8-293 rolls without slipping along the horizontal plane. The angular velocity and acceleration of the wheel are 6 rad per sec clockwise and 5 rad per sec² clockwise, respectively. Determine the angular acceleration of BC.

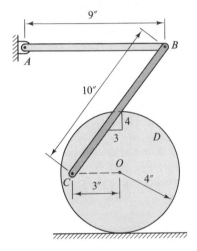

Figure P8-293

*8-294 Crank OA in Fig. P8-294 rotates with a constant angular velocity of 10 rad/s clockwise. Determine the angular velocity and acceleration of member AB when in the position shown.

8-295 Wheel C of Fig. P8-295 rolls without slipping along the horizontal plane. The crank B oscillates about an axis at O and drives the wheel by means of a pin moving in the slot in the wheel. The two lengths are $b = 5$ in. and $c = 10$ in. In the position shown, the angular velocity of B is 5 rad per sec clockwise, and its angular acceleration is zero. Determine

Figure P8-294

(a) The angular acceleration of *C*.
(b) The acceleration of *P* relative to the slot in *C*.

Figure P8-295

Figure P8-296

8-296 A vane-type pump or compressor is illustrated in Fig. P8-296. The smaller circular disk with slots rotates about a fixed axis at *O* with a constant angular velocity of 1700 rpm. The vanes slide in the slots, and the outer ends of the vanes maintain contact with the large stationary cylinder. Determine the velocity and acceleration of point *A* on the vertical vane when it is in the indicated position.

8-297 In the mechanism in Fig. P8-297 block *B* has a velocity of 10 ips down the slot and acceleration of 20 ips² up the slot when in the indicated position. Determine
(a) The angular acceleration of member *CD*.
(b) The acceleration of pin *E* relative to member *CD*.

Figure P8-297

Figure P8-298

8-298 Body A in Fig. P8-298 rotates about the Z axis with a constant angular velocity of $3\mathbf{k}$ rad per sec. The forked member B rotates in bearings in A with a constant relative angular velocity of 5 rad per sec clockwise looking from B toward A. The disk C rotates in the fork with a constant angular velocity relative to B of $10\mathbf{i}$ rad per sec. Determine the absolute angular velocity and acceleration of the disk C.

8-299 Determine the velocity and acceleration of point P at the top of disk C in Problem 8-298.

9

Kinetics–Force, Mass, and Acceleration

9-1 INTRODUCTION

Kinetics is the study of the relationship between the resultant force system and the associated motion of bodies that are not in equilibrium. The parallelogram law for the addition of forces and methods for determining the resultant of force systems were studied in Chapters 1 and 2. When the resultant of a force system acting on a body is zero, the body is in equilibrium. Equations of equilibrium were studied in Chapter 4 using the free-body diagram concept as an aid in the visualization of forces. Kinematics, the geometric study of motion without considering the associated forces, was presented in Chapter 8 as a preliminary to the study of the general kinetics problem. Equations relating the unbalanced force systems, the inertial properties (mass and mass moment of inertia), and the motion of bodies not in equilibrium will be developed in this and succeeding chapters.

The foundation for the classical study of kinetics is based on the laws of motion first presented in a formal manner in 1687 by Sir Isaac Newton (1642–1727). Although called Newton's laws of motion, it should be noted that Newton drew upon the scientific achievements of others over a historical period ranging from the builders of the pyramids to Newton's contemporaries. Perhaps the most significant discoveries were made by Galileo Galilei (1564–1642), who performed experiments on falling bodies and pendulums and disproved earlier false theories about dynamics. Significant developments in science pertinent to kinetics have been achieved since Newton's time, and these developments will certainly continue in the future, but the laws of motion stated by Newton are still recognized as the basic laws of engineering kinetics.

The motion of bodies acted upon by unbalanced force systems can be analyzed by several methods. The most useful method for any particular problem depends on the forces acting (constant or variable) and the results to be obtained (reactions, accelerations, velocities, displacements, and so on). The analysis of kinetics problems by force, mass, and acceleration, by the principle of work and energy, and by the impulse-momentum principle is discussed in the following chapters. Although these principles appear to differ in form, they are all immediately derivable from Newton's laws.

9-2 NEWTON'S LAWS OF MOTION

Newton's laws can be stated as follows:

1. If the resultant force acting on a particle is zero, the particle will remain at rest or move with a constant velocity (constant speed and direction).

2. If the resultant force acting on a particle is not zero, the particle will be accelerated in the direction of the force, and the magnitude of the acceleration will be directly proportional to the force and inversely proportional to the mass of the particle.

3. The force exerted by one particle on a second particle is equal in magnitude and opposite in direction to the force exerted by the second particle on the first.

These three laws were developed from a study of planetary motion, that is, the motion of particles. As stated, Newton's laws apply only to the motion of particles. Bodies are made up of particles, however, and these laws have been extended to apply to bodies.

The first law indicates that a particle has some property called *inertia* and that as a result a force is required to produce a change in the motion of the particle.

The second law is a quantitative statement of the relationship which must exist between the force acting on the particle, the inertia (mass) of the particle, and the resulting acceleration.

A mathematical statement of Newton's second law is

$$\mathbf{a} = K \frac{\mathbf{F}}{m},$$ (9-1)

where $\mathbf{a}$ is the absolute acceleration, $\mathbf{F}$ is the resultant force, m is the mass of the particle, and K is a dimensionless constant. If units of acceleration, force, and mass are chosen arbitrarily, the constant K must be determined experimentally. It is convenient to have the constant K be unity, which is possible if the units of either acceleration, force, or mass are selected in terms of the other two quantities. Only those cases in which K is unity are considered here, so that Eq. (9-1) becomes

$$\mathbf{a} = \frac{\mathbf{F}}{m}.$$ (9-1a)

A suitable frame of reference must be used for the measurement of absolute accelerations in order to satisfy Newton's second law. Such a reference frame is called a *newtonian* or *inertial reference frame* and is ideally fixed in space. Suppose that an inertial reference frame does exist, and as a result of a force $\mathbf{F}$, an observer in that frame measures an absolute acceleration $\mathbf{a}_P$ satisfying the second law,

$$\mathbf{a}_P = \frac{\mathbf{F}}{m}.$$ (a)

Let $\mathbf{a}_{P/M}$ be the acceleration observed in a moving frame due to the same force, and let $\mathbf{a}_M$ be the acceleration of the moving frame itself, so that

$$\mathbf{a}_P = \mathbf{a}_M + \mathbf{a}_{P/M}.$$

When this expression is substituted into Eq. (a) it becomes

$$\mathbf{a}_M + \mathbf{a}_{P/M} = \frac{\mathbf{F}}{m}$$

or

$$\mathbf{a}_{P/M} = \frac{\mathbf{F}}{m} - \mathbf{a}_M.$$ (b)

The acceleration $\mathbf{a}_{P/M}$ *observed* in the moving frame satisfies the second law only when $\mathbf{a}_M$ is equal to zero, that is, when the velocity $\mathbf{v}_M$ is a constant, where $\mathbf{v}_M$ is the velocity of the moving frame. Thus, a true *inertial reference frame* is one that is fixed in space, or translates with constant velocity and has no rotation. In most engineering applications, however, motion is measured relative to a reference frame fixed on the rotating earth, since the error introduced by the $\mathbf{a}_M$ term in Eq. (b) is negligible. (For example, the maximum error in calculating the acceleration of a freely falling particle near the earth's surface is approximately 0.34 percent when the rotation of the earth about its axis is neglected.)

Newton's third law makes it possible to extend the second law for particles to bodies, since bodies are made up of a system of particles and the forces between the particles always occur in collinear pairs which balance each other.

A fourth law formulated by Newton is Newton's law of universal gravitation, which states that the force of attraction **F** between two masses m_1 and m_2 separated by a distance r has the magnitude

$$F = \frac{Gm_1m_2}{r^2}.$$

G is a universal gravitational constant, which has an experimentally determined value of

$$G = 3.438(10)^{-8}\,\frac{\text{ft}^4}{\text{lb-sec}^4} = 6.670(10)^{-11}\,\text{m}^3/(\text{kg} \cdot \text{s}^2).$$

9-3 UNITS AND DIMENSIONS

Physical quantities have both *qualitative measures* called *dimensions* and *quantitative* measures called *units*. The application of Newton's second law requires the use of a homogeneous set of dimensions. Since the second law relates the three quantities of force, mass, and acceleration, the dimensions of only two of the quantities can be specified. The dimensions of the third quantity are then derived from Eq. (9-1a). It is commonly agreed that the dimensions of length (L) and time (T) are fundamental, so that the dimensions of acceleration are specified (LT^{-2}). One now has the choice of specifying either the mass (m) or the force (F) as a fundamental dimension.

When length, time, and mass are selected as fundamental dimensions the dimensions of force are derived from Eq. (9-1a). A *unit force* can then be defined as the force necessary to produce a unit acceleration of a particle having a unit mass. Systems of units in which the dimensions of mass, length, and time are fundamental and the dimension of force is the derived quantity are called *absolute systems* of units. The *International System* of Units (SI) is an absolute system in which acceleration is measured in meters per second per second (m/s^2) and mass is measured in kilograms (kg). The derived unit of force is then the force which will produce an acceleration of $1\,\text{m/s}^2$ when applied to a mass of 1 kg and is called a *newton* (N). One newton is the same as $1\,\text{m-kg/s}^2$.

When length, time, and force are selected as fundamental dimensions the dimension of mass is derived from Eq. (9-1a). A *unit mass* can thus be defined as the mass of a particle which will have a unit acceleration when acted upon by a unit force. Systems of units in which the dimensions of force, length, and time are fundamental and the dimension of mass is the derived quantity are called *gravitational systems* of units. In gravitational systems, the unit of force is defined in terms of the gravitational attraction of the earth for a standard mass. If the location of the standard mass is changed, the unit of force is changed due to variations in the earth's gravitational field. Thus, the term *gravitational* implies that the units change when the standard mass is moved in the gravitational field, whereas the term *absolute* implies that the units are independent of the position of the standard mass.

In the *British gravitational system* of units, acceleration is measured in feet per second per second and force in pounds. The derived unit of mass will be 1 lb divided by 1 fps², or 1 lb-sec²-per ft. The unit of mass of 1 lb-sec²-per ft is frequently called a *slug* or a *g-pound*. Both the *SI* and *British gravitational systems* of units will be used in this text and are summarized in Table 9-1. When the British gravitational system is used, the unit of mass will be called a slug.

The mass of a body in British gravitational units can be derived from Eq. (9-1a) if the acceleration produced by some known force can be determined. Experiments show that if the attraction of the earth for the body (the *weight* of the body) is the only force acting on the body, the acceleration will be approximately $g = 32.2$ fps² downward. When the y axis is vertical, Newton's second law is

$$\mathbf{a} = \frac{\mathbf{F}}{m},$$

and when values are substituted in this expression it becomes

$$-g\mathbf{j} = \frac{-W\mathbf{j}}{m}$$

from which

$$m = \frac{W}{g}. \tag{9-2}$$

In the *British gravitational system* of units, the weight is measured in pounds, and the acceleration of gravity is approximately 32.2 fps². In

TABLE 9-1 SYSTEMS OF UNITS

Dimension	British Gravitational Units	x Factor =	International System of Units
length	foot (ft)	0.3048	meter (m)
time	second (sec)	1.000	second (s)
force	pound (lb)	4.448	newton (N)
mass	slug (lb-sec²/ft)	14.59	kilogram (kg)

this text the acceleration of gravity on or near the surface of the earth will be taken as 32.2 fps^2 or 9.81 m/s^2.

9-4 EQUATIONS OF MOTION OF A PARTICLE

When a force system acts on a particle, the system must be concurrent (a particle has no dimensions), and the resultant is a single force acting through the particle. Newton's second law of motion for a particle can be expressed mathematically as

$$\mathbf{R} = m\mathbf{a}, \qquad (9\text{-}3)$$

where $\mathbf{R}$ is the resultant force, m is the mass of the particle, and $\mathbf{a}$ is its acceleration. Equation (9-3) is a vector equation of motion for particles which expresses the equality of the resultant force vector $\mathbf{R}$ and the $m\mathbf{a}$ vector as shown in Fig. 9-1. The $m\mathbf{a}$ vector, which is the effect of the resultant force, is frequently called the *effective* force of the particle because of this equality.

Figure 9-1

In the solution of kinetics problems (as in kinematic analysis) it is generally necessary to resolve the vector equation into a set of component equations suitable to the particular problem being analyzed. Resolving both sides of Eq. (9-3) into rectangular components, the vector equation of motion can be written in component form as

$$R_x\mathbf{i} + R_y\mathbf{j} + R_z\mathbf{k} = \sum F_x\mathbf{i} + \sum F_y\mathbf{j} + \sum F_z\mathbf{k} = m(a_x\mathbf{i} + a_y\mathbf{j} + a_z\mathbf{k}) \quad (9\text{-}4a)$$

or in scalar form as

$$R_x = \sum F_x = ma_x, \qquad R_y = \sum F_y = ma_y, \qquad R_z = \sum F_z = ma_z. \quad (9\text{-}4b)$$

When using these equations, a sign convention should be established for the forces and acceleration. A convenient sign convention is to show a set of axes and consider that the signs of all force and acceleration components are the same as the signs of the corresponding axes. The direction of unknown forces should be assumed and shown on the free-body diagram. Unknown accelerations should generally be assumed in the directions of the positive axes.

The kinematics of a problem, discussed in Chapter 8, generally indicate which components are most convenient for the kinetic analysis. As an example, normal and tangential components are particularly useful in problems where the particle is constrained to move in a circular path. In this case, when there are forces perpendicular to the plane of the circle, the resultant of these forces must be zero since there is no acceleration component in this direction.

Section 9-5
Motion of the mass
center of a
system of particles

9-5 MOTION OF THE MASS CENTER
OF A SYSTEM OF PARTICLES

The motion of the mass center of a system of particles can be obtained from Newton's equation of motion,

$$\mathbf{R} = m\mathbf{a},$$

for a single particle. The system of particles may be considered either as discrete particles, the particles of matter which make up a solid body, rigid or nonrigid, or the individual particles of a gas or fluid. Consider the collection of particles indicated by Fig. 9-2(a). The ith particle has a mass m_i and a position vector from the origin of coordinates $\mathbf{r}_i$. As shown in Fig. 9-2(b), each particle may be subjected to an external force $\mathbf{F}_i$ plus internal forces from the other particles of the system shown as $\mathbf{F}_{i_1}, \mathbf{F}_{i_2}, \mathbf{F}_{i_3}, \mathbf{F}_{ij}, \cdots$. These internal forces arise from such sources as magnetic fields, electrical charges, and molecular and elastic forces. Note that, according to Newton's third law, on mass m_j there is a force $\mathbf{F}_{ji}$ exerted by the mass m_i which is equal in magnitude to and collinear with $\mathbf{F}_{ij}$ but in an opposite direction so that

$$\mathbf{F}_{ij} + \mathbf{F}_{ji} = 0.$$

According to Newton's second law, the equation of motion for each particle is

$$\mathbf{F}_i + \sum_{j=1}^{n} \mathbf{F}_{ij} = m_i \mathbf{a}_i, \tag{9-5}$$

where n is the number of particles in the system. An equation of motion for the entire system may be obtained by adding all the equations of motion for the individual particles as follows:

$$\sum_{i=1}^{n} \mathbf{F}_i + \sum_{i=1}^{n} \left(\sum_{j=1}^{n} \mathbf{F}_{ij} \right) = \sum_{i=1}^{n} m_i \mathbf{a}_i.$$

The second term is the sum of all internal forces of the system which

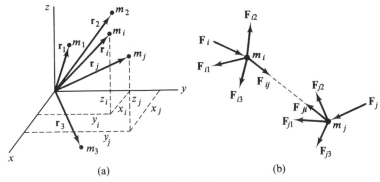

(a)　　　　　　　　(b)

Figure 9-2

must be zero, since they occur in equal, opposite, collinear pairs.[1] If **R** represents the resultant external force acting upon the system, this equation may be written as

$$\mathbf{R} = \sum_{i=1}^{n} \mathbf{F}_i = \sum_{i=1}^{n} m_i \mathbf{a}_i. \qquad (9\text{-}6)$$

Thus the *resultant* (magnitude, direction, and a point on the line of action) *of the external forces acting on a system of particles is equal to the resultant of the* **ma** *vectors* (effective forces) *of the particles of the system.* The preceding statement, a corollary of Newton's second and third laws, is called *d'Alembert's principle*, in honor of Jean LeRond d'Alembert (1717–1783), a French mathematician, who was the first to publish it.

The position of the mass center of a system of particles can be located by Eq. (3-3a). Thus, for the system of particles in Fig. 9-2(a),

$$m\mathbf{r}_G = \sum_{i=1}^{n} m_i \mathbf{r}_i. \qquad (9\text{-}7)$$

Equation (9-7) can be differentiated with respect to time to give

$$m\dot{\mathbf{r}}_G = m\mathbf{v}_G = \sum_{i=1}^{n} m_i \dot{\mathbf{r}}_i = \sum_{i=1}^{n} m_i \mathbf{v}_i \qquad (9\text{-}8)$$

and

$$m\ddot{\mathbf{r}}_G = m\mathbf{a}_G = \sum_{i=1}^{n} m_i \ddot{\mathbf{r}}_i = \sum_{i=1}^{n} m_i \mathbf{a}_i. \qquad (9\text{-}9)$$

Equations (9-6) and (9-9) can be combined to give

$$\mathbf{R} = \sum \mathbf{F} = m\mathbf{a}_G, \qquad (9\text{-}10)$$

which can also be written in vector component or scalar form similar to Eq. (9-4). The G subscript should be retained (for example, $\mathbf{a}_{Gx}$) to emphasize that $\mathbf{a}_G$ is the acceleration of the mass center. Equation (9-10) expresses the relationship between the resultant of the external forces acting on any system of particles and the *acceleration of the mass center* of the system regardless of the type of motion. This relationship is commonly called the *principle of motion of the mass center* of a system of particles. Equation (9-10) for a system of particles is the same as the equation of motion for a single particle located at the mass center of the system and having a mass equal to the total mass of the system. Thus, any body can be considered to be a particle in applying the equations of motion just developed. One very important difference should be noted, however. With a single particle, the line of action of the resultant of the force system acting on the particle passes through the particle, since the force system is concurrent. *With a system of particles, the external force system is not necessarily concurrent, and the*

---

[1] The $\mathbf{F}_{ii}$ terms will be zero, since a particle cannot exert a force upon itself.

Section 9-6
*Procedure for the solution of
problems in kinetics*

resultant usually does not pass through the mass center of the system of particles. To locate the resultant force acting on a body or system of particles, it is necessary to determine the moment of the force system with respect to one or more axes. Methods of determining these moments are developed in the articles for translation, rotation, and plane motion of rigid bodies. If Eq. (9-3) or (9-10) does not supply sufficient information to solve the problem, the body cannot be treated as a particle, and moment equations (which take into account the dimensions of the body and its angular motion) must be used.

In some instances, bodies with negligible mass, such as pulleys and rollers, may be involved in dynamics problems. Force and moment equations for such bodies can be written as in statics since terms involving accelerations will also involve zero mass.

9-6 PROCEDURE FOR THE SOLUTION OF PROBLEMS IN KINETICS

Problems can usually be analyzed most readily and with less chance of error if a definite procedure is followed. The following steps have been found to be helpful in problems in kinetics and will be emphasized in the succeeding articles:

1. Determine carefully what data are given and what is required in the problem.

2. Draw a free-body diagram of one of the particles or bodies involved. (It is usually desirable to include only one body in a free-body diagram when using the method of force, mass, and acceleration, instead of combining several bodies as was frequently done in statics.)

3. Decide what type of motion is involved (particle motion, translation, rotation, or plane motion of a rigid body), select suitable component directions for the problem, and write the general equations of motion for the particle or body being considered.

4. From the free-body diagram (step 2) and the general equations of motion (step 3), determine the number of unknown quantities involved (such as forces, acceleration components, and distances). A list of the unknowns may prove helpful.

5. (a) If the number of equations in step 3 is equal to (or greater than) the number of unknowns, proceed to step 8.

(b) If step 4 indicates more unknowns than the number of equations from step 3, additional equations must be obtained as indicated in step 6.[2]

---

[2] Special cases may arise in which certain unknowns (but never all of them) may be evaluated even though there are more unknowns than independent equations.

493

6. Obtain additional equations in one or more of the following ways:

(a) By using the kinematic relationships between the motions of the various particles, lines, or bodies.

(b) By drawing other free-body diagrams and repeating step 3.

(c) By using the equations of friction if sliding or impending motion exists between two rough surfaces.

7. When as many independent equations are available as the total number of unknown quantities, the analysis is complete.

8. Substitute specific values in the equations of motion from steps 3 and 6 (if more than one free body is necessary). When two or more free-body diagrams are used, the directions of the reference axes (positive directions) must be consistent. Finally, solve for the desired unknowns as determined in step 1.

The above procedure will be used in this text primarily to determine unknown quantities (such as the forces acting on a particle or its acceleration) at a particular instant. Kinetics can also be used to calculate these quantities as a function of time or position along the path of motion, or even to calculate the path of motion itself if it is unknown. In the latter case, Newton's laws of motion are used to derive one or more *differential equations of motion* which must be integrated to obtain equations for the position of the body as a function of time or some other parameter.

Deriving the correct differential equations of motion is an essential step in the calculation of the path of motion of a body. Since the engineer is the one most familiar with the particular problem being studied, he is usually the best qualified person to derive the differential equations of motion. The resulting differential equations of motion are often quite difficult to solve, and an analytical solution in terms of elementary algebraic functions might not be possible. In such cases, engineering solutions can usually be obtained by numerically integrating the differential equations of motion. Digital or analog computers are often necessary, and the services of trained computer programmers and mathematicians might be required. The most essential feature of such an analysis, however, is a properly derived set of differential equations of motion which, together with the appropriate *initial conditions*, accurately represent the physical problem. As the term implies, initial conditions are the conditions which are necessary to describe mathematically the position and velocity of the particle at the beginning of the analysis (usually at $t = 0$).

It is beyond the scope of this text to give a detailed presentation of the calculation of paths of motion including the solution of various types of differential equations; consequently, the application of Newton's laws in deriving the differential equations of motion will be emphasized. It will frequently be found that a first integration of the differential equations can be obtained by using some of the techniques presented in Chapter 8. As will be shown later, a first integration of the differential equations can also frequently be obtained by applications of the work-energy principle (Chapter 10) or the impulse-momentum principle (Chapter 11).

Section 9-6
Procedure for the solution of
problems in kinetics

EXAMPLE 9-1

An airplane pilot completes a dive by pulling out in a circular path in the vertical plane. The speed of the airplane at the bottom of the path is 300 mph. The pilot weighs 170 lb, and the maximum acceleration he can withstand is 8 g. Determine the minimum radius of the circle and the *apparent* weight of the pilot at the bottom of the path.

SOLUTION

The trajectory of the airplane is shown in Fig. 9-3(a). At the time of interest the trajectory is a circular path so that the indicated n and t directions will provide suitable coordinates for the problem.

The minimum radius of the circle can be determined from kinematics. The magnitude of the normal acceleration at the bottom of the path is

$$a_n = \frac{v^2}{r}, \qquad \text{(a)}$$

where

$$v = 300 \text{ mph} = 300\left(\frac{5280}{3600}\right) = 440 \text{ fps}$$

and

$$a_n = 8g = 8(32.2) = 258 \text{ fps}^2.$$

Substituting the above values, Eq. (a) becomes

$$258 = \frac{(440)^2}{r},$$

from which

$$r = \underline{750 \text{ ft.}} \qquad \text{Ans.}$$

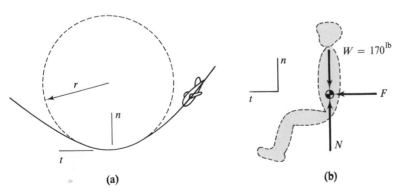

(a)   (b)

Figure 9-3

A free-body diagram of the pilot is shown in Fig. 9-3(b). The force **N** is the pilot's apparent weight, i.e., the force that would be recorded on a spring scale located in the seat cushion. The equation of motion is

$$(N - W)\mathbf{e}_n + F\mathbf{e}_t = m(a_n\mathbf{e}_n + a_t\mathbf{e}_t)$$

or in scalar form

$$\mathbf{e}_n: \quad N - W = \frac{W}{g}(8g), \qquad\qquad (b)$$

$$\mathbf{e}_t: \quad F = \frac{W}{g}a_t \qquad\qquad (c)$$

The force N can be calculated from Eq. (b) as

$$N = 9W$$

or

$$N = 9(170) = \underline{1530\,\text{lb.}}$$

Ans.

Equation (c), which relates the force and acceleration in the tangential direction, is not required in the solution for N, and it should be noted that N is independent of the tangential acceleration.

EXAMPLE 9-2

The small body A in Fig. 9-4(a) has a mass of 0.15 kg and swings in a horizontal plane on the end of cord AB. The line AO has a constant angular speed of 25 rpm. Determine the tension in the cord and the angle ϕ.

SOLUTION

The free-body diagram of body A is shown in Fig. 9-4(b). The circular path of the body suggests that normal and tangential components

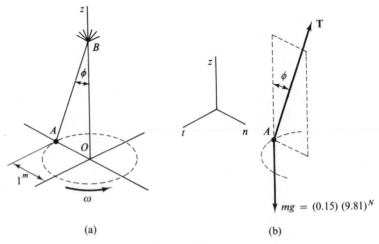

(a)   (b)

Figure 9-4

would be appropriate for this problem. The positive directions of the axes are shown in Fig. 9-4(b).

The force $\mathbf{T}$ in the free-body diagram has two unknown characteristics, the magnitude T and the angle ϕ. The equation of motion is

$$\mathbf{R} = T(\cos \phi \mathbf{e}_z + \sin \phi \mathbf{e}_n) - mg\mathbf{e}_z = m(a_n\mathbf{e}_n + a_t\mathbf{e}_t + a_z\mathbf{e}_z).$$

There are no forces in the t direction, and

$$a_t = 0$$

since the angular speed ω of line OA is constant. The vector equation of motion thus gives two scalar equations:

$$\mathbf{e}_n: \quad T \sin \phi = ma_n,$$

$$\mathbf{e}_z: \quad T \cos \phi - mg = ma_z.$$

The body moves in a horizontal plane so that

$$a_z = 0,$$

and from kinematics

$$a_n = r\omega^2 = 1.00\left[\frac{25(2\pi)}{60}\right]^2 = 6.85 \text{ m/s}^2.$$

Substituting values for m, g, and the acceleration components into the scalar equation gives

$$T \sin \phi = 0.15(6.85) = 1.028$$

and

$$T \cos \phi - 0.15(9.81) = T \cos \phi - 1.472 = 0$$

from which

$$T = \underline{1.795N}, \qquad \phi = \underline{34.9°}. \qquad \text{Ans.}$$

EXAMPLE 9-3

Block A in Fig. 9-5(a), (p. 499), weighs 10.00 lb, and block B weighs 6.44 lb. The blocks are connected by a flexible inextensible cord which passes over a smooth pulley at C. The weights of the cord and the pulley can be neglected. The coefficient of friction between the plane and block A is 0.30. The blocks are initially at rest. Determine the tension in the cord and the velocity of A after it has traveled 5.00 ft.

SOLUTION

The velocity of A can be determined from the principles of kinematics once the acceleration is known. A free-body diagram of A is shown in Fig. 9-5(b). The frictional force $\mathbf{F}$ must be in a direction opposite to the direction of the velocity. Since the blocks are released from rest, the velocity a short time after release will be in the same direction as the acceleration. The direction of the acceleration is unknown and must be assumed. In this case, the acceleration is assumed to be up the plane, which means that the velocity will also be up the plane. Therefore, the frictional force will be down the plane.

Coordinate axes are chosen parallel and perpendicular to the plane because block A slides along the plane, thus, $\mathbf{a}_{A_y} = 0$ and $\mathbf{a}_{A_x} = \mathbf{a}_A$. The equation of motion for body A is

$$\mathbf{R} = \sum \mathbf{F} = m\mathbf{a}_A$$

or

$$T\mathbf{i} - F\mathbf{i} + N\mathbf{j} + 5.00(0.8\mathbf{i} - 0.6\mathbf{j}) + 10.00(-0.6\mathbf{i} - 0.8\mathbf{j}) = \frac{10.00}{32.2}\, a_A \mathbf{i}.$$

This vector equation of motion for body A gives the two scalar equations

$$\mathbf{i}: \quad T - F + 4.00 - 6.00 = 0.310 a_A \tag{a}$$

and

$$\mathbf{j}: \quad N - 3.00 - 8.00 = 0. \tag{b}$$

Equations (a) and (b) contain four unknowns, T, F, a_A, and N. An additional equation is available from friction since the block is assumed to slide on the plane, and the magnitude of the friction force is therefore equal to the limiting value

$$F = \mu N = 0.30N. \tag{c}$$

This still leaves four unknowns and three equations.

Another equation can be obtained by drawing a free-body diagram of body B [Fig. 9-5(c)], and applying Newton's second law:

$$\mathbf{R} = \sum \mathbf{F} = m\mathbf{a}_B.$$

Note that $\mathbf{a}_B = \mathbf{a}_{B_y}$ since block B has motion only in the y direction. It should also be noted that the positive y direction is down for the free-body diagram of block B. The logic for this selection will be explained later.

The cord tension $\mathbf{T}$ has the same magnitude in both free-body diagrams. This fact can be demonstrated by drawing a free-body diagram of the pulley as in Fig. 9-5(d). Since the pulley has negligible mass, the equations of equilibrium are valid, and from the equation $\sum M_O = 0$ the two tensions are equal. (It will be shown later that when the pulley mass is not zero, the two tensions are not equal.) The equation of motion for block B thus becomes

$$-T\mathbf{j} + 6.44\mathbf{j} = \frac{6.44}{32.2}\, a_B \mathbf{j}$$

or

$$-T + 6.44 = 0.20 a_B \tag{d}$$

Equation (d) gives a fourth equation but introduces a fifth unknown, a_B. Since the cord is inextensible, however, kinematics gives

$$a_B = a_A \tag{e}$$

Section 9-6
Procedure for the solution of
problems in kinetics

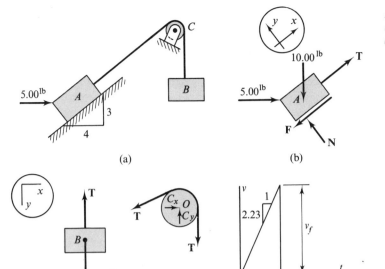

(a) (b)

(c) (d) (e)

Figure 9-5

for the positive directions shown in Fig. 9.5(b) and 9.5(c). Consistent positive directions must be selected for these accelerations to avoid the use of a negative sign when they are equated. If $\mathbf{a}_A$ were assumed to be down the plane, $\mathbf{a}_B$ should be assumed upward to be consistent. Note that the friction force $\mathbf{F}$ should also be reversed in this case.

There are now five equations (a), (b), (c), (d), and (e) and five unknowns. The magnitude

$$N = 11 \text{ lb}$$

is obtained from Eq. (b) and substituted into Eq. (c) so that

$$F = 0.3N = 0.3(11) = 3.30 \text{ lb}.$$

When the value of F is substituted in Eq. (a), it becomes

$$T - 3.30 + 4.00 - 6.00 = 0.310a_A,$$

and Eq. (d) becomes

$$T + 6.44 = 0.20a_B = 0.20a_A.$$

The above two equations are solved simultaneously to give

$$T = 5.99 \text{ lb}, \qquad \mathbf{a}_A = 2.23\mathbf{i} \text{ fps}^2. \qquad \text{Ans.}$$

A v-t diagram for block A is shown in Fig. 9-5(e). The area of the triangle is 5.0 ft, and the slope is 2.23 fps^2. Thus,

$$\text{area} = \frac{v_f t}{2} = 5.0$$

499

and

$$\text{slope} = \frac{v_f}{t} = 2.23,$$

from which

$$\mathbf{v}_f = \underline{4.72\mathbf{i}\,\text{fps}.}$$

Ans.

If the acceleration had been assumed in the wrong direction, the value of a_A would have had a negative sign, indicating that the resulting velocity and hence the frictional force were assumed to be in the wrong direction. As a result it would be necessary to solve the problem again with the frictional force in the correct direction.

Another method of solution is to assume that the bodies are in equilibrium and determine the frictional force necessary to maintain equilibrium. If the friction thus determined is less than the limiting value (μN), the system will remain at rest. If the frictional force for equilibrium is more than the limiting value, the system will be accelerated in a direction opposite to the direction of the friction (since the blocks are initially at rest), and the direction of the velocity and acceleration will then be known.

If the force system acting on block A produces a reversal of velocity during the time interval, the direction of the frictional force will also reverse, and the acceleration will change in magnitude though not necessarily in direction.

EXAMPLE 9-4

The 2.00-lb body A and 4.00-lb body B in Fig. 9-6(a) are initially at rest. The coefficients of friction are $\mu_A = 0.70$ between bodies A and B, and $\mu_B = 0.20$ between body B and the horizontal surface. Determine the accelerations of the two bodies due to the 5.00-lb force applied to body A.

SOLUTION

Three different combinations of motion are possible for the system shown in Fig. 9-6(a):

(a) Body A accelerates, and body B remains at rest.

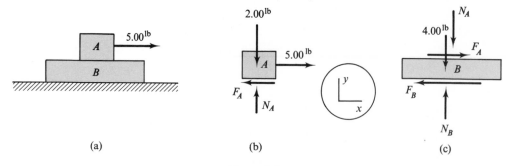

(a) (b) (c)

Figure 9-6

Section 9-6
Procedure for the solution
of problems in kinetics

(b) Bodies A and B both move with the same acceleration.

(c) Bodies A and B both move but with different accelerations.

Free-body diagrams of bodies A and B are shown in Fig. 9-6(b) and 9-6(c). Note that the friction forces must oppose any slipping or tendency to slip. The equations of motion for bodies A and B are, respectively,

$$\mathbf{R}_A = (5.00 - F_A)\mathbf{i} + (N_A - 2.00)\mathbf{j} = \frac{2.00}{32.2}\, a_A\mathbf{i},$$

$$\mathbf{R}_B = (F_A - F_B)\mathbf{i} + (N_B - N_A - 4.00)\mathbf{j} = \frac{4.00}{32.2}\, a_B\mathbf{i}.$$

The scalar equations for body A are

$$\mathbf{i}: \quad 5.00 - F_A = \frac{2.00}{32.2}\, a_A, \tag{a}$$

$$\mathbf{j}: \quad N_A - 2.00 = 0, \tag{b}$$

and for body B the equations are

$$\mathbf{i}: \quad F_A - F_B = \frac{4.00}{32.2}\, a_B, \tag{c}$$

$$\mathbf{j}: \quad N_B - N_A - 4.00 = 0. \tag{d}$$

The four equations (a), (b), (c), and (d) contain six unknowns, N_A, N_B, F_A, F_B, a_A, and a_B. The force N_A as determined from Eq. (b) is

$$N_A = 2.00\,\text{lb},$$

and the force N_B, determined from Eq. (d), is

$$N_B = N_A + 4.00 = 2.00 + 4.00 = 6.00\,\text{lb}.$$

Two equations, (a) and (c), and four unknowns, F_A, F_B, a_A, and a_B, remain. The two additional equations necessary to complete the solution are determined by considering the three possible combinations of motion.

(a) Assume that body A accelerates and that body B remains at rest. In this case the two additional equations are

$$a_B = 0$$

and

$$F_A = F'_A = \mu_A N_A = 0.70(2.00) = 1.400\,\text{lb}.$$

In addition, since body B remains at rest, the friction force F_B must be less than its limiting value F'_B, where

$$F'_B = \mu_B N_B = (0.20)(6.00) = 1.200\,\text{lb}.$$

When $a_B = 0$ is substituted into Eq. (c) it becomes

$$F_A - F_B = 0$$

or

$$F_B = F_A = 1.400\,\text{lb}.$$

Since F_B is greater than F_B', the assumption that body B remains at rest is invalid.

(b) Assume that bodies A and B move with the same acceleration. In this case

$$a_A = a_B$$

and

$$F_B = F_B' = 1.200 \text{ lb}$$

since B slides on the horizontal surface. In addition, since there is no relative motion between A and B, the friction force F_A must be less than its limiting value; that is,

$$F_A < F_A'$$

or

$$F_A < 1.400 \text{ lb}.$$

Equations (a) and (c) now become

$$5.00 - F_A = \frac{2.00}{32.2} a_A = \frac{2.00}{32.2} a_B$$

and

$$F_A - F_B' = F_A - 1.20 = \frac{4.00}{32.2} a_B.$$

The above two equations can be solved simultaneously to give

$$a_B = a_A = 20.4 \text{ fps}^2$$

and

$$F_A = 3.73 \text{ lb}.$$

Since F_A is greater than F_A', the assumption that bodies A and B move with the same acceleration is invalid.

(c) Assume that bodies A and B both move but with different accelerations. In this case

$$F_A = F_A' = 1.400 \text{ lb}$$

and

$$F_B = F_B' = 1.200 \text{ lb},$$

and in addition a_A must be greater than a_B. Substituting the above values into Eq.(a) and (c) yields

$$5.00 - F_A' = 5.00 - 1.400 = \frac{2.00}{32.2} a_A,$$

$$F_A' - F_B' = 1.400 - 1.200 = \frac{4.00}{32.2} a_B$$

from which

$$a_A = 58.0 \text{ fps}^2, \qquad a_B = 1.610 \text{ fps}^2$$

and

$$\mathbf{a}_A = \underline{58.0\mathbf{i} \text{ fps}^2}, \qquad \mathbf{a}_B = \underline{1.610\mathbf{i} \text{ fps}^2}. \qquad \text{Ans.}$$

Since $\mathbf{a}_A$ is greater than $\mathbf{a}_B$, block A will slip to the right on B, and the limiting force will oppose motion as shown.

PROBLEMS

Note: In the following problems unless otherwise specified, all cords, ropes, and cables are assumed to be flexible, inextensible, and of negligible weight. Pulleys are assumed to have negligible mass, and pins, axles, and pegs are assumed to be smooth unless otherwise stated.

9-1 A spring-operated platform scale is placed on the floor of an elevator. Neglect the weight of the platform scale. When the elevator is at rest, a man stands on the scale, which indicates 180 lb. Determine
(a) The acceleration of the elevator when the scale indicates a load of 225 lb.
(b) The load indicated by the scale when the elevator has an acceleration of 15 fps² downward.

9-2 An astronaut weighs 170 lb on the surface of the earth. Assume the earth is a sphere with a 4000-mile radius. Determine
(a) The astronaut's mass and weight when he occupies a space station which has a circular orbit 500 miles above the surface of the earth.
(b) The satellite speed required for the astronaut to experience a weightless sensation.

9-3 The 50-lb block A in Fig. P9-3 is initially at rest on a horizontal surface. The coefficients of static and kinetic friction between the block and surface are, respectively $\mu_s = 0.40$ and $\mu_k = 0.30$. The applied force **P** is slowly increased from zero until the block moves. Determine the minimum magnitude of the force **P** required to move the block and the resulting acceleration.

Figure P9-3

9-4 Block A in Fig. P9-4 weighs 75 lb, and the coefficient of friction between A and the plane is 0.50. Determine the acceleration of A under the action of the 25-lb force when the velocity of A is (a) 8 fps up the plane, (b) 15 fps down the plane, (c) zero.

9-5 A 7-lb box is given a velocity of 10 fps up a plane which is inclined 30° to the horizontal. The coefficient of friction between the box and

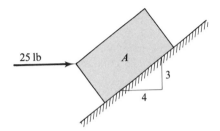

Figure P9-4

the plane is 0.60. Determine the velocity of the box (a) after 0.25 sec, (b) after 0.50 sec.

9-6 Solve Problem 9-5 if the coefficient of friction is 0.40.

9-7. Body A in Fig. P9-7 weighs W lb and is initially moving to the right under the action of the force **P**, which has a constant magnitude. The coefficient of friction between A and the horizontal plane is μ. Determine
(a) The angle θ for maximum acceleration.
(b) The magnitude of the resulting acceleration.

Figure P9-7

9-8 Solve Problem 9-7 if body A is initially moving to the left.

*9-9 Block A in Fig. P9-9 has a mass of 5 kg, and block B has a mass of 2 kg. The coefficient of friction between A and the plane is 0.4. The blocks are initially at rest. Determine the acceleration of B and the tension in the cord.

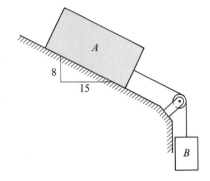

Figure P9-9

*9-10 The masses of blocks A, B, and C in Fig. P9-10 are 15, 25, and 10 kg, respectively, and the system is initially at rest. The coefficient of friction between B and the horizontal surface is 0.10. Determine
(a) The tension in each cord.
(b) The acceleration of body B.

Figure P9-10

9-11 The three bodies A, B, and C in Fig. P9-11 weigh 20, 25, and 15 lb, respectively. The horizontal plane is smooth. Determine
(a) The acceleration of body C.
(b) The tension in the cord connecting B and C.
(c) The tension in the cord connecting A and B.

Figure P9-11

9-12 Solve Problem 9-11 if the coefficient of friction between the horizontal plane and bodies A and B is 0.2. The blocks start from rest.

9-13 Bodies A and B in Fig. P9-13 weigh 50 and 30 lb, respectively. The coefficient of friction between A and the inclined plane is 0.3. Body A has an initial velocity of 10 fps up the plane. Determine
(a) The acceleration of block A.
(b) The tension in the cord.

Figure P9-13

9-14 Body A in Fig. P9-13 has a velocity of 12 fps down the plane. Determine the weight of body B required to produce an acceleration of body A of 8.05 fps² up the plane. All other data are the same as in Problem 9-13.

*9-15 Block A in Fig. P9-15 has a mass of 2.6 kg, and block B has a mass of 1.0 kg. The coefficient of friction between A and the plane is 0.25. Block A has an initial velocity of 3 m/s down the plane. Determine
(a) The acceleration of block A.
(b) The tension in the cable.

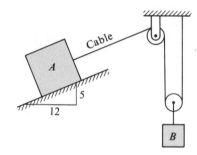

Figure P9-15

9-16 The two blocks A and B in Fig. P9-16 weigh 15 and 10 lb, respectively. The coefficient of friction is 0.50 between A and the plane and 0.20 between B and the plane. Determine the magnitude of the resultant force between the blocks after they have been released from rest.

Figure P9-16

9-17 In Fig. P9-17 body A weighs 15 lb and body B weighs 35 lb. The bodies are initially at rest on the smooth horizontal plane. The coefficient of friction between A and B is 0.25. Determine the acceleration of each of the bodies if the magnitude of the force P is (a) 10 lb, (b) 20 lb.

Figure P9-17

9-18 Blocks A and B in Fig. P9-18 are released from rest in the position shown. The weight of B is twice the weight of A. The coefficient of static friction for all surfaces is 0.40, and the coefficient of kinetic friction is 0.30. Determine the acceleration of block A.

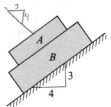

Figure P9-18

9-19 Block A in Fig. P9-19 weighs W lb and slides along a vertical circular track 10 ft in diameter. The coefficient of friction between A and the track is 0.30. When in the position shown, block A has a velocity of 10 fps down and to the left. Determine the acceleration of A for this position.

Figure P9-19

9-20 Particle A in Fig. P9-20 weighs 5.0 lb and swings in a vertical plane at the end of an 8-ft cord. When the angle θ is 30°, the magnitude of the velocity of A is 4 fps. Determine the tension in the cord and the angular acceleration of the cord for this position.

Figure P9-20

9-21 Particle A of mass m travels on a circular path, as shown in Fig. P9-21, with a speed of 14 fps. Determine the radius r of the path.

Figure P9-21

9-22 In Fig. P9-21, particle A is connected to point B by a 9-ft cord and to point O by a 15-ft cord.
(a) Derive an expression for the tension in AB in terms of the mass of the body and the angular velocity of AB.
(b) At what angular velocity will the tension in the cord AB become zero?
(c) As the angular velocity is decreased, the angle AOB will also decrease. Determine the angular velocity at which the angle will become zero.

*9-23 Body A in Fig. P9-23 has a mass of 9 kg and rests on a smooth plane. The force P applied to body A is increased from zero at a very slow rate. Determine the maximum force P that can be applied without causing the smooth 6-kg cylinder to leave body A at point C.

Figure P9-23

*9-24 A racing car travels around a curve at a constant speed of 250 km/hr. The radius of curvature of the track is 500 m, measured horizontally, and the super elevation of the track, angle θ in Fig. P9-24, is 30°. Determine the minimum coefficient of friction between the track and tires if there is to be no tendency for the car to slip.

Figure P9-24

9-25 The turntable in Fig. P9-25 rotates in a horizontal plane with a constant angular acceleration of 3.0 rad per sec² counterclockwise when looking downward. The small block A on the disk weighs 2.0 lb. Determine the magnitude of the frictional force of the disk on the block when the angular velocity of the disk is 2.0 rad per sec counterclockwise. Assume that the block does not slip on the disk.

Figure P9-25

9-26 The coefficient of friction between the block and disk in Problem 9-25 is 0.60. Determine the angular velocity of the disk when the block starts to slip. The angular acceleration of the disk is the same as in Problem 9-25.

9-27 A cardboard box weighing 50 lb comes out the end of a chute with a horizontal velocity of 15 fps and lands without bouncing on the platform of a 100-lb pushcart as shown in Fig. P9-27. The coefficient of friction between the cart and the box is 0.4. The inertia and the rolling resistance of the cart wheels can be neglected. Assume the box comes to rest relative to the cart before impacting the end of the cart. Determine

(a) The final velocity of the box and cart after the box has ceased to slide with respect to the cart.

(b) The distance the cart slides along the top of the cart.

(c) The total distance the cart moves before the box and cart have the same velocity.

Determine the time required for block A to travel the length of block B and the corresponding displacement of block B.

Figure P9-27

9-28 Block A in Fig. P9-28 weighs 10.00 lb and is 5.00 in. long. Block B weighs 20.00 lb and is 15.0 in. long. Block B is on a smooth plane, and the coefficient of friction between A and B is 0.40. Both blocks are initially at rest when the 8.00-lb force is applied to block A. Determine the time required for block A to travel the length of block B and the corresponding displacement of block B.

Figure P9-28

9-29 Bodies A, B, and C in Fig. P9-29 weigh 40, 30, and 20 lb, respectively. The system is released from rest. Determine the acceleration of body A and the tension in the cord connecting bodies B and C.

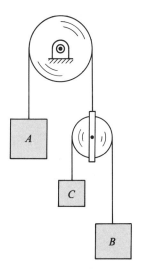

Figure P9-29

9-30 A *top view* of three bodies is shown in Fig. P9-30 as they slide on a smooth horizontal plane. Bodies *A*, *B*, and *C* weigh, respectively, 32.2, 64.4, and 96.6 lb. Determine the accelerations of bodies *A*, *B*, and *C*.

Top view

Figure P9-30

9-31 In Fig. P9-31, body *A* weighs 5.0 lb and body *B* weighs 10.0 lb. The coefficients of friction are 0.4 between *A* and *B* and 0.2 between *B* and the horizontal surface. The weight of body *C* can be neglected. Determine
(a) The maximum value of *P* for which there is no motion.

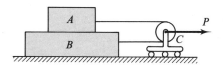

Figure P9-31

(b) The minimum value of *P* for which *A* slides on *B*.

9-32 Ring *B* in Fig. P9-32 slides along rod *OA*, which rotates in the vertical plane. In the position shown, rod *OA* has an angular velocity of 5 rad per sec counterclockwise and an angular acceleration of 3 rad per sec² clockwise. The distance *r* is 1 ft and is increasing at a rate of 2 fps. The coefficient of friction between the ring and rod is 0.2. Determine the acceleration of the ring relative to the rod.

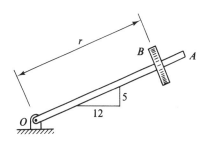

Figure P9-32

9-33 The cone in Fig. P9-33 has a central angle *θ* and rotates about the vertical *z* axis. A small body which has a mass *m* is connected to the vertex of the cone by a cord of length *L* and rotates with the cone. Derive, in terms of the length *L*, the angle *θ*, the mass *m*, and the angular velocity of the cone, an expression for
(a) The tension in the cord.
(b) The reaction of the cone on the particle.

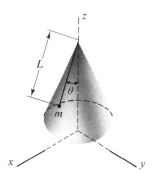

Figure P9-33

9-34 The 4-in. bar AB in Fig. P9-34 rotates in the horizontal (xy) plane around the z axis with an angular speed of ω rad per sec. Particle P weighs 2 lb and is suspended from AB by two 7-in. cords.

(a) Determine the angular speed of AB and the tension in each cord when the distance $d = 6$ in.

(b) As ω decreases the distance d will also decrease. Determine the value of ω for which d approaches zero.

9-36 Bodies A and B in Fig. P9-36 weigh 10 lb each. The cable attached to body B passes around a smooth peg inside body A. In the position shown, A has a velocity of 6 fps to the right on the smooth horizontal surface. Determine

(a) The tension in the cord.

(b) The acceleration of A.

(c) The acceleration of B.

Figure P9-34

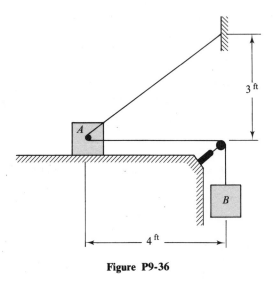

Figure P9-36

9-35 The 8.05-lb body A in Fig. P9-35 is connected to the 16.1-lb body B by a 7.0-ft cord which passes over the smooth peg at C. The horizontal slot is smooth. When the distance x is 4.0 ft the velocity of B is 5.0 fps to the right and the force P is 10.0 lb to the right. Determine the acceleration of B and the tension in the cord.

9-37 In Fig. P9-37, body A is connected to body B by a cord which goes through smooth rollers at C. Body A weighs 15 lb and body B weighs 12 lb. The coefficient of friction between body A and the horizontal surface is 0.6. At the instant shown, body B has a velocity of 6 fps up and to the right. If the cord connecting body A to the wall is cut at this instant, determine the acceleration of body A.

Figure P9-35

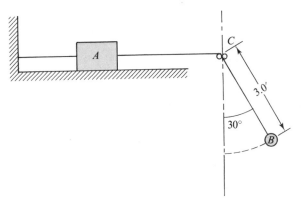

Figure P9-37

9-7 REVERSED EFFECTIVE FORCES

It was first noted by Jean LeRond d'Alembert (1717–1783) that dynamics problems may be treated by the methods of statics if Newton's second law of motion is written as

$$\mathbf{F} + (-m\mathbf{a}) = 0,$$

where the $(-m\mathbf{a})$ term is known as the *reversed effective force*. A corollary of d'Alembert's principle [see Art. 9-5] can be stated as follows: *If a force equal in magnitude to and collinear with, but opposite in sense to, the resultant of the m**a** quantities of the particles is added to the external forces acting on the particles, the resulting force system will be in equilibrium.* Although equivalent to forces, reversed effective forces are not true forces since they do not represent the action of one body upon another.

The concept of reversed effective forces does not present any new information and is therefore only an alternative to the method previously presented in this chapter. It is important to decide which method will be used early in the solution of any problem since if the two methods are mixed in a single solution, the m**a** terms are often used once as *reversed effective forces* and again as *effective forces*, resulting in an incorrect solution. When applying d'Alembert's principle, it is desirable to show the reversed effective forces as dashed lines to distinguish them from true forces.

The following example illustrates the use of reversed effective forces.

EXAMPLE 9-5

A small body weighing 2.0 lb is suspended from the top of a bus by a 52-in. cord. As the bus is accelerated to the left the body swings toward

509

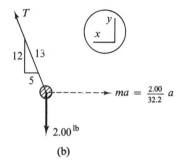

(a)

(b)

Figure 9-7

the rear of the bus as shown in Fig. 9-7(a). Assume that the acceleration is constant and that the pendulum does not oscillate. When the distance d is 20.0 in., determine (a) the acceleration of the bus and (b) the magnitude of the tension in the cord.

SOLUTION

A free-body diagram of the body is illustrated in Fig. 9-7(b). The 2.00-lb weight **W** and tension **T** are the external forces acting on the body. The reversed effective force is shown as a dashed line to distinguish it from the true forces. It should be noted that since the acceleration is known to be toward the left, the reversed effective force is known to act toward the right. The equilibrium equation for the body is

$$\mathbf{R} = 0$$

$$T\left(\frac{5}{13}\mathbf{i} + \frac{12}{13}\mathbf{j}\right) - 2.00\mathbf{j} - \frac{2.00}{32.2}a\mathbf{i} = 0.$$

The associated scalar equations are

$$\mathbf{i}: \quad \frac{5}{13}T - \frac{2.00}{32.2}a = 0 \tag{a}$$

$$\mathbf{j}: \quad \frac{12}{13}T - 2.00 = 0. \tag{b}$$

The tension T, from Eq. (b), is

$$T = \underline{2.17\,\text{lb},} \qquad\qquad \text{Ans.}$$

and when this value is substituted into Eq. (a) the value of a becomes

$$a = 13.42$$

and

$$\mathbf{a} = \underline{13.42\mathbf{i}\,\text{fps}^2.} \qquad\qquad \text{Ans.}$$

PROBLEMS

*9-38 Solve Problem 9-9 by the reversed effective-force method.

9-39 Solve Problem 9-20 by the reversed effective-force method.

*9-40 Solve Problem 9-23 by the reversed effective-force method.

*9-41 Solve Problem 9-24 by the reversed effective-force method.

9-42 Solve Problem 9-25 by the reversed effective-force method.

9-43 A package rests on the front seat of an automobile as shown in Fig. P9-43. The coefficient of friction between the package and the seat cushion is 0.6. When stopping at a traffic signal, at what deceleration will the package begin to slide from the seat?

Figure P9-43

9-44 In Fig. P9-44 the homogeneous 5-lb ball C is suspended from the 20-lb body B by a cord attached at A. The motion of the system is such that the angle θ remains constant at 30° when the velocity of B is to the left. The coefficient of friction between B and the plane is 0.4. Determine the magnitude of the force **P**.

Figure P9-44

9-45 The shaft in Fig. P9-45 rotates in bearings A and B at a speed of 1200 rpm. The weights at C and D are 2 and 5 lb, respectively. Neglect the effects of gravity. Determine the forces exerted on the shaft by the bearings when in the position shown.

9-46 The 5-in. radius disk in Fig. P9-46 spins about the vertical axis at a constant speed. The pendulum arms are 10 in. long. Determine the rate of spin when the angle θ is 30°.

Figure P9-45

Figure P9-46

*9-47 In Fig. P9-47, blocks A and B have masses of 10 and 40 kg, respectively. Determine the acceleration of A and the tension in the cord if the fixed drum is smooth.

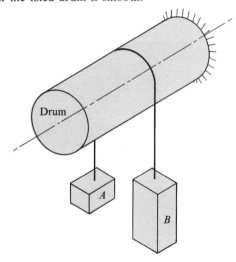

Figure P9-47

*9-48 Solve Problem 9-47 if the coefficient of friction between the cord and the fixed drum is 0.15 and the bodies are released from rest. [*Hint:* Review belt friction in Art. 5-6 if necessary.]

9-49 Derive an equation relating the large and small tensions in the flat belt in Fig. P9-49 when slipping impends between the belt and rotating pulley. The belt weighs w lb per ft of length, and the pulley has a constant angular velocity of ω rad per sec. Neglect the force that the earth exerts on the belt (the weight). (Review belt-friction equations for stationary belts or band brakes in Art. 5-6 if necessary.)

Figure P9-49

PART B–KINETICS OF RIGID BODIES

9-8 INTRODUCTION

The application of Newton's laws of motion to the kinetics of rigid bodies will be developed in Part B of this chapter. Kinetics has been defined as the relationship between the resultant force system and the associated motion of bodies that are not in equilibrium. When the size of a body can be neglected in kinetic analysis, the body can be treated as a *particle* with its mass concentrated at a single point, and the line of action of the resultant force must pass through the particle. Consequently, a single vector equation obtained from Newton's second law, $\mathbf{R} = m\mathbf{a}$, is sufficient to describe the relation of the unbalanced force system acting on a particle to the inertia or mass of the particle and the resulting acceleration.

It was shown in Art. 9-5 that the relationship between the resultant of the external forces acting on any *system of particles* and the acceleration of the mass center of the system is also governed by a single vector equation, $\mathbf{R} = m\mathbf{a}_G$. For a system of particles, however, the external force system is not necessarily concurrent, and the resultant usually does not pass through the mass center of the system. In addition, the equation of motion for the mass center of a system of particles gives no information regarding the motion of the individual particles *relative* to the mass center.

A *rigid body* is a system of particles in which all particles remain at fixed distances from each other. The remainder of this chapter will be devoted to the effects of the points of application of forces on the motion of a rigid body relative to its center of mass, as well as on the motion of the mass center itself. It should be obvious that additional equations of motion will be required. These equations can be obtained from moment equations which take into account the point of application of forces, the dimensions of the body, and its angular motion. Methods of determining these equations will be developed for translation, rotation, and plane motion of rigid bodies in the following articles.

9-9 TRANSLATION OF A RIGID BODY

When every line in a rigid body remains parallel to its initial position the motion of the body is defined as *translation*. If a rigid body has translation, all particles of the body have the same acceleration.

Figure 9-8 represents a rigid body which has a motion of translation so that each particle of the body has an acceleration **a** as indicated. The resultant of all external forces acting on the body is shown as **R**. From d'Alembert's principle [Eq. (9-6)] the resultant of the external forces **R** must be equal to the sum of the $m_i \mathbf{a}_i$ values for any system of particles which includes rigid bodies. Let the body be made up of n particles and m_i be the mass of the ith particle. The force **R** is

$$\mathbf{R} = \sum_{i=1}^{n} m_i\, \mathbf{a}_i = \left(\sum_{i=1}^{n} m_i \right) \mathbf{a} = m\mathbf{a},$$

where m is the mass of the body and **a** is the acceleration of each point in the body, including the mass center.

Let Q be any point on the line of action of the resultant of the

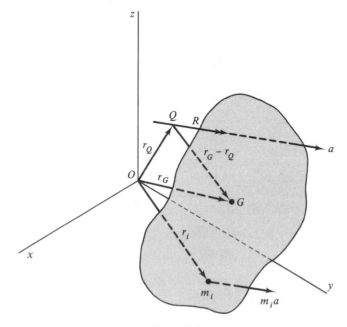

Figure 9-8

513

external forces, $\mathbf{R}$. The moment of $\mathbf{R}$ with respect to the origin O is

$$\mathbf{M}_O = \mathbf{r}_Q \times \mathbf{R} = \mathbf{r}_Q \times m\mathbf{a}. \tag{a}$$

The moment with respect to O must also be equal to the sum of the moments of the $m_i\mathbf{a}_i$ values; that is,

$$\mathbf{M}_O = \sum_{i=1}^{n} (\mathbf{r}_i \times m_i\mathbf{a}_i) = \sum_{i=1}^{n} (\mathbf{r}_i \times m_i\mathbf{a})$$

$$= \sum_{i=1}^{n} (m_i\mathbf{r}_i) \times \mathbf{a} = m\mathbf{r}_G \times \mathbf{a}. \tag{b}$$

When the expressions for the moment $\mathbf{M}_O$ in Eq. (a) and (b) are equated, the result is

$$\mathbf{r}_Q \times m\mathbf{a} = m\mathbf{r}_G \times \mathbf{a},$$

which reduces to

$$m(\mathbf{r}_G - \mathbf{r}_Q) \times \mathbf{a} = 0.$$

The last equation is satisfied if $\mathbf{r}_Q = \mathbf{r}_G$, in which case the point Q is at the mass center G, and $\mathbf{R}$ passes through G. The equation is also satisfied provided that $\mathbf{r}_Q - \mathbf{r}_G$ is parallel to $\mathbf{a}$. However, from Eq. (9-10), $\mathbf{R}$ has the same sense and slope as $\mathbf{a}$, which means that if $\mathbf{r}_G - \mathbf{r}_Q$ is parallel to $\mathbf{a}$, $\mathbf{R}$ must pass through G. Consequently, *if a rigid body has a motion of translation, the resultant of the external forces applied to the body must pass through the mass center of the body and the resultant moment of the external forces about the mass center must be zero.* The equations of motion for translation are

$$\sum \mathbf{F} = m\mathbf{a},$$
$$\sum \mathbf{M}_G = 0, \tag{9-11a}$$

which are equivalent to

$$\sum F_x = ma_x, \qquad \sum F_y = ma_y, \qquad \sum F_z = ma_z,$$
$$\sum M_{Gx} = 0, \qquad \sum M_{Gy} = 0, \qquad \sum M_{Gz} = 0. \tag{9-11b}$$

In many problems, a body may have coplanar translation, and if the external forces are all in the plane of motion, only the force components in the plane of motion and the moment equation about an axis perpendicular to the plane of motion are relevant. If the mass center moves in the xy plane, Eq. (9-11b) becomes

$$\sum F_x\mathbf{i} + \sum F_y\mathbf{j} = m(a_x\mathbf{i} + a_y\mathbf{j}),$$
$$\sum M_{Gz}\mathbf{k} = 0. \tag{9-12}$$

When a translating body has rectilinear motion, it is usually desirable to select either the x or y axis parallel to the acceleration, in which case the other acceleration component is zero. When a body has plane curvilinear translation, it is frequently convenient to use Eq.

(9-12) with the x and y axes replaced by n and t axes in the directions of the normal and tangential components of acceleration. The following examples illustrate the principles just discussed.

EXAMPLE 9-6

A man wants to slide the homogeneous 100-lb box in Fig. 9-9(a) across the floor by pushing on it with a force **P** as indicated. The coefficient of friction between the box and the floor is 0.20. Can an acceleration of 8.0 fps² to the right be produced without tipping the box? If so, determine the magnitude of the force **P**.

SOLUTION

Figure 9-9(b) is a free-body diagram of the box. Since the body has a motion of translation, the equations of motion are

$$\sum F_x \mathbf{i} + \sum F_y \mathbf{j} = m(a_{G_x}\mathbf{i} + a_{G_y}\mathbf{j}),$$

$$\sum M_{G_z}\mathbf{k} = \mathbf{0}.$$

Assume that the box will slide without tipping. A check for the validity of this assumption can be obtained by determining the distance x; if x is less than 1.5 ft, the box will not tip. The force **P** is large enough to cause the box to slip; therefore, the friction will equal the limiting value. That is,

$$F = F' = \mu N = 0.20N. \tag{a}$$

The equations of motion for the axes shown on the free-body diagram are

$$P\mathbf{i} - F\mathbf{i} + N\mathbf{j} - 100\mathbf{j} = \frac{100}{32.2}(8.0\mathbf{i} + 0\mathbf{j}) = 24.8\mathbf{i}$$

and

$$\sum M_{G_z}\mathbf{k} = xN\mathbf{k} - 2P\mathbf{k} - 2.5F\mathbf{k} = 0.$$

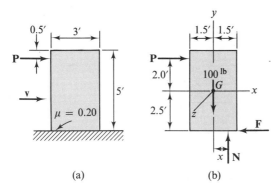

(a)   (b)

Figure 9-9

The scalar equations are

$$\mathbf{i}: \quad P - F = 24.8, \tag{b}$$

$$\mathbf{j}: \quad N - 100 = 0, \tag{c}$$

$$\mathbf{k}: \quad xN - 2P - 2.5F = 0. \tag{d}$$

Equations (a)–(d) can be used to determine three unknown forces and one unknown distance; thus all unknown values can be determined. The value for N is determined from Eq. (c) and substituted into Eq. (a) and (b). The results are

$$\mathbf{N} = 100\mathbf{j}\,\text{lb}, \qquad \mathbf{F} = -20\mathbf{i}\,\text{lb}, \qquad \mathbf{P} = \underline{44.8\mathbf{i}\,\text{lb}.} \quad \text{(Ans.)}$$

Substituting these values into Eq. (d), the value for x is found to be

$$x = 1.397\,\text{ft}.$$

Since x is less than 1.5 ft, tipping does not impend. If x were equal to 1.5 ft, the box would be on the verge of tipping. If x were more than 1.5 ft, the box would tip over before $\mathbf{P}$ increased to 44.8 lb, and the solution would not be valid.

EXAMPLE 9-7

The assembly in Fig. 9-10(a) has translation with the acceleration of the particles parallel to the x axis. The homogeneous 1600-lb body A is supported above body B by bearings C and D on a smooth rod. Cord EF, which is parallel to the xz plane, is subjected to a tensile force of 2000 lb. Determine the bearing reactions at C and D on body A and the acceleration of the assembly.

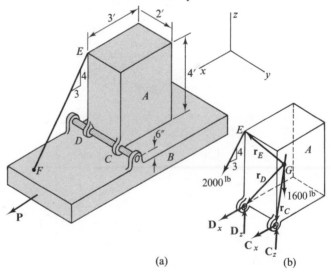

(a)   (b)

Figure 9-10

The weight of A and the cord tension are given. The reactions at C and D on A and the acceleration of A are required.

Figure 9-10(b) is a free-body diagram of body A. Since the supporting rod is smooth, there are no components of force in the y direction at C and D. Body A has translation, but not all forces are in one plane; therefore Eq. (9-11) are required. The equations are

$$\sum \mathbf{F} = m\mathbf{a}, \qquad \sum \mathbf{M}_G = 0.$$

There are five unknown quantities on the free-body diagram and in the equation of motion, namely $\mathbf{C}_x$, $\mathbf{C}_z$, $\mathbf{D}_x$, $\mathbf{D}_z$, and $\mathbf{a} = a_x\mathbf{i}$.

The two vector equations are, in general, equivalent to six algebraic equations. Since there are no forces or accelerations in the y direction, however, the component equation $\sum \mathbf{F}_y = m\mathbf{a}_y$ will not contribute any useful information; thus only five equations are available to determine the five unknowns.

The force in the cable is parallel to the xz plane, and it can be written as

$$\mathbf{T} = 2000(0.6\mathbf{i} - 0.8\mathbf{k}) = (1200\mathbf{i} - 1600\mathbf{k}) \text{ lb.}$$

The positive directions are indicated by the coordinate axes. The force equation gives

$$\sum \mathbf{F} = \mathbf{T} + \mathbf{C} + \mathbf{D} + \mathbf{W} = m(a_x)\mathbf{i},$$

which becomes

$$(1200\mathbf{i} - 1600\mathbf{k}) + (C_x\mathbf{i} + C_z\mathbf{k}) + (D_x\mathbf{i} + D_z\mathbf{k}) + (-1600\mathbf{k}) = \left(\frac{1600}{32.2}\right)(a_x\mathbf{i}).$$

In component form, the equation becomes

$$1200 + C_x + D_x = 49.7a_x \qquad \text{(a)}$$

and

$$-1600 + C_z + D_z - 1600 = 0. \qquad \text{(b)}$$

To write the moment equation of motion in vector form, the position vectors from G to points on the action lines of forces **C**, **D**, and **T** are required. Points C, D, and E are convenient and will be used. Since the block is homogeneous, G lies at the geometrical center of the block, and the position vectors are

$$\mathbf{r}_C = (1.5\mathbf{i} + \mathbf{j} - 2\mathbf{k}) \text{ ft}, \qquad \mathbf{r}_D = (1.5\mathbf{i} - \mathbf{j} - 2\mathbf{k}) \text{ ft},$$

$$\mathbf{r}_E = (1.5\mathbf{i} - \mathbf{j} + 2\mathbf{k}) \text{ ft}.$$

The moment equation is

$$\sum \mathbf{M}_G = \mathbf{r}_C \times \mathbf{F}_C + \mathbf{r}_D \times \mathbf{F}_D + \mathbf{r}_E \times \mathbf{T} = 0,$$

or

$$(1.5\mathbf{i} + \mathbf{j} - 2\mathbf{k}) \times (C_x\mathbf{i} + C_z\mathbf{k}) + (1.5\mathbf{i} - \mathbf{j} - 2\mathbf{k}) + (D_x\mathbf{i} + D_z\mathbf{k})$$

$$+ (1.5\mathbf{i} - \mathbf{j} + 2\mathbf{k}) \times (1200\mathbf{i} - 1600\mathbf{k}) = 0,$$

which expands to

$$[C_z\mathbf{i}-(2C_x+1.5C_z)\mathbf{j}-C_x\mathbf{k}]+[-D_z\mathbf{i}-(2D_x+1.5D_z)\mathbf{j}+D_x\mathbf{k}]$$
$$+[1600\mathbf{i}+(2400+2400)\mathbf{j}+1200\mathbf{k}]=0.$$

Since the $\mathbf{i}$, $\mathbf{j}$, and $\mathbf{k}$ components must each be zero, this is equivalent to the following three scalar equations:

$$\mathbf{i}: \quad C_z-D_z+1600=0, \tag{c}$$

$$\mathbf{j}: \quad -2(C_x+D_x)-1.5(C_z+D_z)+4800=0, \tag{d}$$

$$\mathbf{k}: \quad -C_x+D_x+1200=0. \tag{e}$$

Equations (b) and (c) may be solved simultaneously for the values of C_z and D_z:

$$C_z=+800, \qquad D_z=+2400.$$

Since, from Eq. (b), $C_z+D_z=3200$, Eq. (d) may be written as

$$-2(C_x+D_x)-1.5(3200)+4800=0$$

or

$$C_x+D_x=0. \tag{f}$$

The simultaneous solution of Eq. (e) and (f) gives

$$C_x=+600, \qquad D_x=-600.$$

The bearing reactions on body A, are, therefore,

$$\mathbf{F}_C=(600\mathbf{i}+800\mathbf{k})\ \text{lb}=\underline{1000(0.600\mathbf{i}+0.800\mathbf{k})\ \text{lb}} \quad \text{Ans.}$$

and

$$\mathbf{F}_D=(-600\mathbf{i}+2400\mathbf{k})\ \text{lb}=\underline{2470(-0.243\mathbf{i}+0.970\mathbf{k})\ \text{lb}.}$$
$$\text{Ans.}$$

The acceleration can be found from Eq. (a) as

$$1200+600-600=49.7a_x \quad \text{or} \quad a_x=24.15$$

from which

$$\mathbf{a}=\underline{24.2\mathbf{i}\ \text{fps}^2.} \qquad \text{Ans.}$$

PROBLEMS

9-50 The resultant of the force system acting on body E in Fig. P9-50 passes through the mass center G. The resultant moment of external forces about point H is found to be zero. What conclusion can be drawn from these facts concerning the motion of the body and the acceleration of point P?

*9-51 The 90,000-kg jet airliner of Fig. P9-51 is being stopped by brake action only. The engines are at idle and producing negligible thrust,

Figure P9-50

and the velocity of the aircraft is too slow to produce measurable aerodynamic lift or drag.

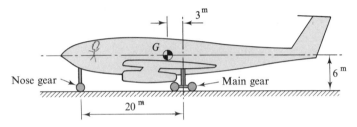

Figure P9-51

Only the main landing gear is equipped with brakes, and the coefficient of friction between the tires and runway is 0.7. If under these conditions the brakes are applied hard enough that the main wheels are on the verge of slipping, determine the vertical force in the nose wheel strut.

9-52 The uniform bar AB in Fig. P9-52 weighs 10.00 lb and is connected to two collars of negligible weight. The collars slide freely on parallel shafts which are in a vertical plane. Determine

(a) The acceleration of the bar.

(b) The reactions at A and B.

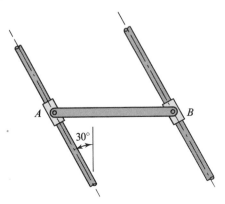

Figure P9-52

9-53 The homogeneous bar in Fig. P9-53 weighs 90.0 lb and slides on the smooth horizontal plane. It is connected to the 50.0-lb body B by a cord which passes over a smooth peg at C. Determine the force P which will result in translation of body A.

9-54 The crate in Fig. P9-54 weighs 100 lb and is lowered by three vertical cables connected at

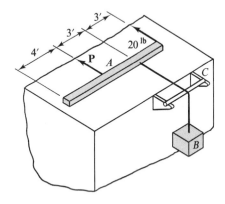

Figure P9-53

points A, B, and C. A constant tension of 30.0 lb is maintained in cable A. If the crate is to have translation, determine

(a) The tension required in the other two cables.

(b) The acceleration of the crate.

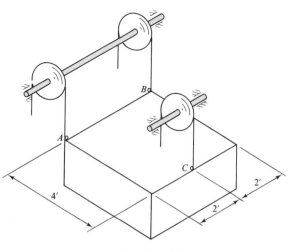

Figure P9-54

519

9-55 Determine the minimum accelerations of the blocks in Fig. P9-55 if A is not to move relative to B. The coefficient of friction between A and B is 0.20, and the horizontal plane is smooth.

Figure P9-55

Figure P9-56

9-56 The homogeneous cylindrical drum D in Fig. P9-56 is to be pushed along the horizontal surface by the bumper B. The drum weighs 42.0 lb, and the coefficients of static and kinetic friction between the drum and all surfaces are, respectively, $\mu_s = 0.60$ and $\mu_k = 0.50$. The drum is to have translation. Determine
(a) The minimum force P required.
(b) The resulting acceleration of the drum.

9-57 Bar BC of Fig. P9-57 weighs 161 lb and has a radius of gyration of mass with respect to a horizontal axis through G of 3.0 ft. The bar is supported by flexible cords AB and CD which rotate in the same vertical plane. The speed of the mass center, when in the position shown, is 12 fps. Determine the tension in cable AB for this position.

9-58 The homogeneous body in Fig. P9-58 weighs 0.50 lb-per ft³ and is supported by three 6-ft flexible cords attached to the body at A, B, and C. The body moves parallel to the yz plane and, in the position shown, has a velocity of 12.0 fps. Determine the tension in each of the cords and the acceleration of the body.

9-59 The 32.2-lb homogeneous rod BC is connected to the 96.6-lb frame D in Fig. P9-59 by the pin at C and the cord AB. Determine the maximum magnitude of P for which no relative motion will take place between BC and D.

Figure P9-57

Figure P9-58

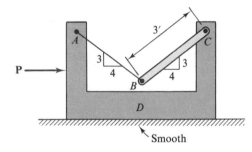

Figure P9-59

9-60 The sliding door in Fig. P9-60 weighs 900 lb. The mass and friction of the rollers may be considered negligible. Counterweight *B* weighs 100 lb and is connected to the door by a cable. Determine the acceleration of the door and the reactions of the track on the rollers.

9-61 Blocks *A* and *B* in Fig. P9-61 are connected by cable *CD*. Both blocks are symmetrical about the plane of motion, and cable *CD* and the force *P* are in the plane of motion.

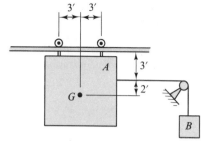

Figure P9-60

Block *A* weighs 6.0 lb, and block *B* weighs 5.0 lb. Both blocks are to move in translation. Determine the maximum allowable force *P*.

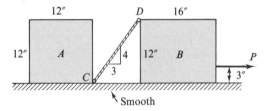

Figure P9-61

9-62 A long slender rod is connected to a block by a smooth pin at *A* in Fig. P9-62. The coefficient of friction between the block and the slot is 0.5. When the block is moving down the slot, determine the acceleration of the system and the corresponding angle θ between the rod and the vertical.

Figure P9-62

521

9-63 The uniform bars AB and CD in Fig. P9-63 weigh 64.4 and 128.8 lb, respectively. The 48.3-lb ball E is suspended by a cord as shown. The acceleration of F is gradually increased from zero to a constant value of 12 fps² to the right. For this acceleration, determine
(a) The angle θ.
(b) The components of the pin reaction at B on CD.

Figure P9-63

9-64 The homogeneous bar AB in Fig. P9-64 weighs 5 lb and is connected to the 25-lb block C by means of a smooth pin at B. Determine the components of the reaction at B on bar AB.

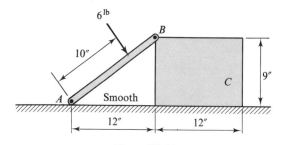

Figure P9-64

*9-65 Wheel C in Fig. P9-65 rolls along the plane without slipping. The homogeneous 8-kg bar AB is pinned to body C at A. The coefficient of friction between the bar at B and the plane is 0.6. The angular velocity and acceleration of wheel C at this instant are 4 rad per sec clockwise and 2 rad per sec² counterclockwise, respectively. Determine, for this instant, the components of the pin reaction at A on body AB.

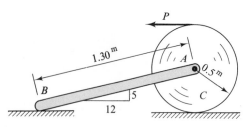

Figure P9-65

*9-66 The crate on the truck in Fig. P9-66 has a mass of m kg and is restrained from tipping by the cable. The coefficient of friction between the crate and truck is 0.30. Determine the maximum acceleration of the truck to the left if the crate is to remain upright. Would such an acceleration be reasonable?

*9-67 The truck in Problem 9-66 has a velocity of 50 km/hr to the left when the brakes are applied. Determine the shortest stopping distance if the crate is not to tip or slide into the truck cab.

9-68 The homogeneous wedge-shaped block in Fig. P9-68 has a mass of m units and is mounted on the frame E by means of smooth bearings at A and B and balls at C and D. A small part of the wedge is cut away at the bearings to eliminate axial thrust along the pins. The assembly is given an acceleration $a\mathbf{i}$ in the x direction. Determine the reactions of body E on the block.

9-69 The homogeneous bar AB in Fig. P9-69 is 25 in. long and weighs 32.2 lb. The carriage E weighs 60 lb and slides along a smooth horizontal plane. AB is connected to E by a ball and socket at B, a smooth surface at A, and a flexible cord from O to C. Point C is 10 in. from A (15 in. from B). The force $P = 30$ lb results in translation parallel to the xy plane. Determine all unknown forces acting on AB.

9-70 In Fig. P9-70 a 25-lb container of nitroglycerine is being moved in transporter B, which weighs 50 lb. The explosive is suspended inside the transporter by wires at C, D, and E and can be moved in translation only. The breaking strength of the wires is 90 lb. Determine the maximum force $\mathbf{P}$ that can be applied to the transporter.

Figure P9-66

Figure P9-68

Figure P9-70

9-71 In Fig. P9-71, block *A* weighs 10 lb, and block *B* weighs 15 lb. The constant force *P* has the maximum possible magnitude without causing *A* to tip.
(a) Determine the force *P*.
(b) If *B* has an initial velocity of 32.2 fps to the left, determine its velocity after moving a total distance of 50 ft.

Figure P9-69

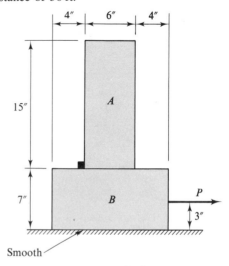

Figure P9-71

9-72 The 500-lb homogeneous block A in Fig. P9-72 is placed on a frame as shown. The support at C is smooth, and the coefficient of friction at D is 0.20. Members EF and GH rotate at a constant angular speed of 2 rad per sec. Determine the components of the reactions at C and D on A when the system is in the configuration shown.

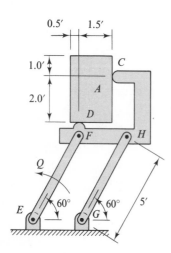

Figure P9-72

*9-73 Member BC in Fig. P9-73 has a mass of 4 kg, and A has a mass of 2 kg. The surface between A and BC is smooth. The system is held in equilibrium by the four cords. The angle θ is 30° when the horizontal cord at B is cut. Show that the tension in the cord D will become zero and determine the acceleration of A just after the cord is cut.

Figure P9-73

*9-74 Cord D in Problem 9-73 is removed, and the smooth surface between A and BC is replaced by a rough one. Determine the minimum coefficient of friction if A is not to slip on BC either before or after the horizontal cord is cut.

9-10 ROTATION OF A RIGID BODY

When a rigid body has a motion of rotation, all particles of the body travel in circular paths about the fixed axis of rotation. Figure 9-11(a) shows a rigid body rotating about the fixed Z axis. The XYZ coordinate system is fixed in space with its origin O at the intersection of the axis of rotation and the plane in which the mass center moves. The xyz coordinate system also has its origin at point O and is fixed in the body and rotates with it. The z and Z axes are collinear, and for convenience the x and y axes are so oriented that the x axis passes through G, the mass center of the body. The angular velocity $\boldsymbol{\omega}$ and the angular acceleration $\boldsymbol{\alpha}$ will lie along the z axis, and $\boldsymbol{\omega} = \omega \mathbf{k}$ and $\boldsymbol{\alpha} = \alpha \mathbf{k}$. The mass center travels on a circular path of radius r_G in the XY plane with a normal acceleration of $-r_G \omega^2 \mathbf{i}$ and a tangential acceleration of $r_G \alpha \mathbf{j}$. Equation (9-10), when applied to a rotating body, becomes

$$\mathbf{R} = \sum \mathbf{F} = m\mathbf{a}_G = m(a_{Gx}\mathbf{i} + a_{Gy}\mathbf{j}) \qquad (9\text{-}13a)$$

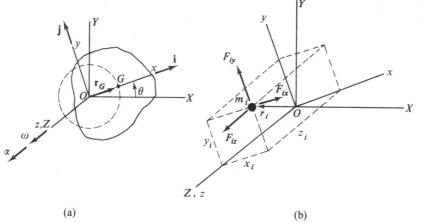

(a)

(b)

Figure 9-11

or in scalar form

$$\mathbf{i}: \quad \sum F_x = ma_{Gx} = m(-r_G\omega^2),$$

$$\mathbf{j}: \quad \sum F_y = ma_{Gy} = m(r_G\alpha), \qquad (9\text{-}13\text{b})$$

$$\mathbf{k}: \quad \sum F_z = 0.$$

Equations (9-13a) and (9-13b) were derived with the x axis through G. Sometimes, particularly for composite bodies, it is not convenient or necessary to locate the mass center of the body and place the x axis through it. The only requirement is that the directions of the components of $\mathbf{a}_G$ (or of $m\mathbf{a}_G$) for each component part of the body be consistent with the axes used to express the components of the forces.

The moment equation of motion for rotation can be developed by consideration of a particle of the rotating body as shown in Fig. 9-11(b). The position of the particle m_i relative to the x, y, z axes is $\mathbf{r}_i = x_i\mathbf{i} + y_i\mathbf{j} + z_i\mathbf{k}$. Since m_i is a particle of the rigid body, its acceleration is, from Eq. (8-22),

$$\mathbf{a}_i = \boldsymbol{\alpha} \times \mathbf{r}_i + \boldsymbol{\omega} \times (\boldsymbol{\omega} \times \mathbf{r}_i)$$

$$= (\alpha\mathbf{k}) \times (x_i\mathbf{i} + y_i\mathbf{j} + z_i\mathbf{k}) + (\omega\mathbf{k}) \times [\omega\mathbf{k} \times (x_i\mathbf{i} + y_i\mathbf{j} + z_i\mathbf{k})]$$

$$= -(x_i\omega^2 + y_i\alpha)\mathbf{i} + (x_i\alpha - y_i\omega^2)\mathbf{j}.$$

From Newton's second law, the resultant force (both internal and external) acting on the particle is

$$\mathbf{R}_i = m_i\mathbf{a}_i,$$

and the moment of the resultant force about point O is

$$\mathbf{M}_{Oi} = \mathbf{r}_i \times \mathbf{R}_i = \mathbf{r}_i \times m_i \mathbf{a}_i.$$

When $\mathbf{r}_i$ and $\mathbf{a}_i$ are expressed in component form the moment becomes

$$\mathbf{M}_{Oi} = (x_i \mathbf{i} + y_i \mathbf{j} + z_i \mathbf{k}) \times m_i [-(x_i \omega^2 + y_i \alpha) \mathbf{i} + (x_i \alpha - y_i \omega^2) \mathbf{j}]$$
$$= \mathbf{i} m_i (y_i z_i \omega^2 - x_i z_i \alpha) - \mathbf{j} m_i (x_i z_i \omega^2 + y_i z_i \alpha) + \mathbf{k} m_i (x_i^2 + y_i^2) \alpha.$$

When the moments about O of all the forces applied to all the particles of a body are added together, the moments of the internal forces are eliminated since they always occur in balancing pairs, and thus the sum of the moments about O of the external forces becomes

$$\sum \mathbf{M}_O = \sum_{i=1}^n \mathbf{M}_{Oi} = \mathbf{i} \sum_{i=1}^n m_i (y_i z_i \omega^2 - x_i z_i \alpha)$$

$$- \mathbf{j} \sum_{i=1}^n m_i (x_i z_i \omega^2 + y_i z_i \alpha) + \mathbf{k} \sum_{i=1}^n m_i (x_i^2 + y_i^2) \alpha.$$

For a rigid body with a continuous mass distribution, the particle mass m_i can be replaced by the mass dm of a small element of the body and the summation $\sum_{i=1}^n$ can be replaced by the integration $\int_m$, and the moment becomes

$$\sum \mathbf{M}_O = \int_m d\mathbf{M}_O = \mathbf{i} \left(\omega^2 \int_m yz \, dm - \alpha \int_m xz \, dm \right)$$

$$- \mathbf{j} \left(\omega^2 \int_m xz \, dm + \alpha \int_m yz \, dm \right) + \mathbf{k} \alpha \int_m (x^2 + y^2) \, dm.$$

The integrals represent products and moments of inertia of the mass of the body with respect to the xyz coordinate system which is fixed in the body. Using the notation developed in Chapter 6, the moment equation becomes

$$\sum \mathbf{M}_O = (\omega^2 I_{yz} - \alpha I_{xz}) \mathbf{i} - (\omega^2 I_{xz} + \alpha I_{yz}) \mathbf{j} + \alpha I_z \mathbf{k}, \qquad (9\text{-}14a)$$

which is equivalent to

$$\mathbf{i}: \quad \sum \mathbf{M}_{Ox} = (\omega^2 I_{yz} - \alpha I_{xz}),$$

$$\mathbf{j}: \quad \sum \mathbf{M}_{Oy} = -(\omega^2 I_{xz} + \alpha I_{yz}), \qquad (9\text{-}14b)$$

$$\mathbf{k}: \quad \sum \mathbf{M}_{Oz} = \alpha I_z.$$

Figure 9-11(a) was drawn with the x axis through the mass center. This limitation, however, was not involved in deriving Eq. (9-14), and the only restrictions on the coordinate system are that it be a right-hand system and that the z axis be the axis of rotation.

If the axis of rotation, the z axis, is parallel to an axis of revolution of a homogeneous body (the axis of revolution will pass through G) and if the xy plane is the plane of motion of the body, the products of inertia I_{xz} and I_{yz} will be zero. They are zero because $I_{xz'}$ and $I_{y'z'}$—where y' and z' are axes through G parallel to y and z, respectively—are zero due to symmetry, and the transfer terms are zero because z_G is zero.

If the xz plane is a plane of symmetry, $I_{yz} = 0$ as shown in Chapter 6, and if the yz plane is a plane of symmetry, $I_{xz} = 0$.

If the force system and body are both symmetrical about the xy plane, the equations for $\sum M_{Ox}$, $\sum M_{Oy}$, and $\sum F_z$ provide no useful information (both sides of the equations vanish), and Eq. (9-13b) and (9-14b) become

$$\sum F_x = ma_{Gx},$$

$$\sum F_y = ma_{Gy}, \tag{9-15}$$

$$\sum M_{Oz} = I_z\alpha.$$

When using Eq. (9-13), (9-14), and (9-15), a consistent sign convention is essential. In all derivations, the positive directions were the same as the positive directions of the coordinate axes. The vector representations for positive $\boldsymbol{\omega}$ and $\boldsymbol{\alpha}$ have the same direction as the z axis.

Because the normal and tangential components of the acceleration of the mass center of a rotating rigid body are generally the most convenient components to determine, it is usually easier to determine the normal and tangential components of the pin or axle reaction on the body than to determine the horizontal and vertical components. It is often convenient to show normal and tangential components of the pin reaction on the free-body diagram. In the preceding derivation, the x axis is the normal direction for the acceleration of the mass center, and the y axis is the tangential direction.

In Art. 9-8 it was shown that, for a body with translation, the resultant external force acts through the mass center of the body. When a rigid body has rotation, however, the resultant external force does not pass through the mass center. If the mass center is on the axis of rotation, the resultant of the external force system is a couple because $\mathbf{R}$ is zero from Eq. (9-13) and $\mathbf{M}$ is different from zero in Eq. (9-14) unless α and ω are zero.

The location of a point on the action line of the resultant external force, when it is a force, can be determined in the following manner. For simplicity, consider a body which is symmetrical about the plane of motion (the xy plane) and assume that the forces are also symmetrical with respect to this plane. Figure 9-12 represents the plane of motion of the mass center of the body, and the components of the resultant external force (including the reaction at O) are shown as $\mathbf{R}_x$ and $\mathbf{R}_y$ at point P. The moment of the resultant force about O is

$$\mathbf{M}_O = (q\mathbf{i}) \times (R_x\mathbf{i} + R_y\mathbf{j}) = qR_y\mathbf{k}.$$

Figure 9-12

From Eq. (9-13b), $R_y = mr_G\alpha$, and from Eq. (9-14b), $M_O = \alpha I_z$; therefore,

$$\alpha I_z = qmr_G\alpha$$

and

$$q = \frac{I_z}{mr_G}.$$

The parallel-axis theorem for I_z gives

$$I_z = (I_G)_z + mr_G^2 = mk_G^2 + mr_G^2,$$

where k_G is the radius of gyration of mass of the body about an axis through G parallel to the z axis. When this expression is substituted into the equation for q it becomes

$$q = \frac{mk_G^2 + mr_G^2}{mr_G} = \frac{k_G^2}{r_G} + r_G.$$

This equation shows that q is always more than r_G.

The point P at which the resultant of the external forces acting on a rotating rigid body (symmetrical about the xy plane) *intersects the x axis is called the center of percussion of the body.* Since the moment of the external forces about the center of percussion is zero, a force applied to the body through the center of percussion will not affect the tangential component of the reaction of the pin or axle at the axis of rotation. With ballistic pendulums, impact-testing machines, and similar apparatus, the load is usually applied to the body through the center of percussion to eliminate the bearing reaction due to the applied load. In baseball, the bat will sting the batter's hands unless the ball is hit at the center of percussion of the swinging bat.

EXAMPLE 9-8

The unbalanced wheel A in Fig. 9-13(a) weighs 64.4 lb and has a radius of gyration of mass relative to the horizontal axis of rotation

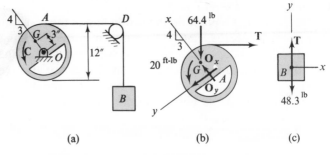

(a)   (b)   (c)

Figure 9-13

through O of 0.40 ft. The wheel is symmetrical about the plane of motion. The block B weighs 48.3 lb and is fastened to A by an inextensible cord which passes over the smooth drum at D and is wrapped around A. The moment of the couple $\mathbf{C}$ applied to A is 20.0 ft-lb counterclockwise. When A is in the position indicated, it has an angular velocity of 6.0 rad per sec clockwise. Determine the normal and tangential components of the reaction at O on body A. Neglect bearing friction at O.

SOLUTION

The free-body diagram of body A is shown in Fig. 9-13(b). The reaction at O is resolved into normal and tangential components along the x and y axes, respectively. Since the body is symmetrical about the plane of motion, Eq. (9-15) applies, and the three equations of motion for body A are

$$\sum F_x = ma_{Gx} = -mr_G\omega^2,$$
$$O_x - 0.60T - 0.80(64.4) = -2.0(\tfrac{3}{12})(-6.0)^2; \tag{a}$$

$$\sum F_y = ma_{Gy} = mr_G\alpha,$$
$$O_y - 0.80T + 0.60(64.4) = 2.0(\tfrac{3}{12})\alpha; \tag{b}$$

$$\sum M_{Oz} = I_z\alpha = mk_z^2\alpha,$$
$$20.0 - \tfrac{6}{12}T + 0.60(\tfrac{3}{12})(64.4) = 2.0(0.40)^2\alpha. \tag{c}$$

The three equations of motion for body A contain four unknowns, O_x, O_y, T, and α. An additional equation can be obtained from the free-body diagram of body B in Fig. 9-13(c). The equation of motion in the y direction is

$$\sum F_y = ma_y,$$
$$T - 48.3 = 1.5a_{By}. \tag{d}$$

This equation contains no additional unknown forces, and the unknown acceleration a_{By} can be expressed in terms of the angular acceleration of A. Noting the direction of the positive y axis for body B and the direction of $\boldsymbol{\alpha}$ for body A, this kinematical relationship becomes

$$a_{By} = \tfrac{6}{12}\alpha. \tag{e}$$

There are now five equations and five unknowns. When Eq. (e) is substituted into Eq. (d), Eq. (d) and (c) can be solved simultaneously for T, which is

$$T = 54.2 \text{ lb},$$

and for α, which is

$$\alpha = 7.93 \text{ rad per sec}^2.$$

These results can then be substituted into the first two equations to determine the components of the reactions. The results are

$$\mathbf{O}_x = \underline{66.1\mathbf{i}\,\mathrm{lb}} \qquad\qquad \text{Ans.}$$

and

$$\mathbf{O}_y = \underline{8.7\mathbf{j}\,\mathrm{lb.}} \qquad\qquad \text{Ans.}$$

EXAMPLE 9-9

The 36-lb thin, homogeneous, right triangular plate A in Fig. 9-14(a) rotates about the z axis with an angular velocity of $4\mathbf{k}$ rad per sec and an angular acceleration of $6\mathbf{k}$ rad per sec^2 in the position shown. The plate is mounted on a smooth, horizontal circular rod and supported by smooth bearings at B and C. Determine the torque $\mathbf{T}$ applied to the rod and the bearing reactions on the rod when the plate is in the given position. Neglect the weight of the circular rod.

SOLUTION

The angular velocity and acceleration of the plate, together with its weight and dimensions, are given. The couple $\mathbf{T}$ and the bearing reactions are required.

A free-body diagram of the plate and rod is shown in Fig. 9-14(b). Since the bearing rod is smooth, there are no z components of force at B and C.

Body A has rotation, but neither the body nor the force system is symmetrical with respect to the plane of motion (the xy plane), and although Eq. (9-15) are valid for this problem, they do not provide enough information to solve the problem. The general force and moment equations, Eq. (9-13) and (9-14), can be used to solve this problem. The equations are

$$\sum \mathbf{F} = m\mathbf{a}_G$$

and

$$\sum \mathbf{M}_O = (\omega^2 I_{yz} - \alpha I_{xz})\mathbf{i} - (\omega^2 I_{xz} + \alpha I_{yz})\mathbf{j} + \alpha I_z\mathbf{k}.$$

These equations involve four unknown force components and an unknown moment $\mathbf{T}$. The acceleration of G can be obtained from the given data and the principles of kinematics. The moment and products

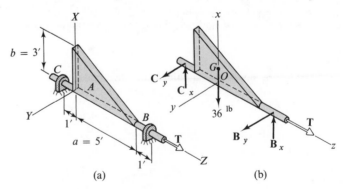

(a)   (b)

Figure 9-14

of inertia can be determined from the data, either by integration or from the table on p. 296 or Appendix B.

The two vector equations of motion are equivalent to six component equations. As previously noted, however, there are no forces in the z direction, and the force equation

$$\sum F_z = ma_{Gz}$$

will not provide any useful information; therefore, the two general motion equations are equivalent to only five scalar equations for this problem. Since there are only five unknown force and couple components, the analysis is complete.

The values of the moments and products of inertia are obtained from the parallel-axis theorem. Thus (see Chapter 6 or Appendix B)

$$I_z = I_{zG} + mx_G^2 = m(2b^2 + 3c^2)/36 + md^2$$

$$= \frac{36}{32.2}\left[\frac{2(3)^2 + 3(0)^2}{36}\right] + \frac{36}{32.2}(1)^2$$

$$= 1.677 \text{ slug-ft}^2,$$

$$I_{xz} = (I_{xz})_G + mx_G z_G = -mab/36 + m\left(\frac{a}{3}\right)\left(\frac{c}{2}\right)$$

$$= -\frac{36}{32.2}\left[\frac{5(3)}{36}\right] + 0$$

$$= -0.466 \text{ slug-ft}^2.$$

Since the xz plane is a plane of symmetry, I_{yz} is zero.

The acceleration of G is

$$\mathbf{a}_G = -r_G\omega^2\mathbf{i} + r_G\alpha\mathbf{j} = -(1)(4)^2\mathbf{i} + (1)(6)\mathbf{j}$$

$$= (-16\mathbf{i} + 6\mathbf{j}) \text{ fps}^2.$$

Equation (9-13a) becomes

$$\sum \mathbf{F} = m\mathbf{a}_G$$

$$B_x\mathbf{i} + B_y\mathbf{j} + C_x\mathbf{i} + C_y\mathbf{j} - 36\mathbf{i} = \frac{36}{32.2}(-16\mathbf{i} + 6\mathbf{j}),$$

which is equivalent to

$$\mathbf{i}: \quad B_x + C_x = 36 - \frac{36}{32.2}(16) = 18.11 \qquad \text{(a)}$$

and

$$\mathbf{j}: \quad B_y + C_y = 6.71. \qquad \text{(b)}$$

Equation (9-14a) for moments becomes

$$\sum \mathbf{M}_O = \mathbf{r}_B \times \mathbf{F}_B + \mathbf{r}_C \times \mathbf{F}_C + \mathbf{r}_G \times \mathbf{W} + \mathbf{T} = -\alpha I_{xz}\mathbf{i} - \omega^2 I_{xz}\mathbf{j} + \alpha I_z\mathbf{k},$$

or

$$(4.33\mathbf{k}) \times (B_x\mathbf{i} + B_y\mathbf{j}) + (-2.67\mathbf{k}) \times (C_x\mathbf{i} + C_y\mathbf{j}) + (1\mathbf{i}) \times (-36\mathbf{i}) + T\mathbf{k}$$

$$= -6(-0.466)\mathbf{i} - (4)^2(-0.466)\mathbf{j} + 6(1.677)\mathbf{k},$$

which, when expanded, is equivalent to

$$\mathbf{i}: \quad -4.33B_y + 2.67C_y = 2.80, \tag{c}$$

$$\mathbf{j}: \quad 4.33B_x - 2.67C_x = 7.46, \tag{d}$$

$$\mathbf{k}: \qquad\qquad T = 10.06. \tag{e}$$

Equations (a) and (d) can be solved simultaneously for B_x and C_x, Eq. (b) and (c) can be solved for B_y and C_y, and Eq. (e) gives T. The components are

$$B_x = 7.97 \text{ lb}, \qquad B_y = 2.16 \text{ lb},$$

$$C_x = 10.15 \text{ lb}, \qquad C_y = 4.55 \text{ lb},$$

and

$$T = 10.06 \text{ ft-lb.}$$

The couple and bearing reactions are

$$\mathbf{T} = \underline{10.06\mathbf{k} \text{ ft-lb,}} \qquad\qquad \text{Ans.}$$

$$\mathbf{F_B} = 7.97\mathbf{i} + 2.16\mathbf{j} = \underline{8.25(0.966\mathbf{i} + 0.262\mathbf{j}) \text{ lb,}} \qquad \text{Ans.}$$

and

$$\mathbf{F_C} = 10.15\mathbf{i} + 4.55\mathbf{j} = \underline{11.12(0.913\mathbf{i} + 0.409\mathbf{j}) \text{ lb.}} \quad \text{Ans.}$$

PROBLEMS

9-75 The slender, homogeneous rod in Fig. P9-75 is 3 ft long and weighs 2 lb. In the configuration shown, the rod rotates about point A with a counterclockwise angular velocity of 5 rad per sec. Determine
(a) The angular acceleration of the rod.
(b) The normal and tangential components of the reaction at A.

Figure P9-75

9-76 In Fig. P9-76, a 50-lb homogeneous drum is at rest against a small step at A. In an effort to push the drum over the step, a force of 20 lb is applied as shown. Determine
(a) The initial angular acceleration of the drum.
(b) The corresponding normal and tangential components of the reaction at A.

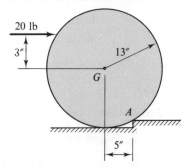

Figure P9-76

*9-77 The solid homogeneous cylinder in Fig. P9-77 has a mass of 30 kg and is rotating at 1200 rpm clockwise about a fixed horizontal axis through O. The coefficient of kinetic friction between the brake and the cylinder is 0.20. If the tension in the spring when the brake is applied is 100 N, determine the time required for the cylinder to stop rotating. Neglect the thickness of the vertical members.

Figure P9-77

9-78 Body A in Fig. P9-78 weighs 32.2 lb and is rotating about a horizontal axis at O with an angular velocity of 3.0 rad per sec counterclockwise when in the position indicated. The coefficient of friction between the brake drum and the brake BCD is 0.50, and body A has a radius of gyration of mass with respect to a horizontal axis through G of 6.0 in. Assume symmetry about the plane of motion. Determine the angular acceleration of A.

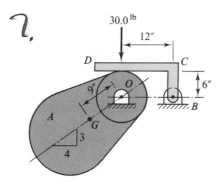

Figure P9-78

9-79 The chain-drive system in Fig. P9-79 consists of a 4-lb sprocket A and a 1-lb sprocket B. Sprocket A is driven by a 30-in-lb torque as indicated. Determine the angular acceleration of sprocket B.

Figure P9-79

9-80 The pinion, gear, and flywheel in Fig. P9-80 have moments of inertia about their axes of rotation of 0.030, 0.050, and 0.750 slug-ft², respectively. A torque of 120 ft-lb is applied to the pinion in the direction indicated. Determine the angular acceleration of the flywheel. Neglect bearing friction.

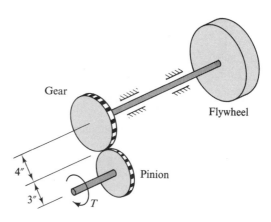

Figure P9-80

***9-81** The 5-kg block in Fig. P9-81 slides on a smooth horizontal surface. The pulley has a mass moment of inertia of 0.5 kg · m² about its axis of rotation, and there is no slipping between the pulley and the cable. Determine the angular acceleration of the pulley and the forces exerted by the cable on the blocks.

Figure P9-81

533

9-82 The 12-lb rigid bar in Fig. P9-82 rotates in the vertical plane about the fixed point O. In the position shown, the mass center G of the bar has a tangential acceleration of 7 ft per sec² down. Determine the radius of gyration of mass of the bar with respect to G.

Figure P9-82

9-83 The slender homogeneous rod in Fig. P9-83 is supported by a cord at A and a horizontal pin at B. The cord is suddenly cut. Determine, for the instant of release,
(a) The location of pin B which will result in the maximum angular acceleration of the rod.
(b) The resulting angular acceleration of the rod.

Figure P9-83

***9-84** The uniform homogeneous rocker arm ACE in Fig. P9-84 rotates about a smooth horizontal axis through C with an angular velocity of 8 rad/s counterclockwise and an angular acceleration of 24 rads/s² clockwise in the position shown. Part AC has a mass of 20 kg, and part CE has a mass of 30 kg. Determine, for this position, the force P and the components of the pin reaction at C.

Figure P9-84

9-85 The 15-lb bar in Fig. P9-85 has a radius of gyration of mass with respect to a horizontal axis through G of 3 in. and rotates about a horizontal axis at A. The 5-lb body C is restrained from sliding in the smooth slot by the cord CD. When AB is in the position shown, the tension in the cord is 4 lb. Determine the normal component of the reaction at A on the bar and the angular acceleration of AB.

Figure P9-85

9-86 Figure P9-86 represents the pendulum of an impact-testing machine used to measure the strength of the test specimen used as a cantilever beam. The rod AB weighs 20.0 lb, and each of the cylindrical disks weighs 60.0 lb. Determine the distance the striking edge should be placed from the center of the disks to eliminate any horizontal bearing reaction at the instant of impact.

Figure P9-86

9-87 The homogeneous cylinder in Fig. P9-87 has an initial angular velocity of 800 rpm counterclockwise. The coefficient of friction between the cylinder and the horizontal and vertical surfaces is 0.2. Determine the time required for the cylinder to come to rest.

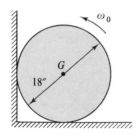

Figure P9-87

*9-88 The assembly in Fig. P9-88 rotates about a fixed horizontal axis through A. The solid homogeneous disk has a mass of 30 kg and is connected to the rigid A-frame ABC by the cord and a pin at C. When the assembly is in the position shown, it has an angular velocity of 2 rad per sec clockwise. The weight of the A-frame can be neglected. Determine the tension in the cord.

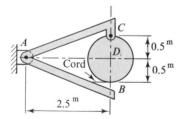

Figure P9-88

9-89 The center O of the homogeneous cylinder in Fig. P9-89 is restrained by a smooth slot.

Figure P9-89

The cylinder is at rest when it is lowered onto the belt, which has a constant velocity of 20 fps as shown. The coefficient of friction between the belt and the cylinder is 0.4. Determine
(a) The maximum angular velocity of the cylinder.
(b) The distance traveled by the belt during the time required for the cylinder to reach its maximum angular velocity.

9-90 The homogeneous cylinder A weighs 161 lb and rotates about a vertical axis as shown in Fig. P9-90. A small disk B weighing 16.1 lb, which may be considered a particle, is placed on the cylinder. The coefficient of static friction between the disk and cylinder is 0.50. The system is at rest when the 63-lb force is applied. Determine the angular velocity of the cylinder when the disk starts to slip.

Figure P9-90

9-91 Cylinder A in Fig. P9-91 weighs 8 lb and has an initial angular velocity of 300 rpm clockwise. The 2-lb cylinder B is initially at rest when

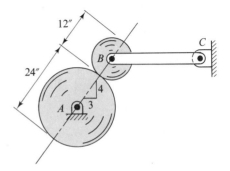

Figure P9-91

it is lowered onto cylinder A by the arm BC. The weight of arm BC is negligible, and the coefficient of friction between the cylinders is 0.2. Determine the final angular velocity of the cylinders and the contact time required to reach those velocities.

*9-92 The 20-kg cage A in Fig. P9-92 is raised and lowered by a cable and drum mounted on gear B. The drum and gear B have a mass of 30 kg and a radius of gyration of mass of 0.40 m with respect to the axis of rotation. The 15-kg gear C has a radius of gyration of mass of 0.20 m with respect to its axis of rotation. Determine the acceleration of A when the 120-N force is applied to C as shown.

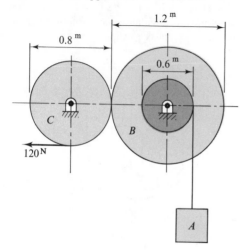

Figure P9-92

9-93 The unbalanced 32.2-lb cylinder in Fig. P9-93 has a radius of gyration of mass about a horizontal axis through G of 0.70 ft and rotates

about a parallel axis through E. In the position shown, the velocity of O is 2.0 fps downward. Determine

(a) The acceleration of point O.
(b) The normal component (along line EG) of the pin reaction at E on the cylinder.

9-94 The symmetrical body A in Fig. P9-94 weighs 644 lb, and body B weighs 64.4 lb. Body B has a velocity of 16 fps upward at the beginning of a certain time interval. During the time interval, body B travels 104 ft and has a displacement of 40 ft downward. Determine the radius of gyration of mass of body A with respect to its axis of rotation.

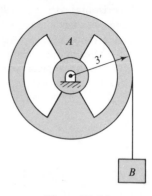

Figure P9-94

9-95 The drum in Fig. P9-95 weighs 322 lb and has a radius of gyration of mass with respect to a horizontal axis through G of 1.50 ft. Body A weighs 64.4 lb and has a velocity of 10.0 fps upward in the position indicated. Determine
(a) The acceleration of body A.
(b) The reaction of the axle on the drum.

Figure P9-93

Figure P9-95

9-96 Body A in Fig. P9-96 weighs 161 lb and has a radius of gyration of mass with respect to a horizontal axis through G of 3.0 ft. The homogeneous body B weighs 95 lb, and the coefficient of friction between A and B is 0.50. Body A has an angular velocity of 3.0 rad per sec clockwise in the position shown. Determine (for this position)

(a) The angular acceleration of body A.
(b) The reaction at O on body A.

Figure P9-96

9-97 The thin rod AB in Fig. P9-97 slides along the inside of the smooth fixed cylinder. The rod weighs 2 lb and in the configuration shown has an angular velocity of 5.0 rad per sec clockwise. Determine the angular acceleration of the rod and the reactions at A and B.

Figure P9-97

9-98 The slender homogeneous bar AB in Fig. P9-98 has a mass m and a length $\sqrt{2}l$ and is supported by three cords as shown. Determine the tensions in cords OA and OB just after cord C is cut.

9-99 The slender homogeneous bar in Fig. P9-99 weighs 20 lb and is held in equilibrium by three cords as shown. Determine the tensions in cords A and B just after cord C is cut.

Figure P9-98

Figure P9-99

9-100 The system in Fig. P9-100 consists of three slender homogeneous rods pinned at points A, B, and C to form an equilateral triangle. Each rod is 21 in. long and weighs 8 lb. Determine the reaction on the frame at A at the instant the system is released from rest with member AC aligned with the vertical direction.

Figure P9-100

9-101 A torque **T** of 24 in-lb applied as shown in Fig. P9-101 produces an angular acceleration of disk A of 4 rad per sec². Disk A and its power shaft have a weight of 18 lb and a radius of gyration of mass of 2.5 in. with respect to the geometric axis through its mass center. Disk B weighs 30 lb and has a radius of gyration of mass with respect to the vertical axis through its mass center of 4.0 in. Assume that the thrust bearing is frictionless and that A does not slip on B. Determine the radial distance x from the center of B to the point of contact of A on B.

Figure P9-101

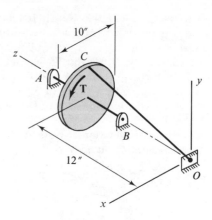

Figure P9-102

9-102 The system in Fig. P9-102 consists of a 9-lb slender homogeneous rod and a 25-lb homogeneous disk. The disk rotates in smooth bearings at A and B, and the rod rotates in a smooth ball and socket joint at O. The disk and rod are rigidly connected at C. A 24-in-lb torque **T** is applied to the disk as shown. Determine the angular acceleration of the system when in the indicated configuration.

9-104 The system in Fig. P9-104 consists of two slender vertical bars BC and AD connected by the slender horizontal rod AB. The vertical bars weigh 4 lb each and rotate in smooth bearings at C and D. The horizontal rod weighs 6 lb and is rigidly connected to the vertical bars. The 20-lb force applied to the horizontal rod at P lies in the xy plane. In the configuration shown,

Figure P9-103

9-103 Rod AB in Fig. P9-103 weighs 20 lb and rotates with its ends in horizontal smooth bearings. The torque **T** causes member OA to rotate about the z axis with a constant angular velocity of 10 rad per sec. Determine the bearing reactions on AB at A and B when in the position shown.

the angular velocity of the system is 10**k** rad per sec. Determine
(a) The angular acceleration of the system.
(b) The bearing reactions at C and D.

*9-105 The eccentric circular cone in Fig. P9-105 has a mass of 600 kg and is mounted on a horizontal shaft whose mass can be neglected.

Figure P9-104

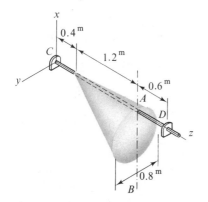

Figure P9-105

The cone has an angular velocity of $5\mathbf{k}$ rad/s and an angular acceleration of $-15\mathbf{k}$ rad/s² when in the position shown. Determine the bearing reactions at C and D and the couple which must be applied to the shaft.

9-106 Figure P9-106 represents a slender rod A and a small ball B, each weighing 0.60 lb. The bodies rotate about the vertical shaft and are supported by a smooth step or thrust bearing at

D (a bearing which can exert a force in the z direction) and by the cord C. Determine the tension in the cord and the bearing reaction when the angular velocity of the system is $20\mathbf{k}$ rad per sec.

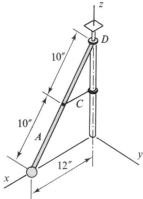

Figure P9-106

9-107 Rod AB and body C in Fig. P9-107 each have a mass m. The system is released from rest when $\theta = \theta_0$. Determine the distance x which will result in the maximum angular acceleration of rod AB and the resulting angular acceleration when $\theta_0 = 30°$

Figure P9-107

9-11 PLANE MOTION OF A RIGID BODY

When every particle of a moving rigid body remains a constant distance from a fixed reference plane, the body has plane motion. All lines in the body in, or parallel to, the plane of motion (the plane in which the mass center moves) have the same angular velocity and the same angular acceleration at any instant. The particles of a rigid body having plane motion all travel along plane curves, and the relations between the angular motion of the body and the accelerations of various particles of the body can be obtained from kinematics as indicated in Chapter 8.

The general equations of plane motion can be conveniently developed by selecting an XYZ coordinate system fixed in space and an xyz coordinate system fixed in, and moving with, the body with its origin at the mass center. The XY and xy planes are placed in the plane of motion as shown in Fig. 9-15. Vectors representing the angular velocity and the angular acceleration of the body will be parallel to the Z axis. The xyz coordinate system moves with the body, and in general the x and y axes will not be parallel to the X and Y axes. Since $\mathbf{R}_G$ is the position vector for the mass center, G, the equation of motion of the mass center is [see Eq. (9-10)]

$$\sum \mathbf{F} = m\mathbf{a}_G = m\ddot{\mathbf{R}}_G \qquad (9\text{-}16\text{a})$$

or

$$\sum F_x = ma_{Gx} = m\ddot{R}_{Gx},$$
$$\sum F_y = ma_{Gy} = m\ddot{R}_{Gy}, \qquad (9\text{-}16\text{b})$$
$$\sum F_z = ma_{Gz} = 0.$$

Figure 9-15

The angular velocity and angular acceleration of the x and y axes, as well as of any other line in the body parallel to the plane of motion, are $\omega\mathbf{k}$ and $\alpha\mathbf{k}$, respectively. The acceleration of any particle m_i in Fig. 9-15 can be expressed in terms of its position in the moving coordinate system (see Art. 8-21). The position of the particle relative to the xyz axes is $\mathbf{r}_i = x_i\mathbf{i} + y_i\mathbf{j} + z_i\mathbf{k}$, and since x, y, and z are constant for any given particle, $\mathbf{v}_{P/xyz}$ and $\mathbf{a}_{P/xyz}$ are zero. Thus, the acceleration of m_i, from Eq. (8-28), is

$$\mathbf{a}_i = \mathbf{a}_G + \boldsymbol{\alpha} \times \mathbf{r}_i + \boldsymbol{\omega} \times (\boldsymbol{\omega} \times \mathbf{r}_i)$$

$$= \mathbf{a}_G + (\alpha\mathbf{k}) \times (x_i\mathbf{i} + y_i\mathbf{j} + z_i\mathbf{k}) + (\omega\mathbf{k}) \times [(\omega\mathbf{k}) \times (x_i\mathbf{i} + y_i\mathbf{j} + z_i\mathbf{k})]$$

$$= \mathbf{a}_G - (\alpha y_i + \omega^2 x_i)\mathbf{i} + (\alpha x_i - \omega^2 y_i)\mathbf{j}.$$

The resultant force, both internal and external forces, applied to the particle is

$$\mathbf{F}_i = m_i\mathbf{a}_i,$$

and the moment of this resultant, with respect to the mass center of the

body, is

$$\mathbf{M}_{Gi} = \mathbf{r}_i \times \mathbf{F}_i = \mathbf{r}_i \times m_i \mathbf{a}_i$$

$$= (x_i \mathbf{i} + y_i \mathbf{j} + z_i \mathbf{k}) \times m_i [\mathbf{a}_G - (\alpha y_i + \omega^2 x_i)\mathbf{i} + (\alpha x_i - \omega^2 y_i)\mathbf{j}],$$

which, when expanded, becomes

$$\mathbf{M}_{Gi} = \mathbf{r}_i \times m_i \mathbf{a}_G + m_i(\omega^2 y_i z_i - \alpha x_i z_i)\mathbf{i}$$

$$- m_i(\omega^2 x_i z_i + \alpha y_i z_i)\mathbf{j} + m_i \alpha (x_i^2 + y_i^2)\mathbf{k}.$$

The resultant moment of the external force system with respect to the mass center is

$$\mathbf{M}_G = \sum_{i=1}^{n} \mathbf{M}_{Gi}$$

since the internal forces are eliminated in the summation. As in Art. 9-9, the mass of the particle m_i can be replaced by the mass of the element dm, and the summation can be replaced by integration for a rigid body with a continuous mass distribution. The resultant moment of the external force system about the center of mass then becomes

$$\mathbf{M}_G = \left(\int_m \mathbf{r}\, dm \right) \times \mathbf{a}_G + \omega^2 \mathbf{i} \int_m yz\, dm - \alpha \mathbf{i} \int_m xz\, dm$$

$$- \omega^2 \mathbf{j} \int_m xz\, dm - \alpha \mathbf{j} \int_m yz\, dm + \alpha \mathbf{k} \int_m (x^2 + y^2)\, dm.$$

The quantity $\int_m \mathbf{r}\, dm$ must be zero since $\mathbf{r}$ is measured from G [see Eq. (3-3b)]. The other integrals are moments and products of inertia with respect to the xyz coordinate system, and the moment equation reduces to

$$\mathbf{M}_G = (\omega^2 I_{yz} - \alpha I_{xz})\mathbf{i} - (\omega^2 I_{xz} + \alpha I_{yz})\mathbf{j} + \alpha I_z \mathbf{k}. \qquad (9\text{-}17a)$$

Equation (9-17a) yields the three scalar equations:

$$\mathbf{i}: \quad M_{Gx} = \omega^2 I_{yz} - \alpha I_{xz},$$

$$\mathbf{j}: \quad M_{Gy} = -\omega^2 I_{xz} - \alpha I_{yz}, \qquad (9\text{-}17b)$$

$$\mathbf{k}: \quad M_{Gz} = \alpha I_z.$$

It should be noted that in the preceding equations the moments and products of inertia are computed with respect to axes which have an origin at the mass center. Equations (9-17) are identical to Eq. (9-14) except that the moment center is on the axis of rotation for Eq. (9-14) and at the mass center for Eq. (9-17). Hence, plane motion of a rigid body can be analyzed in two steps: (1) the motion of the mass center and (2) the rotation of the body with respect to an axis through the mass center. The motion of the mass center can be analyzed by Newton's second law just as if the rigid body were a mass particle, and the rotation of the body about an axis through the mass center is treated as if the body were rotating about a fixed axis through G.

In many engineering problems involving plane motion, the body has an axis of revolution (the axis of revolution will pass through the mass center of a homogeneous body) or is symmetrical with respect to the plane of motion (the xy plane). In either case, I_{xz} and I_{yz} are zero, and Eq. (9-17b) reduce to

$$M_{Gx} = 0,$$

$$M_{Gy} = 0, \qquad\qquad (9\text{-}18)$$

$$M_{Gz} = \alpha I_z.$$

Furthermore, many plane motion problems are essentially two-dimensional in that the body is symmetrical with respect to the plane of motion and the applied forces can be assumed to be in this plane, in which case Eq. (9-16) and Eq. (9-18) reduce to

$$\sum F_x = ma_{Gx},$$

$$\sum F_y = ma_{Gy}, \qquad\qquad (9\text{-}19)$$

$$\sum M_{Gz} = I_z\alpha.$$

The other three equations yield no useful information.

In the preceding analysis, the origin of the moving coordinate system was placed at the mass center because the equations developed in this manner are in general the most useful. There are some problems, however, for which other points within the body may be used for the origin of the moving coordinate system and result in a considerable saving of effort.

Figure 9-16 shows a rigid body with plane motion similar to Fig. 9-15 except that the origin of the moving coordinate system is at some point A which does not coincide with the mass center G but does lie in the plane of motion of the mass center. The development of the moment equation of the body with respect to A proceeds as before, so that

$$\mathbf{M}_{Ai} = \mathbf{r}_i \times m_i \mathbf{a}_A + m_i(\omega^2 y_i z_i - \alpha x_i z_i)\mathbf{i}$$

$$-m_i(\omega^2 x_i z_i + \alpha y_i z_i)\mathbf{j} + m_i\alpha(x_i^2 + y_i^2)\mathbf{k}.$$

From Fig. 9-16,

$$\mathbf{r}_i = \mathbf{r}_G + \boldsymbol{\rho}_i,$$

and the first term in the moment equation can be written as

$$\mathbf{r}_i \times m_i \mathbf{a}_A = \mathbf{r}_G \times m_i \mathbf{a}_A + \boldsymbol{\rho}_i \times m_i \mathbf{a}_A.$$

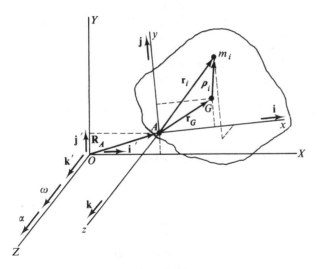

Figure 9-16

The summation of moments about A gives

$$\mathbf{M}_A = \sum_{i=1}^{n} \mathbf{M}_{Ai} = \sum_{i=1}^{n} \mathbf{r}_G \times m_i \mathbf{a}_A + \sum_{i=1}^{n} \boldsymbol{\rho}_i \times m_i \mathbf{a}_A$$

$$+ \sum_{i=1}^{n} m_i(\omega^2 y_i z_i - \alpha x_i z_i)\,\mathbf{i}$$

$$- \sum_{i=1}^{n} m_i(\omega^2 x_i z_i + \alpha y_i z_i)\mathbf{j} + \sum_{i=1}^{n} m_i\alpha(x_i^2 + y_i^2)\mathbf{k}.$$

Upon replacing the summation by integration, the first term of the preceding equation becomes

$$\int_m \mathbf{r}_G \times \mathbf{a}_A \, dm = \mathbf{r}_G \times \mathbf{a}_A \int_m \, dm = (\mathbf{r}_G \times \mathbf{a}_A)m.$$

Similarly, the second term becomes

$$\int_m \boldsymbol{\rho} \times \mathbf{a}_A \, dm = \int_m \boldsymbol{\rho} \, dm \times \mathbf{a}_A,$$

which must be zero as in the previous discussion, since $\int \boldsymbol{\rho} \, dm$ is the first moment of the mass with respect to G. The last three terms produce moment of inertia terms similar to those in Eq. (9-17a), and the moment of the external forces about point A becomes

$$\mathbf{M}_A = (\mathbf{r}_G \times \mathbf{a}_A)m + (\omega^2 I_{yz} - \alpha I_{xz})\mathbf{i} - (\omega^2 I_{xz} + \alpha I_{yz})\mathbf{j}$$

$$- \alpha I_z \mathbf{k}. \tag{9-20}$$

Note that in Eq. (9-20) the moment of inertia terms are computed with respect to axes which have an origin at A, whereas in Eq. (9-17) the origin of the axes is at G. Equation (9-20) reduces to the form of Eq. (9-17a) if the first term on the right is zero and this will occur if

1. $\mathbf{r}_G = 0$, in which case A coincides with G.

2. $\mathbf{a}_A = 0$, which corresponds to the rotation about a fixed axis (see Art. 9-9).

3. $\mathbf{r}_G \times \mathbf{a}_A = 0$, which implies that $\mathbf{r}_G$ and $\mathbf{a}_A$ are parallel.

The third case provides the only new information. Equation (9-20) reduces to the same form as Eq. (9-17a), provided the acceleration of the moment center is along the line AG. For example, when a wheel is rolling without slipping along a fixed plane, the acceleration of the point of contact (the instantaneous center of zero velocity) is normal to the plane, and if the mass center is on the line through the point of contact normal to the plane, Eq. (9-17) can be applied with respect to an axis through the instantaneous center. It should be noted, however, that this procedure is valid only when the acceleration of the moment axis is toward or away from the mass center.

When any of the equations of motion for plane motion is used, a consistent sign convention must be followed. In the derivations, the positive directions for velocities, accelerations, and forces were the positive directions of the coordinate axes. Any given quantity in the opposite direction must be used as a negative number in the equations of motion. When two or more bodies are involved in a single problem, different orientations of the axes and different sign conventions can be used for different bodies, provided related kinematical equations are properly written.

In problems involving wheels, cylinders, and similar bodies rolling on a plane, the data may not always indicate whether or not the body slides as it rolls along the plane. There are three possibilities for problems of this type: (1) The problem may state that the wheel *rolls without slipping* or that *slipping impends*. (2) The problem may state that the body *rolls and slips*. (3) The problem may not specify which of these conditions exists. *The method of solution for each of these situations is discussed assuming that the body and force system are symmetrical about the plane of motion and that the mass center is at the geometrical center of the wheel.* A similar analysis applies if the mass center of the wheel is not at the geometrical center except that the kinematic relationships are somewhat more involved. Note that Eq. (9-19) are the only equations needed for the problem as stated.

1. When the problem states that the body *rolls without slipping* the acceleration of the mass center is $a_G = r\alpha$—see Art. 8-19. This equation of kinematics can be used with the equations of motion to solve the problem. The maximum friction ($F' = \mu N$) is usually *not* developed in this type of problem. The friction is a force unknown in magnitude and direction and can be determined from the equations of

motion. A special case of this type of problem is illustrated in Example
9-10 in which *slipping impends*. In this case, both $F = F' = \mu N$ and $a_G = r\alpha$ are valid.

2. When the problem states that the body *rolls and slips*, the friction force must equal its limiting value of μN. The direction of the velocity of the point of contact must be determined from the velocity data in order to show the friction with the correct sense (opposite to this velocity). *Note that a_G does not equal $r\alpha$.*

3. When the problem does not state that one of these conditions exists (that is, no slipping, slipping impends, or rolls and slips), one of the conditions (usually no slipping) must be assumed, and a check made to indicate the validity of the assumption. The following steps are suggested:

a. Assume that no slipping occurs (write "assume no slip" as part of the solution) and use $a_G = r\alpha$.
b. Solve the equations of motion for the friction and normal forces. This friction is the force which must be developed to prevent slipping.
c. Compute the maximum friction available from $F' = \mu N$.
d. Check the assumption; that is, compare F and F'.
   If $F < F'$, assumption is correct and $a_G = r\alpha$.
   If $F = F'$, slipping impends, assumption is correct, and $a_G = r\alpha$.
   If $F > F'$ the body rolls and slips and a_G *is not equal to* $r\alpha$.
e. If the check indicates that the body rolls and slips, the preceding solution is not valid. The friction force is equal to F' in the direction of F found in step b. Values of a_G and α can be determined from the equations of motion.

EXAMPLE 9-10

Body A in Fig. 9-17(a) is a 100-lb solid homogeneous cylinder with a narrow slot cut in it as indicated by the dashed circle. The effect of the slot on the moment of inertia of the cylinder can be neglected. The coefficient of friction between the cylinder and plane is 0.40. Block B is connected to A by the flexible cord which passes over the smooth drum and is wrapped around the cylinder in the slot. Determine the maximum weight B can have if the cylinder is to roll without slipping along the inclined plane. The point on A in contact with the plane has impending motion down the plane.

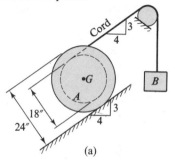

(a)

Figure 9-17

545

SOLUTION

The free body diagrams for the two bodies are shown in Fig. 9-17(b) and (c). For impending motion, the frictional force **F** opposes the motion or impending motion of the point of contact; therefore, the friction on A is up the plane. Body A has plane motion and body B has translation. The general equations of motion for A are

$$\sum F_x = ma_{Gx}, \qquad \sum F_y = ma_{Gy}, \qquad \sum M_{Gz} = I_z\alpha,$$

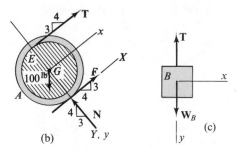

Figure 9-17 (*Continued*)

and the equation of motion for B is

$$\sum F_y = ma_{By}.$$

The x, y, and z axes for body A are selected parallel and perpendicular to the plane as shown, and a_{Gy} is zero. The axes for Fig. 9-17(b) are selected as shown so that a positive angular acceleration, $\alpha\mathbf{k}$, will result in a positive acceleration of G, $a_{Gx}\mathbf{i}$. In the two free-body diagrams there are four unknown forces ($\mathbf{W}_B$, $\mathbf{T}$, $\mathbf{F}$, and $\mathbf{N}$), and the four motion equations contain three unknown acelerations (a_{Gx}, α, and a_{By}), making seven unknowns and four equations of motion. Three additional relationships must be obtained from friction and kinematics.

Since slipping impends, the magnitude of the frictional force is

$$F = F' = 0.40N.$$

The wheel does not slip; therefore, from kinematics, the acceleration of G is

$$a_{Gx}\mathbf{i} = r\alpha\mathbf{i} = \alpha\mathbf{i}.$$

Note: $\alpha = \alpha\mathbf{k}$ is assumed clockwise and $(\mathbf{a}_G)_x$ is assumed up the plane. The x component of the acceleration of E (equal in magnitude to the acceleration of B) can be expressed in terms of the angular acceleration of A by relative motion. Thus, with α assumed to be clockwise,

$$\mathbf{a}_E \qquad = \qquad \mathbf{a}_G \qquad + \qquad \mathbf{a}_{E/G}$$

from which

$$(a_E)_x = (a_B)_y = +1.00\alpha + 0.75\alpha = +1.75\alpha,$$

where $(\mathbf{a}_E)_x$ is up the plane and $(\mathbf{a}_B)_y$ is downward.

When specific values are substituted in the equations of motion for A, they become

$$\sum F_x = T + F - \tfrac{3}{5}(100) = \frac{100}{32.2} a_{Gx} = \frac{100}{32.2} \alpha,$$

$$\sum F_y = \tfrac{4}{5}(100) - N = \frac{100}{32.2} a_{Gy} = 0,$$

and

$$\sum M_{Gz} = 0.75T - 1.00F$$

$$= \frac{1}{2}\frac{100}{32.2}(1.00)^2 \alpha.$$

Likewise, for B,

$$\sum F_y = W_B - T = \frac{W_B}{32.2} a_{By}$$

$$= \frac{W_B}{32.2} 1.75\alpha,$$

and from friction,

$$F = 0.40N.$$

Solution of these equations gives

$$W_B = \underline{313 \text{ lb.}} \qquad\qquad \text{Ans.}$$

EXAMPLE 9-11

The wheel in Fig. 9-18(a) weighs 96.6 lb and rolls along the horizontal plane. The coefficient of friction between the wheel and plane is 0.20. The radius of gyration of the mass of the wheel with respect to its geometric axis is 1.30 ft. Determine the acceleration of the center of the wheel and the angular acceleration of the wheel.

SOLUTION

The free-body diagram of the wheel is drawn in Fig. 9-18(b). *Assume that the wheel does not slip* since the statement of the problem did not specify either slipping or no slipping. The positive axes are shown on the diagram, and the accelerations are assumed to be in the positive

(a)

Figure 9-18

547

(b)

Figure 9-18 (*Continued*)

directions of the axes. The equations of motion are

$$\sum F_x = 40 - 30 - F = ma_{Gx} = 3.0a_{Gx}, \qquad (a)$$

$$\sum F_y = 96.6 - N = ma_{Gy} = 0, \qquad (b)$$

and

$$\sum M_{Gz} = 1.5F - 1.5(30) = I_{Gz}\alpha = 3.0(1.30)^2\alpha. \quad (c)$$

Since rolling without slipping is assumed, the accelerations are related by

$$a_{Gx} = r\alpha = 1.5\alpha.$$

Note: This equation is valid only for the given orientation of the axes. If y had been chosen upward and z to the front, a_{Gx} would equal $-r\alpha$ for a right-hand system of axes.

Simultaneous solution of these equations gives

$F = 21.42$ lb	$\mathbf{F} = -21.42\mathbf{i}$ lb,
$N = 96.6$ lb,	$\mathbf{N} = -96.6\mathbf{j}$ lb,
$\alpha = -2.54$ rad per sec²,	$\boldsymbol{\alpha} = -2.54\mathbf{k}$ rad per sec²,
$a_G = 1.5(-2.54) = -3.81$ fps²,	$\mathbf{a}_G = -3.81\mathbf{i}$ fps².

The force $F = 21.42$ lb is the friction required to prevent slipping. The maximum friction which can be developed is

$$F' = \mu N = 0.20(96.6) = 19.32 \text{ lb}.$$

Since F' is less than F, the assumption that the wheel does not slip is wrong, and the solution is incorrect. The value obtained for N is correct, however, since the equation for N did not involve the assumption. When $F = F' = 19.32$ lb is substituted into Eq. (a) and (c) they become

$$\sum F_x = 40 - 30.0 - 19.32 = 3.00a_{Gx}$$

and

$$\sum M_{Gz} = 1.50(19.32) - 1.50(30) = 5.07\alpha.$$

Solution of these equations gives

$$a_{Gx} = -3.11 \text{ fps}^2, \qquad \underline{\mathbf{a}_{Gz} = -3.11\mathbf{i} \text{ fps}^2}, \qquad \text{Ans.}$$

and

$$\alpha = -3.16 \text{ rad per sec}^2, \qquad \underline{\boldsymbol{\alpha} = -3.16\mathbf{k} \text{ rad per sec}^2}. \quad \text{Ans.}$$

EXAMPLE 9-12

Bar AB in Fig. 9-19(a) is 5.0 m long and has a mass of 40 kg. The ends of the rod are connected to rollers of negligible mass which move in the smooth slots as indicated. When the bar is in the position shown, the

Figure 9-19

moment of the couple C is $150\,\text{N}\cdot\text{m}$ counterclockwise, and the velocity of A is 3.5 m/s down the slot. Determine

(a) the angular acceleration of AB.

(b) The reactions on the bar at A and B.

SOLUTION

Figure 9-19(b) is a free-body diagram of the bar. The body is symmetrical about the plane of motion and the forces are all in this plane; therefore, Eq. (9-19) are the only motion equations which provide useful information. The motion equations are

$$\sum F_x = ma_{Gx}, \qquad \sum F_y = ma_{Gy}, \qquad \sum M_{Gz} = I_{Gz}\alpha.$$

The free-body diagram has two unknown forces, and the three equations of motion contain two unknown linear acceleration components and the unknown angular acceleration of the rod. It is therefore necessary to obtain two additional equations relating some of the five unknown quantities before the problem can be solved.

The ends of the bar are constrained to move along the slots, and thus they have rectilinear motion with accelerations and velocities along the slots. With this information, it is possible to express the acceleration of the mass center in terms of the angular velocity and angular acceleration of the bar. The angular velocity of the bar can be obtained from the instantaneous center as shown in Fig. 9-19(c), and the acceleration terms can be related by means of relative acceleration equations. The velocity of A is 3.5 m/s downward, and from Fig. 9-19(c) it is also equal to 1.75ω; thus

$$1.75\omega = 3.5$$

from which $\omega = 2\,\text{rad/s}$ and $\boldsymbol{\omega} = 2\,\text{rad/s}$. The angular acceleration is assumed to be clockwise, and the accelerations of A and B are related by the equation

$$\mathbf{a}_B = \mathbf{a}_A + \mathbf{a}_{B/A}.$$

549

When horizontal components are equated the equation becomes

$$\tfrac{4}{5}a_B = 0 + \tfrac{3}{5}(5\alpha) - \tfrac{4}{5}(5\omega^2),$$

$$a_B = 3.75\alpha - 5(2)^2 = 3.75\alpha - 20,$$

and

$$\mathbf{a}_B = (3.75\alpha - 20).$$

The relative acceleration equation for points G and B is

$$\mathbf{a}_G = \mathbf{a}_B \qquad\qquad + \mathbf{a}_{G/B}$$

$$= (3.75\alpha - 20) +$$

$$= \overleftarrow{3.0}\alpha + \overrightarrow{2.25}\alpha + \overrightarrow{16} + \overrightarrow{12} + \overrightarrow{1.5}\alpha + \overrightarrow{2\alpha} + \overleftarrow{8} + \overrightarrow{6}$$

$$= (\overleftarrow{1.5}\alpha - 8) + (\overrightarrow{0.25}\alpha - 18).$$

When this expression for $\mathbf{a}_G$ is substituted into the force equations of motion they become

$$\sum F_x = ma_{Gx},$$

$$R_A - 0.6R_B = 40(1.5\alpha - 8) = 60\alpha - 320; \qquad\qquad (a)$$

$$\sum F_y = ma_{Gy},$$

$$0.8R_B - 40(9.81) = 40(0.25\alpha - 18) = 10\alpha - 720. \qquad\qquad (b)$$

The moment equation can be written conveniently by resolving $\mathbf{R}_B$ into components and applying the principle of moments; thus

$$\sum M_{Gz} = I_{Gz}\alpha.$$

$$0.8R_B(2.0) - 0.6R_B(1.5) - R_A(1.5) - 150 = \tfrac{1}{12}(40)(5)^2\alpha$$

or

$$0.7R_B - 1.5R_A = 83.3\alpha + 150. \qquad\qquad (c)$$

When Eq. (a)–(c) are solved simultaneously the results are

$$R_A = -408\text{N}$$

$$R_B = -380\text{N}$$

$$\alpha = 2.34 \text{ rad/s}^2.$$

The negative signs indicate that $\mathbf{R}_A$ and $\mathbf{R}_B$ were assumed to be In the wrong direction. The results are

$$\mathbf{R}_A = \underline{408\text{N} \rightarrow}, \qquad\qquad \text{Ans.}$$

$$\mathbf{R}_B = \underline{380\text{N}}\,, \qquad\qquad \text{Ans.}$$

$$\alpha = \underline{2.34 \text{ rad/s}^2}. \qquad\qquad \text{Ans.}$$

EXAMPLE 9-13

Two 2-ft-diameter homogeneous disks weighing 96.6 lb each are connected by a 4-ft rod of negligible weight. Welded to this rod are two homogeneous 1- by 2-ft rectangular plates as shown in Fig. 9-20(a). Each plate weighs 48.3 lb. The assembly rolls without slipping along a straight path on a horizontal plane. When in the position indicated, the angular velocity is 10 rad per sec as shown, and the magnitude of the couple **C** is such that slipping impends. The coefficient of friction between the wheels and plane is 0.25. Determine the angular acceleration and the couple **C**.

SOLUTION

Figure 9-20(b) is a free-body diagram of the assembly showing the orientation of the moving xyz coordinate system with its origin at the mass center, G. The friction forces are assumed to have moments about the z axis which oppose **C**. Due to symmetry, G lies midway between the disks on the center line of the rod, and the total weight of the assembly is

$$W = 2(96.6) + 2(48.3) = 289.8 \text{ lb.}$$

The body is not symmetrical about the plane of motion, and the forces do not all lie in this plane; therefore, the general moment equation, [Eq. (9-17a)] will be needed, and it will be necessary to obtain values for I_{xz}, I_{yz}, and I_z.

The body is symmetrical with respect to the yz plane; therefore, I_{xz} must be zero. The xz plane is a plane of symmetry for the two disks, so I_{yz} for the disks is also zero. The product of inertia of each plate with respect to planes of symmetry is zero (see Art. 6-10 or Appendix B), but the transfer terms are different from zero, and I_{yz} becomes

$$I_{yz} = \sum [(I_{yz})_G + m y_G z_G]$$

$$= 0 + \frac{48.3}{32.2}\left(-\frac{1}{2}\right)(1) + 0 + \frac{48.3}{32.2}\left(\frac{1}{2}\right)(-1)$$

$$= -1.500 \text{ slug-ft}^2.$$

(a)   (b)

Figure 9-20

The value of I_z is also obtained by the parallel axis theorem (see the table in Art. 6-14 or Appendix B).

The moment of inertia is

$$I_z = \sum (I_z)_{\text{disks}} + \sum [(I_z)_G + md^2]_{\text{plates}}$$

$$= 2\left[\frac{1}{2}\frac{96.6}{32.2}(1.00)^2\right] + 2\left[\frac{1}{12}\frac{48.3}{32.2}(1.00)^2 + \frac{48.3}{32.2}(0.50)^2\right]$$

$$= 4.00 \text{ slug-ft}^2.$$

The force equation of motion is

$$\sum \mathbf{F} = m\mathbf{a}_G,$$

which, from Fig. 9-20(b), becomes

$$-F_1\mathbf{i} - F_2\mathbf{i} + N_1\mathbf{j} + N_2\mathbf{j} - 289.8\mathbf{j} = \frac{289.8}{32.2}\mathbf{a}_G.$$

The moments of the forces about G can be obtained by vector multiplication, but in this example the moments can be determined more readily by inspection of Fig. 9-20(b). The equation is

$$\sum \mathbf{M}_G = (\omega^2 I_{yz} - \alpha I_{xz})\mathbf{i} - (\omega^2 I_{xz} + \alpha I_{yz})\mathbf{j} + \alpha I_z \mathbf{k},$$

which becomes

$$(2N_2 - 2N_1)\mathbf{i} + (2F_2 - 2F_1)\mathbf{j} + (C - 1F_1 - 1F_2)\mathbf{k}$$
$$= (10)^2(-1.500)\mathbf{i} - \alpha(-1.500)\mathbf{j} + \alpha(4.00)\mathbf{k}.$$

The two general equations of motion are equivalent to five algebraic equations (there are no forces or acceleration component in the z direction). There are five unknown forces and couples in addition to the linear acceleration of G and the angular acceleration of the body. The problem stated that the wheels roll without slipping; therefore,

$$\mathbf{a}_G = r\alpha(-\mathbf{i}) = -\alpha\mathbf{i}.$$

The algebraic equations of motion are

$$-F_1 - F_2 = 9a_G = -9\alpha, \qquad \text{(a)}$$

$$N_1 + N_2 - 289.8 = 0, \qquad \text{(b)}$$

$$2N_2 - 2N_1 = -150, \qquad \text{(c)}$$

$$2F_2 - 2F_1 = 1.500\alpha, \qquad \text{(d)}$$

and

$$C - F_1 - F_2 = 4.00\alpha \qquad \text{(e)}$$

These five equations contain six unknowns, and one additional equation can be obtained from friction, according to the statement of the problem. As C is increased, F_1 and F_2 will also increase until one of them is equal to the limiting value. Since there are two possibilities, an assumption must be made and checked. If F_1 is assumed to equal μN_1, a simultaneous solution of the equations gives

$$N_1 = 182.4 \text{ lb},$$

$$N_2 = 107.4 \text{ lb},$$

$$F_1 = \mu N_1 = 0.25(182.4) = 45.6 \text{ lb},$$

and

$$F_2 = 53.9 \text{ lb}.$$

Since F_2 is greater than μN_2, which is $0.25(107.4)$ or 26.8 lb, the right-hand wheel will slip first, and the problem must be solved assuming that F_2 is equal to μN_2. This solution gives

$$N_1 = 182.4 \text{ lb}, \qquad N_2 = 107.4 \text{ lb},$$

$$F_2 = \mu N_2 = 26.85 \text{ lb}$$

$$F_1 = 22.72 \text{ lb},$$

$$\alpha = 5.51 \text{ rad/sec}^2,$$

and

$$C = 71.6 \text{ ft-lb}.$$

Since F_1 is less than μN_1, this is the correct assumption, and slipping impends at the right wheel. The results are

$$\alpha = 5.51\mathbf{k} \text{ rad per sec}^2, \qquad\qquad \text{Ans.}$$

and

$$\mathbf{C} = 71.6\mathbf{k} \text{ ft-lb.} \qquad\qquad \text{Ans.}$$

PROBLEMS

9-108 The 16.1-lb wheel in Fig. P9-108 is a homogeneous cylinder with a narrow slot in which the cord is wrapped. Neglect the effect of the slot on the moment of inertia of the mass of the wheel. The wheel rolls without slipping on the horizontal plane. Determine
(a) The acceleration of G.
(b) The coefficient of friction required between the wheel and the plane to prevent slipping.

*9-109 Figure P9-109 represents a 5-kg solid homogeneous cylinder. Determine
(a) The least coefficient of friction between the cylinder and the plane for which the cylinder will roll down the plane without slipping starting from rest.
(b) The acceleration of the center of the cylinder when it rolls without slipping.

Figure P9-108

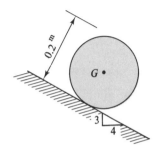

Figure P9-109

9-110 The 9-lb homogeneous sphere in Fig. P9-110 has a narrow slot cut in it for the cord. Neglect the effect of the slot on the moment of inertia of mass of the sphere, and assume that the cord acts in the vertical plane of symmetry. The sphere is released from rest with G 80 in. above the floor. Determine

(a) The angular acceleration of the sphere.

(b) The velocity of G immediately prior to impact with the floor.

Figure P9-110

9-111 The 32.2-lb homogeneous sphere in Fig. P9-111 has a narrow slot cut in it for the cord. Neglect the effect of the slot on the moment of inertia of mass of the sphere, and assume that both cords act in a vertical plane of symmetry. Determine the force **P** when the angular acceleration of the sphere is 12 rad per sec² counterclockwise.

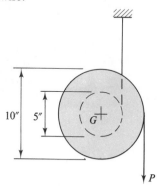

Figure P9-111

9-112 Cylinder *A* in Fig. P9-112 weighs 32.2 lb and rolls along the inclined plane without slipping. Neglect the effect of the slot on the moment of inertia of the disk. Body *B* weighs 16.1 lb. Determine the angular acceleration of the disk.

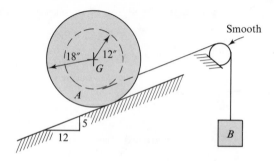

Figure P9-112

9-113 Wheels *A* and *B* in Fig. P9-113 are identical 16.1-lb solid homogeneous cylinders with smooth axles at their geometric centers. The 32.2-lb carriage *C*, with mass center at *G*, is supported on the axles of *A* and *B*, and the wheels roll without slipping on the inclined plane. Determine the acceleration of the carriage *C*.

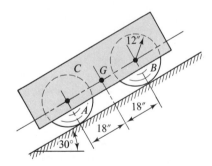

Figure P9-113

9-114 In Fig. P9-114 the homogeneous cylinder *A* is mounted on smooth bearings inside box *B* and connected by a cable to body *C*. Bodies *A*, *B*, and *C* weigh 50, 20, and 70 lb, respectively. All surfaces are smooth. Determine the acceleration of bodies *B* and *C*.

*****9-115** The solid homogeneous cylinder in Fig. P9-115 has a mass of 13 kg. The force *P* is applied to a cord wrapped around a thin slot (neglect the effect of the slot on the moment of inertia of the cylinder). The coefficient of friction between the cylinder and the inclined plane is 0.30. Determine the angular acceleration of

Figure P9-114

the cylinder and the linear acceleration of G when
(a) $P = 24$ N.
(b) $P = 60$ N.

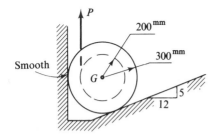

Figure P9-115

***9-116** The coefficient of friction between the cylinder and the inclined plane in Problem 9-115 is increased to 0.60. Determine the angular acceleration of the cylinder and the linear acceleration of G when
(a) $P = 60$ N.
(b) $P = 75$ N.

9-117 The 12.88-lb homogeneous disk A in Fig. P9-117 is connected by a rope to the 32.2-

Figure P9-117

lb homogeneous cylinder B. Determine the angular acceleration of body B.

***9-118** Body A in Fig. P9-118 is a 60-kg homogeneous cylinder with a narrow slot cut for the flexible cable. Neglect the effect of the slot on the moment of inertia. When in the position shown, the acceleration of the mass center of the cylinder is 2 m/s² upward. Determine the mass of body B.

Figure P9-118

9-119 In Fig. P9-119 the homogeneous rod CD is suspended from homogeneous cylinders A and B by cables which are free to unwind from the cylinders without slipping. The weights of cylinders A and B and the rod are 10, 6, and 30 lb, respectively. The system is released from rest in the configuration shown. Determine the angular acceleration of the rod.

Figure P9-119

9-120 Body *A* in Fig. P9-120 rolls without slipping, weighs 322 lb, and has a radius of gyration of mass with respect to *G* of 2 ft. The angular acceleration of *A* is 10 rad per sec² clockwise, and the velocity of *G* is 8 fps to the right. Body *B* weighs 96.6 lb and is acted on by a horizontal force **P** of proper magnitude to keep the cord supporting *B* vertical. Determine

(a) The force **P**.

(b) The couple **C**.

Figure P9-120

9-121 The solid homogeneous sphere in Fig. P9-121 weighs 64.4 lb and rolls without slipping on the horizontal plane. Determine the distance *x* for which the frictional force will be zero and the corresponding angular acceleration of the sphere.

*9-122 The coefficient of friction between the homogeneous 145-kg cylinder in Fig. P9-122 and the horizontal plane is 0.30. Neglect the effect of the slot on the moment of inertia of the cylinder. The cylinder is acted upon by a 700-N force as shown. Determine

(a) The frictional force on the cylinder.

(b) The acceleration of the mass center of the cylinder.

(c) The angular acceleration of the cylinder.

Figure P9-121

Figure P9-122

9-123 The 4-lb homogeneous rod in Fig. P9-123 is supported on a horizontal surface by friction pads at *A* and *B*. The coefficient of friction between the pads and the surface is 0.5. The rod is at rest, and the force *P* is 2 lb as shown. At the instant the cord is cut, determine

(a) The angular acceleration of the rod.

(b) The acceleration of the mass center *G*.

9-124 Solve Problem 9-123 when the force *P* is 8 lb as shown.

9-125 The homogeneous rod *AB* in Fig. P9-125 is held in equilibrium by the smooth horizontal plane and the cord *BC*. Determine the acceleration of the mass center of the rod and the angular acceleration of the rod at the instant the cord is cut.

Figure P9-123

Figure P9-125

9-126 Solve Problem 9-125 if the coefficient of friction between the rod and the horizontal plane is 0.6.

9-127 Solve Problem 9-125 if the coefficient of friction between the rod and the horizontal plane is 0.2.

9-128 The wheel in Fig. P9-128 weighs 322 lb and has a radius of gyration of mass of 1.6 ft with respect to a horizontal axis through O. The wheel rolls and slips on the horizontal plane. In the position shown, the velocity of O is 4.0 fps to the right, the angular velocity is 3.0 rad per sec clockwise, and the angular acceleration is 2.0 rad per sec² clockwise. The coefficient of friction between the wheel and the plane is 0.20. Determine

(a) The couple C.
(b) The acceleration of point O.

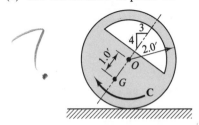

Figure P9-128

*9-129 The 150-kg wheel in Fig. P9-129 has a radius of gyration of mass with respect to an axis through the mass center G of 0.30 m. A couple

T of 300 N · m acts on the wheel. The coefficient of static friction between the horizontal surface and the wheel is 0.40. Determine the magnitude of the maximum angular velocity the wheel can have in the position shown if the wheel does not slip.

Figure P9-129

9-130 A homogeneous sphere with a 9-in. radius weighs W lb and is placed on an inclined plane with a slope of 3 vertical to 4 horizontal. The coefficient of friction between the sphere and plane is 0.20. If the sphere is released from rest, determine the distance the center of the sphere will move in 2.0 sec.

9-131 The homogeneous cylinder C in Fig. P9-131 is initially at rest when it is placed on a belt moving with a constant velocity of 40 fps to the left. The coefficient of friction between the belt and the cylinder is 0.5. Determine the angular velocity of the cylinder when it strikes the wall.

Figure P9-131

9-132 Solve Problem 9-131 if the belt moves with a constant velocity of 30 fps to the left.

*9-133 The homogeneous cylinder in Fig. P9-133 (p. 558) has a mass of 45 kg and is initially at rest on cart B. A force **P** is applied to the cart, giving it an acceleration of 5 m/s² to the right. The cylinder does not slip on the cart. Determine.

(a) The acceleration of the mass center of the cylinder.
(b) The friction force acting on the cylinder.

Figure P9-133

9-134 Block A and homogeneous cylinder B in Fig. P9-134 weigh 32.2 lb each. The inclined plane is smooth, and the coefficient of friction between A and B is 0.5. The bodies are released from rest when the 20-lb force is applied to A. Determine the acceleration of body A and the angular acceleration of cylinder B.

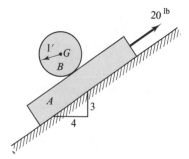

Figure P9-134

9-135 Body A in Fig. P9-135 weighs 128.8 lb, and the coefficient of friction between it and the plane is 0.2. Body B weighs 644 lb, has a radius of gyration of mass of 1.5 ft with respect to a horizontal axis through G, and rolls without slipping. The moment of the couple $\mathbf{C}$ is 100 ft-lb. Determine the acceleration of body A when it is being pulled to the right by body B.

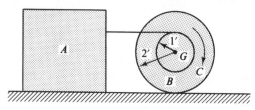

Figure P9-135

9-136 The 7-lb homogeneous rod AB in Fig. P9-136 is supported by two thin inextensible wires AD and BC. At the instant wire BC is cut, determine

(a) The angular acceleration of the rod.
(b) The tension in wire AD.

Figure P9-136

9-137 The slotted bar AB in Fig. P9-137 has a velocity of 4.0 fps to the left in the position shown when the cord breaks. The radius of gyration of the mass of the bar about its mass center is 12 in. The coefficient of friction between the bar and the pin at O is 0.50. Determine the acceleration of the mass center of the bar just after the cord breaks.

Figure P9-137

9-138 The homogeneous rod AB in Fig. P9-138 is held in equilibrium on the smooth corner C by cord AD. Determine the angular acceleration of the rod at the instant the cord is cut.

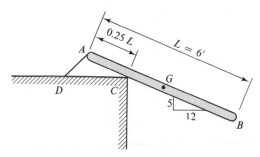

Figure P9-138

9-139 The uniform slender bar BC in Fig. P9-139 weighs 193.2 lb. It has an angular velocity of 2 rad per sec and an angular acceleration of 6 rad per sec², both clockwise, in this position. Determine the components of all unknown forces acting on BC.

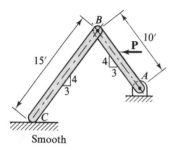

Smooth

Figure P9-139

*9-140 The velocity of point B on the 40-kg homogeneous rod in Fig. P9-140 is 3 m/s down. End B of the rod is smooth, and the coefficient of friction between end A and the horizontal plane is 0.5. Determine the angular acceleration of the rod.

Figure P9-140

9-141 The 10-ft bar AB in Fig. P9-141 weighs 322 lb and has a radius of gyration of mass of 3 ft with respect to a horizontal axis through G. The angular velocity of the bar is 2 rad per sec clockwise in the position shown. Determine the couple C that will cause the bar to have an angular acceleration of 5 rad per sec² counterclockwise. Neglect the weights of the smooth guide blocks at A and B.

9-142 The homogeneous cylinder A in Fig. P9-142 weighs 193.2 lb and in the position shown is acted on by the torque $T = 40$ ft-lb counterclockwise. The wheel has a constant angular velocity of 8 rad per sec counterclockwise. Rods BC and DE each weigh 64.4 lb. Determine the reaction of the pin at B on member BC.

Figure P9-141

Figure P9-142

9-143 In Fig. P9-143, body A is a homogeneous cylinder weighing 64.4 lb, and B is a block weighing 8.05 lb. The flexible cord connecting A and B passes over two small pulleys whose weights and diameters may be neglected. When the angle θ is 90°, the velocity of A is 10 fps to the right. Body A rolls along the plane without slipping. Determine all unknown forces acting on A when $\theta = 90°$.

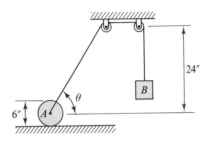

Figure P9-143

559

9-144 The unbalanced wheel in Fig. P9-144 weighs 64.4 lb and rolls without slipping along the inclined plane. The center of the wheel C has a velocity of 2.0 fps and an acceleration of 10.0 fps², both up the plane, when in the position shown. Determine the radius of gyration of the mass of the wheel with respect to a horizontal axis through G.

Figure P9-144

9-12 REVERSED EFFECTIVE FORCES AND COUPLES

A corollary of d'Alembert's principle for a system of particles was presented in Art. 9-7. The corollary states the following: *If a force equal in magnitude to and collinear with, but opposite in sense to, the resultant of the ma quantities of the particles is added to the external forces acting on the system, the resulting force system will be in equilibrium.* d'Alembert's reasoning can be extended to rigid bodies. Thus, by d'Alembert's principle, the equation of motion of the mass center of a rigid body becomes

$$\mathbf{F} + (-m\mathbf{a}_G) = 0. \qquad (9\text{-}21)$$

When the body has translation, the resultant external force passes through the mass center so the $m\mathbf{a}_G$ vector also must pass through the mass center. For translation, therefore, the problem may be solved by the methods of statics if the reversed effective force $(-m\mathbf{a}_G)$ is placed at the mass center in the free-body diagram. The equations $\sum \mathbf{F} = 0$ and $\sum \mathbf{M} = 0$ can be applied using all forces on the diagram including the reversed effective force $(-m\mathbf{a}_G)$. This method offers the advantage that the moment equation can be written with respect to any point or axis, thereby making it possible to select moment centers which eliminate the greatest number of unknowns from the moment equation.

The application of *d'Alembert's principle* becomes somewhat more involved when the body has angular motion. For example, if the equations of motion for a rigid body with plane motion are examined—see Eq. (9-16) and (9-17)—it is evident that by *d'Alembert's principle* they become

$$\mathbf{F} + (-m\mathbf{a}_G) = 0$$

and

$$\sum \mathbf{M}_G + [-(\omega^2 I_{yz} - \alpha I_{xz})\mathbf{i} + (\omega^2 I_{xz} + \alpha I_{yz})\mathbf{j} - \alpha I_z \mathbf{k}] = 0, \qquad (9\text{-}22)$$

where the second term of the last equation represents the *reversed effective couple.* Reversed effective couples are not true couples any more than reversed effective forces are true forces. By placing the reversed effective force $(-m\mathbf{a}_G)$ at the mass center and the reversed effective couple anywhere on the body, the problem may be treated as one of statics. Care must be taken to make sure that the moments and products of inertia are computed with respect to axes through the mass center of the body. Once the reversed effective forces and couples are shown on the free-body diagram, the equations of motion are simply $\sum\mathbf{F} = 0$ and $\sum\mathbf{M} = 0$, where all forces and moments, including the reversed effective forces and couples, are used. Any convenient point may be used as a moment center.

When solving a problem involving a rigid body rotating about a fixed axis, it may be more convenient to use Eq. (9-14) to compute the moment of the reversed effective couple. The *xyz* axes are used to compute moments and products of inertia. If this is done, the reversed effective force $(-m\mathbf{a}_G)$ must be placed through the origin used to determine the reversed effective couple.

As indicated in Art. 9-7, the concept of reversed effective forces and couples does not present any new information and is therefore only an alternative to the methods previously presented. It is important to decide which method will be used early in the solution of any problem since if the two methods are mixed in a single solution, the *m***a** terms are often used once as reversed effective forces and again as acceleration terms, resulting in incorrect results. When applying d'Alembert's principle, it is desirable to show the reversed effective forces and couples as dashed lines to distinguish them from true forces and couples.

Reversed effective forces are helpful in the study of the banking or superelevation of highway and railroad curves. *Superelevation* is defined as the difference in elevation between the outer and inner edges of a roadway or between the outer and inner rails of railroad tracks. The analysis is similar for highway and railroad curves; hence, the following discussion of superelevation of highways is also applicable to railroads.

The tendency for an automobile to slide away from the center of curvature of a curve or to overturn while traveling on a curve can be overcome by banking the highway. The car in Fig. 9-21(a) (p. 562) is traveling along a horizontal highway curve which has a radius of R ft. The speed of the car is v fps. The free-body diagram of the car is shown in Fig. 9-21(b). The mass of the car is obviously not symmetrical with respect to the plane of motion, the horizontal plane in which G moves, and therefore the reversed effective force should not pass through G. The exact location of the reversed effective force would be difficult to determine, however, and it is found that the error introduced by placing the force through G is comparatively small; therefore, the

Figure 9-21

reversed effective force is shown acting through G. Moreover, the position of the reversed effective force does not affect the force equations and is important only in an investigation of the tendency of the car to overturn. The ideal angle of bank, θ, is one which will eliminate the frictional forces $\mathbf{F}_1$ and $\mathbf{F}_2$. This ideal angle can be determined from the following equations of equilibrium when F_1 and F_2 are set equal to zero:

$$\sum F_x = N_1 \sin \theta + N_2 \sin \theta - m(a_G)_x = 0,$$

$$\sum F_y = N_1 \cos \theta + N_2 \cos \theta - W = 0.$$

The value of $(a_G)_x$ is v^2/R. Therefore,

$$\sin \theta (N_1 + N_2) = \frac{W}{g} \frac{v^2}{R} \quad \text{and} \quad \cos \theta (N_1 + N_2) = W$$

from which

$$\tan \theta = \frac{v^2}{gR}. \tag{a}$$

The angle θ is seen to be a function of the speed of the car and the radius of the curve.

When the speed of a car is greater than the value of v, from Eq. (a) the car will tend to slide outward along the radius of curvature, and when the speed is less than the value from Eq. (a), the car will tend to slide in toward the center of curvature. This tendency for slower-moving vehicles to slide toward the center of the curve limits the permissible amount of superelevation because the amount of friction that can be developed is rather small when the highway is icy.

When the car is on the verge of overturning, the reaction N_1 of Fig. 9-21(b) will be zero. The overturning velocity can be shown to be a function of d, h, θ, and R by use of the equations of equilibrium for the free-body diagram of Fig. 9-21(b) when the car is assumed not to slip. Whether the car will slip or overturn first will depend on the coefficient of friction as well as the quantities d, h, θ, and R.

The following example illustrates the use of reversed effective forces and couples.

EXAMPLE 9-14

The uniform slender bar *BE* in Fig. 9-22(a) weighs 64.4 lb and is fastened to the vertical shaft by a smooth pin at *E* and the cord *CD*. The mechanism rotates about the vertical axis with a constant angular velocity of 50 rpm. Determine the force of the cord and the components of the pin reaction at *E* on *BE*.

SOLUTION

The *xyz* coordinate system is attached to the body and rotates with it so that the body is always in the *xz* plane. The angular-velocity is $\omega\mathbf{k} = 50(2\pi/60)\mathbf{k}$ rad per sec. The angular acceleration is zero, and the acceleration of the mass center is

$$\mathbf{a}_G = -x_G\omega^2\mathbf{i} = -1.8\left(50\frac{2\pi}{60}\right)^2\mathbf{i} = -49.35\mathbf{i}\,\text{fps}^2.$$

Therefore, the reversed effective force is

$$-m\mathbf{a}_G = \frac{64.4}{32.2}(49.35)\mathbf{i} = 98.7\mathbf{i}\,\text{lb}.$$

The bar is not symmetrical with respect to the plane of motion, the *xy* plane; therefore, Eq. (9-14a) must be used to determine the *reversed*

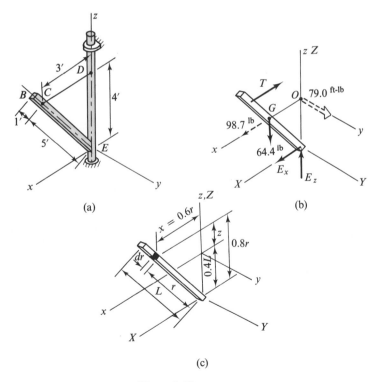

(a)

(b)

(c)

Figure 9-22

effective couple. Note that the reversed effective force must act through the origin at O if Eq. (9-14a) is used to compute the moment of the couple. The moment and products of inertia are obtained using Fig. 9-22(c). The mass of the element is

$$dm = \rho A \, dr,$$

and the mass of the bar is

$$m = \rho A L = 2.00 \text{ slugs.}$$

The moment of inertia of the element about the z axis is

$$dI_z = x^2 \, dm = (0.6r)^2 \rho A \, dr,$$

and the resultant moment of inertia is

$$I_z = \rho A (0.36) \int_0^L r^2 \, dr = 0.12 \rho A L^3 = 0.12 m L^2$$

$$= 0.12(2.00)(6.0)^2 = 8.64 \text{ slug-ft}^2.$$

The bar is symmetrical about the xz plane; therefore, I_{yz} is zero. The product of inertia of the element about the xy and yz planes is

$$dI_{xz} = xz \, dm = (0.6r)(0.8r - 0.4L)\rho A \, dr,$$

and the resultant product of inertia is

$$I_{xz} = 0.6\rho A \int_0^L (0.8r^2 - 0.4Lr) \, dr$$

$$= 0.04\rho A L^3 = 0.04 m L^2 = 0.04(2.00)(6.0)^2$$

$$= 2.88 \text{ slug-ft}^2.$$

From Eq. (9-14a), the reversed effective couple is

$$\mathbf{C}_R = 0\mathbf{i} + \left(50\frac{2\pi}{60}\right)^2 (2.88)\mathbf{j} - 0\mathbf{k}$$

$$= 79.0\mathbf{j} \text{ ft-lb.}$$

The reversed effective force and couple are shown by dashed lines on Fig. 9-22(b). The equations of equilibrium are

$$\Sigma F_z = E_z - 64.4 = 0$$

from which

$$E_z = 64.4 \text{ lb}$$

and

$$\mathbf{E}_z = 64.4\mathbf{k} \text{ lb}; \qquad\qquad \text{Ans.}$$

$$\Sigma M_y = -4T + 0.8(3)(98.7) + 0.6(3)64.4 + 79.0 = 0$$

from which

$$T = 107.9 \text{ lb}$$

and

$$\mathbf{T} = -107.9\mathbf{i} \text{ lb}; \qquad\qquad \text{Ans.}$$

(b)

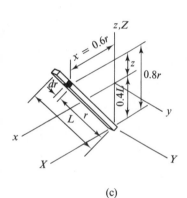

(c)

Figure 9-22(b) and (c)

and

$$\sum F_x = E_x + 98.7 - 107.9 = 0$$

from which

$$E_x = 9.2 \text{ lb}$$

and

$$\mathbf{E}_x = \underline{9.2\mathbf{i}\,\text{lb}}. \qquad\qquad \text{Ans.}$$

PROBLEMS

9-145 Solve Problem 9-62 by the reversed effective-force method.

9-146 Solve Problem 9-71 by the reversed effective-force method.

9-147 Solve Problem 9-72 by the reversed effective-force method.

*9-148 Solve Problem 9-88 by the reversed effective-force method.

9-149 Solve Problem 9-93 by the reversed effective-force method.

9-150 Solve Problem 9-128 by the reversed effective-force method.

9-151 Solve Problem 9-141 by the reversed effective-force method.

9-152 Block A in Fig. P9-152 weighs 128.8 lb, slides in a smooth track, and is connected by a smooth pin D to the uniform bar B, which weighs 64.4 lb. In the position shown, the acceleration of the mass center of B is 15 fps² at a slope of 4 horizontally to the left and 3 vertically upward, and the angular acceleration of B is 8 rad per sec² clockwise. Determine the force P and the couple C which act on bar B.

Figure P9-152

9-153 The rigid frame $ABCD$ in Fig. P9-153 slides along the horizontal plane with an acceleration of 6 fps² to the right. The solid homogeneous cylinder F weighs 161 lb and is fastened to the frame at E with a smooth pin. When in the position shown, the cylinder has an angular velocity of 5 rad per sec and an angular acceleration of 10 rad per sec², both clockwise. Determine the force Q and the components of the force on the body at E.

Figure P9-153

*9-154 Figure P9-154 represents a 30-kg solid homogeneous cylinder and two 15-kg slender rods. The cylinder rolls without slipping on the horizontal surface. Determine the acceleration of the center of the cylinder.

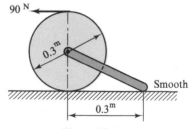

Figure P9-154

9-155 The 64.4-lb homogeneous bar AB in Fig. P9-155 (p. 566) is pinned at A and rests against the 96.6-lb body D at B. Determine the maximum weight body C can have without causing AB to rotate. All surfaces of contact are smooth, and it is assumed that body D will not tip.

9-156 The homogeneous disk A and pulley B in Fig. P9-156 are supported by and keyed to a shaft which is mounted in bearings at C and D. The disk weighs 50 lb, body W weighs 20 lb, and the weight of the pulley and shaft may be neglected. In the position shown, the magnitude of the angular velocity of A is 2.0 rad per sec. Determine the components of the bearing reactions on the shaft at C and D.

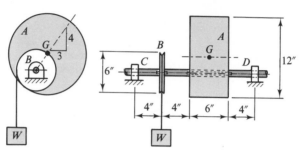

Figure P9-156

9-13 CLOSURE

An orderly method of procedure is a definite aid in solving problems in kinetics. The main essentials of the systematic procedure of Art. 9-6 include (1) a complete free-body diagram; (2) a careful analysis of the unknown forces, accelerations, and so on; (3) the timely use of kinematic and friction relations to reduce the number of unknowns involved; and (4) the use of the appropriate equations of motion.

When a rigid body has pure rotation, it is usually desirable to resolve the pin reaction into components in the directions of the normal and tangential components of the acceleration of the mass center of the body. Moments are usually computed with respect to an axis through the mass center when the body has either translation or plane motion and with respect to the axis of rotation when the body has rotation. When the reversed effective-force method is used, moments can be taken with respect to any axis. Whether or not to use reversed effective forces is a matter of individual preference.

Instantaneous values of forces and accelerations are obtained by the methods of force, mass, and acceleration. Velocities, displacements, and distances traveled can be determined by means of v-t or ω-t diagrams when accelerations are constant. If some of the forces acting on a body are variable, the acceleration is variable, and the differential equations of kinematics must be used to determine velocities, displacements, and so on. In this case, the principle of work and energy or of impulse and momentum as discussed in the next two chapters may provide a more direct solution.

PROBLEMS

Note: Unless otherwise specified, all cords, ropes, and cables are assumed to be flexible, inextensible, and of negligible weight. All pins and axles are assumed to be smooth unless otherwise stated.

9-157 Body A in Fig. P9-157 weighs 32.2 lb, and body B weighs 48.3 lb. The coefficient of friction between A and B is 0.40, and the plane is smooth. Determine the maximum horizontal force which can be applied at C without causing A to tip or slide on B
(a) If the force is to the left.
(b) If the force is to the right.

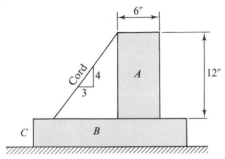

Figure P9-157

9-158 Body A in Fig. P9-158 weighs 20.0 lb, and body B weighs 10.0 lb. The coefficient of friction is 0.20 at C, 0.50 at D, and zero at E. The bodies are initially at rest when the force $P = 8.00$ lb is applied. Determine the acceleration of A with respect to B immediately after the force is applied.

Figure P9-158

9-159 The 64.4-lb homogeneous triangular block B in Fig. P9-159 is connected to frame A by a smooth pin at O. Determine the maximum acceleration up the plane that can be given frame A without causing B to rotate.

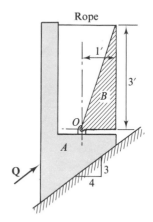

Figure P9-159

*9-160 The two slender homogeneous bars in Fig. P9-160 are rigidly connected at B and rotate in a vertical plane. The mass of bar AB is 45 kg, and the mass of bar CD is 15 kg. In the position shown, the angular velocity is 3.0 rad/s clockwise. Determine
(a) The angular acceleration of the bars.
(b) The reaction of the pin at A on AB.

Figure P9-160

9-161 Body A in Fig. P9-161 is a 96.6-lb slender homogeneous bar which is rigidly connected to the 193.2-lb homogeneous half-cylinder B. Bodies A and B rotate about a horizontal axis at O. Block C is attached to a rope which wraps around B. In the position shown, the angular velocity and acceleration of A and B are 3.0 rad per sec and 2.0 rad per sec², both clockwise. Determine

(a) The weight of block C.

(b) The magnitude of the bending moment at the junction of A and B.

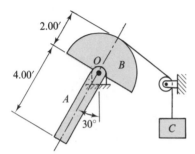

Figure P9-161

9-162 The homogeneous body A in Fig. P9-162 rolls and slips on the plane. It has a radius of gyration of mass with respect to a horizontal axis through G of 2.0 ft. The weight of A is 644 lb. The coefficient of friction between A and the plane is 0.2. The initial velocity of G is 4.5 fps down the plane, and the cable is taut. Determine the acceleration of point G.

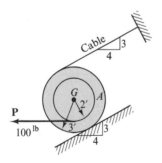

Figure P9-162

*__9-163__ The homogeneous cylinder A in Fig. P9-163 has a mass of 60 kg, and the slender homogeneous bar BC has a mass of 15 kg. The

bar is welded to the cylinder in its plane of symmetry. The assembly moves on a smooth horizontal surface. In the position shown, the body has an angular velocity of 2.0 rad/s clockwise. Determine the acceleration of point C at the instant shown.

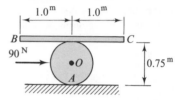

Figure P9-163

9-164 Determine the components of the force exerted by the piston pin A on the connecting rod AB of the reciprocating engine in Fig. P9-164 for the crank position of 90°. The crank has a constant angular velocity of 100 rad per sec clockwise. The connecting rod weighs 8.05 lb and has a centroidal radius of gyration of 2.77 in. The pressure of the expanding gases on the 7.00-lb piston at this position is 85.0 psi. Neglect friction between the piston and the cylinder.

Figure P9-164

9-165 Body A in Fig. P9-165 weighs 128.8 lb (including brackets) and is pinned at O to the 64.4-lb solid homogeneous drum C. The cord which unwinds from C passes over a smooth peg as indicated. Determine the acceleration of the 96.6-lb block B.

9-166 The assembly in Fig. P9-166 consists of a 32.2-lb solid homogeneous cylinder C rigidly connected to the 64.4-lb slender homogeneous bar AB. The assembly rotates in a vertical plane about the fixed axis at A and, in the position shown, has an angular velocity of 4 rad per sec

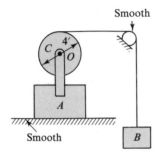

Figure P9-165

(a) The angular acceleration of rod BC.
(b) The pin reaction on rod BC.

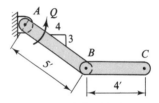

Figure P9-167

clockwise. The magnitude of the moment of the couple **T** is 100 ft-lb. Determine, for this position,

(a) The angular acceleration of the assembly.
(b) The bearing reaction at A on the bar.

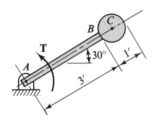

Figure P9-166

9-167 Both rods in Fig. P9-167 move in the vertical plane. When in the position shown, rod AB has an angular velocity of 2 rad per sec clockwise and an angular acceleration of 3 rad per sec^2 counterclockwise, and the homogeneous 64.4-lb rod BC has an angular velocity of 3 rad per sec clockwise. Determine

9-168 The 50-lb homogeneous cylinder in Fig. P9-168 is connected to the wall by cord AB and is acted upon by the vertical force **P**. The cord and the force **P** both lie in the plane of motion, and the effect of the slot can be neglected. The coefficient of friction between the cylinder and the floor is 0.20. Determine the acceleration of G and the tension in the cord when the force P is (a) 4.0 lb, (b) 10.0 lb.

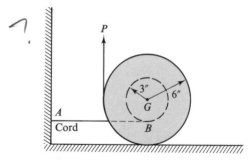

Figure P9-168

9-169 Solve Problem 9-168 for the case when the force P is 20.0 lb.

10

Kinetics–Work and Energy

PART A–PARTICLES

10-1 INTRODUCTION

In Chapter 9, Newton's laws of motion relating force, mass, and acceleration were used in the solution of problems in kinetics. Newton's laws are concerned with the forces acting on a particle or body at any instant and the resulting acceleration. If the position or velocity is required as a function of time, the acceleration obtained by the methods of Chapter 9 can be integrated as discussed in Chapter 8. This integration is usually straightforward if the acceleration is found to be constant or a function of the time. When the acceleration is obtained as a function of the position of the particle or body the problem can frequently be solved more readily by applying the principle of work and kinetic energy, which equates the work done on a particle or body to the change in kinetic energy of the particle or body. When any of the forces acting on a particle vary with the position of the particle the method of work and energy is particularly effective.

10-2 WORK DONE BY A FORCE OR A SYSTEM OF FORCES

The term *work*, as used in mechanics, should not be confused with such common terms as yardwork, roadwork, and housework. The definition of *work* and its calculation are presented in Art. 7-3. This material is included here as a review and for a ready reference.

The work done by a constant force **F** during a displacement **q** is defined as the product of the rectangular component of the force in the direction of the displacement and the magnitude of the displacement. If the angle between **F** and **q** is θ, the work done can be expressed as

$$U = Fq \cos \theta, \qquad (10\text{-}1)$$

where U is the work done. Figure 10-1 shows that the work done by the force **F** during the displacement **q** as the block moves from position A to position B is the product of the magnitude of the rectangular component of **F** in the direction of the displacement, $F \cos \theta$, and the magnitude of the displacement. Thus U is equal to $(F \cos \theta)q$. The work done may also be thought of as the product of the magnitude of **F** and the magnitude of the component of the displacement in the direction of **F**; that is, $U = F(q \cos \theta)$.

Since the definition of the work done by the force **F** during the displacement **q** is the same as the definition of the scalar or dot product of the vectors **F** and **q**, it is convenient to express the work done as

$$U = \mathbf{F} \cdot \mathbf{q}. \qquad (10\text{-}1a)$$

Work is the scalar product of two vectors; therefore it is a scalar quantity having only magnitude and algebraic sign. The quantities **F** and **q** can be expressed in vector notation as

$$\mathbf{F} = (F_x \mathbf{i} + F_y \mathbf{j} + F_z \mathbf{k})$$

and

$$\mathbf{q} = (q_x \mathbf{i} + q_y \mathbf{j} + q_z \mathbf{k}).$$

The work done by **F** can be obtained by means of the dot product which determines both the magnitude and algebraic sign; thus,

$$U = \mathbf{F} \cdot \mathbf{q} = F_x q_x + F_y q_y + F_z q_z.$$

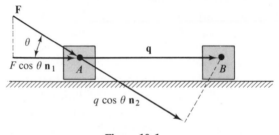

Figure 10-1

From the definition of work [see Eq. (10-1)] it may be noted that Section 10-2
Work done by a force
or a system of
forces the work done by a force is positive when the displacement and the component of the force along the displacement have the same sense ($0 \le \theta < 90°$). When the displacement and the component of the force in the direction of the displacement are in opposite directions ($90° < \theta \le 180°$), the force does negative work. When $\theta = 90°$, the force has no component along the displacement and the work done by the force is zero. Since work is the product of force and displacement, it has dimensions of FL and common units are ft-lb, in-lb, and $N \cdot m$ (Newton·meter or joule).

Note that, by definition, the work done by a *constant* force does not depend on the path traveled by the point of application of the force during the displacement. Thus, the work done by the constant force **F** in Fig. 10-2 is the same if it is moved from position A to position B along path a, along path b, or along the straight line from A to B. Since the weight of a particle, W, is constant, the work done by the weight is Wh, where h is the vertical distance the particle moves. This work is positive when the particle moves downward and negative when it moves upward.

When the force varies either in magnitude or direction, as in Fig. 10-3, Eq. (10-1) is valid only for an infinitesimal change in position, and the work done by the force as it moves from A to B is

$$U = \int_{r_A}^{r_B} \mathbf{F} \cdot d\mathbf{r} = \int_{s_A}^{s_B} F \cos \theta \, ds. \tag{10-2}$$

Note that ds is the magnitude of the infinitesimal displacement $d\mathbf{r}$ and that θ is the angle between **F** and $d\mathbf{r}$. Equation (10-2) is valid for any general force and displacement, and it can be used to compute the work done if relationships between F, θ, and s can be established. Equation (10-2) can be written in a form which is sometimes more convenient to use; that is,

$$U = \int \mathbf{F} \cdot d\mathbf{r} = \int (F_x \mathbf{i} + F_y \mathbf{j} + F_z \mathbf{k}) \cdot (dx \mathbf{i} + dy \mathbf{j} + dz \mathbf{k})$$

$$= \int F_x \, dx + \int F_y \, dy + \int F_z \, dz. \tag{10-2a}$$

Figure 10-2

Figure 10-3

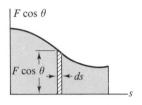

Figure 10-4

Although **F** and $d\mathbf{r}$ are vector quantities, the dot product is a scalar quantity, and ordinary scalar integration can be used to determine the work done.

Sometimes, a graphical or semigraphical method for determining the work done by a force is useful, especially when the mathematical relation between F, θ, and s in Eq. (10-2) is either unknown or cumbersome. When the component of the force in the direction of the displacement, $F \cos \theta$, is plotted against the position function s as indicated in Fig. 10-4, the area under the curve represents the work done—see Eq. (10-2). The indicator card used to determine the work done by the steam on the piston of a steam engine is an example of an F-s diagram.

A spring is a body which deforms when acted on by a force. The relationship of the force and deformation depends on the material used and on the dimensions and shape of the spring (helical, spiral, conical, cantilever, or other). Only springs which exhibit a linear relationship between force and displacement or deformation are considered in this section. In this case, the magnitude, F, of the force is given by the expression

$$F = ks,$$

where s is the deformation of the spring from its unloaded position and k is a constant known as the *modulus* of the spring. The force-displacement diagram for a spring is shown in Fig. 10-5. As the spring is stretched an amount s from its unstretched length l_o, the force increases from zero to $F = ks$. The spring modulus, k, is the slope of the F-s diagram and has dimensions of force divided by length, often expressed as pounds per foot, pounds per inch, or newtons per meter.

The work done in stretching the spring in Fig. 10-5 from an initial length $l_o + s_1$ to a final length $l_o + s_2$ can be determined from Eq. (10-2a). Since the force and displacement are in the same direction,

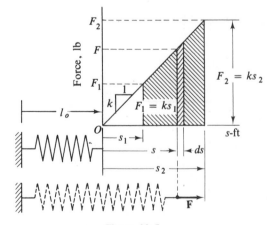

Figure 10-5

Section 10-2
Work done by a force
or a system of
forces

the angle θ is zero and the equation becomes

$$U = \int_{s_1}^{s_2} F \cos \theta \, ds = \int_{s_1}^{s_2} ks \, ds = \frac{k(s_2^2 - s_1^2)}{2}.$$

The force $\mathbf{F}$ is the force exerted by a body on the spring. The spring must exert the same force on the body except in the opposite direction. Consequently,

$$U_{\text{on body}} = -U_{\text{on spring}} = -\int_{s_1}^{s_2} ks \, ds.$$

The preceding expression can also be obtained as the area under the F-s diagram between s_1 and s_2, the shaded trapezoidal area. This area is

$$U = \text{area} = \frac{F_1 + F_2}{2}(s_2 - s_1).$$

When $F_1 = ks_1$ and $F_2 = ks_2$ are substituted into the expression, it becomes

$$U = \frac{ks_1 + ks_2}{2}(s_2 - s_1) = \frac{k(s_2^2 - s_1^2)}{2}$$

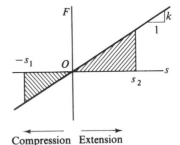

Compression Extension

Figure 10-6

as before. When the spring is initially unstretched, the trapezoid becomes a triangle and the work done on the spring is $ks_2^2/2$.

It should be noted that as a spring is being deformed (either stretched or compressed) the force on the spring and the displacement are in the same direction and the work done *on the spring* is positive. If a spring is initially deformed and released gradually, the force and displacement are in opposite directions and the work done *on the spring* is negative. The force of the spring on a body or particle is opposite the force of the body on the spring, but both the particle and the end of the spring have the same displacement; therefore, the work done *by a particle on a spring* and the work done *by the spring on the particle* will have the same magnitudes but opposite signs.

When a spring is constructed so that it can act in either tension or compression, such as an automobile suspension spring, it is usually assumed that the spring modulus is the same in tension and compression. The force-displacement diagram will be similar to Fig. 10-6, and it is apparent that, during a change of length from tension to compression (or the reverse), the force will change direction while the displacement will continue in the same direction. Consequently, the area on one side of the s axis will represent positive work and that on the other side will give negative work.

When two bodies are connected by a flexible inextensible cable the resultant work done on the bodies by the force in the cable is zero. The two forces at the ends of the cable have the same magnitude (the mass of the cable is assumed to be negligible), and if the cable is inextensible, the components of the displacements of the two ends in the direction of the forces must have the same magnitude; therefore the resultant work done by the cable is zero. The following examples demonstrate the calculation of the work done by various force systems.

EXAMPLE 10-1

The coefficient of friction between the 100-lb body A in Fig. 10-7(a) and the inclined plane is 0.20. The spring has a modulus of 20 lb per ft and is stretched 1 ft in the position shown. Determine the total work done on body A while it is moving 5 ft down the plane.

SOLUTION

A free-body diagram of body A is shown in Fig. 10-7(b). There are four forces acting on the body, and three will do work as body A moves 5 ft down the plane. The body has no acceleration normal to the plane; therefore $N = W \cos \theta = 100(\frac{4}{5}) = 80$ lb. The frictional force opposes the motion and thus acts up the plane, and since slipping occurs, $F = \mu N = 0.20(80) = 16$ lb. The work done by the friction is negative since it is opposite the displacement; thus

$$U_F = -F \, \Delta s = -16(5) = -80 \text{ ft-lb.}$$

The work done by the weight is positive since the vertical component of the displacement is in the direction of the weight; that is,

$$U_W = W(\Delta s_y) = 100[(\tfrac{3}{5})5] = 300 \text{ ft-lb.}$$

The work done by the spring force is negative since it is always directed up the plane and the movement is down the plane. The spring force is equal to the modulus (20 lb per ft) multiplied by the stretch, s, of the spring. The work may be computed by drawing an F-s diagram or making use of an integral. Since the initial stretch is 1 ft and the final stretch is 6 ft, the work done by the spring is

$$U_{\text{spring}} = -\int_1^6 20s \, ds = -10s^2 \big]_1^6 = -350 \text{ ft-lb.}$$

The total work is

$$U = -80 + 300 - 350 = -\underline{130 \text{ ft-lb.}} \qquad \text{Ans.}$$

(a)　　　　　　　　　　　(b)

Figure 10-7

Section 10-2
Work done by a force
or a system of
forces

EXAMPLE 10-2

The 200-lb body A in Fig. 10-8(a) slides in the smooth slot and is connected to the 300-lb body B by an inextensible cord. The spring has a modulus of 20 lb per ft and is arranged so that it can act in either compression or tension. The unstressed length of the spring is 7 ft, and the spring is initially perpendicular to the slot. Determine the work done by the forces acting on the two bodies as A moves 12 ft down the slot from the position shown. The slope and magnitude and the 100-lb force remain constant throughout the motion.

SOLUTION

A free-body diagram of the system in an intermediate position is drawn in Fig. 10-8(b). The force in the cord is an internal force of the system, and since the cord is inextensible, the work done by the cord is zero. There are six forces on the free-body diagram, and two of these forces, **N** and **R**, are normal to their corresponding displacements; therefore they do no work. The resultant work on the system is thus

$$U = U_{300\text{lb}} + U_{200\text{lb}} + U_{100\text{lb}} + U_{\text{spring}}.$$

The work done by the 300-lb force is

$$U_{300\text{lb}} = 300(12) = 3600 \text{ ft-lb}.$$

The work done by the 200-lb force is

$$U_{200\text{lb}} = 200(\tfrac{4}{5})(12) = 1920 \text{ ft-lb}.$$

The work done by the 100-lb force is

$$U_{100\text{lb}} = \mathbf{F} \cdot \mathbf{q} = 100(0.6\mathbf{i} + 0.8\mathbf{j}) \cdot 12(0.6\mathbf{i} - 0.8\mathbf{j})$$

$$= 1200(0.36 - 0.64) = -336 \text{ ft-lb}.$$

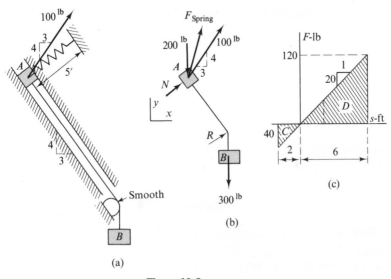

Figure 10-8

The work done by the spring can be computed from the F-s diagram in Fig. 10-8(c). As body A moves 12 ft down the slot the length of the spring changes from 5 to 13 ft. Since the spring has an unstressed length of 7 ft, it is initially compressed 2 ft and exerts a 40-lb compressive force on A. In the final position, the spring is stretched 6 ft and exerts a tensile force of 120 lb on A. The area C of the F-s diagram represents negative work done by A on the spring since the spring is in compression but is increasing in length; however, area C represents positive work of the spring on body A since the component of the force on body A along the slot and the displacement of body A are both down the slot. In a similar manner, area D represents negative work done by the spring on body A. The work done by the spring on A is

$$U_{\text{spring}} = C - D = \frac{40(2)}{2} - \frac{120(6)}{2} = -320 \text{ ft-lb}.$$

The total work done on the two bodies is

$$U = 3600 + 1920 - 336 - 320 = \underline{4864 \text{ ft-lb}}.$$
$$\text{Ans.}$$

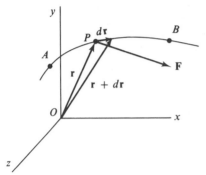

Figure 10-9

10-3 PRINCIPLE OF WORK AND KINETIC ENERGY

The particle P in Fig. 10-9 has a mass m and travels along the path from A to B. Let F be the resultant force acting on the particle, and let the axes shown be a newtonian set of axes. The work done by the force as it moves through a differential displacement, $d\mathbf{r}$, is

$$dU = \mathbf{F} \cdot d\mathbf{r} = m\ddot{\mathbf{r}} \cdot \dot{\mathbf{r}} \, dt = \tfrac{1}{2}m \frac{d}{dt}(\dot{\mathbf{r}} \cdot \dot{\mathbf{r}}) \, dt = \tfrac{1}{2}m \, d(v^2),$$

where $v = |\dot{\mathbf{r}}|$. The work done in traveling from the initial position, A, to the final position, B, is

$$U = \tfrac{1}{2}m \int_{v_i}^{v_f} d(v^2) = \tfrac{1}{2}m(v_f^2 - v_i^2),$$

where the subscripts i and f indicate initial and final velocities. The term $\tfrac{1}{2}mv^2$ is called the *kinetic energy* of the particle and is denoted by the symbol T. The kinetic energy of the particle is its capacity to do work by virtue of its velocity. The above equation can be written

$$U = \Delta T = T_f - T_i. \qquad (10\text{-}3)$$

This equation leads to the *principle of work and kinetic energy*, which can be stated as follows: *The work done on a particle* (by all forces acting on the particle) *is equal to the change of the kinetic energy of the particle.*

EXAMPLE 10-3

Section 10-3
Principle of work and
kinetic energy

The 20-lb body in Fig. 10-10(a) is constrained to slide along the smooth rod BC which lies in a vertical plane. When the body is in position A its velocity is 30 fps to the right, and the spring (modulus of 40 lb per ft) is stretched 6 ft. Determine the speed of the body when it reaches the position A'.

SOLUTION

A free-body diagram of the body is drawn in an intermediate position in Fig. 10-10(b). The equation relating the work and kinetic energy is

$$U = \Delta T$$

$$= \tfrac{1}{2}m(v_f^2 - v_i^2).$$

Since the force N is the reaction of the smooth rod on the particle and is always normal to the motion of the particle, N does no work. The length of the spring changes from 21 to 17 ft. Since it is initially stretched 6 ft, it is stretched 2 ft in the final position. The spring is under tension; therefore the component of the spring force along the path is always in the direction of the movement of the body, and the spring will do positive work on the body. This work on the body can be evaluated from an integral as follows:

$$U_{\text{spring}} = -\int_{s_i}^{s_f} ks\, ds = -\int_6^2 40s\, ds = -20s^2 \Big]_6^2$$

$$= 640 \text{ ft-lb.}$$

The work done by the weight on the body is

$$U_{20\text{lb}} = W(\Delta s_y) = -(20)(6) = -120 \text{ ft-lb.}$$

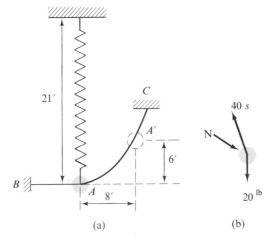

(a)   (b)

Figure 10-10

The total work done on the body is

$$U = U_{spring} + U_{20lb}$$

$$= 640 - 120 = 520 \text{ ft-lb}.$$

The work and kinetic energy equation becomes

$$U = \tfrac{1}{2}m(v_f^2 - v_i^2)$$

$$520 = \frac{1}{2}\left(\frac{20}{32.2}\right)(v_f^2 - 30^2)$$

from which

$$v_f = \underline{50.7 \text{ fps.}} \qquad \text{Ans.}$$

EXAMPLE 10-4

A rocket is fired vertically upward, and after burnout at position 1 it has a mass m and a vertical velocity v_1. From position 1 to position 2, as shown in Fig. 10-11, the only force acting on it is due to the inverse square law of gravitation, GMm/r^2, where G is the universal gravitation constant, M is the mass of the earth, and r is the distance from the center of the earth to the rocket. Determine the speed of the body when it reaches position 2.

Figure 10-11

SOLUTION

Since the gravitational force is a function of the position of the rocket, the principle of work and kinetic energy is used:

$$U = \tfrac{1}{2}m(v_f^2 - v_i^2).$$

The gravitational force is downward, and the movement of the mass m is upward; therefore, the work done by gravity is negative. Hence

$$U = -\int_{r_1}^{r_2} \frac{GMm}{r^2}\,dr = \frac{GMm}{r}\bigg]_{r_1}^{r_2} = GMm\left(\frac{1}{r_2} - \frac{1}{r_1}\right)$$

and

$$= GMm\frac{r_1 - r_2}{r_1 r_2} = -GMm\frac{r_2 - r_1}{r_1 r_2}$$

$$\Delta T = \tfrac{1}{2}m(v_2^2 - v_1^2) = U = -GMm\left(\frac{r_2 - r_1}{r_1 r_2}\right)$$

from which

$$v_2^2 = v_1^2 - 2GM\left(\frac{r_2 - r_1}{r_1 r_2}\right)$$

or

$$v_2 = \underline{\left[v_1^2 - 2GM\left(\frac{r_2 - r_1}{r_1 r_2}\right)\right]^{1/2}}. \qquad \text{Ans.}$$

PROBLEMS

Note: Unless otherwise specified, the mass of each spring and cord may be neglected. All cords are to be considered flexible and inextens-

ible; all pins, pulleys, and fixed drums are assumed to be smooth.

10-1 A stone weighing 0.25 lb is thrown with a velocity of 100 fps at an angle of 60° above the horizontal. By making use of the principle of work and kinetic energy, determine the maximum height it will rise above the point of throwing.

10-2 The force P acting on the 40-lb block A in Fig. P10-2 has a constant direction, as shown, and a magnitude that varies according to the equation $P = 6s^2$ lb, where s gives the position of A in feet. When $s = 2$ ft body A has a velocity of 4 fps to the right. Determine the velocity of A when $s = 5$ ft. The plane is smooth.

Figure P10-2

10-3 Solve Problem 10-2 if the coefficient of friction between body A and the horizontal plane is 0.10.

***10-4** Block A in Fig. P10-4 has a mass of 2.0 kg and has a velocity of 5 m/s up the plane. The coefficient of friction between the block and surface is 0.20. There is a gradual transition from the horizontal to the inclined surface. Locate the position where the block will finally come to rest.

Figure P10-4

10-5 The 10-lb block A in Fig. P10-5 is projected from the 10-ft tube by a spring whose unstressed length is 2.5 ft. The modulus of the spring is 400 lb per in. The friction between the block and tube is 2.0 lb. Determine the velocity A as it leaves the tube.

10-6 The 64.4-lb collar shown in Fig. P10-6 is released from rest in position A where the spring is unstretched. Determine the modulus of

Figure P10-5

the spring if the velocity of the collar is 6 fps after it slides downward 5 in. The rod is smooth, and the 50-lb force maintains the same slope throughout the motion.

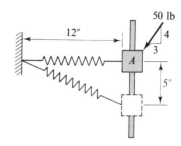

Figure P10-6

10-7 The 3.22-lb particle A in Fig. P10-7 is moving on a smooth horizontal surface and is connected to point O by an elastic cord which has a modulus of 18.7 lb per ft. In the position shown, the elastic cord is stretched 1 ft, and the velocity of A is as indicated. Determine the magnitude of the velocity of A when it reaches its maximum distance from point O of 10 ft.

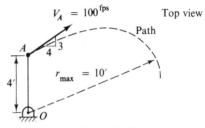

Figure P10-7

10-8 The smooth rod in Fig. P10-8 has a negligible mass and rotates about a vertical axis through O with a clockwise angular velocity of 6 rad per sec. Block B weighs 32.2 lb and is held by a string in contact with the uncompressed spring. The modulus of the spring is 27 lb per ft. The string breaks, and in the resulting motion the spring is compressed a maximum of 1 ft. Determine the speed of the block when the spring reaches its maximum compression.

Figure P10-8

10-9 The 64.4-lb body A in Fig. P10-9 slides in the smooth slot. In the position shown, the spring is compressed 1 ft, and body A has a velocity of 10 fps up the slot. The velocity of the body is reduced to 8 fps up the slot after moving 5 ft. The 130-lb force maintains a constant slope and magnitude during the motion. The spring is attached to the block and to point C. Determine the modulus of the spring.

Figure P10-9

*10-10 The 20-kg block A in Fig. P10-10 slides along the smooth slot. Spring B has a modulus of 1000 N/m, and spring C has a modulus of 500 N/m. In the position shown, spring B is stretched 0.80 m, spring C is unstressed, and the block has a velocity of 10 m/s down the slot. Determine the velocity of A when it reaches position D.

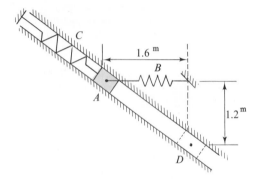

Figure P10-10

10-11 The 10-lb body A in Fig. P10-11 is 6 in. from the spring when its velocity is 6 fps down the plane. The modulus of the spring is 10 lb per inch, and the coefficient of friction between A and the inclined plane is 0.20. Determine the distance traveled down the plane by the body before coming to rest.

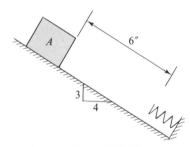

Figure P10-11

10-12 The 32.2-lb block W in Fig. P10-12 is released from rest in the position shown when the spring is stretched 1 ft. The spring constant is 4 lb per ft, and the coefficient of friction between the block and the inclined plane is 0.20. Determine
(a) The expression for the velocity of the block in terms of its displacement, x, during its motion down the plane.
(b) The maximum compression of the spring.

*10-13 The 30-kg block shown in Fig. P10-13 slides on a smooth plane and is connected to a spring with a modulus of 150 N/m. When in the position shown, the block has a velocity of 10 m/s up the plane, and the spring is stretched

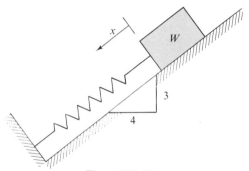

Figure P10-12

0.30 m. Determine the velocity of the block after it moves

(a) Two meters up the plane.

(b) Four meters up the plane.

Figure P10-13

*10-14 The block in Fig. P10-14 has a mass of 2 kg and slides along the rough ($\mu = 0.20$) plane. Spring A has a modulus of 15 N/m, and spring B has a modulus of 60 N/m. When the block has a velocity of 0.60 m/s to the right spring A is compressed 0.40 m and spring B is stretched 0.50 m. Determine the velocity of the block after it has moved 0.40 m.

Figure P10-14

10-15 The body in Fig. P10-15 weighs 32.2 lb and is constrained to slide along the smooth rod which lies in a vertical plane. The tension in the nonlinear spring is given by the equation $F = 9s^2$, where F is in pounds and s is the distance in

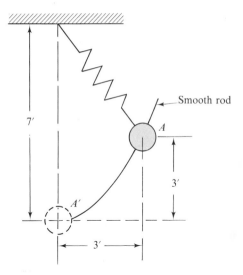

Figure P10-15

feet the spring is stretched. When the body is in position A it has a velocity of 5.0 fps down along the rod, and the spring is stretched 1.0 ft. Determine the velocity of the body when it reaches position A'.

10-16 A pendulum which weighs 3.86 lb and has a cord length of 10 in. is released from rest when the cord is horizontal as shown in Fig. P10-16. When the mass reaches the bottom of its path it strikes a spring with a modulus of 100 lb per in. Determine

(a) The maximum kinetic energy of the mass.

(b) The maximum compression of the spring.

(c) The tension in the cord just before the mass contacts the spring.

Figure P10-16

583

Figure P10-17

10-17 Body *A* in Fig. P10-17 weighs 25 lb, and the modulus of the spring is 10 lb per ft. Initially the system is in equilibrium, and the tension in the rope is zero. The drum starts to rotate and produces a constant tension in the rope of 50 lb. Determine the velocity of *A* after it has moved 5.0 ft up the smooth slot.

10-18 The particle *C* in Fig. P10-18 weighs 40 lb and is constrained to move along a smooth circular path which lies in a vertical plane. The particle is released from rest. The spring tension is 25 lb when in the position shown. Determine the modulus of the spring if the velocity of *C* is 8.0 fps to the left as it passes through point *B*.

Figure P10-18

Figure P10-19

***10-19** The 50-kg block *A* and the 20-kg block *B* in Fig. P10-19 are held in equilibrium by the force *P* and the spring. The spring has a modulus of 150 N/m and has a tension of 50 N in the position shown. The coefficient of friction between the blocks and the planes is 0.10. The force **P** is removed, allowing the blocks to move. Determine the velocity of *A* after it has moved 2.0 m.

10-20 The 322-lb body *A* in Fig. P10-20 slides on the smooth horizontal surface, and body *B* weighs 128.8 lb. In the position shown, the spring is stretched 1 ft, and body *A* has a velocity of 10 fps to the right. When body *A* has moved 6 ft its velocity is 5 fps to the right. Determine the modulus of the spring.

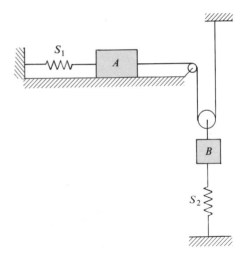

Figure P10-20

10-21 The 322-lb body A in Fig. P10-21 slides on the smooth horizontal surface, and the 64.4-lb body B is suspended from a pulley of negligible weight. The spring modulus is 200 lb per ft. In the position shown, body A has a velocity of 10 fps to the right, and the spring is stretched 2 ft. Determine the velocity of A when it has moved 5 ft to the right.

Figure P10-21

10-22 In Fig. P10-22 block A weighs 64.4 lb, block B weighs 32.2 lb, and the weight of the pulley is negligible. The modulus of spring S_1 is 50 lb per ft. When in the position shown, the tension in S_1 is 50 lb, the tension in S_2 is 300 lb, and the velocity of A is 10 fps to the right along the smooth plane. Determine the modulus of spring S_2 if body A comes to rest after moving 4 ft.

***10-23** The 50-kg body A in Fig. P10-23 has a velocity of 3 m/s to the right when the spring

Figure P10-22

(modulus of 50 N/m) is stretched 0.3 m. The coefficient of friction between A and the horizontal plane is 0.3. Body B has a mass of 30 kg. Determine the distance A will have moved when its velocity is 1 m/s to the right.

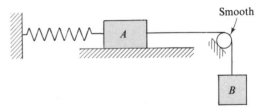

Figure P10-23

10-24 The 644-lb body A in Fig. P10-24 slides on a smooth horizontal plane. In the position shown, the 128.8-lb body B has a velocity of 8 fps downward, and the spring is stretched 2 ft. The modulus of the spring is 20 lb per ft. Determine the velocity of A when it passes under the smooth drum at C.

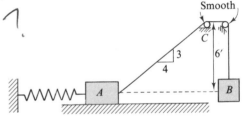

Figure P10-24

10-25 The 20-lb block A in Fig. P10-25 is connected to the 30-lb block B by the 20-in. bar of negligible weight. The blocks slide in smooth slots and are released from rest in the position shown. The spring has a modulus of 2 lb per in. and is stretched 6 in. in the indicated position. Determine the velocity of B when it has moved downward 4 in.

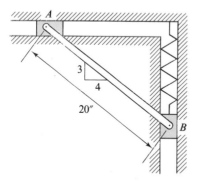

Figure P10-25

10-26 In Fig. P10-26, the two blocks A and B, each weighing 40 lb and connected by the rigid 25-in. bar of negligible mass, slide in smooth slots. In the position shown block A has velocity of 8 ips down the slot, and the spring (modulus of 16 lb per in.) is compressed 5 in. The constant force F has a magnitude of 65 lb and remains at the indicated slope throughout the motion. Determine the velocity of block A when it has moved 15 in. downward to point A'.

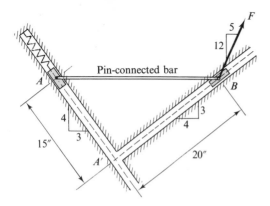

Figure P10-26

*10-27 In Fig. P10-27 body A has a mass of 15 kg, body B has a mass of 7 kg, and the modulus of the spring is 8.0 N/m. The plane is smooth. In the position shown, body A is at rest, and the tension in the spring is 60 N. Determine

(a) The velocity of A as it passes under the small peg at C.

(b) The tension in the cord as A passes under C.

Figure P10-27

Figure P10-28

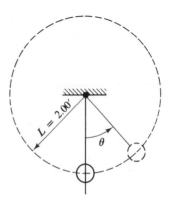

Figure P10-29

10-28 A particle of weight W starts from rest in position A in Fig. P10-28 and slides on the smooth cylindrical surface. Determine
(a) The velocity of the particle as a function of θ.
(b) The value of the angle θ when the particle leaves the surface.

10-29 The small body in Fig. P10-29 weighs 3.0 lb and swings in a vertical plane at the end of a 2.00-ft cord. When the angle θ is 20° the

velocity of the body is 9.00 fps. Determine
(a) The tension in the cord when $\theta = 60°$.
(b) The tension in the cord when $\theta = 0°$.

10-30 Determine the minimum speed that the body in Problem 10-29 must have when $\theta = 0$ if it is to travel in a full circle.

10-31 The speed of the body in Problem 10-29 is $2\sqrt{gL}$ when $\theta = 0$. Determine the angle θ when the cord becomes slack.

10-4 POTENTIAL ENERGY—CONSERVATION OF ENERGY

The potential energy of a body or system of bodies is the capacity of the body to do work as a result of the relative positions of the various particles of the system. If a spring is compressed, the material of the spring is stressed and energy is stored in the spring. The stresses in the spring tend to restore the spring to its original unstressed length, and the force exerted by the spring can do work as the spring returns to its original length. As long as the stresses in the spring remain below the elastic limit[1] for the material of the spring, the *elastic potential energy* or strain energy stored in the spring is equal to the work done on the spring in compressing it. Steam compressed in a boiler has potential energy due to its pressure, and this energy is transformed into work as the steam drives the piston. Steam also has thermal energy due to the temperature difference between the steam and the atmosphere. These two forms of stored energy in steam are closely related and are usually combined in describing the energy of the steam.

When a body is raised above any reference or datum plane, it is able to do work as it returns to the reference plane, as a result of the attraction of the earth for the body. The hammer of a pile driver is an example of such a body. Many pile drivers do work as a result of the attraction of the earth for the hammer. The *gravitational potential energy* is not stored in the hammer alone but rather in the system of bodies consisting of the hammer and the earth.

The work done by the weight **W** of particle P as it moves from position P_1 to position P_2, as shown in Fig. 10-12, (p. 588) is

$$U = -W(y_2 - y_1) = Wy_1 - Wy_2. \qquad (10\text{-}4)$$

When the work done by a force depends only on the initial and final positions of the particle and is independent of the path the force is called a conservative force. The force W in Fig. 10-12 is an example of

---

[1] The greatest stress which can be developed in a material without a permanent deformation remaining upon complete release of the stress is defined as the *elastic limit* of the material.

Figure 10-12

a conservative force. The scalar function Wy, which is equal to the work done by the force W in moving from the displaced position to the datum plane, is called the potential energy, V, of the gravity force with respect to the datum plane.

Equation (10-4) may be written in the form

$$U = V_1 - V_2 = -(V_2 - V_1) = -\Delta V. \qquad (10\text{-}4a)$$

From this equation it is observed that the work done by a conservative force is equal to the negative of its change in potential energy. Since only the change in the potential energy is involved in Eq. (10-4a), the choice of a datum plane is arbitrary because the potential energy varies only by a constant for different datum planes, and the *change* of the potential energies in moving from the initial to the final positions is always the same.

Body A of weight W, shown in Fig. 10-13(a), slides on a smooth horizontal plane and is attached to a spring having a modulus k and an unstretched length L_O. A free-body diagram of A is shown in Fig. 10-13(b). The potential energy of the spring force, ks, acting on body A is equal to the work done by the spring on body A as it moves from a stretched position (length $= L_O + s$) to a datum or reference position (length $= L_O$). The spring force on body A is to the left, and the movement is also to the left; therefore the work done during the displacement $\mathbf{s}$, which is the potential energy, must be positive. Thus

$$V = -\int_s^0 ks\, ds = -\frac{k}{2}\left[s^2 \right]_s^0 = \frac{ks^2}{2}, \qquad (10\text{-}5)$$

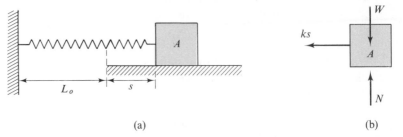

(a)

(b)

Figure 10-13

which is the potential energy of the spring force when the spring is stretched (or compressed) a distance s.

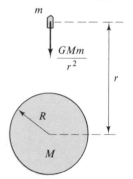

In the discussion of Fig. 10-12 the weight of the body was considered to be constant. When bodies move in space the variation of the gravity force with the distance from the center of the earth must be considered. A body of mass m is shown in Fig. 10-14 at a distance r from the center of the earth. The gravity force, GMm/r^2, from Newton's law of universal gravitation, is also shown. The potential energy of this force is the work done by it as it moves from the position shown to the reference position, which, for convenience, is selected at $r = \infty$. Since the force on the body is always toward the earth and the movement is away from the earth, the work done by the force on the body is negative. The potential energy is

$$V = -\int_r^\infty \frac{GMm}{r^2}\,dr = GMm\left[\frac{1}{r}\right]_r^\infty = -\frac{GMm}{r}. \qquad (10\text{-}6)$$

Figure 10-14

When a particle is acted upon by conservative forces, such as the weight and spring forces, the work and kinetic energy relationship as given by Eq. (10-3),

$$U = \Delta T = T_f - T_i,$$

can be written in a form involving potential energy. Since the work done by conservative forces is equal to the negative of the change in potential energy of these forces, this equation becomes

$$-(V_f - V_i) = T_f - T_i,$$

where V_f and V_i are the potential energies in the final and initial positions. This equation can be rewritten as

$$V_i + T_i = V_f + T_f, \qquad (10\text{-}7)$$

which states that the sum of the initial potential and kinetic energies must equal the sum of the final potential and kinetic energies. This statement is known as the principle of conservation of mechanical energy.

EXAMPLE 10-5

The small 2-lb collar of Fig. 10-15 is attached to a spring which has a modulus of 4 lb per ft and an unstretched length of 3 ft. The collar slides along the fixed smooth circular wire which is in a vertical plane, and when in the position A the collar has a velocity of 3 fps to the

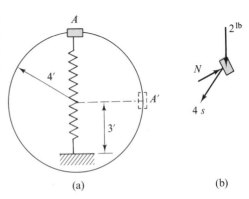

(a)

(b)

Figure 10-15

589

right. Determine the velocity of the collar when it reaches the position
A'.

SOLUTION

A free-body diagram of the particle in an intermediate position is
shown in Fig. 10-15(b). The weight and spring forces are conservative,
and the normal force, N, does no work and so will not enter into the
conservation of energy equation which will be used in solving this
problem. The datum plane for the weight is chosen as the horizontal
diameter of the circle, and the unstretched length of the spring ($s = 0$) is
used as the reference position for the spring.
The conservation of energy equation is

$$V_i + T_i = V_f + T_f. \tag{a}$$

The initial stretch of the spring is $s_i = 4$ ft, and the
final stretch is $s_f = 2$ ft. Thus

$$V_i = \frac{1}{2} ks_i^2 + Wh_i = \frac{1}{2}(4)(4)^2 + 2(4) = 40 \text{ ft-lb},$$

$$T_i = \frac{1}{2} mv_i^2 = \frac{1}{2}\left(\frac{2}{32.2}\right)(3)^2 = 0.28 \text{ ft-lb},$$

$$V_f = \frac{1}{2} ks_f^2 + Wh_f = \frac{1}{2}(4)(2)^2 + 0 = 8 \text{ ft-lb},$$

$$T_f = \frac{1}{2} mv_f^2 = \frac{1}{2}\left(\frac{2}{32.2}\right)v_f^2 = 0.0311v_f^2.$$

(a)

(b)

Figure 10-15

When these values are substituted into Eq. (a) it becomes

$$40 + 0.28 = 8 + 0.0311v_f^2,$$

which gives

$$v_f = 32.2 \text{ fps}$$

and

$$\mathbf{v}_f = 32.2 \text{ fps downward.} \qquad \text{Ans.}$$

PROBLEMS

10-32 The small 6.00-lb body in Fig. P10-32
swings in a vertical plane at the end of a 10-ft
cord. The angular velocity of the cord is 0.50
rad per sec clockwise in the position shown.
(a) Use the principle of conservation of mechanical energy to determine the velocity of the body
when in its lowest position.
(b) Determine the tension in the cord when the
body is in its lowest position.

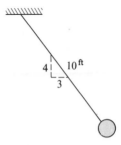

Figure P10-32

*10-33 The 150-kg block in Fig. P10-33 slides
down the smooth plane. In the position shown

the block has a velocity of 3 m/s down the plane
and the spring (modulus = 25 N/m) is stretched

0.3 m. Use the conservation of mechanical energy equation to determine the velocity of the block after it has moved 1.5 m down the plane.

Figure P10-33

10-34 The 128.8-lb body in Fig. P10-34 slides along the smooth rod BC and is connected to springs S_1 and S_2. The modulus of S_1 is 40 lb per ft and its unstretched length is 7 ft. The modulus of S_2 is 20 lb per ft, and its unstretched length is 8 ft. The body has a velocity of 10 fps down the rod when in position A. Determine its velocity when it reaches A'.

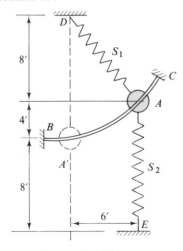

Figure P10-34

10-35 The 12.88-lb body A in Fig. P10-35 slides in the smooth slot and is attached to a spring having a modulus of 4.0 lb per ft. When the spring is compressed 2 ft body A has a

velocity of 3 fps down the slot. The 13-lb force is constant throughout the motion. Determine the velocity of A after it has moved 10 ft down the slot.

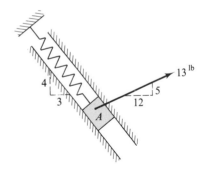

Figure P10-35

10-36 A satellite of mass m describes an orbit about the earth as shown in Fig. P10-36. When the satellite is at position A it has a speed of v_1. Use the principle of conservation of mechanical energy to show that the speed at position B is

$$v_2 = \left[v_1^2 + \frac{2gR^2(r_1 - r_2)}{r_1 r_2} \right]^{1/2}$$

where g is the acceleration of gravity at the earth's surface.

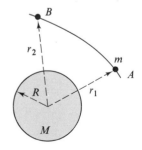

Figure P10-36

*__10-37__ Solve Problem 10-27(a) by using the conservation of mechanical energy equation.

10-38 Solve Problem 10-28 by using the conservation of mechanical energy equation.

10-5 · CONSERVATIVE FORCES

As indicated in Art. 10-4, if the work done by a force as it moves from an initial to a final position is independent of the path traveled, the force is called a conservative force. When the potential energy of a conservative force can be expressed as a function of its position, that is, when $V = V(x, y, z)$, the differential of the potential energy is

$$dV = \frac{\partial V}{\partial x}\, dx + \frac{\partial V}{\partial y}\, dy + \frac{\partial V}{\partial z}\, dz.$$

The work done by the force F during a displacement $d\mathbf{r} = \mathbf{i}\, dx + \mathbf{j}\, dy + \mathbf{k}\, dz$ is

$$dU = \mathbf{F} \cdot d\mathbf{r} = F_x\, dx + F_y\, dy + F_z\, dz.$$

From Eq. (10-4a) $dU = -dV$; therefore

$$F_x\, dx + F_y\, dy + F_z\, dz = -\frac{\partial V}{\partial x}\, dx - \frac{\partial V}{\partial y}\, dy - \frac{\partial V}{\partial z}\, dz$$

or

$$\left(F_x + \frac{\partial V}{\partial x}\right) dx + \left(F_y + \frac{\partial V}{\partial y}\right) dy + \left(F_z + \frac{\partial V}{\partial z}\right) dz = 0.$$

Since the force is conservative and the work done is independent of the path, the preceding equation must be satisfied for any arbitrary choice of dx, dy, and dz. This requires that the coefficients of the differentials must be zero. Hence

$$F_x = -\frac{\partial V}{\partial x}, \qquad F_y = -\frac{\partial V}{\partial y}, \qquad F_z = -\frac{\partial V}{\partial z}. \qquad (10\text{-}8)$$

These equations state that if a force is conservative, its components are given by the negative partial derivatives of its potential function as indicated by Eq. (10-8).

When the partial derivatives of F_x and F_y are taken with respect to y and x, respectively, they become

$$\frac{\partial F_x}{\partial y} = -\frac{\partial^2 V}{\partial y\, \partial x} \quad \text{and} \quad \frac{\partial F_y}{\partial x} = -\frac{\partial^2 V}{\partial x\, \partial y}.$$

Since the order of the partial derivatives is immaterial, it follows that

$$\frac{\partial F_x}{\partial y} = \frac{\partial F_y}{\partial x}. \qquad (10\text{-}9a)$$

In a similar manner it can be shown that

$$\frac{\partial F_y}{\partial z} = \frac{\partial F_z}{\partial y} \qquad (10\text{-}9b)$$

and

$$\frac{\partial F_z}{\partial x} = \frac{\partial F_x}{\partial z}. \qquad (10\text{-}9c)$$

These three equations must be satisfied by any force which is conservative. Thus Eq. (10-9) provide a convenient method to determine whether or not a force is conservative.

EXAMPLE 10-6

Show that the force $\mathbf{F} = x^2\mathbf{i} + y^2\mathbf{j} + z^2\mathbf{k}$ is conservative and determine the potential function of the force.

SOLUTION

Since $F_x = x^2$, $\partial F_x/\partial y = \partial F_x/\partial z = 0$. Similarly, $\partial F_y/\partial x = \partial F_y/\partial z = 0$ and $\partial F_z/\partial x = \partial F_z/\partial y = 0$. Consequently, Eq. (10-9a, b, c) are satisfied, and the force is conservative.

From Eq. (10-8),

$$F_x = x^2 = -\frac{\partial V}{\partial x}$$

from which

$$V = -\frac{x^3}{3} + f_1(y, z),$$

where $f_1(y, z)$ is a function of y and z but not of x. In a similar manner,

$$F_y = y^2 = -\frac{\partial V}{\partial y} \quad \text{and} \quad F_z = z^2 = -\frac{\partial V}{\partial z},$$

which give

$$V = -\frac{y^3}{3} + f_2(x, z)$$

and

$$V = -\frac{z^3}{3} + f_3(x, y).$$

It can be shown by direct substitution in these three equations for V that

$$f_1(y, z) = -\frac{y^3 + z^3}{3} + C_1,$$

$$f_2(x, z) = -\frac{x^3 + z^3}{3} + C_2,$$

and

$$f_3(x, y) = -\frac{x^3 + y^3}{3} + C_3,$$

and thus the potential energy function can be written as

$$V = -\frac{x^3 + y^3 + z^3}{3} + C, \qquad \text{Ans.}$$

where C is an arbitrary constant which can be determined if the value of the potential energy is known for any specified set of values of x, y, and z.

PROBLEMS

10-39 The force $\mathbf{F} = [xy\mathbf{i} + (x^2/2)\mathbf{j}]$ lb acts on a particle which moves in the xy plane.

(a) Show that the force is conservative.
(b) Determine the potential energy function.

Figure P10-40

10-40 Determine the work done by the force of Problem 10-39 as it moves from O to B along the paths shown in Fig. P10-40
(a) From O to A to B.

(b) From O to B along the straight line OB.
(c) From O to B along the parabola $y^2 = x$.

10-41 Show that the force $\mathbf{F} = (x^2\mathbf{i} + xy\mathbf{j})$ lb is nonconservative and determine the work done by the force in moving from O to B along the straight line and along the parabola $y^2 = x$ as shown in Fig. P10-40.

10-42 Which of the following forces are conservative?
(a) $\mathbf{F} = (2x\mathbf{i} + 3xy\mathbf{j} - 6\mathbf{k})$ lb.
(b) $\mathbf{F} = (x\mathbf{i} + y\mathbf{j} + z\mathbf{k})$ lb.
(c) $\mathbf{F} = (4\mathbf{i} + 3y\mathbf{j} + xy\mathbf{k})$ lb.
Determine the potential energy function of the forces which are conservative.

PART B—RIGID BODIES

10-6 WORK OF COUPLES AND FORCES ON RIGID BODIES

The work done by a force was shown to be given by the dot product of two vectors, force and linear displacement. The work done by a couple can be obtained by applying this definition to each force of the couple.

Consider the work done by the couple in Fig. 10-16(a) as it rotates through the infinitesimal angle $d\theta$. The work done by the two forces is

$$U = \mathbf{P}_1 \cdot d\mathbf{r}_1 + \mathbf{P}_2 \cdot d\mathbf{r}_2,$$

where $\mathbf{r}_1$ and $\mathbf{r}_2$ are the position vectors to forces $\mathbf{P}_1$ and $\mathbf{P}_2$, respectively, and $d\mathbf{r}_1$ and $d\mathbf{r}_2$ are the displacements of the two forces. Since $\mathbf{P}_1$ and $d\mathbf{r}_1$ are in the same direction, $\mathbf{P}_1$ does positive work, while $\mathbf{P}_2$

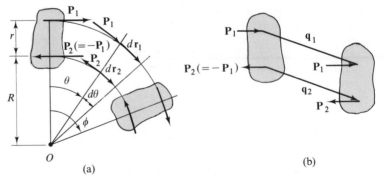

Figure 10-16

does negative work by the same reasoning. Thus the work can be written as

Section 10-6
Work of couples and
forces on rigid bodies

$$dU = P_1 \, dr_1 - P_2 \, dr_2.$$

From the definition of a couple, P_1 is equal to P_2, and from the geometry of the figure,

$$dr_1 = (R + r) \, d\theta \quad \text{and} \quad dr_2 = R \, d\theta.$$

Therefore, the work of the couple is

$$dU = P_1(R + r) \, d\theta - P_1 R \, d\theta = P_1 r \, d\theta = T \, d\theta.$$

where T is the moment of the couple.

The work done during a finite angular displacement can be obtained by integration of the expression

$$U = \int_{\theta_i}^{\theta_f} T \, d\theta, \qquad (10\text{-}10)$$

and if T is constant, the work is

$$U = T(\theta_f - \theta_i) = T \, \Delta\theta,$$

where $\Delta\theta$ is the change of θ, that is, the angular displacement of the couple. The work is positive if the angular displacement is in the same direction as the sense of rotation of the couple and negative if the displacement is in the opposite direction.

If the body is translated as shown in Fig. 10-16(b), the work done by the couple is

$$U = \mathbf{P}_1 \cdot \mathbf{q}_1 + \mathbf{P}_2 \cdot \mathbf{q}_2 = \mathbf{P}_1 \cdot \mathbf{q}_1 - \mathbf{P}_1 \cdot \mathbf{q}_1 = 0$$

since $\mathbf{P}_2 = -\mathbf{P}_1$ and $\mathbf{q}_2 = \mathbf{q}_1$. Thus a couple does no work due to a translation of the body on which it acts. When a couple acts on a body which is simultaneously translated and rotated the couple does work only as a result of the rotation.

When the magnitude of T is a variable, Eq. (10-10) must be integrated to obtain the work done. Moment-angular position diagrams (similar to F-s diagrams) are frequently useful for problems involving variable couples, particularly when the T-θ diagram is a straight line. Linear variations occur for some torsion springs (such as a watch spring), in which case the moment and angular position are related by

$$T = k\theta,$$

where k is the torsional spring constant and is often expressed in such units as foot-pounds per radian, inch-ounces per radian, or newton meters per radian.

The work done on a *rigid* body by a system of external forces and couples is the algebraic sum of works done by the individual forces and couples. The internal forces of a body occur in equal and opposite collinear pairs, and if the body is rigid, the component of the displacement of each force of the pair in the direction of the forces must be equal in magnitude; therefore the resultant work done by the internal forces is zero. *If the body is not rigid, this statement is not valid since work will be done by the internal forces.* When two or more rigid bodies are connected by smooth pins or by flexible, inextensible cables the resultant work done by the connecting members of the system is also zero. The two forces at the ends of a cable have the same magnitude (the mass of the cable is assumed to be negligible), and if the cable is inextensible, the components of the displacements of the two ends in the direction of the forces must have the same magnitudes; therefore, the resultant work done by the cable must be zero. When two bodies are connected by an *extensible* cable the bodies will do work on the cable when its length is changed, and the cable will do an equal amount of work on the bodies with an opposite sign. Thus if the cable is stretched, the bodies will do positive work on the cable, while the cable will do negative work on the bodies. The following examples demonstrate the calculation of the work done by various forces and force systems.

EXAMPLE 10-7

The bar AB in Fig. 10-17(a) weighs 20 lb and is connected to blocks at the ends which slide in smooth horizontal and vertical slots as indicated. The spring OA is arranged so that it can act in either tension or compression. It has a modulus of 5.0 lb per in. and an unstressed length of 7 in. Determine the work done by all forces acting on the bar as it moves to the horizontal position $A'B'$ as indicated by dashed lines. The weights of the blocks at A and B may be neglected.

SOLUTION

A free-body diagram of the bar in an intermediate position is shown in Fig. 10-17(b). There are four forces and a couple acting on the body. Two of the forces, $\mathbf{R}_1$ and $\mathbf{R}_2$, are normal to their corresponding displacements, and as a result they do no work. The resultant work on the bar is thus

$$U = U_{\text{spring}} + U_{\text{weight}} + U_{\text{couple}}.$$

An F-s diagram for the spring is shown in Fig. 10-17(c). The length of the spring changes 8 in., from 5 to 13 in. The spring is initially compressed 2 in. and exerts a 10-lb force on the block A. In the final position, the spring is stretched 6 in. and has a tensile force of 30 lb. The area C represents negative work *on the spring* since the spring is in compression but is increasing in length. The area C represents positive work on block A since the vertical component of the force on A and the displacement are in the same (downward) direction. In a similar manner, the area D represents negative work of the spring on block A.

Section 10-6
Work of couples and
forces on rigid bodies

(a)   (b)

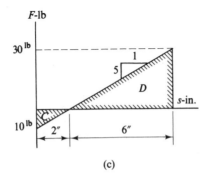

(c)

Figure 10-17

The work of the spring on AB is

$$U_{\text{spring}} = \frac{10(2)}{2} - \frac{30(6)}{2} = -80 \text{ in-lb on } AB.$$

The work done by the weight is

$$U_{\text{weight}} = 20(6) = 120 \text{ in-lb on } AB,$$

and the work done by the couple is

$$U_{\text{couple}} = 60\frac{\pi}{2} = 94.2 \text{ in-lb on } AB.$$

The total work done on AB is

$$U = -80 + 120 + 94.2 = \underline{134.2 \text{ in-lb.}} \qquad \text{Ans.}$$

597

EXAMPLE 10-8

The 20-lb wheel A in Fig. 10-18(a) is connected to the 50-lb block B by means of an inextensible cord of negligible weight which passes over a smooth peg at C. The wheel rolls on plane E without slipping, and the coefficient of friction between B and plane H is 0.30. Determine the work done on the system of bodies (A and B) while body A moves 10 ft up the plane E.

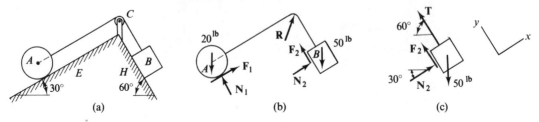

Figure 10-18

SOLUTION

The free-body diagram of the system of bodies is drawn in Fig. 10-18(b). The forces N_1, N_2, and R are perpendicular to the motion of their points of application, and so they do no work. Since the wheel rolls without slipping, the frictional force F_1 also does no work because the particle of the wheel on which F_1 acts has no velocity and hence no displacement while the force is acting on it.[2] Thus the two weights and the frictional force F_2 are the only forces that do work on the system of bodies.

As body A rolls 10 ft up to the right on plane E, block B slides 10 ft down to the right on plane H. From the free-body diagram of block B in Fig. 10-18(c)

$$\sum \mathbf{F}_x = 0,$$

since there is no acceleration in the x direction. Therefore,

$$[N_2 - 0.50(50)]\mathbf{i} = 0$$

or

$$\mathbf{N}_2 = 25\mathbf{i} \text{ lb},$$

and

$$\mathbf{F}_2 = \mu N_2 \mathbf{j} = 0.30(25)\mathbf{j} = 7.5\mathbf{j} \text{ lb}.$$

---

[2] Forces N_1 and F_1 do no work because they are the components of the force exerted by one particle (or, more exactly, one group of adjacent particles) of the plane on one particle of the periphery of the wheel at an instant when neither particle is moving. At the next instant, the force will be exerted by another particle of the plane on an adjacent particle of the wheel. The work done is zero because none of the forces moves. Each acts for an instant as a holding force and becomes zero as a new force exerted by the next particle becomes the holding force or reaction of the plane on the wheel. If a wheel rolls and slips on a fixed plane or rolls without slipping on a moving plane, the frictional force will do work.

The work done by $\mathbf{F}_2$ is

$$U_{F_2} = \mathbf{F}_2 \cdot \mathbf{q}_B = (7.5\mathbf{j}) \cdot (-10\mathbf{j}) = -75 \text{ ft-lb.}$$

The work done by the weight of A is

$$U_{20\text{lb}} = (-10.0\mathbf{i} - 17.32\mathbf{j}) \cdot (10\mathbf{i}) = -100 \text{ ft-lb.}$$

The work done by the weight of B is

$$U_{50\text{lb}} = (-25.0\mathbf{i} - 43.3\mathbf{j}) \cdot (-10\mathbf{j}) = 433 \text{ ft-lb.}$$

The total work done on the two bodies is

$$U = -75 - 100 + 433 = \underline{258 \text{ ft-lb.}} \qquad \text{Ans.}$$

Section 10-7
Kinetic energy of a rigid
body—plane motion

10-7 KINETIC ENERGY OF A RIGID BODY—PLANE MOTION

The kinetic energy of any body, whether rigid or not, is the sum of the kinetic energies of the separate particles. If the body is not rigid, there is no simple equation relating the velocities of the different particles. The velocities of the various particles in rigid bodies, however, are directly related to each other and to the angular motion of the body. An expression for the kinetic energy of a rigid body which has plane motion is developed here. An expression for the kinetic energy of a rigid body with general motion will

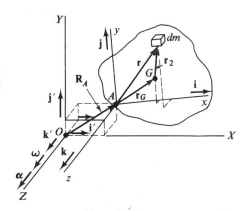

Figure 10-19

be developed in Art. 11-9 in terms of the angular momentum of the body. Consider the body in Fig. 10-19, which has plane motion parallel to the XY plane. Point A is any point fixed in the body and is the origin of the moving xyz axes which move with the body. The xy plane is parallel to the XY plane. The velocity of any particle of mass dm of the body is (see the figure)

$$\mathbf{v} = \dot{\mathbf{R}}_A + \dot{\mathbf{r}} = \dot{\mathbf{R}}_A + \boldsymbol{\omega} \times \mathbf{r},$$

where $\boldsymbol{\omega}$ is the angular velocity of the body. The kinetic energy of the particle is

$$dT = \frac{dm}{2} v^2 = \frac{dm}{2} \mathbf{v} \cdot \mathbf{v}$$

$$= \frac{dm}{2} [\dot{\mathbf{R}}_A + \boldsymbol{\omega} \times \mathbf{r}] \cdot [\dot{\mathbf{R}}_A + \boldsymbol{\omega} \times \mathbf{r}]$$

$$= \frac{dm}{2} \dot{R}_A^2 + dm \dot{\mathbf{R}}_A \cdot (\boldsymbol{\omega} \times \mathbf{r}) + \frac{dm}{2} (\boldsymbol{\omega} \times \mathbf{r}) \cdot (\boldsymbol{\omega} \times \mathbf{r}).$$

The body has plane motion; therefore, $\boldsymbol{\omega} = \omega \mathbf{k}$ and $\mathbf{r} = x\mathbf{i} + y\mathbf{j} + z\mathbf{k}$, from which

$$\boldsymbol{\omega} \times \mathbf{r} = \omega(x\mathbf{j} - y\mathbf{i})$$

and

$$(\boldsymbol{\omega} \times \mathbf{r}) \cdot (\boldsymbol{\omega} \times \mathbf{r}) = \omega^2 (x^2 + y^2).$$

From Fig. 10-19, $\mathbf{r} = \mathbf{r}_G + \mathbf{r}_2$, and the kinetic energy of the particle can be written

$$dT = \frac{dm}{2} \dot{R}_A^2 + dm \dot{\mathbf{R}}_A \cdot (\boldsymbol{\omega} \times \mathbf{r}_G) + dm \dot{\mathbf{R}}_A \cdot (\boldsymbol{\omega} \times \mathbf{r}_2) + \frac{dm}{2} \omega^2 (x^2 + y^2).$$

Since $\dot{\mathbf{R}}_A$, $\mathbf{r}_G$, and $\boldsymbol{\omega}$ do not vary from point to point on the body, the kinetic energy of the body can be determined by integration over the mass as

$$T = \frac{\dot{R}_A^2}{2} \int_m dm + \dot{\mathbf{R}}_A \cdot (\boldsymbol{\omega} \times \mathbf{r}_G) \int_m dm + \dot{\mathbf{R}}_A \cdot \left(\boldsymbol{\omega} \times \int_m \mathbf{r}_2 \, dm \right)$$

$$+ \frac{\omega^2}{2} \int_m (x^2 + y^2) \, dm$$

$$= \frac{m \dot{R}_A^2}{2} + m \dot{\mathbf{R}}_A \cdot (\boldsymbol{\omega} \times \mathbf{r}_G) + 0 + \frac{I_z \omega^2}{2}. \tag{10-11}$$

Note that the quantity $\int_m \mathbf{r}_2 \, dm$ is zero because $\mathbf{r}_2 \, dm$ is the first moment of mass with respect to the mass center (see Chapter 3).

Since A is any point in the body, Eq. (10-11) can be simplified by the proper choice of point A. If A is selected at the mass center, $\mathbf{r}_G$ becomes zero, and the equation reduces to

$$T = \frac{m v_G^2}{2} + \frac{I_z \omega^2}{2}, \tag{10-11a}$$

where $\dot{R}_A$ is v_G and the z axis passes through G.

If point A is a point of zero velocity (the instantaneous axis of zero velocity or a point on the axis of rotation for a body which has rotation), $\dot{R}_A$ is zero, and the equation becomes

$$T = \frac{I_z \omega^2}{2}, \tag{10-11b}$$

where the z axis is the instantaneous axis of zero velocity or the fixed axis of rotation for bodies having rotation.

When the body has translation, $\boldsymbol{\omega}$ is zero, and Eq. (10-11) reduces to

$$T = \frac{m v^2}{2}, \tag{10-11c}$$

where v is the magnitude of the velocity of any point on the body since all points have the same velocity.

Since the kinetic energy of a rigid body is a positive scalar quantity, the kinetic energy of a system of bodies is the sum of the kinetic energies of the bodies of the system.

Section 10-7
Kinetic energy of a rigid
body—plane motion

EXAMPLE 10-9

The slender homogeneous bar AC in Fig. 10-20 weighs 25 lb, and body B weighs 100 lb. Bar AC rotates about a horizontal axis at A, and B is pinned to AC at C. The radius of gyration of the mass of B with respect to a horizontal axis through its mass center G is 1.50 ft. In the position shown, AC has an angular velocity of 2.0 rad per sec clockwise, and B has an angular velocity of 5.0 rad per sec clockwise. Determine the kinetic energy of the system.

SOLUTION

The bar rotates about a fixed axis at A, and body B has plane motion. The kinetic energy of the system is the sum of the kinetic energies of the two bodies; thus

Figure 10-20

$$T = T_{AC} + T_B.$$

The velocity of G can be determined by writing a relative velocity equation for points G and C; that is,

$$\mathbf{v}_G = \mathbf{v}_C \qquad + \qquad \mathbf{v}_{G/C}$$

$$\begin{array}{c}\omega = 2 \\ 6\,(2)\end{array} \qquad + \qquad \begin{array}{c}1\,(5) \\ \omega = 5\end{array}$$

$$\downarrow \qquad\qquad \nearrow$$
$$= 12 + 5 \quad = 13 \text{ fps}$$

The moments of inertia of mass are determined by the procedures discussed in Chapter 6. The kinetic energy of AC, from Eq. (10-11b), is

$$T_{AC} = \frac{1}{2} I_A \omega_{AC}^2$$

$$= \frac{1}{2}\left(\frac{1}{3} mL^2\right)\omega_{AC}^2$$

$$= \frac{1}{2}\left[\frac{1}{3}\left(\frac{25}{32.2}\right)(6)^2\right](2)^2$$

$$= 18.63 \text{ ft-lb.}$$

The kinetic energy of body B, from Eq. (10-11a), is

$$T_B = \frac{1}{2} m_B v_G^2 + \frac{1}{2} I_G \omega_B^2$$

$$= \frac{1}{2} m_B v_G^2 + \frac{1}{2}(m_B k_G^2)\omega_B^2$$

$$= \frac{1}{2}\left(\frac{100}{32.2}\right)(13)^3 + \frac{1}{2}\left[\left(\frac{100}{32.2}\right)(1.5)^2\right](5)^2$$

$$= 350 \text{ ft-lb.}$$

The kinetic energy of the system is

$$T = 18.63 + 350$$

$$= 369 \text{ ft-lb.} \qquad \text{Ans.}$$

10-8 PRINCIPLE OF WORK AND KINETIC ENERGY—PLANE MOTION

The principle of work and kinetic energy, as applied to particles, was developed in Art. 10-3. The principle states that the work done on a particle (by all forces acting on the particle) is equal to the change of kinetic energy of the particle; that is,

$$U = \Delta T = T_f - T_i.$$

The principle of work and kinetic energy can be extended to include bodies and systems of bodies, since bodies are composed of particles. The total work done by all the forces acting on the particles of a body is equal to the sum of the changes of the kinetic energy of the particles, which is the same as the change of the kinetic energy of the body. The internal forces between the particles can do work as well as the external forces due to outside bodies. If the body is rigid, however, the resultant work done by the internal forces is zero, as was indicated in Art. 10-6. Furthermore, if two or more bodies are connected by inextensible cords or cables, the net work done by the connecting members is zero, and the work done by the external forces on the entire system is equal to the change of the kinetic energy of the system of rigid bodies. Thus the principle of work and kinetic energy *for a system of rigid bodies* is expressed

$$U_{\text{ext}} = \Delta T = T_f - T_i. \qquad (10\text{-}12)$$

The general equation can be written in the following special forms for the particular type of motion indicated:

translation: $\quad U = \frac{1}{2}m(v_f^2 - v_i^2),$ $\qquad (10\text{-}12a)$

rotation: $\quad U = \frac{1}{2}I_z(\omega_f^2 - \omega_i^2),$ $\qquad (10\text{-}12b)$

plane motion: $\quad U = \frac{1}{2}m[(v_G)_f^2 - (v_G)_i^2] + \frac{1}{2}I_z(\omega_f^2 - \omega_i^2).$ $\qquad (10\text{-}12c)$

Note that the z axis used to calculate I_z is the axis of rotation in Eq. (10-12b) and the axis through the mass center perpendicular to the plane of motion in Eq. (10-12c). The work and kinetic energy equation for plane motion can also be written in terms of the moment of inertia of mass with respect to the instantaneous axis of zero velocity. Since the moment of inertia is not necessarily constant with respect to the axis through the instantaneous center, the equation for plane motion is

$$U = \frac{1}{2}(I_O)_f\omega_f^2 - \frac{1}{2}(I_O)_i\omega_i^2, \qquad (10\text{-}12d)$$

where $(I_O)_i$ and $(I_O)_f$ are the moments of inertia of the mass of the body with respect to the initial and final instantaneous axes of zero velocity.

Equation (10-12) is not limited to any specific types of motion and can be used when Eq. (10-12a)–(10-12d) do not apply. Since work and energy are scalar quantities, free-body diagrams involving two or more bodies connected by smooth pins, flexible cables, and so on, can be used without encountering the difficulties which arise with vector quantities when working with force, mass, and acceleration.

EXAMPLE 10-10

The unbalanced wheel A in Fig. 10-21(a) weighs 100 lb and has a radius of gyration of mass with respect to a horizontal axis through O of 0.80 ft. The wheel rolls along the horizontal plane without slipping and is connected to the 50.0-lb body B by means of an inextensible cord which passes over the smooth pulley at C and is wrapped around A. Neglect the mass of the pulley at C. The modulus of the spring S is 20.0 lb per ft, and its mass may be neglected. The spring is attached to the wheel at O. When the mass center G of the wheel is in the position indicated on the diagram, the tension in the spring is 50.0 lb, and the velocity of O is 4.0 fps to the left. Determine the velocity of O when the wheel has rolled 90° counterclockwise.

SOLUTION

Figure 10-21(b) is a free-body diagram of the system of bodies. The equation relating the work and kinetic energy of this system is

$$U = T_f - T_i$$

$$U_{100} + U_{50} + U_P = \tfrac{1}{2}m_A[(v_G)_f^2 - (v_G)_i^2]$$

$$+ \tfrac{1}{2}(I_z)_A[(\omega_A)_f^2 - (\omega_A)_i^2]$$

$$+ \tfrac{1}{2}m_B[(v_B)_f^2 - (v_B)_i^2].$$

The forces $\mathbf{F}$, $\mathbf{N}$, and $\mathbf{Q}$ do no work, as was shown in Art. 10-6. The

(a)   (b)

Figure 10-21

final positions of the two bodies are indicated by dashed lines. The displacement of point O is

$$q_O = -r(\Delta\theta)\mathbf{i} = -1\frac{\pi}{2}\mathbf{i} = -1.571\mathbf{i}\,\text{ft}.$$

The displacement of body B has the same magnitude as the displacement of point E on the cord, which can be obtained by relative motion. Thus,

$$q_E = q_O + q_{E/O} = q_O - r(\Delta\theta)\mathbf{i}$$

$$= -1.571\mathbf{i} - \left(\frac{\pi}{2}\right)\mathbf{i}$$

$$= -3.142\mathbf{i}\,\text{ft}$$

and

$$q_B = 3.14\mathbf{j}\,\text{ft}.$$

The displacement of the mass center G has both a horizontal and vertical component, but since the force at G (the weight) is vertical, only the vertical component of the displacement of G (0.50 ft downward) is needed. The work done on the system by the two weights is

$$U_{100} = F(q_G)_y = 100(0.50) = 50.0 \text{ ft-lb},$$

$$U_{50} = Fq_B = -50(3.142) = -157.1 \text{ ft-lb}.$$

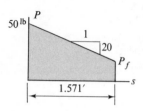

Figure 10-22

The force of the spring on the body varies as the wheel rolls, and the work done can be determined by integration or from the F-s diagram for the spring (Fig. 10-22). The final force is determined from the slope of the diagram as follows:

$$k = 20 = \frac{50.0 - P_f}{1.571}$$

from which

$$P_f = 50.0 - 20(1.571) = 18.58 \text{ lb}.$$

The work done by the spring is proportional to the area under the diagram and is

$$U_P = \frac{50.0 + 18.58}{2}(1.571) = 53.9 \text{ ft-lb}.$$

The total work on the system of bodies is

$$U = 50 - 157.1 + 53.9 = -53.2 \text{ ft-lb}.$$

Equations relating the various velocities can be determined from kinematics using the instantaneous center of zero velocity of A. Thus,

$$(\omega_A)_i = \frac{(v_O)_i}{r} = \frac{4.00}{1.00} = 4.00 \text{ rad per sec},$$

$$(v_G)_i = (1.00^2 + 0.50^2)^{1/2}(\omega_A)_i = 1.118(4.00) = 4.47 \text{ fps},$$

$$(v_B)_i = 2.00(\omega_A)_i = (2.00)(4.00) = 8.00 \text{ fps}.$$

Similarly,

$$(v_G)_f = 0.50(\omega_A)_f, \qquad (v_B)_f = 2.00(\omega_A)_f, \qquad (v_O)_f = 1.00(\omega_A)_f.$$

The moment of inertia $(I_z)_A$, where the z axis passes through G, as obtained by the parallel-axis theorem, is

$$(I_z)_A = I_O - md^2$$

$$= \frac{100}{32.2}(0.80)^2 - \frac{100}{32.2}(0.50)^2 = 1.211 \text{ slug-ft}^2.$$

When these values are substituted in the work and kinetic energy equation, it becomes

$$-53.2 = \frac{1}{2}\frac{100}{32.2}[0.50^2(\omega_A)_f^2 - 4.47^2] + \frac{1}{2}(1.211)[(\omega_A)_f^2 - 4.00^2]$$

$$+ \frac{1}{2}\frac{50}{32.2}[2.00^2(\omega_A)_f^2 - 8.00^2]$$

from which

$$(\omega_A)_f = 3.01 \text{ rad per sec}$$

and

$$(\boldsymbol{\omega}_A)_f = 3.01\mathbf{k} \text{ rad per sec.}$$

The magnitude of the final velocity of O is

$$(v_O)_f = 1.00(\omega_A)_f = 3.01 \text{ fps}$$

and

$$(\mathbf{v}_O)_f = -3.01\mathbf{i} \text{ fps.} \qquad \qquad \text{Ans.}$$

Since $(\omega_A)_f$ is obtained as a square root, it can be either positive or negative, and the direction is determined from the statement of the problem. In this example, the wheel is turning counterclockwise initially, and the velocity is to be determined after it turns 90° counterclockwise; therefore, the final angular velocity must also be counterclockwise.

PROBLEMS

Note: Unless otherwise specified, the mass of all springs and cords may be neglected. All cords are to be considered flexible and inextensible, and all pins are smooth. Where cords or cables are shown tangent to a circular body, it is assumed that they are wrapped around it and that they wind up on the body or unwind from it as the body moves unless the statement of the problem indicates otherwise.

10-43 The angular velocity of the 3.86-lb homogeneous bar in Fig. P10-43 is 10 rad per sec clockwise in the vertical position. After the bar rotates 90° clockwise, it strikes the two springs and continues to rotate until the spring

Figure P10-43

S_1 is compressed 0.20 in. The modulus of S_1 is twice that of S_2, and the springs are unstressed when the bar first strikes them. Determine the modulus of S_1.

*10-44 The 50-kg homogeneous bar AB in Fig. P10-44 has an angular velocity of 4 rad/s clockwise in the position shown, and the spring is stretched 0.3 m in this position. The modulus of the spring is 25 N/m. Will the bar reach the horizontal position? If it becomes horizontal, what will be its angular velocity?

Figure P10-44

10-45 In Fig. P10-45, block A weighs 32.2 lb, block B weighs 96.6 lb, and the unbalanced drum weighs 64.4 lb. The radius of gyration of the mass of the drum with respect to the axis of rotation is 1.50 ft. In the position shown, the angular velocity of the drum is 4.0 rad per sec counterclockwise. Determine the angular velocity of the drum the first time G is directly below the axis of rotation. Assume that the cable does not slip on the pulley.

Figure P10-45

10-46 The homogeneous disk B in Fig. P10-46 weighs 322 lb and has a 4-ft-diameter hub of negligible mass. The 128.8-lb body C has a velocity of 20 fps downward in the position shown. Determine the spring modulus required to make the velocity of C 10 fps downward after it has fallen 4 ft. The spring has an initial tension of 200 lb.

Figure P10-46

10-47 In Fig. P10-47, the homogeneous bars AB ($m = 6$ slugs) and BC ($m = 3$ slugs) are welded at B and rotate about a horizontal axis at A. When AB is horizontal, its angular velocity is 2.0 rad per sec clockwise, and the spring force is 20 lb. After the bars have rotated 90° clockwise, their angular velocity is 3.0 rad per sec clockwise. Determine the spring modulus.

Figure P10-47

10-48 The pulley in Fig. P10-48 weighs 96.6 lb and has a radius of gyration of mass with respect to its axis of rotation of 0.60 ft. Body A weighs 32.2 lb, and the modulus of the spring is 10 lb per ft. When A has a velocity of 8.0 fps downward, the spring is stretched 6.0 in. Determine the displacement of A while its velocity is changed to 2.0 fps downward.

Figure P10-48

Figure P10-50

10-49 The 322-lb symmetrical body A in Fig. P10-49 has a radius of gyration of mass of 2 ft with respect to its axis of rotation. The modulus of the spring is 15 lb per in. The 64.4-lb body B has a velocity of 12 fps downward when the spring tension is 300 lb. Determine the distance B will have moved downward when its velocity is 3 fps downward.

(both upward) while moving upward 6 ft. Determine the weight of B.

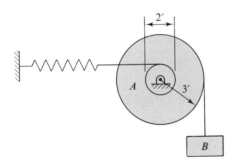

Figure P10-49

10-50 Body A in Fig. P10-50 weighs 64.4 lb and has a radius of gyration of mass with respect to its horizontal axis of rotation of 0.25 ft. The modulus of the spring is 40 lb per ft. The angular velocity of the body is 3 rad per sec clockwise, and the tension in the spring is 2.0 lb in the position shown. Determine the vertical component of the pin reaction on body A when it has rotated 90° clockwise.

10-51 Body A in Fig. P10-51 weighs 60 lb, has a radius of gyration of mass with respect to a horizontal axis through G of 0.60 ft, and rolls without slipping along the inclined plane. The velocity of body B changes from 15 to 6 fps

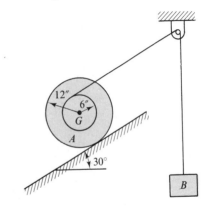

Figure P10-51

*10-52 The 300-kg wheel in Fig. P10-52 rolls without slipping on the horizontal plane and has a radius of gyration of mass with respect to a horizontal axis through G of 0.4 m. The spring, with a modulus of 50 N/m, is stretched 0.6 m in

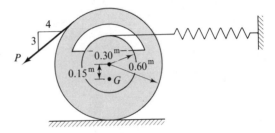

Figure P10-52

the position shown, and the angular velocity of the wheel is 5 rad/s counterclockwise. The constant force **P** is applied at the same slope throughout the motion. Determine the force **P** if the angular velocity of the wheel is 4 rad/s counterclockwise after it rolls 90° counterclockwise.

10-53 The unbalanced wheel in Fig. P10-53 weighs 644 lb and has a radius of gyration of mass with respect to a horizontal axis through G of 1.5 ft. The wheel rolls without slipping, and the modulus of the spring is 20 lb per ft. In the position shown the spring is stretched 1 ft, and the angular velocity of the wheel is 4 rad per sec counterclockwise. Determine if it is possible for the wheel to roll through a 90° counterclockwise angle before it stops. The 100-lb force is applied at the same slope throughout the motion.

Figure P10-53

10-54 The eccentric wheel in Fig. P10-54 weighs 100 lb, has a radius of gyration of mass with respect to a horizontal axis through G of

Figure P10-54

0.40 ft, and rolls without slipping along the horizontal plane. The slope of the 30-lb force remains constant. The angular velocity of A changes from 8 to 4 rad per sec, both clockwise, while the wheel rolls 180° clockwise from the position shown. Determine the weight of body B.

10-55 The wheel in Fig. P10-55 weighs 322 lb and has a radius of gyration of mass with respect to an axis through G of 2.0 ft. Body A weighs 257.6 lb, and the modulus of the spring is 40 lb per ft. Initially the spring is stretched 1.0 ft, and the angular velocity of the wheel is 4 rad per sec counterclockwise. Determine the angular velocity of the wheel after it has rotated 90° counterclockwise.

Figure P10-55

10-56 The coefficient of friction between the 32.2-lb body A in Fig. P10-56 and the plane is 0.50. The 64.4-lb wheel B rolls without slipping and has a radius of gyration of mass with respect to a horizontal axis through G of 2.0 in. The spring modulus is 1.0 lb per in. In the position shown, the velocity of A is 4.0 fps to the left, and the spring is unstretched. Determine the force P which will cause the velocity of A to increase to 5.0 fps to the left while A moves 1.0 ft.

Figure P10-56

*10-57 The 150-kg homogeneous cylinder *A* in Fig. P10-57 rolls without slipping on the horizontal plane. In the position shown, the 30-kg body *B* has a velocity of 5 m/s downward, and the spring is stretched 0.6 m. The spring modulus is 50 N/m. Determine the angular velocity of *A* when it passes directly under the smooth peg *C*.

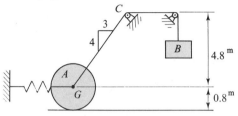

Figure P10-57

10-58 The 644-lb symmetrical body *A* in Fig. P10-58 rolls without slipping and has a radius of gyration of mass of 2 ft with respect to a horizontal axis through *G*. The homogeneous disk *B* weighs 128.8 lb, and body *C* weighs 64.4 lb. The spring (modulus of 20 lb per ft) is stretched 2 ft when body *A* has an angular velocity of 6 rad per sec clockwise. Determine the angular velocity of *A* after *G* has moved 3 ft to the right.

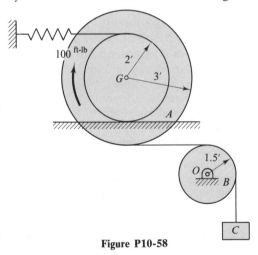

Figure P10-58

10-59 The 161-lb nonhomogeneous bar *AB* in Fig. P10-59 has its mass center at *G* and is pinned at the ends to guide blocks, of negligible weight, which slide in smooth slots. The radius of gyration of the mass of *AB* with respect to a horizontal axis through *G* is 4 ft. In the position

shown the bar has an angular velocity of 3.0 rad per sec counterclockwise, and the spring is compressed 2 ft. Determine the modulus of the spring which will reduce the angular velocity of the bar to 1.5 rad per sec when it reaches the horizontal position.

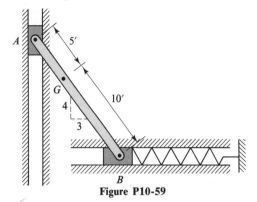

Figure P10-59

10-60 The 322-lb nonhomogeneous bar in Fig. P10-60 has a radius of gyration of mass of 4 ft with respect to a horizontal axis through the mass center *G*. In the position shown the angular velocity of the bar is 4 rad per sec clockwise, and the spring is compressed 2 ft. Determine the modulus of the spring if the angular velocity of the bar is 3 rad per sec clockwise when it reaches the horizontal position. Neglect the mass of the smooth guide blocks at *A* and *B*.

Figure P10-60

*10-61 The semicircular body A in Fig. P10-61 has a mass of 17 kg and rotates about a fixed horizontal axis at O. The homogeneous rod BC has a mass of 3 kg and moves in a vertical plane. When line OB is vertical as shown, the angular velocity of A is 7.0 rad/s clockwise. Determine the angular velocity of A after it has rotated 90° clockwise to the position shown by dashed lines.

Figure P10-63

Figure P10-61

*10-62 If body A in Fig. P10-61 has a mass of 10 kg, BC has a mass of 3 kg, and the system is at rest initially (when B is directly above O), will A start rotating clockwise or counterclockwise? Determine the angular velocity of A when OB is horizontal after starting from rest.

10-63 In Fig. P10-63, A weighs 128.8 lb, B weighs 966 lb, and D weighs 322 lb. Body B is a solid homogeneous cylinder, and body D is a solid homogeneous sphere which rolls without slipping. The couple C_v on body B varies according to the equation $C_v = 30\theta^2 + 90$, where θ, the angular position of B, is in radians and C_v is in foot-pounds. The positive direction is clockwise for C_v and θ. If θ is zero when the velocity of A is 20 fps downward, determine the distance A moves while its velocity is changing to 10 fps downward.

*10-64 The homogeneous 150-kg rod in Fig. P10-64 is released from rest in the position shown. The initial tension in the spring is 500 N. The rod attains an angular velocity of 2.0 rad/s clockwise when it becomes horizontal. Determine the spring constant. Neglect the weights of blocks A and B.

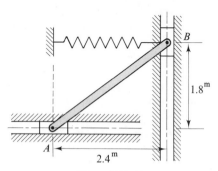

Figure P10-64

10-65 The spring in Fig. P10-65 has a modulus of 10 lb per ft and is stretched 2 ft in the position shown. The homogeneous 96.6-lb bar has an angular velocity of 3 rad per sec clockwise in the vertical position. When the bar has rotated 90° clockwise, determine
(a) The horizontal component of the reaction at O on the bar.
(b) The angular acceleration of the bar.

10-66 The homogeneous 10-ft bar AB in Fig. P10-66 has a mass of 6 slugs and is connected to body C by a flexible cable. Ends A and B move

Figure P10-65

in smooth slots as shown. The system is released from rest when AB is vertical. When C has fallen 8 ft to C' its velocity is 15 fps. Determine the mass of C.

Figure P10-67

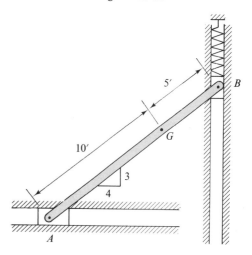

Figure P10-68

Figure P10-66

10-67 The 64.4-lb body A in Fig. P10-67 rolls without slipping and has a radius of gyration of mass of 2 in. with respect to a horizontal axis through G. The homogeneous disk B weighs 32.2 lb, and body C weighs 12.88 lb. The modulus of the spring is 1.0 lb per in., and the spring is stretched 6 in. when the velocity of C is 1.0 fps downward. Determine the angular velocity of A after C has moved 12 in. downward.

10-68 The 322-lb nonhomogeneous bar AB in Fig. P10-68 has a radius of gyration of mass of 4 ft with respect to a horizontal axis through the mass center G. In the position shown the angular velocity of the bar is 4 rad per sec clockwise, and the spring (modulus of 40 lb per ft) is stretched 1 ft. Determine the angular velocity of the bar when it passes through the horizontal position. Neglect the weights of the smooth guide blocks A and B.

*10-69 The homogeneous 30-kg bar AB in Fig. P10-69 is at rest in the position shown. The

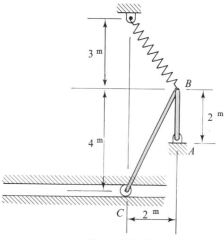

Figure P10-69

homogeneous bar BC has a mass of 90 kg, and the mass of the roller at C may be neglected. The modulus of the spring is 300 N/m, and its unstretched length is 3 m. Determine the angular velocity of AB after it has rotated 90° counterclockwise.

10-70 The 322-lb homogeneous cylinder A in Fig. P10-70 rolls without slipping. The cord from the spring wraps around a hub of negligible mass. The 96.6-lb homogeneous slender rod BC has a small roller of negligible weight at C. Body A has an initial angular velocity of 1 rad per sec counterclockwise, and the spring is initially stretched 2 ft. Determine the spring constant required for the entire assembly to come to rest after body A rotates 90° counterclockwise.

10-72 The homogeneous slender rod AB in Fig. P10-72 weighs 322 lb and rotates in a vertical plane about a horizontal axis through O. In the position shown, the angular velocity of AB is 2.0 rad per sec counterclockwise. After AB has turned through an angle of 90° counterclockwise, its angular velocity is 1.0 rad per sec counterclockwise. Block C weighs 96.6 lb. The spring has an unstretched length of 5.0 ft. Determine the modulus of the spring.

Figure P10-72

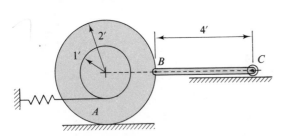

Figure P10-70

10-71 The 96.6-lb cylindrical body in Fig. P10-71 has a radius of gyration of mass with respect to a horizontal axis through the mass center G of 2.5 in. and rolls without slipping. The body has an angular velocity of 1 rad per sec clockwise in the position shown, and the spring is stretched 1 in. Determine the modulus of the spring if the angular velocity of the body is 3 rad per sec clockwise after it has turned 90° clockwise. The 20-lb force remains at the same slope throughout the motion.

10-73 The 161-lb unbalanced wheel A in Fig. P10-73 rolls without slipping on the circular surface and has a radius of gyration of mass with respect to a horizontal axis through G of 1.5 ft. The weight of B can be neglected. The spring has a modulus of 100 lb per ft, has an unstretched length of 4.0 ft, and is attached to member B and to the support at D. The wheel has an angular velocity of 3.0 rad per sec clockwise when in the top position. Determine the angular velocity of A after it has rolled to position A' (shown by dashed lines).

Figure P10-71

Figure P10-73

*10-74 When the 150-kg homogeneous rod *AB* is in the position shown in Fig. P10-74 the *tension* in the spring is 225 N. The weights of the blocks at *A* and *B* and friction can be neglected. The bar is released from rest in the indicated position. When the bar becomes vertical its angular velocity is 1.0 rad/s counterclockwise. Determine the spring modulus.

Figure P10-74

10-75 The slender homogeneous bars *AB* and *BC* in Fig. P10-75 weigh 64.4 and 128.8 lb, respectively. A torsion spring with a modulus of 120 ft-lb per rad is attached to bar *AB* and the support at *A*. The system is at rest, and the torque of the spring on *AB* is zero in the position shown. The end *C* of the bar *BC* is displaced slightly to the right and released. Determine the angular velocity of *AB* when it has rotated through 120°.

10-76 In Fig. P10-76, collars *A* and *B* each weigh 16.1 lb, and the weight of the bar can be neglected. The 25-in. bar is fastened to *A* and *B* by ball and socket joints, and length *b* is 9 in. The system starts from rest when *x* = 0, and a force 5**i** lb is applied to *B*. Determine the velocity of *B* when **x** = 20**i** in. Neglect friction.

10-77 The homogeneous semicircular body in Fig. P10-77 has a uniform thickness and weighs 32 lb. The uniform bar *AB* weighs 20 lb and is pinned to *C*. Body *C* rolls without slipping, and the surface at *A* is smooth. The system is at rest

Figure P10-75

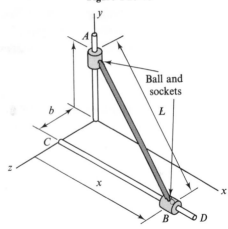

Figure P10-76

in the position shown. When *C* has rotated 90° clockwise it has an angular velocity of 2.0 rad per sec clockwise. Determine the constant force **P**.

Figure P10-77

613

*10-78 The homogeneous body in Fig. P10-78 is released from rest in the position shown. Determine the velocity of A just before AB becomes horizontal.

Figure P10-78

10-9 POWER AND EFFICIENCY

Power, as used in mechanics, is defined as *the time rate of doing work*. From the definition of power, its dimensions are force multiplied by length and divided by time, FL/T, and common units are foot-pounds per second, foot-pounds per minute, horsepower (hp), watts, kilowatts (kW), British thermal units per hour, and so on. The magnitudes of the various units have the relationships indicated in the following equations: 1 hp = 550 ft-lb per sec = 33,000 ft-lb per min = 746 W = 0.746 kW. By definition, 1 British thermal unit (Btu) is equal to 778 ft-lb of energy.

Power is used in rating machines. It is usually more important to know the amount of work the machine can do in a second, hour, or day than merely to know that the machine can do so much work. A large pump and a small pump may both be capable of filling a tank with water from a reservoir, but the time required may differ considerably for the two pumps, and thus their rates of doing the work (power) will not be the same.

Since power is the time rate of doing work, the power delivered to a body by a force $\mathbf{F}$ acting on the body which has a displacement $d\mathbf{r}$ during a time dt is

$$P = \frac{dU}{dt} = \frac{\mathbf{F} \cdot d\mathbf{r}}{dt} = \mathbf{F} \cdot \mathbf{v}. \tag{10-13a}$$

The power delivered by a couple with a moment T acting on a body which has an angular displacement $d\theta$ in the same plane as the couple during a time dt is

$$P = \frac{dU}{dt} = \frac{T \, d\theta}{dt} = T\omega.$$

This expression is equivalent to

$$P = \mathbf{T} \cdot \boldsymbol{\omega}. \tag{10-13b}$$

The units used in expressing power can also be used with an interval of time to express work or energy. Thus a horsepower-hour (hp-hr) is the amount of work done in an hour by a force working at the rate of one horsepower. That is,

$$1 \text{ hp-hr} = \left(33,000 \frac{\text{ft-lb-hr}}{\text{min}}\right)\left(60 \frac{\text{min}}{\text{hr}}\right) = 1,980,000 \text{ ft-lb} = 2545 \text{ Btu}.$$

Similarly, 1 kilowatt-hour (kWh) = 2,655,000 ft-lb = 3413 Btu.

The term *efficiency* can be used to denote a number of different quantities. When referring to machines, *efficiency is defined as the ratio of the energy output of the machine per unit time to the energy supplied to the machine in the same time interval.* Efficiency can be further qualified by calling it mechanical efficiency, electrical efficiency, hydraulic efficiency, overall efficiency, and so on. The mechanical efficiency of a machine is

$$\eta_{mech} = \frac{\text{power output}}{\text{power input}}.$$

It is impossible to eliminate all frictional forces between the moving parts of a machine. Therefore, some of the energy supplied to the machine must be utilized in overcoming the frictional resistance; hence, the power output is less than the energy supplied per unit of time. The kinetic energy of the water as it leaves a water wheel is another example of energy which is not used by the machine to do work. These are only two of many examples which could be cited indicating that the efficiency of any engine is always less than unity, or 100%. Some machines for transforming electric or hydraulic energy to mechanical energy, such as motors and turbines, have efficiencies which may run as high as 90% or more when operating under ideal conditions.

10-10 DISSIPATION OF MECHANICAL ENERGY

Frictional forces are usually undesirable in machines, since they reduce the efficiency of the machines by dissipating some of the energy in unusable forms which are manifested by an increase of temperature, the production of sound, and other phenomena. In some instances, however, the transformation of mechanical energy to thermal energy by means of frictional forces may be desirable, as in the use of brakes to stop an automobile. The kinetic energy of the automobile is dissipated (transformed into thermal energy) through the action of the brake lining sliding on the brake drum. The energy which is transferred to the brake lining by the frictional forces, and which is indicated by an increase in the temperature of the lining, is eventually dissipated into the atmosphere. Frictional forces are also utilized for transmitting

(a)   (b)

Figure 10-23

energy in such apparatus as clutches and belt drives, but the dissipation of energy in such applications is usually small.

A *dynamometer* is a machine or instrument for measuring the force or power developed by an engine. The efficiency can be determined when the power developed is known. The Prony brake illustrated in Fig. 10-23(a) is an example of an absorption-type dynamometer in which the energy transmitted to the drum by the shaft is transformed into thermal energy by frictional forces and dissipated into the atmosphere or into cooling water. The brake drum rotates in a clockwise direction due to the turning moment transmitted by the shaft. When the turning moment is balanced by the frictional moment of the brake on the drum, the angular velocity of the drum remains constant. The work done on the drum by the torque in the shaft is obtained as the product of the turning moment and the angle through which the drum rotates. The power supplied to the drum (and dissipated by the friction) is given by Eq. (10-13b) as

$$P = \mathbf{T} \cdot \boldsymbol{\omega},$$

where P is the power, $\mathbf{T}$ is the turning moment, and $\boldsymbol{\omega}$ is the angular velocity of the drum and shaft.

The turning moment of the shaft on the drum in Fig. 10-23(a) can be obtained from the free-body diagram of the brake and drum in Fig. 10-23(b). The reaction of the shaft on the drum is resolved into a pair of component forces, $\mathbf{R}_x$ and $\mathbf{R}_y$, and a couple, $\mathbf{T}$. The system is in equilibrium; therefore,

$$\sum M_O = 0$$

from which

$$T = Fa,$$

where F is the net load at the end of the brake arm. The expression just developed is not valid for all brake arrangements, and a free-body diagram of either the brake, the drum, or both together should be drawn to help determine the turning moment.

PROBLEMS

10-79 A jet of water flows from a nozzle with a velocity of 200 fps. The diameter of the jet is 3.00 in., and the water weighs 62.4 lb per ft³. Determine
(a) The kinetic energy per pound of water in the jet.
(b) The kinetic energy of the water which flows from the nozzle in 1 sec.
(c) The horsepower of the water in the jet.

*10-80 Determine the torque that must be developed by a motor revolving at 2000 rpm to generate 2.5 kW.

10-81 A 165-lb man walks up a flight of stairs 10.5 ft high in 10 sec. Determine the average horsepower exerted by the man.

10-82 The pumps at the Grand Coulee Dam raise the water to be used in irrigation a distance of 280 ft. Each pump is operated by a 65,000-hp motor and has a capacity of 700,000 gal per min. Water weighs 8.34 lb per gal. Determine the overall efficiency of the pumping system.

10-83 A hoist lifts a load of 3000 lb at a speed of 10 fps when 50 kW is supplied to the motor. Determine the overall efficiency of the motor-hoist unit.

10-84 (a) A 2000-lb load is being lifted with a constant velocity of 10 fps. Determine the power required.
(b) Determine the power required if the 2000-lb load has a velocity of 10 fps upward and is accelerated 4 fps² upward.

*10-85 The moment **M** in Fig. P10-85 causes the drum to have an angular velocity of 150 rpm clockwise. The coefficient of friction between the belt and drum is 0.30. Determine the power dissipated by the brake.

10-86 The Prony brake in Fig. P10-86 is used to measure the power developed by a small impulse-type water wheel. The couple **C** transmitted to the drum by the shaft causes the brake drum to have a clockwise angular velocity of 50 rad per sec. The net load **Q** (neglecting the weight of the brake arm) is 12.5 lb upward. The

All dimensions in millimeters

Figure P10-85

power in the jet of water driving the wheel is 2.40 hp before it strikes the wheel. Determine
(a) The power dissipated by the Prony brake.
(b) The efficiency of the water wheel.

Figure P10-86

10-87 The couple **C** acting on the drum in Fig. P10-87 causes it to rotate with a constant angular velocity of 24 rad per sec clockwise. The coefficient of friction between the brake and drum is 0.50. Determine the power dissipated by the brake.

Figure P10-87

617

*10-88 The brake mechanism in Fig. P10-88 is used to lower body A with a constant velocity of 3.0 m/s. The coefficient of friction between the brake and drum D is 0.40. Neglect the weight of the brake. Determine

(a) The mass of body A.

(b) The power dissipated by the brake.

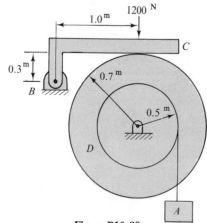

Figure P10-88

10-89 The drum in Fig. P10-89 has a constant angular velocity of **ω** rad per sec clockwise, and the coefficient of friction between the drum and brake band is **μ**. Derive an expression for the power dissipated by the brake in terms of the

Figure P10-89

force **Q**, the angular velocity of the drum, the coefficient of friction between the brake and drum, and the dimensions of the brake.

10-90 The electric generator in Fig. P10-90 is used as a dynamometer to measure the power output of a motor. The frame of the generator (the field) is mounted on smooth bearings and is restrained from turning by the two forces **Q** and **Q′**, which constitute a couple. The armature of the generator is driven by the motor. As the load on the generator is increased, the forces **Q** and **Q′** also increase. When the motor is driving the generator with an angular velocity of 1700 rpm counterclockwise, the magnitudes Q and Q′ are 9.75 lb each, and 8.50 kW is being supplied to the motor. Determine the efficiency of the motor.

Figure P10-90

10-91 A truck-and-trailer unit which weighs 10 tons travels along a highway at 45 mph. The wind and frictional resistance is 50 lb per ton. The efficiency of the power transmission from the engine to the wheels is 65%. Determine the power which must be developed by the engine while moving up a 4% grade (4-ft rise per 100 ft of horizontal distance).

10-11 CLOSURE

The principle of work and kinetic energy equates the work done by all the forces acting on a body or system of bodies to the change in the kinetic energy of the bodies. Work is either a positive or a negative scalar product of force and displacement, whereas kinetic energy is a

The 10-kg body C is connected to block B by a flexible inextensible cord. The system is released from rest when AB is vertical. Determine

(a) The tension in the cord just after the system is released.

(b) The velocity of C after it has moved 1.6 m downward.

Figure P10-98

*10-99 Determine the reactions on the ends of the bar in Problem 10-98 at the instant the bar becomes horizontal.

Figure P10-100

10-100 The homogeneous slender rod in Fig. P10-100 moves with its ends in contact with the two smooth surfaces as indicated. The rod is at rest in the vertical position when the lower end is pulled out slightly to start the bar moving. Determine the angle θ when the upper end leaves the vertical wall.

10-101 The upper end of the rod in Problem 10-100 is constrained by a roller in a smooth slot to move along a vertical line as shown in Fig. P10-101. Determine the angle θ at which the lower end will leave the floor. Motion is started in the same manner as in Problem 10-100.

Figure P10-101

Kinetics–Impulse and Momentum

11

11-1 INTRODUCTION

The principles of force, mass, and acceleration were developed from Newton's laws of motion and their application to kinetics problems was illustrated in Chapter 9. In Chapter 10, the principle of work and kinetic energy was derived from Newton's laws of motion and applied to the analysis of problems in kinetics. The principles of impulse and momentum, which also are based on Newton's laws of motion, will be developed and their use explained in this chapter. Since each of the three methods has certain advantages for some types of problems, it is desirable to understand all three methods.

When an acceleration or a force at an instant is to be obtained, the method of force, mass, and acceleration is usually the most direct. When one or more of the forces involved are variables and can be expressed as functions of the position of the body, the method of work and kinetic energy is generally best. If one or more of the forces involved are variables and can be expressed as functions of time, the principles of impulse and momentum usually provide the most direct solution. The principles of impulse and momentum are particularly effective for problems involving an impact or collision between two bodies, for problems dealing with fluids, and for problems in which the mass varies with time, such as rockets and hoisting cables. Many problems can be solved readily by more than one of the three methods. In such cases, solutions by two methods provide an excellent means of checking the result.

11-2 PRINCIPLE OF LINEAR IMPULSE AND LINEAR MOMENTUM

The linear impulse $\mathbf{I}_L$ of a force $\mathbf{F}$ during a time interval from t_i to t_f is defined as the definite integral of the product of the force and the time dt from t_i to t_f; that is,

$$\mathbf{I}_L = \int_{t_i}^{t_f} \mathbf{F}\, dt. \tag{11-1}$$

The linear impulse is a vector and has fundamental dimensions of FT with common units of pound-seconds and newton-seconds. If the force $\mathbf{F}$ is constant in direction, the linear impulse has the same line of action as the force. When the force varies in direction, the component form of Eq. (11-1) can be used to obtain the components of the linear impulse vector for the time interval involved. The x component is

$$(\mathbf{I}_L)_x = \int_{t_i}^{t_f} \mathbf{F}_x\, dt$$

with similar expressions for the y and z components. The component form is convenient to indicate the relationship between the force and time graphically. In Fig. 11-1, the magnitudes of the x component of the force are plotted against the corresponding values of time. The shaded area under the curve of Fig. 11-1 equals the magnitude of the x component of the linear impulse of the force $\mathbf{F}$ during the time interval from t_i to t_f.

The linear impulse in the x direction can also be written

$$(\mathbf{I}_L)_x = (\mathbf{F}_x)_{\text{avg}}\, \Delta t,$$

where $(\mathbf{F}_x)_{\text{avg}}$ is the time average value of $\mathbf{F}_x$. It should be noted that the time average value of the force is usually not the same as the distance average value of the force used in determining the work done by a force. Problem 11-6 illustrates the preceding statement.

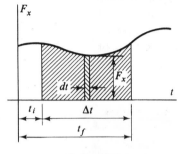

Figure 11-1

In the preceding chapter, it was shown that a force does work only when it has a component in the direction of its displacement. Every force that acts during a time interval does, however, have a linear impulse whether it does any work or not.

The linear impulse of a system of forces during a time interval is the vector sum of the linear impulses of the separate forces of the system during that time interval.

The linear momentum of a particle of mass m at any instant is defined as the product of the mass of the particle and its velocity at that instant; that is,

$$\mathbf{G} = m\mathbf{v}, \tag{11-2}$$

where $\mathbf{G}$ is the linear momentum of the particle.

The linear momentum of a particle is a vector quantity with the same direction as the velocity of the particle. The vector representing the momentum of a particle has a line of action which passes through the particle, and thus it is a localized vector. The dimensions of linear momentum are obtained as follows:

$$mv = \frac{FT^2}{L}\left(\frac{L}{T}\right) = FT.$$

Thus the dimensions, and consequently the common units, of linear momentum and linear impulse are the same.

The linear momentum of a system of n particles at any instant is the vector sum of the linear momentums of the particles of the system at that instant. This sum, as given by Eq. (9-8), is

$$\mathbf{G} = \sum_{i=1}^{n} m_i \mathbf{v}_i = m\mathbf{v}_G, \tag{11-3}$$

where m is the total mass of the system of particles and $\mathbf{v}_G$ is the velocity of the mass center of the system of particles.

Equation (11-3) gives the magnitude and direction of the linear momentum of any system of particles. The linear-momentum vector usually does not pass through the mass center of the system of particles. The position of the resultant linear momentum vector of a rigid body can be obtained by the principle of moments as explained in Art. 11-9.

From the principle of motion of the mass center for a system of particles having a total mass m the equation of motion given by Eq. (9-10) may be written as

$$\mathbf{R} = m\mathbf{a}_G = m\frac{d}{dt}(\mathbf{v}_G),$$

where $\mathbf{R}$ is the resultant of all the external forces acting on the system of particles. This equation may be written as

$$\mathbf{R}\,dt = m\,d\mathbf{v}_G. \tag{11-4}$$

The definite integral of Eq. (11-4) is

$$\left.\begin{array}{c} \int_{t_i}^{t_f} \mathbf{R}\, dt = m \int_{\mathbf{v}_i}^{\mathbf{v}_f} d\mathbf{v}_G = m(\mathbf{v}_{Gf} - \mathbf{v}_{Gi}) \\ \\ \mathbf{I}_L = \mathbf{G}_f - \mathbf{G}_i, \end{array}\right\}$$

or   (11-5)

where the subscripts i and f signify initial and final values of time and velocity.

The principle of linear impulse and linear momentum as expressed mathematically in Eq.(11-5) can be stated in words as follows: *The linear impulse of a force system acting on any system of particles during a time interval is equal to the change in the linear momentum of the system of particles during that time interval.*

Equation (11-4) with m constant can also be written as

$$\mathbf{R} = \frac{d}{dt}(m\mathbf{v}_G), \qquad (11\text{-}6)$$

which indicates that the resultant force acting on a system of particles is equal to the time rate of change of the linear momentum of the system of particles. Newton's second law of motion *for a particle* is sometimes stated in this form. Although Eq. (11-6) is generally not directly applicable to a system of particles that is gaining or losing material, there are instances for which it can be correctly used for variable mass problems. If the particles being added to the system have no velocity before the addition takes place or if the velocity of the particles being ejected from the system is zero after being ejected, then Eq. (11-6) can be applied.

Equations (11-5) are vector equations and can be used in component form. However, a consistent sign convention must be used for both impulses (or forces) and momentums (or velocities). Positive directions should be assumed if not given, and errors will frequently be avoided if positive axes are shown on the sketch.

The principle of linear impulse and linear momentum applies to any system of particles including a system of rigid bodies. Because of the vector quantities involved, however, it is usually better to draw separate free-body diagrams of each rigid body than to use combinations of bodies, as is commonly done when using the principle of work and kinetic energy.

The following examples illustrate the application of the principle of linear impulse and momentum.

EXAMPLE 11-1

The coefficient of friction between the 200-lb body A in Fig. 11-2(a) and the horizontal plane is 0.40. Body B weighs 60 lb. The magnitude of the variable force $\mathbf{P}_v$ is given by the equation $P_v = 3t^2$, where P_v is in

(a)   (b)

(c)

Figure 11-2

pounds when t is in seconds. When $t = 0$ the velocity of A is 5.0 fps to the right. Determine the velocity of A when $t = 5$ sec.

SOLUTION

Free-body diagrams of bodies A and B are shown in Fig. 11-2(b) and (c). Positive axes are included with each diagram. Body A is initially moving to the right; therefore the friction will be to the left and will be equal to the limiting value ($F' = \mu N$). The magnitudes of the velocities of A and B are the same, and their final velocities are $\mathbf{v}_{Af} = v_f \mathbf{i}$ and $\mathbf{v}_{Bf} = v_f \mathbf{j}$. There are three unknown quantities, N, T, and v_f, and there are two component linear impulse and linear momentum equations available for body A and one for body B. The component (scalar) equation for body A in the y direction is

$$(I_L)_y = m_A(v_{Ayf} - v_{Ayi})$$

or

$$\int_0^5 (N - 200)\, dt = \frac{200}{32.2}(0 - 0)$$

from which

$$N = 200 \text{ lb.}$$

The component equation in the x direction for body A is

$$(I_L)_x = m_A(v_{Axf} - v_{Axi}),$$

$$\int_0^5 (3t^2 + T - \mu N)\, dt = \frac{200}{32.2}(v_f - 5),$$

$$t^3\Big]_0^5 + \int_0^5 T\, dt - 0.4(200)t\Big]_0^5 = \frac{200}{32.2}(v_f - 5),$$

which reduces to

$$\int_0^5 T\,dt - 275 = \frac{200}{32.2}(v_f - 5). \tag{a}$$

The linear impulse and momentum equation for body B is

$$(I_L)_y = m_B(v_{Bf} - v_{Bi}),$$

$$\int_0^5 (60 - T)\,dt = \frac{60}{32.2}(v_f - 5),$$

which reduces to

$$300 - \int_0^5 T\,dt = \frac{60}{32.2}(v_f - 5). \tag{b}$$

The simultaneous solution of Eq. (a) and (b) gives

$$v_f = 8.10 \text{ fps}$$

and

$$\mathbf{v}_{Af} = 8.10 \text{ fps} \rightarrow. \qquad\qquad \text{Ans.}$$

The tension T is a variable, and this solution does not provide a specific value of T. The integral $\int_0^5 T\,dt$ is equal to the time average value of T multiplied by the time interval, and thus an *average* value for T can be obtained. This solution assumed that the friction force on A was constant, which requires that the velocity not become zero. This assumption can be checked by determining the minimum velocity which will occur when the acceleration is zero. If the velocity of A is to the right when the acceleration is zero, the assumption is correct. If the minimum velocity is found to be to the left, the assumption that the friction is constant (equal to μN) is not valid, and the problem would have to be solved using this information. For this problem the acceleration is zero when $t = 2.58$ sec and the minimum velocity is 0.733 fps to the right. Thus the solution is valid. It is suggested that this minimum velocity, and the time at which it occurs, be checked by the reader.

EXAMPLE 11-2

The coefficient of friction between the 200-lb block of Fig. 11-3(a) and the plane is 0.50. The magnitude of the variable force $\mathbf{P}_v$ is given by the expression $P_v = 10t + 100$, where P_v is in pounds when t is in seconds.

(a)　　　　　　　　　　　(b)

Figure 11-3

The block is at rest when $t = 0$. Determine the velocity of the body when $t = 15$ sec.

SOLUTION

The free-body diagram of the body is shown in Fig. 11-3(b). The impulse and momentum equation in the y direction is

$$(I_L)_y = m(v_{yf} - v_{yi}),$$

$$\int_0^5 \left(N - 200 - \frac{3}{5} P_v\right) dt = \frac{200}{32.2}(0 - 0)$$

or

$$N = 200 + 0.6(10t + 100)$$

$$= (260 + 6t) \text{ lb.}$$

If the horizontal component of P_v is greater than the limiting value of F, the block will start to slide. When $t = 0$, $N = 260$ lb and $\mu N = 130$ lb. The horizontal component of P_v is $0.8(100) = 80$ lb when $t = 0$; therefore the block will not slip. The value of t when motion starts can be obtained by equating the horizontal component of $\mathbf{P}_v$ to the limiting value of F, that is,

$$0.8(10t + 100) = \mu N = 0.5(260 + 6t)$$

from which

$$t = 10 \text{ sec.}$$

The velocity will be zero from $t = 0$ to $t = 10$ sec. For t greater than 10 sec the force F will be

$$F' = \mu N = 0.5(260 + 6t) = 130 + 3t.$$

The linear impulse and momentum equation for the interval from $10 \text{ sec} < t < 15 \text{ sec}$ is

$$(I_L)_x = m(v_{xf} - v_{xi}),$$

$$\int_{10}^{15} (0.8P_v - F') \, dt = m(v_f - 0),$$

$$\int_{10}^{15} [0.8(10t + 100) - (130 + 3t)] \, dt = \frac{200}{32.2}(v_f - 0),$$

which gives

$$\left. \frac{5t^2}{2} - 50t \right]_{10}^{15} = \frac{200}{32.2}(v_f - 0),$$

$$(562.5 - 750) - (250 - 500) = \frac{200}{32.2} v_f,$$

$$v_f = 10.06 \text{ fps}$$

and

$$\mathbf{v}_f = 10.06 \text{ fps.} \rightarrow. \text{ Ans.}$$

The principle of linear impulse and linear momentum is particularly useful in determining the force developed when a jet of water or other fluid is deflected by a blade. Equation (11-5) can be written as

$$\mathbf{R}\,\Delta t = m(\mathbf{v}'' - \mathbf{v}'),$$

where m is the mass of fluid deflected during the time interval Δt. The force $\mathbf{R}$ is the resultant force acting on the fluid, and $\mathbf{v}'$ and $\mathbf{v}''$ are the velocities of the fluid before and after it is deflected by the blade.

The mass of fluid striking the blade in time Δt depends on the cross-sectional area and the velocity of the jet, the density (mass per unit volume) of the fluid, the velocity of the blade, and whether a single blade or a series of blades mounted on the periphery of a wheel is involved. If the jet strikes a stationary blade, as shown in Fig. 11-4(a), the mass of fluid striking the blade in a time Δt is $A_J v_J'(\Delta t)\rho$,

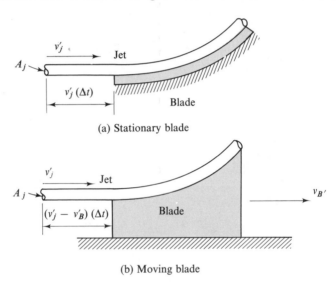

(a) Stationary blade

(b) Moving blade

Figure 11-4

where A_J is the area of the jet, v_J' is the magnitude of its velocity, and ρ is the density of the fluid. If the blade is moving in the same direction as the jet with a speed v_B [see Fig. 11-4(b)], the length of the stream which strikes the blade in a time Δt is $(v_J' - v_B')\,\Delta t$, and the mass striking the blade in the time Δt is $A_J(v_J' - v_B')(\Delta t)\rho$. When a jet of fluid strikes a series of blades on a wheel (such as the runner of an impulse turbine), all the fluid issuing from the nozzle strikes the blades, and the mass of fluid deflected during the time interval Δt is $A_J v_J'(\Delta t)\rho$.

The velocity of the jet as it leaves the blade can be determined from the equation of relative velocity. If the blade is smooth, the force of the blade on the fluid is always normal to the surface and hence normal to the relative velocity of the fluid with respect to the blade. This normal force produces only a normal acceleration of the fluid, and therefore the magnitude of the relative velocity of the fluid leaving the blade is the same as the magnitude of the relative velocity with which the fluid strikes the blade. The direction of the final velocity of fluid relative to the blade is tangent to the blade. The angle through which the relative velocity is changed is called the *deflection angle* of the blade. The final absolute velocity of the water as it leaves the blade can be determined from

$$\mathbf{v}'_{J/B} = \mathbf{v}'_J - \mathbf{v}'_B \quad \text{and} \quad \mathbf{v}''_J = \mathbf{v}''_B + \mathbf{v}''_{J/B},$$

where

$$v''_{J/B} = v'_{J/B}.$$

If the surface of the blade is not smooth, the reaction of the blade on the water will have a component tangent to the surface which will reduce the magnitude of the final relative velocity of the jet with respect to the blade. Thus the final relative velocity might be reduced to 80 or 90% of the initial relative velocity. In a well-designed impulse turbine, the friction between the blade and the jet is usually quite small, and the turbine efficiency may be as much as 90%.

The following problem demonstrates the application of the principle of linear impulse and momentum to a jet.

EXAMPLE 11-3

A jet of water with a cross-sectional area of 2.0 in². has a velocity of 180 fps to the right when it strikes the single curved blade in Fig. 11-5(a). The velocity of the blade is 80 fps to the right, and friction between the blade and water is negligible. The jet remains in a horizontal plane. The water weighs 62.4 lb per ft³. Determine (a) the force of the water on the blade and (b) the power developed.

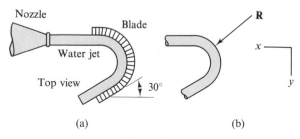

Figure 11-5

SOLUTION

(a) The free-body diagram (top view) of the water which strikes the blade during the time interval Δt is shown in Fig. 11-5(b). In this example, the weight of the fluid is perpendicular to the plane in which the jet moves, but in any case it is customary to neglect the weight of the fluid in comparison with the other forces acting on the jet. The equation of impulse and momentum is

$$\mathbf{R} \, \Delta t = m(\mathbf{v}'' - \mathbf{v}').$$

631

The mass of water which strikes the blade in time Δt is

$$m = A_J(v_J - v_B)\,\Delta t \rho = \frac{2.0}{144}\,(180-80)\,\Delta t\,\frac{62.4}{32.2} = 2.69\,\Delta t.$$

The velocity of the jet relative to the blade before impact is

$$\mathbf{v}'_{J/B} = -180\mathbf{i} - (-80\mathbf{i}) = -100\mathbf{i}\text{ fps.}$$

Since the blade is smooth, the final relative velocity is

$$\mathbf{v}''_{J/B} = 100(0.866\mathbf{i} + 0.500\mathbf{j})\text{ fps,}$$

and the final absolute velocity of the jet is

$$\mathbf{v}''_J = \mathbf{v}''_B + \mathbf{v}''_{J/B} = -80\mathbf{i} + 100(0.866\mathbf{i} + 0.500\mathbf{j})$$

$$= (6.6\mathbf{i} + 50.0\mathbf{j})\text{ fps.}$$

When numerical data are substituted into the equation of impulse and momentum, it becomes

$$\mathbf{R}\,(\Delta t) = 2.69\,(\Delta t)[(6.6\mathbf{i} + 50\mathbf{j}) - (-180\mathbf{i})]$$

from which

$$\mathbf{R} = (502\mathbf{i} + 134.5\mathbf{j})\text{ lb on the water.}$$

The resultant force *on the blade* is

$$\mathbf{R} = -502\mathbf{i} - 134.5\mathbf{j} = \underline{520(-0.965\mathbf{i} - 0.259\mathbf{j})\text{ lb.}} \qquad \text{Ans.}$$

The initial and final absolute velocities of the water used in the equations of impulse and momentum can be replaced by the corresponding velocities of the water relative to the blade because the difference of the two absolute velocities is the same as the difference of the two relative velocities. By the use of relative velocities, the impulse and momentum equation is

$$\mathbf{R}\,(\Delta t) = m(\mathbf{v}''_{J/B} - \mathbf{v}'_{J/B})$$

$$= 2.69(\Delta t)[(86.6\mathbf{i} + 50\mathbf{j}) - (-100\mathbf{i})]$$

and

$$\mathbf{R} = (502\mathbf{i} + 134.5\mathbf{j})\text{ lb on the jet.}$$

This is the same result as that obtained using the absolute velocities.

(b) The reaction on the blade moves in the x direction with the blade; therefore, $\mathbf{R}_y$ does no work. The power delivered to the blade is, from Eq. (10-13a), the product of the force exerted on the blade in the direction of the velocity multiplied by the velocity of the force (and blade); that is,

$$P = R_x v_B = 502(80) = 40{,}200\text{ ft-lb per sec}$$

$$= \frac{40{,}200}{550} = \underline{73.0\text{ hp.}} \qquad \text{Ans.}$$

PROBLEMS

Note: In the following problems, unless otherwise specified, all cords, ropes, and cables are assumed to be flexible, inextensible, and of negligible weight. Pulleys are assumed to have negligible mass, and pins, axles, and pegs are assumed to be smooth unless otherwise stated.

11-1 The coefficient of friction between the 100-lb block A in Fig. P11-1 and the horizontal surface is 0.2. The force $\mathbf{P} = 3t^2\mathbf{i}$ is in pounds when t is in seconds. Determine the resultant linear impulse on the block from $t = 0$ to $t = 5$ sec if the body moves to the right during the entire time interval.

Figure P11-1

*11-2 Block A in Fig. P11-2 moves along the smooth plane. Determine the linear impulse of the force $\mathbf{P}$ in the direction parallel to the inclined plane from $t = 0$ to $t = 17$ s when the magnitude of $\mathbf{P}$ is (a) 13N, (b) $(2t + 20)$N, where t is in seconds.

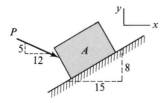

Figure P11-2

*11-3 Solve Problem 11-2 if the magnitude of the force $\mathbf{P}$ is given by the force time diagram in Fig. P11-3.

11-4 The coefficient of friction between the 25-lb body A in Fig. P11-4 and the plane is 0.2. The force $\mathbf{P} = (50 + 30t)\mathbf{i}$ is in pounds when t is in seconds. The body is at rest when $t = 0$. Determine the velocity of the body when $t = 2$ sec.

Figure P11-3

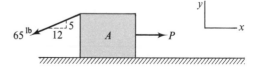

Figure P11-4

11-5 The 250-lb block in Fig. P11-5 is moved to the right along the rough plane ($\mu = 0.2$) for 5 sec by the 90-lb force. Determine the angle θ (between 0 and 90°) for the maximum resultant linear impulse and the magnitude and direction of this impulse.

Figure P11-5

11-6 The 64.4-lb block in Fig. P11-6 moves on a smooth horizontal plane and is acted on by the force $\mathbf{F}$, which varies according to the equation $F = 6t^2$, where F is in pounds when t is in seconds. The block starts from rest at the origin when $t = 0$.

(a) Determine the time average value of the force $\mathbf{F}$ during the interval from $t = 0$ to $t = 3$ sec.

(b) Derive the expression for the position s of the block as a function of time.

(c) Determine the work done by the force $\mathbf{F}$ during the time interval from $t = 0$ to $t = 3$ sec, using the expression $U = \int F \, ds$.

(d) Determine the position average value of the force $\mathbf{F}$ from the expression $U = (F_{avg})(\Delta s)$.

Figure P11-6

*11-7 The three particles in Fig. P11-7 are located in the xy plane. The mass of each particle in kilograms and the velocity in meters per second are indicated on the diagram. Determine
(a) The linear momentum of the system.
(b) The velocity of the mass center.

Figure P11-7

*11-8 Body A in Fig. P11-8 has a mass of 2 kg, and bodies B and C have masses of 3 and 5 kg, respectively. Determine the linear momentum of the system when A has a velocity of 10 m/s downward.

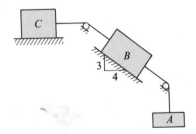

Figure P11-8

11-9 The coefficient of friction between the 100-lb block in Fig. P11-9 and the plane is 0.4. Determine the time elapsed for the 120-lb force **P** to change the velocity of the block from 0 to 64.4 fps to the right.

Figure P11-9

11-10 Solve Problem 11-9 if the magnitude of the force is $P = 80t$, where P is in pounds when t is in seconds, and the velocity of the block is zero when $t = 0$.

11-11 The 10-lb body A in Fig. P11-11 is at rest against the stop on the smooth plane when $t = 0$. The magnitude of the force **P** varies as indicated in the diagram. Determine the velocity of A when $t = 15$ sec.

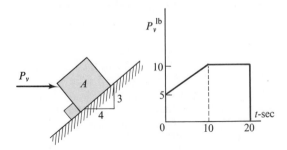

Figure P11-11

11-12 The coefficient of friction between the 322-lb body A in Fig. P11-12 and the plane is 0.2. Body B weighs 64.4 lb. The magnitude of the force **P** in pounds is given by the expression $P = 150 + 30t^2$, where t is in seconds. When $t = 0$, A has a velocity of 2 fps to the right. Determine the velocity of A when $t = 3$ sec.

Figure P11-12

*11-13 The 5-kg body in Fig. P11-13 slides on a smooth horizontal surface. The magnitude of the force **P** is given by $P = 3t^2 + 20$, where P is in newtons when t is in seconds, and the force maintains the constant direction indicated in the plane of motion. When $t = 0$ the velocity of the mass center G is $\mathbf{v}_G = (-6\mathbf{i} + 4\mathbf{j})$ m/s. Determine the velocity of G when $t = 8$ s.

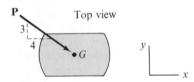

Figure P11-13

*11-14 The magnitude of the force **P** in Problem 11-13 varies as shown in Fig. P11-14, while its direction remains constant. The mass and initial velocity of the body are the same as in Problem 11-13. Determine the velocity of *G* when $t = 8$ s.

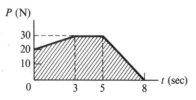

Figure P11-14

*11-15 The masses of bodies *A* and *B* in Fig. P11-15 are 2 and 4 kg, respectively. The coefficient of friction is 0.2 between body *A* and the horizontal plane and 0.4 between body *B* and the inclined plane. Body *B* has an initial velocity of 2 m/s down the plane. Determine the velocity of *A* 3 s later.

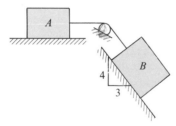

Figure P11-15

*11-16 A box having a mass of 10 kg is given a velocity of 5 m/s down an unloading chute that makes an angle of 20° with the horizontal. The box comes to rest in 3 s. Determine the coefficient of friction between the box and the chute.

11-17 The coefficient of friction between the block *A* in Fig. P11-17 and the horizontal plane is 0.4. Determine the minimum value of the force **P**, in terms of the weight *W* of the block, required to give the block a velocity of 12 fps in 2 sec starting from rest.

11-18 Block *A* in Fig. P11-18 weighs 40 lb and slides along the rough plane ($\mu = 0.25$). The magnitude of the force **P** in pounds is given by the equation $P = 2t$, where *t* is in seconds. Body *A* has a velocity of 15 fps to the left when $t = 0$.

Figure P11-17

Determine
(a) The time *t* when the block comes to rest.
(b) The velocity of *A* when $t = 8$ sec.

Figure P11-18

11-19 The force **P** acting on the 10-lb body in Fig. P11-19 varies as shown. The coefficient of friction between the body and the plane is 0.40, and the body is at rest when $t = 0$. Determine
(a) The maximum velocity of the body.
(b) The velocity of the body when $t = 3$ sec.
(c) The velocity of the body when $t = 5$ sec.

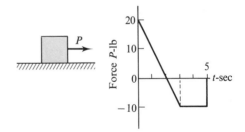

Figure P11-19

11-20 The 6.50-lb block *A* in Fig. P11-20 rests at the bottom of the smooth inclined plane. The magnitude of the force **P** is given by the equation $P = 20(1 - e^{-0.1t})$, where *P* is in pounds when *t* is in seconds. Determine the velocity of *A* when $t = 4$ sec.

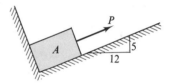

Figure P11-20

11-21 Determine the time t in Problem 11-20 at which the velocity of body A will be 40 fps up the plane.

11-22 The 257.6-lb block A in Fig. P11-22 is connected to the 96.6-lb block B by a flexible wire. The coefficient of friction is 0.20 between A and the plane and 0.25 between B and the plane. Determine the velocity of A 3 sec after it starts from rest.

Figure P11-22

11-23 The 5.0-lb body A in Fig. P11-23 is pulled along the plane by a flexible cord which passes over a drum mounted in the body. The coefficient of friction is 0.40 between the cord and the drum and between the body and the plane. The velocity of the body is 5.0 fps to the right when $t = 0$. *Note:* The body does not stop during the 2-sec interval. Determine
(a) The velocity of A when $t = 1$ sec.
(b) The velocity of A when $t = 2$ sec.
(c) The maximum velocity of A.

Figure P11-23

11-24 Body A in Fig. P11-24 weighs 10 lb, and B weighs 5 lb. The bodies are connected by the cord which passes over smooth pulleys at A and C. The coefficient of friction between B and the plane is 0.20. The magnitude of the force **P** is indicated by the diagram. When $t = 0$ the velocity of B is 30 fps to the right.
(a) Does the velocity of B become zero during the interval $0 < t < 4$ sec?
(b) Determine the velocity of B when $t = 4$ sec.

11-25 Solve Problem 11-24 if the pulley on body A is locked and the coefficient of friction between the pulley and cord is 0.20. All other data are unchanged.

Figure P11-24

11-26 Bodies A, B, and C in Fig. P11-26 have weights of 3, 8, and 5 lb, respectively. The planes on which the bodies slide and the pulleys are smooth. The magnitude of the force **P** varies according to the equation $P = 0.5 + 1.2t$, where P is in pounds when t is in seconds. The system is at rest when $t = 0$. Determine the velocity of each body when $t = 3$ sec.

Figure P11-26

11-27 Bodies A, B, and C in Fig. P11-27 weigh 60, 30, and 10 lb, respectively. The coefficient of friction is 0.50 between A and B and 0.10 between B and the plane. The magnitude of the force **P** is equal to $0.6t^2$, where P is in pounds when t is in seconds. The system is at rest when $t = 0$. Determine
(a) The value of t when motion impends.
(b) The velocity of A when $t = 10$ sec.

11-28 A jet of water 2 in. in diameter impinges on a fixed blade as shown in Fig. P11-28. The velocity of the jet is 60 fps to the right. Determine the force exerted on the smooth blade by the water.

Figure P11-27

Figure P11-28

11-29 A jet of water having a cross-sectional area of a in². strikes a fixed blade and is deflected 90° as shown in Fig. P11-29. The forces required to hold the blade in equilibrium are shown on the figure. The speed of the jet as it strikes the blade is 100 fps. Determine the percentage reduction of the jet speed due to blade friction.

Figure P11-29

11-30 A jet of water 2 in. in diameter strikes the vane shown in Fig. P11-30. The spring has a modulus of 400 lb per ft. Determine the speed of the jet which will cause the spring to be compressed 1 ft.

11-31 The jet of water in Fig. P11-31 has a 0.2-ft² cross-sectional area, leaves the nozzle at 200 fps, and strikes the curved blade as shown. The blade has a constant velocity of 50 fps to the right. Determine

(a) The horsepower delivered to the blade by the water.

Figure P11-30

(b) The velocity of the water as it leaves the blade.

Figure P11-31

11-32 A jet of water having a cross-sectional area of 3 in.² and a velocity of 170 fps to the right strikes a single curved blade and is divided into two equal streams as shown in Fig. P11-32. The blade has a constant velocity of 50 fps to the right, and the friction on the blade reduces the magnitude of the relative velocity of the water with respect to the blade by 6%. Determine

(a) The force exerted on the blade by the water.

(b) The horsepower delivered to the blade by the water.

(c) The kinetic energy remaining in each pound of water as it leaves the blade.

Figure P11-32

11-33 A jet of water flows from a 2-in.-diameter pipe with a velocity of 100 fps and strikes a smooth plunger as indicated in Fig. P11-33. Determine

(a) The force which must be applied to the plunger to keep it stationary.

(b) The force required to move the plunger 2 ft to the left in 0.05 sec with a constant velocity.

(c) The work done in moving the plunger against the water as required in part (b).

11-34 A 4.0-in.-diameter jet of water impinges on the single vane shown in Fig. P11-34. The velocities of the jet and vane are 72 and 20 fps, respectively, both to the right. Determine the angle ϕ for which 25 hp will be developed by the vane.

Figure P11-33

Figure P11-34

11-3 CONSERVATION OF LINEAR MOMENTUM

In Art 11-2, it was shown that the linear impulse of a force system acting on any system of particles during a time interval is equal to the change in the linear momentum of the system of particles during that time interval. *The principle of conservation of linear momentum states that when the linear impulse exerted on a system of particles is zero the change in the linear momentum of the system of particles is zero, and therefore the linear momentum of the system is conserved.* Furthermore, when there is no linear impulse on a system of particles in a given direction the linear momentum is unchanged in that direction even though the total linear momentum may not be constant.

When two particles or bodies, A and B, collide, the linear impulse of A on B is equal in magnitude and opposite in sense to the linear impulse of B on A. Thus, if no other forces are acting on the two bodies in a given direction, the linear momentum of the system composed of the two bodies is conserved in that direction during the impact. The two blocks A and B in Fig. 11-6 are moving along the same path on a smooth plane when they collide. Since there are no horizontal forces external to the system composed of blocks A and B acting on the blocks during the collision, the linear momentum of the system in the

Figure 11-6

horizontal direction is constant. That is,

$$m_A(\mathbf{v}_A)_i + m_B(\mathbf{v}_B)_i = m_A(\mathbf{v}_A)_f + m_B(\mathbf{v}_B)_f, \qquad (11\text{-}7)$$

where the subscripts i and f denote initial and final velocities, respectively. When Eq. (11-7) is used, a positive direction must be selected, and all velocities in the opposite direction must be treated as negative quantities.

When the linear impulse in a given direction is not zero but is known to be relatively small, it can frequently be neglected in order to obtain an approximate solution sufficiently accurate for many purposes. If the linear momentum of the system composed of A and B in Fig. 11-6 is large compared to the linear impulse of the frictional force, Eq. (11-7) is approximately true even though friction exists between the plane and the blocks. The frictional forces could not exceed μN, and the time of impact would be small; therefore, the impulse of the friction on the blocks *during the impact period* would not change the large linear momentum of the blocks materially.

Conservation of linear momentum should not be confused with conservation of kinetic energy, which was discussed in Art. 10-4. The linear momentum of a body or system of bodies is conserved when the linear impulse acting on the body is zero. The kinetic energy of a body is conserved when no work is done on it. It is possible to have the linear momentum of two bodies conserved even though most or all of the kinetic energy is dissipated by producing deformation of the bodies, increasing the temperature, and producing sound and other vibrations. On the other hand, when an elastic steel ball strikes and rebounds from a large steel plate, most of the kinetic energy is conserved, while the linear momentum of the ball may change by a complete reversal in sense with little change in magnitude.

EXAMPLE 11-4

A 10.00-lb block is suspended by a long cord. The block is at rest when a 0.050-lb bullet traveling horizontally to the left strikes the block and is embedded in it. The impact causes the block to swing upward 0.50 ft measured vertically from its lowest position. Determine

(a) The velocity of the bullet just before it strikes the block.
(b) The loss of kinetic energy of the system during impact.

SOLUTION

(a) The velocity of the block and bullet immediately after impact can be found by the principle of work and kinetic energy. Since no dimensions are given, the block is treated as a particle. There are only two forces acting on the block and bullet after impact as indicated in the free-body diagram in Fig. 11-7, and the force **P** does no work because there is no displacement in the direction of **P**. When values are

Figure 11-7

substituted into

$$U = \tfrac{1}{2}m[v_f^2 - v_i^2],$$

the result is

$$-10.05(0.50) = \frac{1}{2}\frac{10.05}{32.2}[0 - v^2]$$

from which

$$\mathbf{v} = 5.67\mathbf{i} \text{ fps,}$$

where $\mathbf{v}$ is the velocity of the block and bullet immediately after impact. During the time of the collision, no forces external to the block and bullet are acting in the horizontal direction. Consequently, the horizontal linear momentum is conserved. The positive x axis has been selected to the left for this example, and the conservation of momentum principle gives

$$\frac{10.00}{32.2}(0) + \frac{0.05}{32.2}\mathbf{v}_B = \frac{10.05}{32.2}(5.67\mathbf{i})$$

from which

$$\mathbf{v}_B = \underline{1140\mathbf{i} \text{ fps,}} \qquad\qquad \text{Ans.}$$

where $\mathbf{v}_B$ is the velocity of the bullet before impact.

(b) The kinetic energy of the system before impact is

$$\frac{1}{2}\left(\frac{0.05}{32.2}\right)(1140)^2 = 1009 \text{ ft-lb.}$$

The kinetic energy of the system after impact is

$$\frac{1}{2}\left(\frac{10.05}{32.2}\right)(5.67)^2 = 5.03 \text{ ft-lb.}$$

The loss of kinetic energy during the collision is $1009 - 5.03 = 1004$ ft-lb, or 99.5%. It should be noted that even though the momentum is conserved during impact, nearly all of the kinetic energy is dissipated as heat and internal work in deforming the block and the bullet.

EXAMPLE 11-5

Crate A in Fig. 11-8(a) slides down the chute with a velocity of 10 m/s and falls on the stationary cart B. The mass of A is 5 kg, and the mass of B is 15 kg. The coefficient of friction between A and B is 0.40. The crate slides along the cart with no rebound. Neglect the mass of the wheels and bearing friction.

(a) Determine the velocity of the cart after the bodies attain a common velocity.
(b) Will the crate hit the end of the cart?
(c) If so, determine the impulse of the cart on the crate.

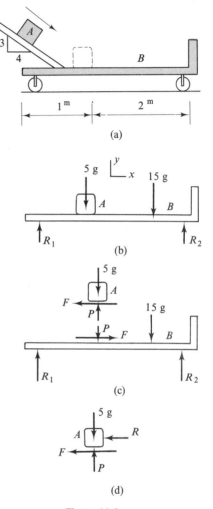

(a)

(b)

(c)

(d)

Figure 11-8

SOLUTION

(a) Figure 11-8(b) is a free-body diagram of the system. There are no horizontal forces external to the system; therefore the linear momentum is conserved in the horizontal direction:

$$\Sigma \, mv_{xi} = \Sigma \, mv_{xf},$$

$$m_A v_{Axi} + m_B v_{Bxi} = m_A v_{Axf} + m_B v_{Bxf} = (m_A + m_B) v_{xf},$$

$$5(\tfrac{4}{5})10 + 0 = (5 + 15) v_{xf}$$

from which

$$\mathbf{v}_{xf} = \underline{2.0 \text{ m/s}} \rightarrow. \qquad \text{Ans.}$$

(b) Various procedures are available for answering part (b). For example, assume that A strikes B after sliding 2 m relative to B. The velocity of A will be more than the velocity of B if the assumption is correct. The linear momentum of the system will be conserved, but the friction force between A and B will do work on the system. From the free-body diagram of body A in Fig. 11-8(c) the normal force P is 5g N, and the friction (as long as sliding occurs) is $\mu P = 0.4(5g) = 2g$ N. The work done on the two bodies during the relative motion is

$$U = -F \, \Delta s_A + F \, \Delta s_B = -F(\Delta s_A - \Delta s_B),$$

and since A slides 2 m farther than B, the work done is

$$U = -F(\Delta s_{A/B}) = -2g(2) = -4g \text{ N·m.}$$

The work and energy principle gives

$$U = \Delta T$$

$$-4g = \tfrac{1}{2}m_A(v_{A2}^2 - v_{A1}^2) + \tfrac{1}{2}m_B(v_{B2}^2 - v_{B1}^2),$$

where the subscript 1 refers to velocities when slipping starts and 2 refers to velocities just before impact. Note that some of the kinetic energy of A was lost when A first hit the cart and that its velocity changed from 10 m/s down the chute to 8 m/s horizontally. The work and energy equation becomes

$$-4g = \tfrac{1}{2}(5)(v_{A2}^2 - 8^2) + \tfrac{1}{2}(15)(v_{B2}^2 - 0). \tag{a}$$

The conservation of linear momentum equation from 1 to 2 gives

$$m_A v_{A1} + m_B v_{B1} = m_A v_{A2} + m_B v_{B2}$$

or

$$5(8) + 0 = 5v_{A2} + 15v_{B2}. \tag{b}$$

Equation (a) becomes

$$v_{A2}^2 + 3v_{B2}^2 = -1.6g + 64, \tag{c}$$

and from Eq. (b),

$$v_{A2} = 8 - 3v_{B2}. \tag{d}$$

When Eq. (d) is substituted into Eq. (c) it becomes

$$12v_{B2}^2 - 48v_{B2} + 1.6(9.81) = 0$$

from which

$$v_{B2} = 3.64 \text{ m/s} \quad \text{or} \quad 0.36 \text{ m/s},$$

and from Eq. (d),

$$v_{A2} = -2.92 \text{ m/s} \quad \text{or} \quad 6.92 \text{ m/s}.$$

The first pair of values ($v_{B2} = 3.64$ m/s and $v_{A2} = -2.92$ m/s) does not satisfy the assumption that $v_A > v_B$. The second set of values, however, does satisfy the assumption; therefore the crate will strike the end of the cart.

(c) The free-body diagram of the crate during the impact period is shown in Fig. 11-8(d). Since the duration of the impact is short, the impulse of the friction force is neglected as compared with the impulse of the force R. The impulse-momentum equation for A is

$$-\int R \, dt = m(v_{Af} - v_{A2}) = 5(2 - 6.92)$$

or

$$\int \mathbf{R} \, dt = \underline{24.6 \text{ N·s}} \leftarrow \text{on } A. \qquad \text{Ans.}$$

PROBLEMS

*11-35 The 10-kg body *A* in Fig. P11-35 slides on a smooth horizontal surface with a velocity of 8 m/s to the right and collides with the 5-kg body *B*, which has a velocity of 6 m/s to the left just before impact. The two bodies lock together on impact.
(a) Determine their common velocity after impact.
(b) If the time of impact is 0.1 s, determine the time average value of the force exerted by body *A* on *B*.

Figure P11-35

11-36 A 150-lb man running with a velocity of 12 fps to the right jumps on a 300-lb cart moving with a velocity of 4 fps to the right. Determine
(a) The common velocity of the man and cart.
(b) The loss of kinetic energy during the impact.

11-37 A 400-lb body has a velocity of 800 fps in free flight when an explosion causes a separation into two parts. Part *A* weighs 300 lb, and *B* weighs 100 lb. Both parts continue in motion along the original line of flight. After the explosion *B* is moving 120 fps faster than *A*. Determine
(a) The speed of *A*.
(b) The linear impulse on *B* caused by the explosion.

*11-38 The block in Fig. P11-38 has a mass of *M* kg and is supported by an inextensible cord *L* m long. The block is at rest when a bullet of

Figure P11-38

mass *m* kg and traveling horizontally to the right strikes the block and is embedded in it. The impact causes the block to move so that the cord makes a maximum angle θ with its initial position. Derive, as a function of θ, *L*, *M*, and *m*, a relation for the velocity of the bullet just before impact.

*11-39 In Fig. P11-39 block *A* has a mass of 9 kg and is moving toward the left with a velocity of 1.5 m/s just before it is struck by the bullet *B*, which has a mass of 50 g and a velocity of 800 m/s to the right. The bullet is embedded in the block. The coefficient of friction between body *A* and the plane is 0.2. How long will the block and bullet continue to move after the impact?

Figure P11-39

11-40 The 4.00-lb body *A* in Fig. P11-40 is suspended by means of an inextensible cord. The body is swinging to the left through its lowest position with a velocity of 6 fps when a bullet weighing 0.01 lb having a velocity of 1500 fps as shown is embedded in the block. How far, measured vertically, will body *A* rise above its lowest position?

Figure P11-40

11-41 The 26-lb block *A* in Fig. P11-41 rests upon a smooth plane as shown. A bullet weighing 0.1 lb is fired horizontally with a velocity of 1300 fps and is embedded in the block. Determine the maximum distance the block will move up the plane.

Figure P11-41

11-42 In Fig. P11-42, the projectile weighs 60 lb and has a velocity of 1500 fps as shown as it leaves the gun. The gun and carriage weigh 3000 lb. Neglecting the mass of the gas, determine

(a) The velocity of the gun and carriage just after the projectile leaves the gun.

(b) The required modulus of the spring if it is not to be deflected more than 18 in. when the gun and carriage hit it. Assume that the spring is initially uncompressed. Neglect the mass of the spring and the small rollers.

Figure P11-42

11-46 A frame of weight $6W$ is at rest on a smooth plane in the position shown in Fig. P11-46. A particle of weight W is attached to the end of rod AB, which is pinned to the frame at A. Neglect the mass of the rod. The rod is released from rest in the horizontal position. Determine the velocity of the frame when the rod is vertical.

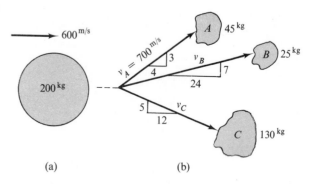

(a)   (b)

Figure P11-43

*11-43** A body having a mass of 200 kg is traveling in free flight with a velocity of 600 m/s as shown in Fig. P11-43(a). An instant later an explosion separates the body into three parts A, B, and C with masses of 45, 25, and 130 kg, respectively. The velocity of A and the directions of the velocities of B and C are indicated in Fig. P11-43(b). Determine the magnitudes of the velocities of B and C.

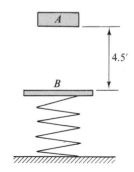

Figure P11-44

11-44 The 100-lb body A in Fig. P11-44 drops 4.5 ft from rest on the 50-lb plate B and remains in contact with it. Before contact body B is at rest on the spring, which has a modulus of 200 lb-per ft. Determine the maximum downward displacement of body B.

11-45 A falling weight of 400 lb is used to drive a 600-lb pile into the ground. Determine the penetration of the pile when the weight is dropped 10 ft. Assume an average resisting force of 15,000 lb and no rebound of the weight after impact.

Figure P11-46

11-47 An open cart of mass M_0 is rolling with a velocity v_0 to the right along a smooth surface. Rain begins to fall with a velocity **u** in the direction shown in Fig. P11-47. The rate (mass per unit of time) at which the mass of water accumulates in the cart is r. Derive an expression for the velocity of the cart in terms of the given data and time.

*11-48 The 1.0-kg block A in Fig. P11-48 slides on a smooth horizontal plane with a velocity of 30**i** m/s when the bullet B strikes and is embedded in the block. The mass of the bullet is 10 g, and it has a velocity of 500 m/s parallel to

Figure P11-47

the yz plane as shown. Determine the final velocity of A.

Figure P11-48

*11-49 Block A in Fig. P11-49 has a mass of 2.0 kg and slides on the smooth inclined plane. Body B has a mass of 1.0 kg and falls on block A and sticks to A. Just before the impact A has a horizontal velocity of 30 m/s as shown, and B has a vertical velocity of 40 m/s downward.

Determine

(a) The final velocity of body B.

(b) The average force of A on B during the impact assuming the duration of the impact to be 0.10 s.

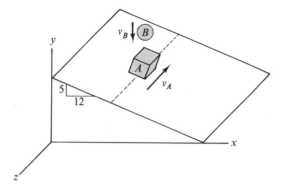

Figure P11-49

11-4 ELASTIC IMPACT

A collision between two bodies which occurs in a very small interval of time and where relatively large reaction forces exist is called *impact.* When two elastic bodies collide, they are compressed until their mass centers attain a common velocity, and then they move apart as the forces within the bodies act to restore the bodies to their original shapes. There has been considerable study of the forces acting during impact and of the resulting stresses and strains in the bodies. The primary concern in elementary engineering mechanics, however, is to obtain relations between the velocities of the bodies before and after impact. For this purpose, some additional definitions will be helpful.

When two bodies collide the straight line normal to the striking surfaces and passing through the point of contact is called the line of impact. When contact occurs over an area instead of at a point, the line of impact is defined as the line of action of the resultant normal force exerted by the bodies on each other. When the centers of gravity of the two colliding bodies are on this line the impact is defined as *central impact. Eccentric impact* occurs when the center of gravity of one or of both bodies is not on the line of impact. When the velocities of the points of contact of the two colliding bodies are along the line of impact *the impact is direct.* Direct impact implies a head-on collision as distinguished from the case of one body striking another body with a glancing blow, which is defined as *oblique impact.* Figure 11-9 shows top views of various types of impact of smooth circular disks sliding on horizontal surfaces and rods rotating on the same surfaces about vertical axes. Only central impact, either direct or oblique, can occur between particles. Eccentric impact is discussed in Art. 11-12.

When two bodies collide, the principle of conservation of linear momentum can frequently be applied for one or more directions to give one or more relations between the velocities of the bodies before and after impact. The velocities of the bodies after impact also depend

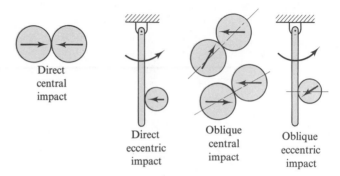

Figure 11-9

on the material properties of the bodies. Experiments indicate that the magnitude of the relative velocity of departure of two spheres which collide with direct central impact depends primarily on their relative velocity of approach and the material properties of the spheres. The preceding statement is not valid if the forces of impact are so large that the bodies are excessively deformed or shattered. *The ratio of the magnitude of the relative velocity of departure of the points of contact of two bodies that collide with direct impact to their relative velocity of approach is defined as the coefficient of restitution for the two bodies.* This ratio is a measure of the elastic properties of the bodies and must be determined experimentally. In mathematical form,

$$e = -\frac{(v_{A/B})_f}{(v_{A/B})_i} = -\frac{(v_A)_f - (v_B)_f}{(v_A)_i - (v_B)_i}, \tag{11-8}$$

where e is the coefficient of restitution, A and B are the points of contact of the two colliding bodies, and i and f indicate initial and final velocities, respectively. The relative velocities of approach and departure are always of opposite sense; therefore, the negative sign before the fraction must be included if the value of e is to be positive.

For perfectly elastic bodies, the coefficient of restitution is approximately one, and for inelastic bodies which remain together after impact the coefficient is zero. A collision in which the final relative velocity is zero is often called plastic impact. If the coefficient of restitution is unity, the kinetic energy of the system is conserved during impact (see Problem 11-68), but since some energy is always used in producing sound, vibrations of the bodies, permanent deformations of the bodies, and possibly in other ways, the coefficient is always less than one.[1] The value of e for any two bodies is therefore between the limits of zero and unity.

The table, following, of approximate values of coefficients of restitution for direct, central impact of spheres of various materials gives an indication of the range of values.

APPROXIMATE COEFFICIENTS OF RESTITUTION

Glass on glass	0.93–0.95
Ivory on ivory	0.88–0.89
Steel on steel	0.5–0.8
Cast iron on cast iron	0.4–0.7
Lead on lead	0.12–0.18
Iron on lead	0.11–0.15
Cork on cork	0.5–0.6
Wood on wood	0.4–0.6
Clay on clay (moist)	0
Putty on putty (moist)	0

---

[1] See Werner Goldsmith, "The Coefficient of Restitution," *Bulletin of the Mechanics Division of the American Society for Engineering Education* (May 1952). Vol. 2, no. 2, p. 10.

When the impact of two bodies is oblique the components of the velocities of the points of contact normal to the surface of contact are used in Eq. (11-8), and the coefficient of restitution is assumed to be the same as for direct impact. The coefficient of restitution for nonspherical bodies depends on the shape of the bodies and on the positions of the bodies relative to the line of impact as well as on the material of the bodies. For oblique impact of *smooth* bodies, the components of the velocities tangent to the contacting surfaces are not altered by the impact because there is no linear impulse on either body in the tangential direction.

EXAMPLE 11-6

Two smooth 4-oz pucks are sliding on smooth ice when they collide as indicated in the top view in Fig. 11-10(a). The velocity of A before impact was 40 fps and the velocity of B before impact was 50 fps in the directions indicated on the figure. The coefficient of restitution for the bodies is 0.56. Determine the velocity of each puck after impact.

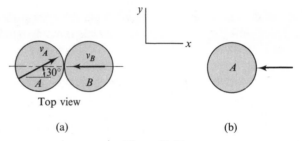

Top view

(a)   (b)

Figure 11-10

SOLUTION

The line of impact is in the x direction and the x components of the velocities of the pucks are changed. The principle of conservation of linear momentum for this direction, with positive to the right, gives

$$[m_A(v_A)_x + m_B(v_B)_x]_i = [m_A(v_A)_x + m_B(v_B)_x]_f,$$

which, in scalar form, becomes

$$\frac{4}{16g}(40)(0.866) + \frac{4}{16g}(-50) = \frac{4}{16g}(v_A)_x + \frac{4}{16g}(v_B)_x, \qquad \text{(a)}$$

where $\mathbf{v}_A$ and $\mathbf{v}_B$ are the final velocities of the pucks. The x components of the final velocities are both assumed to be to the right. Since the impact is oblique, the components of the velocities *along the line of impact* must be used in Eq. (11-8), and, for this position, the equation becomes

$$e = 0.56 = -\frac{(v_A)_x - (v_B)_x}{40(0.866) - (-50)}. \qquad \text{(b)}$$

Equations (a) and (b) can be solved simultaneously to give

$$(v_A)_x = -31.4 \text{ fps}, \qquad (v_B)_x = 16.02 \text{ fps}.$$

The free-body diagram of puck A in Fig. 11-10(b) shows that there is no impulse on the body in the y direction; therefore the linear momentum of A is conserved in the y direction, and the y component of its velocity does not change. A similar analysis shows that the impulse on puck B in the y direction is also zero and that the y component of the velocity of B does not change during the impact. Therefore, the final velocity of puck A is

$$\mathbf{v}_A = -31.4\mathbf{i} + 0.5(40)\mathbf{j} = \underline{37.2 \text{ fps } 32.5°} \qquad , \quad \text{Ans.}$$

and the final velocity of B is

$$\mathbf{v}_B = 16.02\mathbf{i} = \underline{16.02 \text{ fps} \rightarrow}. \qquad \text{Ans.}$$

PROBLEMS

*11-50 Two bodies are moving on a smooth horizontal plane when they collide with direct central impact. Before impact the velocity of the 40-kg body A is 20 m/s to the right, and the 20-kg body B has a velocity of 10 m/s to the left. The coefficient of restitution is 0.6. Determine

(a) The velocity of each body after impact.

(b) The time average value of the force exerted by body A on body B if the contact time is 0.05 s.

11-51 The 40-lb body A in Fig. P11-51 has a velocity of 20 fps to the right when it collides with the 100-lb body B, which is at rest before impact. The spring (modulus 40 lb-per ft) is unstressed before impact. The coefficient of restitution is 0.4. Determine the maximum compression of the spring if the horizontal plane is smooth.

Figure P11-51

11-52 Body A in Fig. P11-51 weighs 32.2 lb, and body B weighs 64.4 lb, and the modulus of the spring is 40 lb per ft. Body A has a velocity of 15 fps to the right when it collides with body

B, which is at rest before impact. Determine the coefficient of restitution if the maximum compression of the spring is 2 ft. The spring is initially uncompressed.

11-53 Body A in Fig. P11-53 weighs 4.0 lb and is released from rest in the position shown. Body A strikes the 2.0-lb body B with direct central impact. Body B is at rest on the smooth horizontal plane before impact. The coefficient of restitution is 0.5. Determine

(a) The velocity of A after impact.

(b) The loss of kinetic energy of body A during impact.

Figure P11-53

11-54 Block A in Fig. P11-54 weighs 50 lb and has a velocity of 30 fps to the left when it

Figure P11-54

strikes the 70-lb body B, which is at rest. The coefficient of restitution between the bodies is 0.6. The horizontal plane is smooth, but the inclined plane and body B have a coefficient of friction of 0.3. Determine the distance body B will travel up the inclined plane before coming to rest.

11-55 The 5-lb smooth disk A and the 15-lb smooth disk B in Fig. P11-55 slide on a horizontal surface and collide with oblique central impact. Before impact the velocities of A and B are 10 and 13 fps, as shown. The coefficient of restitution is 0.6. Determine the velocity of body A after impact.

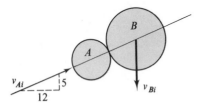

Figure P11-55

11-56 Disks A and B in Fig. P11-56 weigh 25 and 15 lb, respectively, and slide on a smooth horizontal plane. Just before colliding with oblique central impact, the velocities of A and B are 39 and 20 fps, as shown. The striking surfaces are smooth, and the coefficient of restitution is 0.4. Determine the velocity of A after impact.

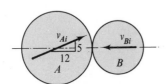

Figure P11-56

11-57 The 5-lb smooth disk A in Fig. P11-57 and the 20-lb smooth disk B slide on a horizontal plane when they collide with oblique central impact. Before impact the velocities are as shown. The coefficient of restitution is 0.5. Determine the kinetic energy of A after impact.

11-58 Disks A and B in Fig. P11-58(a) slide on a smooth horizontal surface and collide with

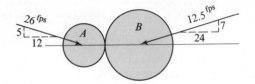

Figure P11-57

direct oblique impact. Disk A weighs 1 lb and has a velocity of 10 fps to the right before impact. Disk B weighs 2 lb and has a velocity of 12 fps upward before impact. The coefficient of restitution for bodies A and B is 0.80. Determine the velocities of A and B after impact if the bodies are in the position shown in Fig. P11-58(b) just before impact.

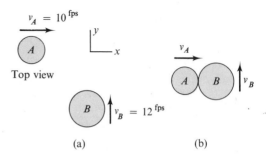

(a)   (b)

Figure P11-58

11-59 Solve Problem 11-58 if disks A and B are in the position shown in Fig. P11-59 at the instant of impact.

Figure P11-59

11-60 Solve Problem 11-58 if disks A and B are in the position shown in Fig. P11-60 just before impact.

11-61 Two spheres A and B in Fig. P11-61 are dropped simultaneously from a height h above a horizontal plane. The coefficients of restitution are e_1 between A and the plane and e_2 between B and the plane. Show that the

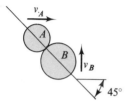

Figure P11-60

velocity of *A* with respect to *B* immediately after impact is $(e_1 - e_2)\sqrt{2gh}$ upward.

Figure P11-61

*11-62 A ball rebounds from a smooth inclined plane as shown in Fig. P11-62. The ball has a velocity of 20 m/s downward just before it strikes the plane, and the coefficient of restitution is 0.8. Determine the velocity just after the impact.

Figure P11-62

*11-63 The 2-kg ball in Fig. P11-63 strikes the smooth horizontal surface with a velocity of 65 m/s. The coefficient of restitution is 0.6. Determine

(a) The velocity of the ball after impact.

Figure P11-63

(b) The maximum height above the surface that the ball will rise.

(c) The time average value of the force exerted by the surface on the ball if the contact time is 0.02 s.

11-64 A ball hits a smooth horizontal plane with a velocity of v_i and rebounds with a velocity of v_f as shown in Fig. P11-64. The coefficient of restitution is *e*. Show that $\tan \theta_f = e \tan \theta_i$.

Figure P11-64

11-65 A 0.322-lb ball hits a rough horizontal plane with a velocity of 60 fps as shown in Fig. P11-65 and rebounds at the indicated angle. The coefficient of restitution is 0.8. Determine the linear impulse of the frictional force on the ball.

Figure P11-65

*11-66 Each of the balls *A* and *B* in Fig. P11-66 has a mass of 1.0 kg. Ball *B* is suspended at rest from a vertical inextensible cord

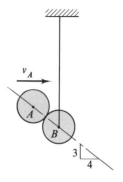

Figure P11-66

when it is struck by A, which has an initial velocity of 25 m/s to the right. The coefficient of restitution for the balls is 0.80. Determine the velocity of each ball after impact.

11-67 The two disks A and B in Fig. P11-67 have identical masses and slide on a smooth

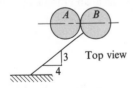

Figure P11-67

horizontal plane. Body B is connected to the wall by an extensible cord. The coefficient of

restitution for disks A and B is 0.80. Initially the velocity of A is 25 fps to the right, and B is at rest. Determine the velocity of each body after impact.

11-68 The two bodies A and B in Fig. P11-68 are moving to the right along the smooth horizontal plane when they collide with direct central impact. Determine the percentage loss of kinetic energy of the two bodies during impact in terms of two masses, the initial velocities, and the coefficient of restitution.

Figure P11-68

11-5 ANGULAR MOMENTUM OF A PARTICLE—CONSERVATION OF ANGULAR MOMENTUM

The linear momentum $\mathbf{G}$ of a particle having a mass m and a velocity $\mathbf{v}$ is $\mathbf{G} = m\mathbf{v}$. The vector $\mathbf{G}$ is localized and passes through the particle as shown in Fig. 11-11. Let the fixed point O be the origin of a Newtonian reference system $Oxyz$. The *angular momentum* of the particle with respect to the fixed point O is the moment of the linear momentum with respect to O. Using the symbol $\mathbf{H}_O$ for angular

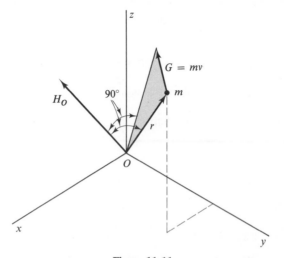

Figure 11-11

momentum, it can be written as

$$\mathbf{H}_O = \mathbf{r} \times m\mathbf{v}, \qquad (11\text{-}9)$$

where $\mathbf{r}$ is the position vector from O to the particle. The angular momentum vector $\mathbf{H}_O$ is perpendicular to the plane formed by the position vector $\mathbf{r}$ and the velocity vector $\mathbf{v}$ as shown in Fig. 11-11.

When Eq. (11-9) is differentiated with respect to time, the time rate of change of the angular momentum is

$$\dot{\mathbf{H}}_O = \dot{\mathbf{r}} \times m\mathbf{v} + \mathbf{r} \times m\dot{\mathbf{v}}$$

or

$$\dot{\mathbf{H}}_O = \mathbf{v} \times m\mathbf{v} + \mathbf{r} \times m\mathbf{a}.$$

The first term on the right is zero, and the second term is the same as $\mathbf{r} \times \mathbf{F}$, where $\mathbf{F}$ is the resultant force acting on the particle. The time rate of change of the angular momentum can thus be written as

$$\dot{\mathbf{H}}_O = \mathbf{M}_O, \qquad (11\text{-}10)$$

where $\mathbf{M}_O$ is the moment about point O of the resultant force acting on the particle. Eq. (11-10) is a vector equation and can be resolved into component scalar equations as

$$\dot{H}_x = M_x,$$
$$\dot{H}_y = M_y, \qquad (11\text{-}10a)$$
$$\dot{H}_z = M_z,$$

where H_x, H_y, H_z and M_x, M_y, M_z are the angular momentum and the moment of the resultant force, respectively, about fixed x, y, and z axes.

Equations (11-10) and (11-10a) are the mathematical form of the *principle of angular momentum for particles, which states that the time rate of change of the angular momentum of a particle with respect to a fixed point O (or a fixed axis) is equal to the moment about point O (or about the fixed axis) of the resultant force acting on the particle.*

When the moment $\mathbf{M}_O$ can be expressed as a function of time in a Newtonian reference system, Eq. (11-10) can be integrated from some initial time t_i to some final time t_f to give

$$\Delta\mathbf{H}_O = (\mathbf{H}_O)_f - (\mathbf{H}_O)_i = \int_{t_i}^{t_f} \mathbf{M}_O \, dt, \qquad (11\text{-}11)$$

where $\int_{t_i}^{t_f} M_O \, dt$ is the *angular impulse*, with respect to O, of the resultant force acting on the particle. Equation (11-11) is the mathematical form of the *principle of angular impulse and angular momentum for a particle and states that during the time interval t_i to t_f the change in the angular momentum about any fixed point O is equal to the angular impulse about point O acting on the particle.* Equation (11-11) can be resolved into three component expressions in the same manner as that explained for Eq. (11-10).

If $\mathbf{M}_O$ is zero, Eq. (11-10) becomes

$$\dot{\mathbf{H}}_O = 0.$$

Since the derivative of $\mathbf{H}_O$ is zero, $\mathbf{H}_O$ must be constant during the time interval that $\mathbf{M}_O$ is zero, and

$$(\mathbf{H}_O)_f = (\mathbf{H}_O)_i.$$

The *principle of conservation of angular momentum for a particle states that if the moment about any fixed point O of the resultant force acting on a particle is zero, the angular momentum of the particle about that point is conserved.*

Equation (11-10) is a vector equation, and although the total moment $\mathbf{M}_O$ may not be zero, it frequently occurs that a component of $\mathbf{M}_O$ is zero along some fixed axis through point O. In such a case, although the total angular momentum $\mathbf{H}_O$ is not conserved, the component of $\mathbf{H}_O$ in the direction of zero moment is conserved. Such a case is illustrated in Example 11-8.

EXAMPLE 11-7

The bob of a simple pendulum has a mass m and is supported by a cord of length L which swings in a vertical plane. Use the principle of angular momentum to determine the differential equation of motion.

SOLUTION

A free-body diagram of the bob swinging in the xz plane is shown in Fig. 11-12. There are two forces acting on the bob, the weight and the

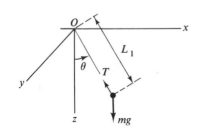

tension in the cord. Since the cord passes through O, the moment of the tensile force with respect to O is zero; therefore the resultant moment of all forces about O is the moment of the weight. That is,

$$\mathbf{M}_O = -mgL \sin\theta\,\mathbf{j}.$$

The magnitude of the linear momentum of the pendulum bob is $mL\dot{\theta}$, and the angular momentum is the moment of the linear momentum; hence

$$\mathbf{H}_O = L(mL\dot{\theta})\mathbf{j}.$$

Figure 11-12

When $\mathbf{H}_O$ is differentiated with respect to time, the principle of angular momentum gives

$$mL^2\ddot{\theta}\mathbf{j} = -mgL \sin\theta\,\mathbf{j}$$

or

$$\ddot{\theta} + \frac{g}{L}\sin\theta = 0. \qquad \text{Ans.}$$

EXAMPLE 11-8

The spherical pendulum in Fig. 11-13(a) is given an initial velocity of v_1 parallel to the y axis when the angle θ is 30°. If θ is to increase to a maximum value of 45°, determine the required value of v_1.

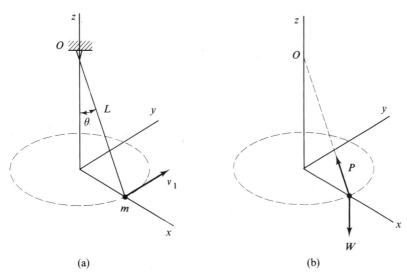

(a)   (b)

Figure 11-13

SOLUTION

A free-body diagram of the particle m is shown in Fig. 11-13(b). The moment of the forces acting on the particle about the fixed pivot O is

$$\mathbf{M}_O = (\mathbf{L} \times \mathbf{P}) + (\mathbf{L} \times \mathbf{W}),$$

where $\mathbf{L}$ is the vector from O to m. The moment due to $\mathbf{P}$ is zero since $\mathbf{P}$ is always directed through O. The moment due to $\mathbf{W}$ is always perpendicular to the plane formed by $\mathbf{L}$ and $\mathbf{W}$ and consequently has no component in the z direction; therefore the component of the angular momentum along the z axis is conserved.

Since the initial velocity $\mathbf{v}_1$ is parallel to the xy plane, the initial angular momentum of the particle about the fixed z axis is

$$(\mathbf{H}_z)_1 = [(\mathbf{L} \times m\mathbf{v}_1) \cdot \mathbf{k}]\mathbf{k} = Lmv_1 \sin 30°\mathbf{k}$$

and

$$(H_z)_1 = Lmv_1 \sin 30°.$$

Although the trajectory of the particle is unknown, when θ reaches a maximum value the velocity v_2 must again be parallel to the xy plane; therefore

$$(\mathbf{H}_z)_2 = [(\mathbf{L} \times m\mathbf{v}_2) \cdot \mathbf{k}]\mathbf{k} = Lmv_2 \sin 45°\mathbf{k}$$

and

$$(H_z)_2 = Lmv_2 \sin 45°.$$

Since the z component of the angular momentum is conserved, $(H_z)_1$ must equal $(H_z)_2$, that is,

$$Lmv_1 \sin 30° = Lmv_2 \sin 45°,$$

from which

$$v_2 = v_1 \left(\frac{\sin 30°}{\sin 45°} \right). \tag{a}$$

A second equation relating v_2 and v_1 can be obtained from the work-energy principle; that is,

$$U_{1-2} = T_2 - T_1$$

or

$$-W(L \cos 30° - L \cos 45°) = \frac{1}{2} \frac{W}{g} v_2^2 - \frac{1}{2} \frac{W}{g} v_1^2.$$

The last equation can be rearranged to the form

$$v_2^2 = v_1^2 - 2gL(\cos 30° - \cos 45°). \tag{b}$$

Substitution of Eq. (a) into Eq. (b) gives

$$v_1^2 \left(\frac{\sin 30°}{\sin 45°}\right)^2 = v_1^2 - 2gL(\cos 30° - \cos 45°)$$

or

$$v_1^2 = \frac{2gL(\cos 30° - \cos 45°)}{1 - (\sin 30°/\sin 45°)^2} = 0.318(2gL).$$

When $L = 3$ ft is substituted into the last equation the solution is obtained as

$$v_1^2 = 61.4$$

or

$$v_1 = 7.84 \text{ fps.} \qquad \text{Ans.}$$

PROBLEMS

11-69 (a) Determine the angular momentum of a 1.61-lb particle with respect to the origin O of a fixed coordinate system when the particle is passing through the point $(4\mathbf{i} + 10\mathbf{j} - 10\mathbf{k})$ ft with a velocity of $(2\mathbf{i} - 6\mathbf{j} + 10\mathbf{k})$ fps.
(b) Solve part (a) if the particle has a mass of 0.5 kg, the coordinates of the point are (4, 10, −5) m, and $\mathbf{v} = (5\mathbf{i} + 2\mathbf{j} - 7\mathbf{k})$ m/s.

11-70 The particles A and B shown in Fig. P11-70 have masses of $3m$ and m, respectively, and are joined by a rigid bar, the mass of which may be neglected. The assembly rotates in the vertical xy plane about a horizontal axis through O. Determine
(a) The angular momentum of the two particles with respect to O.
(b) The differential equation of motion by using the principle of angular momentum.

11-71 Particles A and B shown in Fig. P11-71 have masses of $2m$ and m, respectively, and are

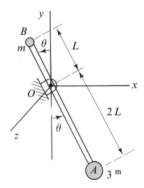

Figure P11-70

joined by rigid bars, the masses of which may be neglected. The assembly rotates in the vertical xy plane. Determine
(a) The angular momentum of the two particles with respect to O.
(b) The differential equation of motion by using Eq. (11-10).

Figure P11-71

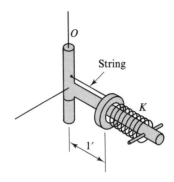

Figure P11-73

11-72 The smooth rod of negligible mass in Fig. P11-72 rotates in a horizontal plane about a vertical axis at O with an angular velocity of 3 rad per sec clockwise. The particle A of mass m can slide along the rod and is held in position by the string. When in the position shown, the string breaks, and the mass is free to slide along the rod. Determine

(a) The velocity of A just before it hits the stop.
(b) The velocity of A just after it hits the stop (assume no rebound).
(c) The impulse of the stop on A during the impact.

Figure P11-72

11-73 A smooth thin rod of negligible mass rotates in a horizontal plane about a vertical axis through O as shown in Fig. P11-73 with a clockwise angular velocity of 6 rad per sec. A 32.2-lb disk is held by a string and is in contact with an unstressed spring. The string then breaks. Determine the value of the spring constant if the maximum compression of the spring is 1 ft.

11-74 The 3.22-lb particle A in Fig. P11-74 slides on the smooth horizontal plane and is

connected to the spring by a cord which passes through a smooth hole in the plane. In the position shown, A has a velocity of 100 fps to the right as indicated, there is no slack in the cord, and the spring is unstretched. During the motion the particle reaches a maximum distance of 10 ft from the hole. Determine the spring constant.

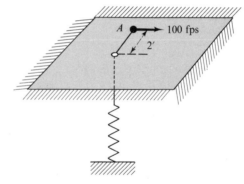

Figure P11-74

*11-75 The 0.1-kg particle A in Fig. P11-75 slides on a smooth horizontal plane and is attached to the fixed point O by means of an elastic cord having a modulus of 20 N/m and an

Figure P11-75

unstretched length of 3 m. When A is in the position shown its velocity is 30 m/s as indicated. Determine

(a) The speed of the particle when the cord becomes slack.

(b) The minimum distance from O to the particle A during the resulting motion.

11-76 The 6.44-lb particle A in Fig. P11-76 is connected to the fixed point O by a spring with a modulus of 20 lb per ft. In the position shown, the spring is stretched 2 ft, and A has the velocity indicated. Determine the maximum distance from O to A during the resulting motion.

Figure P11-76

11-77 A particle weighing 6.44 lb moves on a smooth horizontal plane as shown in Fig. P11-77. Throughout the motion a force of magnitude $F = 4r^3$ acts on the particle and is always directed toward the fixed point O. The force is in pounds when r (the distance of the particle from the fixed point) is in feet. Determine the velocity of the particle at point A if the particle reaches a maximum distance from O of 8 ft.

11-78 A particle travels on the smooth inner

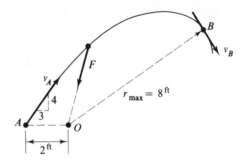

Figure P11-77

surface of a hollow cone with a vertical axis of revolution. When in the position shown in Fig. P11-78, the mass is given an initial velocity in a *horizontal* plane. Determine the radius of the path when the velocity of the mass is again horizontal if the initial speed is

(a) 3 fps.

(b) 30 fps.

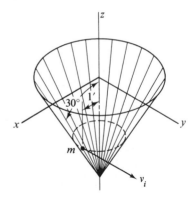

Figure P11-78

11-6 CENTRAL-FORCE MOTION

A central force is defined as a force with a line of action that always passes through a fixed point, known as the center of force. Some common examples of central-force systems include the motion of the planets about the sun, the motion of the moon (as well as artificial satellites) about the earth, the motion of an electron about the nucleus of an atom, and the motion of an α particle in the vicinity of a nuclear charge.

Consider the motion of particle P of mass m in Fig. 11-14 under the action of the central force **F** always directed through the fixed point

O. From Eq. (11-10),

$$\dot{\mathbf{H}}_O = \mathbf{M}_O = 0$$

because $\mathbf{F}$ has no moment about O. Since $\dot{\mathbf{H}}_O$ is zero, the integral of $\dot{\mathbf{H}}_O$ must be a constant; that is

$$\mathbf{H}_O = \mathbf{C}.$$

The angular momentum of a particle about a fixed point was defined as the moment of the linear momentum ($m\mathbf{v}$) about the point; therefore,

$$\mathbf{H}_O = \mathbf{r} \times (m\mathbf{v}) = \mathbf{C}.$$

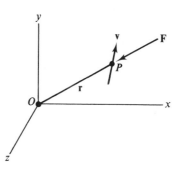

Figure 11-14

Since $\mathbf{H}_O$ is a constant in both magnitude and direction and, from the definition of a cross product, $\mathbf{H}_O$ is perpendicular to the plane of $\mathbf{r}$ and $\mathbf{v}$, it follows that $\mathbf{r}$ and $\mathbf{v}$ must remain in the same plane. Therefore *central-force motion is motion in a plane*. The plane is determined by the center of force, the position of the particle, and its velocity in that position.

Newton's law of universal gravitation states that the force of attraction, $\mathbf{F}$, between two masses M and m separated by a distance r has the magnitude

$$F = \frac{GMm}{r^2},$$

where G is the universal gravitational constant. The experimentally determined value of G is

$$G = 3.438(10)^{-8} \frac{\text{lb-ft}^2}{\text{slug}^2}.$$

This unit could also be written as

$$\frac{\text{ft}^3}{\text{slug-sec}^2} \quad \text{or} \quad \frac{\text{ft}^4}{\text{lb-sec}^4}.$$

In SI units the constant G is

$$G = 6.670(10)^{-11} \, \text{m}^3/\text{kg} \cdot \text{s}^2.$$

If the mass of one of the two bodies is much greater than that of the other body, the larger mass can be considered fixed in space because the smaller mass would have a negligible effect upon the motion of the heavier body. The masses of the sun, earth, and moon are given for reference in this regard. They are

$$M_{\text{sun}} = 1.97(10)^{29} \, \text{slugs} = 2.87(10)^{30} \, \text{kg},$$

$$M_{\text{earth}} = 4.09(10)^{23} \, \text{slugs} = 5.97(10)^{24} \, \text{kg},$$

and

$$M_{\text{moon}} = 5.05(10)^{21} \, \text{slugs} = 7.37(10)^{22} \, \text{kg}.$$

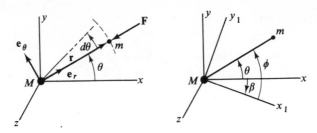

Figure 11-15

Figure 11-15(a) shows a large mass M, a much smaller mass m moving in the xy plane, and the force $\mathbf{F}$ exerted on m by M. Using polar coordinates and considering M fixed in space, the equations of motion of m in component form are, from Eq. (8-17),

$$\sum F_R = ma_R,$$

$$-\frac{GMm}{r^2} = m(\ddot{r} - r\dot{\theta}^2); \tag{a}$$

and

$$\sum F_T = ma_T,$$

$$0 = \frac{m}{r}\frac{d}{dt}(r^2\dot{\theta}). \tag{b}$$

When Eq. (b) is multiplied by r/m and integrated, the result is

$$r^2\dot{\theta} = \text{a constant} = h. \tag{c}$$

Note that the area swept out by the radius vector as it turns through an angle $d\theta$ in time dt is $dA = r(r\,d\theta)/2$; therefore,

$$2\frac{dA}{dt} = r^2\frac{d\theta}{dt} = \text{a constant} = h. \tag{11-12}$$

The quantity $dA/dt = h/2$ is called the *areal speed*, and, as shown here, it is a constant for any central-force system. Equation (11-12) expresses Kepler's second law of planetary motion which states that *the radius vector describes equal areas in equal intervals of time.*

The equation of the path of motion can be obtained by eliminating the parameter t from Eq. (a) and (c) as follows: Let $u = 1/r$, from which $du = -dr/r^2$ and, with Eq. (11-12),

$$\frac{dr}{dt} = \frac{dr}{d\theta}\frac{d\theta}{dt} = \frac{h}{r^2}\frac{dr}{d\theta} = -h\frac{du}{d\theta}.$$

Also

$$\frac{d^2r}{dt^2} = \frac{d}{dt}\left(-h\frac{du}{d\theta}\right) = -h\frac{d^2u}{d\theta^2}\frac{d\theta}{dt} = -h\frac{d^2u}{d\theta^2}\frac{h}{r^2} = -h^2u^2\frac{d^2u}{d\theta^2}. \tag{d}$$

Substitution of Eq (11-12), of Eq. (d), and of u for $1/r$ into Eq. (a) gives

$$-GMmu^2 = m\left[-h^2u^2\frac{d^2u}{d\theta^2} - \frac{1}{u}(hu^2)^2\right]$$

or

$$\frac{d^2u}{d\theta^2} + u = \frac{GM}{h^2}.$$ (e)

Equation (e) is an ordinary, second-order differential equation with constant coefficients and with the right-hand side different from zero. It can be solved by several methods; probably the most common is to assume a solution and verify it by substitution in the differential equation. Since u and its second derivative with respect to θ are involved, a function of θ with a second derivative equal to the negative of the original function plus or minus a constant is desired. Some exponential and trigonometric functions have this property, and the following assumed solution will be used:

$$u = C \cos(\theta + \beta) + D.$$

Substitution into Eq. (e) gives

$$D = \frac{GM}{h^2},$$

and the general solution of the differential equation is

$$u = C \cos(\theta + \beta) + \frac{GM}{h^2} = \frac{1}{r},$$ (f)

where C and β are constants of integration with values determined by the initial conditions. The value of h is given by Eq. (c). By measuring the polar angle from an x_1 axis as shown in Fig. 11-15(b) instead of from the x axis, the angle $\theta + \beta$ can be replaced by an angle ϕ, and Eq. (f) becomes

$$r = \frac{h^2/GM}{1 + (h^2C/GM)\cos\phi}.$$ (11-13a)

Equation (11-13a) is the equation of a conic section in polar form with the origin at one focus. For the simplest case with C equal to zero, r is a constant and the curve is a circle. A conic section is defined as the path of a point P (Fig. 11-16) which moves in such a manner that

$$\frac{OP}{PD} = \text{a constant} = e,$$

where O is a fixed point, AB is a fixed line, OP is the distance from O to P, PD is the perpendicular distance from P to the fixed line, and e is

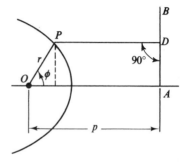

Figure 11-16

the eccentricity. This relation can be written as

$$OP = r = e(PD) = e(p - r \cos \phi)$$

from which

$$r = \frac{ep}{1 + e \cos \phi}.$$ (11-13b)

Equation (11-13b) is the same as Eq. (11-13a) when the eccentricity of the conic section is

$$e = \frac{h^2 C}{GM}$$ (g)

and

$$p = \frac{1}{C}.$$

Therefore

$$ep = \frac{h^2}{GM}.$$

Equation (11-13a) or (11-13b) can now be written as

$$r = \frac{h^2/GM}{1 + e \cos \phi}.$$ (11-14a)

Also Eq. (f) can be written as

$$\frac{1}{r} = C \cos \phi + \frac{GM}{h^2},$$ (11-14b)

which is the equation of the path of the particle in polar form.

The preceding analysis proves Kepler's first law, which states that *the trajectory or orbit of a particle moving under the action of an inverse-square central force is a conic section with the center of force at one of the foci.* When e is less than 1, the locus is a closed path (an ellipse or circle when $e = 0$) since the denominator of Eq. (11-14a) is never zero for any value of ϕ. For values of $e \geq 1$ the path is not closed because, for certain values of ϕ, the denominator becomes zero and r becomes infinite. For $e = 1$ the curve is a parabola, and for $e > 1$ it is a hyperbola.[2]

For planetary motion about the sun and for artificial earth satellites, the orbits are ellipses and e is less than 1. The minimum distance from the center of attraction (at one focus) to the particle, corresponding to $\phi = 0$, is called perigee (r_p), and the maximum distance (for an elliptical orbit), occurring when $\phi = \pi$, is called apogee (r_a). When $\phi = 0$ in Fig. 11-16 the velocity of the particle is perpendicular to the radius vector, and from Eq. (c)

$$r^2 \dot{\theta} = r^2 \dot{\phi} = r_p v_p = h.$$

---

[2] See the discussion of conic sections in any standard text on analytical geometry.

This equation makes it possible to compute the constant h if the velocity and distance at perigee are known.

For an elliptical orbit, Eq. (11-14a) shows that the perigee and apogee distances are

and

$$\left.\begin{array}{l} r_p = \dfrac{h^2/GM}{1+e} \\[4mm] r_a = \dfrac{h^2/GM}{1-e}. \end{array}\right\} \tag{h}$$

These two equations can be solved for e by eliminating h^2/GM, and the result is

$$e = \frac{r_a - r_p}{r_a + r_p}. \tag{i}$$

The semimajor axis of the ellipse is

$$a = \frac{r_p + r_a}{2},$$

which, from Eq. (h), becomes

$$a = \frac{h^2}{2GM}\left(\frac{1}{1+e} + \frac{1}{1-e}\right) = \frac{h^2}{GM}\left(\frac{1}{1-e^2}\right). \tag{j}$$

The semiminor axis is

$$b = a(1-e^2)^{1/2} = \frac{h^2}{GM}(1-e^2)^{-1/2}.$$

The area of an ellipse is

$$A = \pi a b = \pi a^2 (1-e^2)^{1/2}.$$

The areal speed, $dA/dt = h/2$, from Eq. (11-12) gives $A = (h/2)T$, where T is the period (the time for one orbit) and

$$T = \frac{2A}{h} = \frac{2\pi a^2}{h}(1-e^2)^{1/2}.$$

The value of $1-e^2$ can be obtained from Eq. (j), and the period becomes

or

$$\left.\begin{array}{l} T = \dfrac{2\pi a^2}{h}\left(\dfrac{h^2}{GMa}\right)^{1/2} = \dfrac{2\pi a^{3/2}}{(GM)^{1/2}} \\[4mm] \dfrac{T^2}{a^3} = \dfrac{4\pi^2}{GM}. \end{array}\right\} \tag{11-15}$$

Note that the period is independent of the mass of the orbiting body. Equation (11-15) demonstrates Kepler's third law[3] of planetary motion; namely, *the squares of the periods of the planets are proportional to the cubes of the semimajor axes of their orbits.*

The orbit of the moon around the earth is very nearly an ellipse (the sun perturbs it slightly) with $e = 0.056$. Its perigee and apogee distances are 222,000 and 253,000 miles, respectively. Some data on the planets are shown in the accompanying table.

PLANETARY DATA

Planet	e	Mean Distance from Sun in Astronomical Units*	Time for One Revolution About the Sun in Days (d) or Years (yr)†
Mercury	0.206	0.387	88 d
Venus	0.007	0.723	225 d
Earth	0.017	1.000	365 d
Mars	0.093	1.524	687 d
Jupiter	0.048	5.203	11.9 yr
Saturn	0.056	9.539	29.5 yr

* An astronomical unit equals the mean distance from the earth to the sun which is approximately $9.26(10^7)$ miles or $1.49(10^8)$ km.

† The days and years used are the conventional time periods, or solar time, as distinguished from sidereal, or star time, measured by the rotation of the earth in relation to the stars. A sidereal day equals 23 hr, 56 min, and 4.091 sec of mean solar time.

When a satellite is "launched" with a velocity v_p parallel to the earth's surface at a distance r_p from the center of the earth the constant C can be determined by means of Eq. (11-14b). The angle ϕ is zero, and the value of C is

$$C = \frac{1}{r_p} - \frac{GM}{h^2} = \frac{1}{r_p} - \frac{GM}{r_p^2 v_p^2}.$$

When the value of C is substituted into Eq. (g) the eccentricity becomes

$$e = \frac{h^2}{GM}\left(\frac{1}{r_p} - \frac{GM}{h^2}\right)$$

$$= \frac{r_p v_p^2}{GM} - 1. \tag{k}$$

The constant GM for earth satellites can be determined from the force of gravity acting on a particle of mass m located at the surface of

---

[3] The first two laws were published in 1609 and the third in 1619. Newton derived the law of gravitational attraction from Kepler's laws.

the earth. By the law of universal gravitation,

$$\frac{GMm}{R^2} = mg,$$

where R, the radius of the earth, is 3960 miles or 6370 km; therefore

$$GM = gR^2$$
$$= 32.2(3960)^2(5280)^2 = 1.408(10)^{16} \text{ ft}^3/\text{sec}^2.$$

Other values of gR^2 that are commonly used are

$$gR^2 = 1.239(10)^{12} \text{ miles}^3/\text{hr}^2$$
$$= 3.98(10)^{13} \text{ m}^3/\text{s}^2$$
$$= 5.16(10)^{12} \text{ km}^3/\text{hr}^2.$$

Equation (k) can be used to determine the minimum velocity the satellite must have to escape from the earth's gravitational field. The minimum velocity will occur when the satellite is on a parabolic path, that is, when $e = 1$. When $e = 1$ is substituted into Eq. (k) the escape velocity is found to be

$$v_e = \sqrt{\frac{2GM}{r_p}} = \sqrt{\frac{2gR^2}{r_p}}.$$

When the eccentricity is zero the particle will have a circular orbit, and the corresponding velocity, from Eq. (k), is

$$v_{\text{cir}} = \sqrt{\frac{GM}{r_p}} = \sqrt{\frac{gR^2}{r_p}}.$$

The two preceding equations show that

$$v_e = \sqrt{2}v_{\text{cir}}.$$

EXAMPLE 11-9

An earth satellite is launched with a velocity of 18,000 mph parallel to the surface of the earth at an altitude of 540 miles. Determine the maximum altitude reached by the satellite.

SOLUTION

Figure 11-17 shows the satellite P' moving from perigee to apogee. The angular momentum at perigee must equal the angular momentum at apogee; that is,

$$mv_p r_p = mv_a r_a$$

or

$$v_a = \frac{r_p}{r_a} v_p. \tag{a}$$

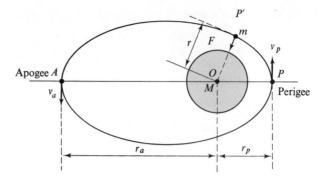

Figure 11-17

The principle of work and energy gives

$$U_{p \to a} = T_a - T_p$$

or

$$-\int_{r_p}^{r_a} \frac{GMm}{r^2}\, dr = \frac{1}{2}\, m(v_a^2 - v_p^2)$$

from which

$$GMm\left(\frac{1}{r}\right)_{r_p}^{r_a} = GMm\left(\frac{r_p - r_a}{r_p r_a}\right) = \frac{1}{2}\, m(v_a^2 - v_p^2). \tag{b}$$

When Eq. (a) is substituted into Eq. (b) the result is

$$GM\left(\frac{r_p - r_a}{r_p r_a}\right) = \frac{1}{2}\left(\frac{r_p^2}{r_a^2}\, v_p^2 - v_p^2\right) = \frac{v_p^2}{2} \frac{r_p^2 - r_a^2}{r_a^2},$$

which can be solved for r_a as

$$r_a = \frac{v_p^2 r_p^2}{2GM - v_p^2 r_p}. \tag{c}$$

The radius of the earth is 3960 miles; therefore

$$r_p = 3960 + 540 = 4500 \text{ miles.}$$

The velocity v_p is given as 18,000 mph, and the value of MG is $1.239(10)^{12}$ miles³/hr² in consistent units. When these values are substituted into Eq. (c) the radius at apogee is

$$r_a = \frac{(18,000)^2(4500)^2}{2(1.239)(10)^{12} - (18,000)^2(4500)}$$

$$= 6430 \text{ miles.}$$

The altitude at apogee is

altitude $= 6430 - 3960 = \underline{2470 \text{ miles above the earth's surface.}}$

Ans.

EXAMPLE 11-10

The satellite of Fig. 11-18 is launched with a velocity of 21,000 mph parallel to the surface of the earth at an altitude of 600 miles.

Determine

(a) The eccentricity of the orbit.
(b) The equation of the orbit.
(c) The time for one orbit.

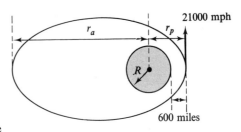

Figure 11-18

SOLUTION

(a) The initial distance of the satellite from the center of the earth, r_p, is

$$r_p = 3960 + 600 = 4560 \text{ miles.}$$

The eccentricity, from Eq. (k), is

$$e = \frac{r_p v_p^2}{GM} - 1$$

$$= \frac{4560(21{,}000)^2}{1.239(10)^{12}} - 1 = \underline{0.623.} \qquad \text{Ans.}$$

(b) The equation of the orbit can be expressed in a variety of forms. If Eq. (11-14a) is used, the value of h^2/GM must be determined. Since h is twice the constant areal speed, its value is

$$h = v_p r_p,$$

and the numerator of Eq. (11-14a) is

$$\frac{h^2}{GM} = \frac{v_p^2 r_p^2}{GM}$$

$$= \frac{(21{,}000)^2 (4560)^2}{1.239(10)^{12}}$$

$$= 7400 \text{ miles.}$$

The equation of the orbit is

$$r = \frac{h^2/GM}{1 + e \cos \phi}$$

$$= \frac{7400}{1 + 0.623 \cos \phi} \qquad \text{Ans.}$$

(c) The period can be obtained from Eq. (11-15) after the semimajor axis, a, has been obtained. Equation (j) for the semimajor axis gives

$$a = \frac{h^2}{GM} \left(\frac{1}{1 - e^2} \right)$$

$$= 7400 \left(\frac{1}{1 - 0.623^2} \right)$$

$$= 12{,}094 \text{ miles,}$$

and the period is

$$T = 2\pi \left(\frac{a^3}{GM}\right)^{1/2}$$

$$= 2\pi \left[\frac{12,094^3}{1.239(10)^{12}}\right]^{1/2}$$

$$= \underline{7.51 \text{ hr.}} \qquad \qquad \text{Ans.}$$

PROBLEMS

11-79 A satellite is launched with a velocity of 20,000 mph parallel to the earth's surface at an altitude of 300 miles. Determine
(a) The altitude at apogee.
(b) The distance from the center of the earth to the center of the elliptical path.

***11-80** A satellite is launched with a velocity of 34,000 km/hr parallel to the surface of the earth at an altitude of 650 km. Determine
(a) The altitude at apogee.
(b) The speed of the satellite when it is at the extremity of the minor axis. (*Note:* From analytical geometry $a^2 = b^2 + c^2$, where a is the semimajor axis, b is the semiminor axis, and c is the distance from the center of the ellipse to the center of the earth).

11-81 At burnout the satellite in Fig. P11-81 reached an altitude of 1200 miles with a velocity of 18,000 mph, making an angle of 80° with the vertical. Determine the perigee and apogee distances.

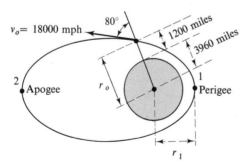

$v_o = 18000$ mph 80° 1200 miles 3960 miles
2 Apogee r_o 1 Perigee r_1

Figure P11-81

11-82 (a) A satellite is traveling in a circular orbit at an altitude of 1000 miles above the earth. Determine the velocity of the satellite.

(b) The velocity of the satellite of part (a) is increased to 19,000 mph by firing booster rockets. Determine the apogee distance for the resulting elliptical orbit.

11-83 A satellite is to be launched with a velocity parallel to the earth's surface at an altitude of 300 miles. Determine
(a) The escape velocity.
(b) The circular orbit velocity.

***11-84** The perigee and apogee distances of a satellite are 7200 and 19,000 km, respectively. Determine
(a) The eccentricity of the orbit.
(b) The velocity at perigee.

11-85 A satellite is launched with a velocity of 21,000 mph parallel to the earth's surface at an altitude of 600 miles. Determine
(a) The equation of the orbit.
(b) The apogee distance.

***11-86** A satellite has an elliptical orbit ($e = 0.2$) about the earth with a minimum distance from the center of the earth of 9600 km. Determine
(a) The distance from the center of the earth to the apogee.
(b) The period of the orbit.

11-87 A particle of mass m is subjected to a central force and travels on a path defined by the expression $r = \theta^2$. Develop the expression for the central force as a function of h and r.

11-88 The central force **F** of Fig. 11-15(a) is $-(km/r^3)\mathbf{e}_r$. For this value of **F**, determine the polar equation of the path of the particle of mass m.

***11-89** A satellite is launched with a velocity of 30,000 km/hr *parallel to the earth* at a distance of 7250 km from the center of the earth.

Determine
(a) The equation of the orbit.
(b) The eccentricity of the orbit.

11-90 The mass of the earth is approximately 81.3 times the mass of the moon. The radius of the moon is approximately 1080 miles. Determine the acceleration of gravity on the moon's surface.

11-91 Figure P11-91 represents the elliptical path of a satellite with one focus at the center of the earth, O, and the other focus at F. An ellipse has the property that the sum of the distances from any point on the ellipse to the two foci is equal to the length of the major axis; that is, $r_p + r_a = 2a$. Show that the length of the semiminor axis, b, is $b = \sqrt{r_p r_a}$, where r_p and r_a are the perigee and apogee distances.

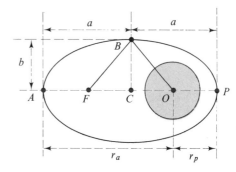

Figure P11-91

11-92 A satellite of the earth has an apogee distance of 10,000 miles and a perigee distance of 5000 miles. Determine
(a) The period of the orbit.
(b) The semiminor axis by using the relationship developed in Problem 11-91.

11-7 SYSTEMS WITH VARIABLE MASS

In previous articles, the mass of the system considered remained constant. There are, however, many dynamics problems such as those involving rockets, hoisting cables, and high-speed particles in which the mass of the system changes with time or in which some parts of the system gain or lose mass from other parts of the system. Variable-mass problems can be divided into two[4] types:

> **1.** Problems in which the mass of the system varies because the number of individual particles, each of constant mass, in the system varies.
> **2.** Problems in which masses of the individual particles change with time, while the number of particles remains constant.

The first variable-mass system is commonly encountered in rocket problems where the system to be studied is defined as the rocket shell plus the unburned fuel.

The second type occurs when the velocities of individual particles of the system become so large that the relativistic changes of mass with velocities must be considered. Relativistic changes in mass are not normally important unless the velocities involved are a significant fraction of the speed of light. Such velocities are attained by charged

---

[4] There could conceivably be a combination of these two types, but such a combination will not be considered.

particles in cyclotrons, X-ray machines, and television picture tubes. In this text, only variable-mass problems of the first type will be treated.[5]

The general procedure to be developed will apply to any body which is either acquiring or expelling mass. The following assumptions are made to simplify the derivation of the differential equation of motion and the subsequent analysis:

1. The motion of the system is along a straight line; consequently, the resultant of the external forces applied to the system is along the path of motion.
2. The acceleration of gravity, g, is constant.

The first assumption excludes any lateral maneuver of the system but permits investigation of the fundamental equations relating acceleration, velocity, and distance traveled along the path. These equations are dependent on the variation of mass of the body. Since the acceleration due to gravity is inversely proportional to the square of the distance from the center of the earth, the second assumption is accurate to within 5% for altitudes less than about 100 miles above the earth.

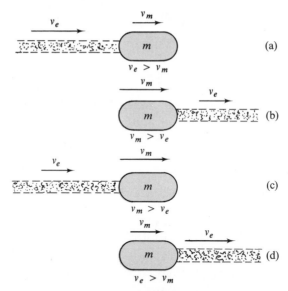

Figure 11-19

Four possible situations in which a large mass m is gaining or losing mass from a stream of particles are illustrated in Fig. 11-19. All velocities are assumed to be to the right, and any velocity to the left

---

[5] Those interested in relativistic mass changes are referred to books on the special theory of relativity, such as M. Russell Wehr and James A. Richards, *Physics of the Atom* (Reading, Mass.: Addison-Wesley Publishing Company, Inc., 1960), Chap. 5.

will be indicated by a minus sign. Figures 11-19(a) and 11-19(b) represent situations in which the large mass m is being increased, first because the stream of particles (with velocity $\mathbf{v}_e$) is to the left of m and $\mathbf{v}_e$ is greater than $\mathbf{v}_m$ and second because the stream is to the right of m and is being overtaken by m ($\mathbf{v}_m$ is greater than $\mathbf{v}_e$). Figures 11-19(c) and 11-19(d) represent cases in which mass is being expelled from m and would occur when m was being accelerated by firing a rocket engine [Fig. 11-19(c)] or decelerated by firing a retrorocket [Fig. 11-19(d)].

Section 11-7
Systems with
variable mass

The equation of motion will be derived for the situation shown in Fig. 11-19(a) but will be valid for any of the four cases when consistent signs are used for velocities and rate of change of mass. Figure 11-20(a) shows the system at time t when the mass m has a velocity $\mathbf{v}_m$ and the increment of mass Δm has a velocity of $\mathbf{v}_e$. The forces $\mathbf{T}_1$ and $\mathbf{T}_2$ are the thrust or mutual reaction between m and Δm, and $\mathbf{R}$ is the resultant of the external forces (such as weight and aerodynamic drag) acting on the system. The diagram in Fig. 11-20(b) illustrates the system at time $(t+\Delta t)$ when Δm has joined m with a common velocity of $\mathbf{v}_m+\Delta\mathbf{v}_m$. The linear momentum of the system at time t is

Figure 11-20

$$\mathbf{G}(t) = m\mathbf{v}_m + \Delta m\mathbf{v}_e,$$

and the momentum at $t+\Delta t$ is

$$\mathbf{G}(t+\Delta t) = (m+\Delta m)(\mathbf{v}_m+\Delta\mathbf{v}_m).$$

The impulse of the external forces (note that $\mathbf{T}_1$ and $\mathbf{T}_2$ are internal to the system) is equal to the change of the momentum of the system; that is,

$$\mathbf{R}\,\Delta t = \mathbf{G}(t+\Delta t) - \mathbf{G}(t)$$

$$= (m\mathbf{v}_m + m\,\Delta\mathbf{v}_m + \Delta m\mathbf{v}_m + \Delta m\,\Delta\mathbf{v}_m) - (m\mathbf{v}_m + \Delta m\mathbf{v}_e)$$

$$= m\,\Delta\mathbf{v}_m - \Delta m(\mathbf{v}_e - \mathbf{v}_m) + \Delta m\,\Delta\mathbf{v}_m.$$

As the time interval approaches zero the incremental quantities Δt, Δm, and $\Delta\mathbf{v}_m$ become the differential quantities dt, dm, and $d\mathbf{v}_m$, respectively, and the equation can be written as

$$\mathbf{R}\,dt = m\,d\mathbf{v}_m - dm(\mathbf{v}_e - \mathbf{v}_m) + dm\,d\mathbf{v}_e.$$

The last term in this expression is a second-order differential and can be neglected. When the equation is divided by dt it becomes

$$\mathbf{R} = m\frac{d\mathbf{v}_m}{dt} - \frac{dm}{dt}(\mathbf{v}_e - \mathbf{v}_m)$$

$$= m\dot{\mathbf{v}}_m - \dot{m}(\mathbf{v}_e - \mathbf{v}_m) = m\dot{\mathbf{v}}_m - \dot{m}\mathbf{v}_{e/m}. \qquad (11\text{-}16)$$

If v_m is greater than v_e (the large mass is overtaking Δm), the equation becomes

$$\mathbf{R} = m\dot{\mathbf{v}}_m + \dot{m}(\mathbf{v}_m - \mathbf{v}_e) = m\dot{\mathbf{v}}_m + \dot{m}\mathbf{v}_{m/e}. \qquad (11\text{-}16a)$$

When Δm is being expelled as in Fig. 11-19(c) or (d), $\dot{m}$ will be negative in Eq. (11-16) and (11-16a).

Note that in general $\mathbf{v}_e$ is not zero, although it may be in special cases, for example, when Δm is at rest before it is "picked up." If Newton's second law of motion is written in the form

$$\mathbf{R} = \frac{d}{dt}(m\mathbf{v}_m) = \dot{m}\mathbf{v}_m + m\dot{\mathbf{v}}_m, \qquad (11\text{-}17)$$

it is seen that this expression agrees with Eq. (11-16) or (11-16a) *only* when $\mathbf{v}_e$ is zero, that is, when the initial velocity of the acquired mass or the final velocity of the expelled mass is zero.

The thrust T can be obtained by applying the impulse-momentum equation to the mass Δm in Fig. 11-20. When $\mathbf{i}$ is used as the unit vector to the right the impulse-momentum equation is

$$-T_1 \Delta t \mathbf{i} = \Delta m (v_m + \Delta v_m - v_e)\mathbf{i}$$

from which

$$-T_1 = \lim_{\Delta t \to 0} \frac{\Delta m}{\Delta t}(v_m - v_e) = \dot{m}v_{m/e} = -\dot{m}v_{e/m}$$

when the higher-order term $\Delta m(\Delta v_m)$ is omitted. Therefore,

$$T_1 = \dot{m}v_{e/m},$$

where T_1 is the magnitude of the thrust force, and since $\dot{m}$ and $v_{e/m}$ are both positive when v_e is greater than v_m, T_1 is positive and is in the direction shown (to the left) in Fig. 11-20(a). The vector expression for the thrust force is $\mathbf{T}_1 = -\dot{m}\mathbf{v}_{e/m}$. The force $\mathbf{T}_2$, which has the same magnitude as $\mathbf{T}_1$ but acts to the right, is

$$\mathbf{T}_2 = \dot{m}v_{e/m}\mathbf{i} = \dot{m}\mathbf{v}_{e/m}. \qquad (a)$$

The last equation can be combined with Eq. (11-16) to give

$$\mathbf{R} = m\dot{\mathbf{v}}_m - \mathbf{T}_2$$

or

$$m\dot{\mathbf{v}}_m = \mathbf{R} + \mathbf{T}_2. \qquad (11\text{-}16b)$$

Again it should be noted that when mass is being expelled from m, as in a rocket, the factor $\dot{m}$ is negative in the preceding equations.

Consider the rocket in Fig. 11-21 with an

y

$+$

$W = (m_o - bt)\,g$

F_d

T

Figure 11-21

initial mass m_0 which burns fuel at a constant rate b; that is, $\dot{m} = -b$. The velocity of the burned gasses with respect to the rocket is $\mathbf{v}_{e/m}$ and is assumed to be constant. The mass of the rocket at time t is

$$m = m_0 - bt,$$

and the magnitude of the thrust is

$$T = \dot{m}v_{e/m} = bv_{e/m}.$$

Note that $\dot{m} = -b$ and that $\mathbf{v}_{e/m}$ is downward; therefore the force $\mathbf{T}_2$ is opposite $\mathbf{v}_{e/m}$; that is $\mathbf{T}_2$ acts upward on the rocket.

The external forces in the vertical direction are the weight $W = mg = (m_0 - bt)g$ and the aerodynamic drag force F_d, both downward. Equation (11-16b) becomes

$$(m_0 - bt)\dot{v}_m = -(m_0 - bt)g - F_d + T \qquad (11\text{-}18a)$$

$$= -(m_0 - bt)g - F_d + bv_{e/m}. \qquad (11\text{-}18b)$$

When any rocket is fired vertically, subject to the assumptions listed, Eq. (11-18a) can be applied if the thrust is known, or Eq. (11-18b) can be used if the relative speed of the exhaust gases is known. If the rocket travels horizontally along a track, the weight and track reaction will be normal to the velocity and thus will be omitted from the equation. The drag force will be zero in the absence of frictional and aerodynamic forces or may have a magnitude kv^n in the presence of an atmosphere where n varies with the speed of the rocket but is generally taken to be 1 or 2. During the analysis, it is usually helpful to use symbols for the constants as well as the variables. Numerical values can be substituted when the formula has been developed to yield the desired results.

EXAMPLE 11-11

The single-stage rocket in Fig. 11-21 has an initial weight of 8050 lb and is fired vertically from rest from the surface of the earth. The rocket contains 6440 lb of fuel, which is consumed in 80 sec. The speed of the exhaust gases relative to the rocket is 5000 fps. At the time of burnout, determine

(a) The acceleration of the rocket in g units.
(b) The velocity of the rocket in miles per hour.
(c) The altitude of the rocket in miles.

Neglect drag forces.

SOLUTION

(a) When air resistance is neglected, the differential equation of the motion, Eq. (11-18b), becomes

$$(m_0 - bt)\dot{v}_m = -(m_0 - bt)g + bv_{e/m}$$

from which

$$\frac{dv}{dt} = \frac{bv_{e/m}}{(m_0 - bt)} - g. \tag{a}$$

From the statement of the problem, $m_0 = 8050/32.2 = 250$ slugs and $b = (6440/32.2)/80 = 2.5$ slugs per sec. Thus, at burnout, when $t = 80$ sec, Eq. (a) gives

$$\frac{dv}{dt} = a = \frac{(2.5)(5000)}{250 - (2.5)(80)} - g = 250 - g$$

or

$$\mathbf{a} = \left[\left(\frac{250}{32.2} \right) g - g \right] \mathbf{j} = \underline{6.76g \uparrow}. \qquad \text{Ans.}$$

(b) When $t = 0$, $v = 0$, and Eq. (a) can be integrated to obtain the velocity. Thus

$$\int_0^v dv = v_{e/m} \int_0^t \frac{b\,dt}{m_0 - bt} - g \int_0^t dt$$

and

$$\left[v \right]_0^v = v_{e/m} \left[-\ln(m_0 - bt) \right]_0^t - \left[gt \right]_0^t$$

from which

$$v = -v_{e/m} \ln\left(\frac{m_0 - bt}{m_0} \right) - gt. \tag{b}$$

When $t = 80$ sec, the speed of the rocket is

$$v = -(5000) \ln\left[\frac{250 - (2.5)(80)}{250} \right] - 32.2(80) = 5471 \text{ fps}$$

and

$$\mathbf{v} = 5471 \left(\frac{3600}{5280} \right) \mathbf{j} = \underline{3730 \text{ mph} \uparrow}. \qquad \text{Ans.}$$

(c) When v in Eq. (b) is replaced by dy/dt, it can be integrated between limits $y = 0$ when $t = 0$ and $y = y$ when $t = t$. That is,

$$\int_0^y dy = -v_{e/m} \int_0^t \ln\left(\frac{m_0 - bt}{m_0} \right) dt - g \int_0^t t\,dt.$$

The first integral to the right of the equal sign can be expressed in the form $\int \ln u\,du$, which is equal to $u(\ln u - 1)$, if dt is multiplied by $-b/m_0$ and the coefficient of the integral is multiplied by $-m_0/b$. The equation becomes

$$\int_0^y dy = \frac{m_0 v_{e/m}}{b} \int_0^t \left[\ln\left(\frac{m_0 - bt}{m_0} \right) \right]\left(\frac{-b}{m_0} dt \right) - g \int_0^t t\,dt.$$

This equation can be integrated to give

$$\left[y \right]_0^y = \frac{m_0 v_{e/m}}{b} \left\{ \left(\frac{m_0 - bt}{m_0} \right)\left[\ln\left(\frac{m_0 - bt}{m_0} \right) - 1 \right] \right\}_0^t - g\left[\frac{t^2}{2} \right]_0^t$$

$$y = \frac{m_0 v_{e/m}}{b} \left\{ \left(\frac{m_0 - bt}{m_0} \right) \left[\ln\left(\frac{m_0 - bt}{m_0} \right) - 1 \right] + 1 \right\} - \frac{1}{2} g t^2.$$

The value of y, when $t = 80$ sec, is

$$y = \frac{(250)(5000)}{(2.5)} \left\{ \left(\frac{250 - (2.5)(80)}{250} \right) \left[\ln\left(\frac{250 - (2.5)(80)}{250} \right) - 1 \right] + 1 \right\}$$

$$- \frac{32.2}{2} (80)^2 = 136{,}000 \text{ ft}$$

from which

$$\mathbf{y} = (136{,}000/5280)\mathbf{j} = \underline{25.8 \text{ miles } \uparrow}. \qquad \text{Ans.}$$

A more complete discussion of the performance of rockets can be found in various advanced texts on rocket performance.[6]

Another example of variable-mass problems arises in the motion of chains or flexible cables. When the velocity of the acquired, or lost, mass is zero before it becomes part of the moving system, or after it has left the system, Eq. (11-17) may be applied directly.

If a chain is allowed to fall link by link through a hole in a support [see Fig. 11-22(a)] or is raised from a fixed plane [see Fig. 11-22(b)], the only external forces acting on the *moving* part of the chain are the weight of the moving part of the chain and any force on the end of the chain away from the support. The floor supports the stationary part of the chain but does not act on the moving links.

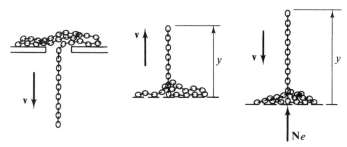

Figure 11-22

When chains are being lowered or dropped onto a fixed plane, as in Fig. 11-22(c), a normal upward force is exerted by the plane on the lowest link of the chain in order to bring it to rest as it strikes the plane. Only that portion of the normal force (the total force includes the weight of the chain on the plane) which changes the velocity of the moving link to zero is shown as N_e. If the chain in Fig. 11-22(c) weighs

---

[6] See, for example, Arthur I. Berman, *The Physical Principles of Astronautics* (New York: John Wiley & Sons, Inc. 1961).

q lb per ft and is moving downward with a speed $\dot{y}$, the mass of the chain brought to rest during a time interval Δt is $(q/g)(\Delta y)$. The velocity of each link changes from $-\dot{y}\mathbf{j}$ to zero during the time Δt. Therefore, the principle of linear impulse and momentum gives

$$(\mathbf{I}_L)_y = N_e\,\Delta t\mathbf{j} = \Delta\mathbf{G} = \frac{q}{g}(\Delta y)[0-(-\dot{y}\mathbf{j})]$$

or

$$N_e = \frac{q}{g}\frac{\Delta y}{\Delta t}\,\dot{y},$$

which becomes, as Δt approaches zero,

$$N_e = \frac{q}{g}\dot{y}^2.$$

The free-body diagram of the moving portion of the chain being lowered onto a plane must include not only the force (if any) applied to the upper end of the chain and the weight of the vertical portion of the chain but also a vertical force, with a magnitude of $(q/g)\dot{y}^2$, applied to the lower end.

EXAMPLE 11-12

A 10-ft cable weighing 4 lb per ft is loosely coiled on the floor. Attached to one end of the cable is a light cord which passes over a smooth pulley to a 16-lb weight as shown in Fig. 11-23(a). A 2-ft length of the cable is initially raised from the floor, and the system is released from rest. Determine the maximum height reached by the upper end of the cable.

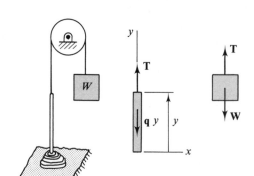

(a)　　　　　(b)　　　　(c)

Figure 11-23

SOLUTION

A free-body diagram of the moving portion of the cable is shown in Fig. 11-23(b). It is assumed that y_{max} is less than the length of the cable (10 ft) so that the weight of the moving portion is given by qy. The resultant external force on the body is $(T-qy)\mathbf{j}$. Since the velocity of the part of the cable on the floor is zero before it leaves the floor, Eq. (11-17) applies, and the equation of motion is

$$\mathbf{R} = \frac{d}{dt}(m\mathbf{v}_m),$$

which becomes, in scalar form,

$$T - qy = \frac{d}{dt}\left(\frac{qy}{g}\,\dot{y}\right). \tag{a}$$

The acceleration of the weight W is $-\ddot{y}\mathbf{j}$, and its equation of motion [see Fig. 11-23(c)] is

$$T - W = \frac{W}{g}(-\ddot{y}) = -\frac{d}{dt}\left(\frac{W}{g}\dot{y}\right). \tag{b}$$

The simultaneous solution of Eq. (a) and (b) gives

$$W - qy = \frac{d}{dt}\left(\frac{qy}{g}\dot{y}\right) + \frac{d}{dt}\left(\frac{W}{g}\dot{y}\right)$$

$$= \frac{1}{g}\frac{d}{dt}[(W + qy)\dot{y}].$$

This differential equation is easier to solve if the right-hand side is rewritten as

$$\frac{du}{dt} = \frac{du}{dy}\frac{dy}{dt} = \dot{y}\frac{du}{dy},$$

where $u = (W - qy)\dot{y}$. Thus,

$$W - qy = \frac{\dot{y}}{g}\frac{d}{dy}[(W + qy)\dot{y}]$$

or

$$(W - qy)\,dy = \left(\frac{\dot{y}}{g}\right)d[(W + qy)\dot{y}]. \tag{c}$$

When Eq. (c) is multiplied by $W + qy$, the right-hand side assumes the form $u\,du$, which is the differential of $u^2/2$. Therefore,

$$(W + qy)(W - qy)\,dy = \frac{1}{g}[(W + qy)\dot{y}]d[(W + qy)\dot{y}]$$

or

$$(W^2 - q^2y^2)\,dy = \frac{1}{2g}\,d[(W + qy)\dot{y}]^2.$$

The integral of this equation is

$$W^2y - \frac{q^2y^3}{3} + C = \frac{[(W + qy)\dot{y}]^2}{2g}. \tag{d}$$

From the given data, when $t = 0$, $y = y_0 = 2$ ft and $\dot{y} = 0$; therefore,

$$W^2y_0 - \frac{q^2y_0^3}{3} + C = 0$$

and

$$C = \frac{q^2y_0^3}{3} - W^2y_0.$$

Consequently Eq. (d) becomes

$$W^2(y - y_0) - \frac{q^2(y^3 - y_0^3)}{3} = \frac{[(W + qy)\dot{y}]^2}{2g}.$$

The maximum height of the cable occurs when $\dot{y} = 0$; therefore,

$$W^2(y_{max} - y_0) - \frac{q^2(y_{max}^3 - y_0^3)}{3} = 0$$

677

or

$$(y_{max} - y_0)\left[W^2 - \frac{q^2(y_{max}^2 + y_{max}y_0 + y_0^2)}{3} \right] = 0.$$

Since $y_{max} - y_0 \neq 0$, the second factor must be zero; that is,

$$y_{max}^2 + y_{max}y_0 + y_0^2 - \frac{3W^2}{q^2} = 0.$$

When the values $y_0 = 2$ ft, $W = 16$ lb, and $q = 4$ lb per ft are substituted, this equation becomes

$$y_{max}^2 + 2y_{max} - 44 = 0$$

from which

$$y_{max} = 5.708 \quad \text{or} \quad -7.708.$$

The negative root has no meaning for this problem; therefore the solution is

$$y_{max} = \underline{5.71 \text{ ft.}} \qquad \text{Ans.}$$

Since y_{max} is less than 10 ft, the assumption that some of the cable remains on the floor is justified. If y_{max} had been more than 10 ft, it would be necessary to determine the velocity when $y = 10$ ft after which the acceleration would be constant and the added distance the cable would rise could be determined directly.

PROBLEMS

11-93 Determine the weight of the fuel consumed in 2 min by a rocket engine producing a thrust of 1,400,000 lb if the exhaust gas has a velocity of 7000 fps relative to the rocket.

***11-94** A booster rocket is used to give additional starting power to an airplane. The fuel is burned at a rate of 4 kg/s and ejected with a relative velocity of 1500 m/s. Determine the additional starting thrust generated by the rocket.

11-95 A fueled rocket with a total mass of 50 slugs, including 40 slugs of fuel, is fired vertically. The fuel is consumed at the rate of 2 slugs per sec and is ejected with a relative velocity of 6000 fps. Neglect the effect of air friction. Determine
(a) The acceleration of the rocket, expressed in terms of g, immediately after ignition.
(b) The velocity of the rocket at time of burnout.

11-96 The two-stage rocket shown in Fig. P11-96 is fired vertically. The shell of rocket A weighs 100 lb and contains 600 lb fuel. The shell

Figure P11-96

of rocket B weighs 50 lb and contains 450 lb of fuel. The payload C weighs 50 lb. When rocket A expends its fuel, shell A is released, and rocket B is immediately fired. Rocket A consumes its fuel at the rate of 20 lb per sec, and rocket B burns its fuel at the rate of 15 lb per sec. The relative velocity of the exhaust for both rockets is 7000 fps. Neglect air friction. Determine

(a) The speed when shell A is released.

(b) The maximum speed attained by the payload C.

11-97 A small CO_2 rocket travels along a horizontal frictionless track. The initial and final weights of the rocket are 12 and 4 lb, respectively. The CO_2 is discharged at the uniform rate of 2 lb per sec with a velocity of 500 fps relative to the rocket. The rocket is subjected to a drag force $F_D = -kv$, where $k = 0.02$ lb-sec per ft and v is the velocity in feet per second. Determine the maximum velocity attained by the rocket. The rocket starts from rest.

11-98 Determine the distance traveled by the rocket. in Problem 11-97 during the time required for the velocity to change from zero to its maximum value.

11-99 A fueled rocket with a total mass of 40 slugs, including 30 slugs of fuel, approaches the moon along a flight path which is perpendicular to the moon's orbital path. The speed of the rocket is 8000 fps when a retrorocket is fired to bring the payload to rest as it touches the moon's surface. If the moon's gravitational attraction is one-sixth that of the earth's, what must be the burning time of the rocket, and at what altitude above the moon's surface must it be fired? The velocity of the exhaust gases with respect to the rocket is 7000 fps. Assume that the attraction of the moon for the rocket does not vary with the distance from the rocket to the moon.

11-100 One end of a pile of a long, uniform chain falls through a hole in its support and pulls the remaining links after it in a steady flow. Assume that the links which are initially at rest acquire the velocity of the falling chain without frictional resistance or interference from the adjacent links. Determine the velocity of the moving links after 10 ft of the chain has fallen through the hole.

*11-101 A body with a mass of 1 kg is dropped from rest and accumulates mass at a rate of 0.1 kg/m of travel. The mass accumulated was at rest before becoming attached to the body. Determine the velocity of the body after it has dropped 10 m. Neglect air resistance.

11-102 A uniform chain 16 ft long and weighing 0.1 lb per ft is coiled neatly on a table. A 21-ft cord is fastened to one end of the chain and passes over a small pulley 18 ft above the coil. The other end of the cord is connected to a 2-lb body A. Determine the velocity of body A just before it hits the table if it is released from rest 15 ft above the table.

11-103 Body A weighs 4 lb and is attached to an 18-ft uniform chain which weighs 0.1 lb per ft. The chain is coiled on the floor. Body A is projected upward from floor level with an initial velocity of 40 fps. Determine the maximum height to which body A will rise.

11-104 Solve Problem 11-103 if the initial velocity of body A is 60 fps upward.

*11-105 A portion of the uniform chain in Fig. P11-105 is coiled on a table, and the other portion passes over a small pulley and hangs

1 m

x

Figure P11-105

down on the other side a distance x. When $x = 2$ m, the chain is released from rest. Determine the relationship between the velocity of the chain at some subsequent time and the length x.

11-106 A uniform chain 20 ft long and weighing 0.1 lb per ft is partly coiled on the floor. A cord is fastened to the upper end of the chain and passes over a small pulley 12 ft above the coiled chain. The other end of the cord is connected to body A, which weighs 0.5 lb. The body is released from rest when 10 ft of the chain is suspended in air. Determine the velocity of body A as a function of the length of the chain suspended above the floor.

The principle of linear impulse and linear momentum equates the linear impulse of the forces acting on a body or system of particles to the change of linear momentum of the body or system of particles. The principle of angular impulse and angular momentum equates the angular impulse of the forces acting on a body or system of particles with respect to a point fixed in space to the change of angular momentum of the particle with respect to the corresponding point. When either the linear or angular impulse is zero, the corresponding linear or angular momentum is unchanged or conserved.

The principle of angular momentum equates the time rate of change of angular momentum of a particle, with respect to a point fixed in space, to the sum of the moments of the external forces acting on the particle with respect to the point. When the resultant moment is zero with respect to any fixed point the angular momentum is constant or conserved with respect to the point.

The concepts of linear and angular impulse and momentum are particularly useful in the analysis of variable-mass problems and of problems involving elastic impact and central-force motion. Since impulse and momentum are vector quantities, it is usually desirable to cut the cords, pull the pins, and perform other acts of liberation so as to work with free-body diagrams of separate bodies.

The coefficient of restitution is the ratio of the magnitude of the relative velocity of departure of two colliding bodies to the magnitude of their relative velocity of approach.

Three methods of solving problems in kinetics have been developed in Chapters 9, 10, and 11. The method of force, mass, and acceleration can be used to obtain instantaneous values of forces and acceleration. The principles of force, mass, and acceleration, together with the equations of kinematics, can also be used when a time interval, a distance traveled, and/or a change in velocity are involved.

The principle of work and kinetic energy equates work done by the forces acting on a body or system of bodies to the change in kinetic energy of the system. Thus it is the logical method when the problem involves forces, distances, and velocities. It is particularly useful when the forces and acceleration of the body vary with the position of the body, for example, a body which is acted on by a spring or a body swinging on the end of a cord.

The method of impulse and momentum is concerned with forces, time intervals, and changes in velocity. It can be applied to problems involving these quantities for either constant or variable forces, although it is particularly useful when an applied force is specified as a function of time. This method is the only one whose application to fluid flow and variable-mass problems was discussed, although the method of force, mass, and acceleration can be used for such problems with necessary modifications.

Neither the method of impulse and momentum nor that of work and kinetic energy gives instantaneous values of acceleration, and the method of work and energy does not give even an average value of forces which do no work. In any problem, it is important that a careful analysis be made to ascertain the given data and the unknown quantities to be obtained before a method of solution is selected. Two of the methods may sometimes be used to supplement each other. For example, in a ballistic pendulum, the principles of work and kinetic energy and of conservation of linear momentum can both be used to good advantage. When the problem can be solved completely by two different methods, they provide a reliable check on the solution.

PROBLEMS

11-107 The coefficient of friction between the 200-lb block in Fig. P11-107 and the horizontal surface is 0.4. The force 20t is in pounds when t is in seconds. When t = 0 the block has a velocity of 40 fps to the left. Determine
(a) The time interval that the block will *not* be moving.
(b) The velocity of the block when t = 6 sec.

Figure P11-107

11-108 The 12.88-lb body A in Fig. P11-108 is connected to the 6.44-lb body B by a rope passing over the two pulleys C and D of negligible mass. The force 2.0t is in pounds when t is

Figure P11-108

in seconds. When t = 0, B has a velocity of 10 fps upward. Determine the velocity of body B when t = 5 sec.

11-109 A bead of mass m, attached to an inextensible cord of negligible mass, is revolving on a smooth horizontal table as shown in Fig. P11-109. The cord passes through a hole in the table and is being pulled downward at a constant rate u by the force P. At the instant shown, the bead is at a distance R from the hole, and the magnitude of the component of the velocity of the bead perpendicular to the cord is v. Determine
(a) The kinetic energy of the bead in the position shown.
(b) The work done by the force P in decreasing the distance of the bead from the hole to R/2.

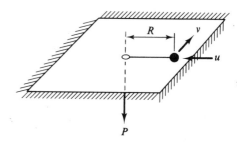

Figure P11-109

11-110 The 9.66-lb body A in Fig. P11-110 slides on a smooth horizontal surface and has a velocity of 20 fps to the right when it strikes the 6.44-lb body B, which is at rest. The coefficient of restitution is 0.5. After impact body B moves up the circular surface. Determine the reaction of the curved surface on body B just as it starts to move up the circular surface (position B in the diagram).

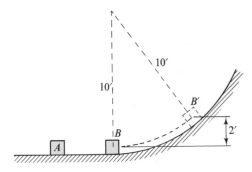

Figure P11-110

11-111 Using the data given in Problem 11-110, determine the reaction of the curved surface on body B when it has moved to position B'.

11-112 Particle A weighs 16.1 lb and moves on a smooth horizontal surface as shown in Fig. P11-112. Throughout the motion a force $F = kr^3$, where F is in pounds when r is in feet, acts on the particle and is always directed toward the fixed point O. When the particle is at its maximum distance from O the magnitude of its velocity is 20 fps. Determine the value of k.

Figure P11-112

*11-113 The two-stage rocket shown in Fig. P11-113 is fired vertically. The shell of each rocket, A and B, has a mass of 50 kg, and each rocket contains 500 kg of fuel. The payload C has a mass of 20 kg. When rocket A consumes its fuel, shell A is released and rocket B is immediately fired. Both rockets burn fuel at a rate of 25 kg/s, and the relative velocity of efflux for both rockets is 2100 m/s. Neglect air friction. Determine

(a) The thrust immediately after rocket A is fired.

(b) The velocity just after rocket A has consumed its fuel.

(c) The maximum speed attained by the payload C.

Figure P11-113

11-114 A snow plow with a rotary blade [Fig. P11-114(a)] is moving with a constant speed of 15 fps and throwing snow with a velocity of 30 fps relative to the plow at an angle of 30° as shown in Fig. P11-114(b). If 300 lb of snow per second is thrown by the plow, determine the magnitude of the forces F and R which must be exerted due to the snow removal.

11-115 A satellite is launched with a velocity of 20,000 mph parallel to the earth's surface at an altitude of 500 miles. Determine

(a) The eccentricity of the orbit.

(b) The equation of the orbit.

(c) The period of the orbit.

11-116 The 8.05-lb body A in Fig. P11-116(a) is moving to the right on a smooth horizontal plane with a velocity of 20 fps when it strikes the dynamic force gage B. The contact force between the gage and body A is shown in Fig. P11-116(b). Determine

Figure P11-114

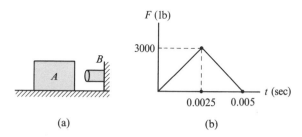

Figure P11-116

(a) The velocity of body A after impact.
(b) The coefficient of restitution between A and the gage.

*11-117 Body A in Fig. P11-117 has a mass of 20 kg, and the cart B has a mass of 80 kg. The coefficient of friction between A and B is 0.20, and the coefficient of restitution is zero. The bodies are at rest when the force P (magnitude = 300 N) is applied. Determine
(a) The velocity of B when it has moved 4.5 m to the left.
(b) The velocity of A when it has moved 4.5 m to the left.
(c) The velocity of A just after it collides with the right end of B.

Figure P11-117

(d) The work done by the friction force on each body while A is sliding on B.
(e) The energy dissipated by the plastic collision.

PART B—RIGID BODIES

11-9 ANGULAR MOMENTUM OF A RIGID BODY OR SYSTEM OF PARTICLES

The angular momentum of a particle with respect to a fixed point was defined in Art. 11-5. In general, *the angular momentum of a*

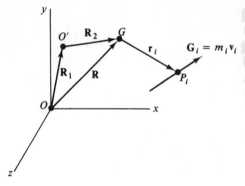

Figure 11-24

particle with respect to any point, fixed or moving, is the moment of the linear momentum of the particle with respect to that point. The angular momentum of a system of particles about any point is the sum of the angular momenta of the individual particles. Figure 11-24 represents a system of n particles (only the ith particle P_i is shown on the diagram) where G is the mass center of the particles. The XYZ axes are fixed in space, and O' is any point in space. The linear momentum of P_i is $m_i\mathbf{v}_i$ as shown, and the angular momentum of P_i with respect to point O' is the vector

$$(\mathbf{H}_{O'})_i = (\mathbf{R}_2 + \mathbf{r}_i) \times \mathbf{G}_i = (\mathbf{R}_2 + \mathbf{r}_i) \times m_i\mathbf{v}_i. \qquad (a)$$

The velocity of P_i in terms of its position vector is

$$\mathbf{v}_i = \frac{d}{dt}(\mathbf{R} + \mathbf{r}_i) = \dot{\mathbf{R}} + \dot{\mathbf{r}}_i.$$

The angular momentum of the system of particles with respect to O' is

$$\mathbf{H}_{O'} = \sum_{i=1}^{n} (\mathbf{R}_2 + \mathbf{r}_i) \times m_i(\dot{\mathbf{R}} + \dot{\mathbf{r}}_i). \qquad (11\text{-}19)$$

When O' is moved to G the vector $\mathbf{R}_2$ becomes zero, and the angular momentum of the system of particles with respect to G becomes

$$\mathbf{H}_G = \sum_{i=1}^{n} \mathbf{r}_i \times m_i(\dot{\mathbf{R}} + \dot{\mathbf{r}}_i)$$

$$= \left(\sum_{i=1}^{n} m_i\mathbf{r}_i\right) \times \dot{\mathbf{R}} + \sum_{i=1}^{n} m_i\mathbf{r}_i \times \dot{\mathbf{r}}_i.$$

The first term in the preceding equation is zero since the summation is the first moment of the mass with respect to its mass center. The vector $\dot{\mathbf{r}}_i$ is the velocity of the ith particle relative to G. The angular momentum of the system with respect to the mass center thus becomes

$$\mathbf{H}_G = \sum_{i=1}^{n} m_i\mathbf{r}_i \times \dot{\mathbf{r}}_i. \qquad (b)$$

When O' is selected as the fixed point O, $\mathbf{R}_2$ is equal to $\mathbf{R}$, and Eq. (11-19) becomes

$$\mathbf{H}_O = \sum_{i=1}^{n} (\mathbf{R} + \mathbf{r}_i) \times m_i(\dot{\mathbf{R}} + \dot{\mathbf{r}}_i)$$

$$= \mathbf{R} \times \dot{\mathbf{R}} \sum_{i=1}^{n} m_i + \mathbf{R} \times \frac{d}{dt}\left(\sum_{i=1}^{n} m_i\mathbf{r}_i\right) + \left(\sum_{i=1}^{n} \mathbf{r}_i m_i\right) \times \dot{\mathbf{R}}$$

$$+ \sum_{i=1}^{n} m_i\mathbf{r}_i \times \dot{\mathbf{r}}_i.$$

684

The summation $\sum_{i=1}^{n} m_i \mathbf{r}_i$ is zero in the second and third terms of the preceding equation, and when Eq. (b) is substituted for the last term the equation becomes

Section 11-9
Angular momentum
of a rigid body or
system of particles

$$\mathbf{H}_O = \mathbf{R} \times \dot{\mathbf{R}} m + \mathbf{H}_G = \mathbf{R} \times m\mathbf{v}_G + \mathbf{H}_G. \qquad (c)$$

Equation (c) demonstrates that the angular momentum of any system of particles with respect to a fixed point O is the sum of the moment of the linear momentum of the mass of the system considered concentrated at the mass center and moving with the velocity of the mass center added to the angular momentum of the particles of the system with respect to the mass center.

Equation (11-19) was developed for any system of particles. When the particles constitute a rigid body the velocities of the various particles can be expressed in terms of the angular velocity of the body and the linear velocity of the mass center. Figure 11-25 represents a rigid body in which P is a particle of mass dm and G is the mass center of the body. The XYZ axes are fixed in space, and the xyz coordinate system is fixed in the body with its origin at G.

From Eq. (8-27) the absolute velocity of particle P is

$$\mathbf{v}_P = \dot{\mathbf{R}} + \mathbf{v}_{P/xyz} + \boldsymbol{\omega} \times \mathbf{r}$$

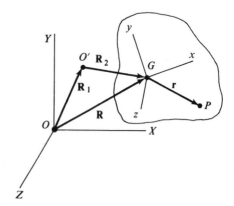

Figure 11-25

where $\mathbf{v}_{P/xyz}$ is the velocity of P relative to the moving axes and is zero in this case since P does not move relative to the xyz axes and $\boldsymbol{\omega}$ is the angular velocity of the body (the xyz axes). The angular momentum of the particle P with respect to point O' (either fixed or moving) is

$$d\mathbf{H}_{O'} = (\mathbf{R}_2 + \mathbf{r}) \times (\mathbf{v}_P \, dm)$$

$$= (\mathbf{R}_2 + \mathbf{r}) \times (\dot{\mathbf{R}} + \boldsymbol{\omega} \times \mathbf{r}) \, dm$$

$$= [\mathbf{R}_2 \times \dot{\mathbf{R}} + \mathbf{R}_2 \times (\boldsymbol{\omega} \times \mathbf{r}) + \mathbf{r} \times \dot{\mathbf{R}} + \mathbf{r} \times (\boldsymbol{\omega} \times \mathbf{r})] \, dm.$$

The angular momentum of the rigid body can be obtained by integrating the preceding expression. The quantities $\mathbf{R}_2$, $\dot{\mathbf{R}}$, and $\boldsymbol{\omega}$ are constant for all elements of mass and can thus be removed from the integrals. The angular momentum of the body is

$$\mathbf{H}_{O'} = \mathbf{R}_2 \times \dot{\mathbf{R}} \int_m dm + \mathbf{R}_2 \times \left(\boldsymbol{\omega} \times \int_m \mathbf{r} \, dm \right) + \left(\int_m \mathbf{r} \, dm \right) \times \dot{\mathbf{R}}$$

$$+ \int_m \mathbf{r} \times (\boldsymbol{\omega} \times \mathbf{r}) \, dm.$$

The quantity $\int_m \mathbf{r} \, dm$ is zero since $\mathbf{r}$ is measured from the mass center of the body and $\mathbf{H}_{O'}$ reduces to

$$\mathbf{H}_{O'} = \mathbf{R}_2 \times \dot{\mathbf{R}}m + \int_m \mathbf{r} \times (\boldsymbol{\omega} \times \mathbf{r}) \, dm$$

$$= \mathbf{R}_2 \times m\mathbf{v}_G + \int_m \mathbf{r} \times (\boldsymbol{\omega} \times \mathbf{r}) \, dm \qquad (11\text{-}20)$$

Equation (11-20) gives the angular momentum of a rigid body at any instant with respect to any point O'. If O' is moved to O, $\mathbf{R}_2$ becomes $\mathbf{R}$, and Eq. (11-20) becomes

$$\mathbf{H}_O = \mathbf{R} \times (m\mathbf{v}_G) + \int_m \mathbf{r} \times (\boldsymbol{\omega} \times \mathbf{r}) \, dm. \qquad (11\text{-}21)$$

If O' is moved to G, $\mathbf{R}_2$ becomes zero, and Eq. (11-20), for moments with respect to the mass center, is

$$\mathbf{H}_G = \int_m \mathbf{r} \times (\boldsymbol{\omega} \times \mathbf{r}) \, dm. \qquad (11\text{-}22)$$

Although Eq. (11-20)–(11-22) provide concise expressions for the moment of momentum of a rigid body, they are more useful when the integral is expressed in terms of moments and products of inertia. The quantity $\mathbf{r}$ can be written as

$$\mathbf{r} = x\mathbf{i} + y\mathbf{j} + z\mathbf{k}$$

and $\boldsymbol{\omega}$ as

$$\boldsymbol{\omega} = \omega_x \mathbf{i} + \omega_y \mathbf{j} + \omega_z \mathbf{k}$$

in which case $\boldsymbol{\omega} \times \mathbf{r}$ becomes

$$\boldsymbol{\omega} \times \mathbf{r} = (\omega_y z - \omega_z y)\mathbf{i} + (\omega_z x - \omega_x z)\mathbf{j} + (\omega_x y - \omega_y x)\mathbf{k}.$$

The vector triple product $\mathbf{r} \times (\boldsymbol{\omega} \times \mathbf{r})$ reduces to

$$\mathbf{r} \times (\boldsymbol{\omega} \times \mathbf{r}) = [\omega_x(y^2 + z^2) - \omega_y xy - \omega_z xz]\mathbf{i}$$
$$+ [-\omega_x xy + \omega_y(z^2 + x^2) - \omega_z yz]\mathbf{j}$$
$$+ [-\omega_x xz - \omega_y yz + \omega_z(x^2 + y^2)]\mathbf{k}.$$

Equation (11-22) for $\mathbf{H}_G$ can now be written as

$$\mathbf{H}_G = \left[\omega_x \int_m (y^2 + z^2) \, dm - \omega_y \int_m xy \, dm - \omega_z \int_m xz \, dm\right]\mathbf{i}$$

$$+ \left[-\omega_x \int_m xy \, dm + \omega_y \int_m (z^2 + x^2) \, dm - \omega_z \int_m yz \, dm\right]\mathbf{j}$$

$$+ \left[-\omega_x \int_m xz \, dm - \omega_y \int_m yz \, dm + \omega_z \int_m (x^2 + y^2) \, dm\right]\mathbf{k}$$

$$= (\omega_x I_x - \omega_y I_{xy} - \omega_z I_{xz})\mathbf{i} + (-\omega_x I_{xy} + \omega_y I_y - \omega_z I_{yz})\mathbf{j}$$

$$+ (-\omega_x I_{xz} - \omega_y I_{yz} + \omega_z I_z)\mathbf{k}. \qquad (11\text{-}22a)$$

Note that Eq. (11-21) can be written as

Section 11-9
Angular momentum
of a rigid body or
system of particles

$$\mathbf{H}_O = \mathbf{R} \times m\mathbf{v}_G + \mathbf{H}_G, \qquad (11\text{-}21\text{a})$$

where $\mathbf{H}_G$ is given by Eq. (11-22a).

Equation (11-22a) for $\mathbf{H}_G$, in terms of the moments and products of inertia of the body and the rectangular components of the angular velocity of the body for general motion, appears rather complicated. For most practical cases, however, many of the terms are zero. Some important special cases are given here for ready reference.

For translation $\boldsymbol{\omega} = 0$, and

$$\mathbf{H}_G = 0. \qquad (11\text{-}23\text{a})$$

When a body has plane motion parallel to the xy plane (note that this includes rotation about an axis parallel to the z axis) ω_x and ω_y are both zero, and Eq. (11-22a) becomes

$$\mathbf{H}_G = \omega_z(-I_{xz}\mathbf{i} - I_{yz}\mathbf{j} + I_z\mathbf{k}). \qquad (11\text{-}23\text{b})$$

This equation will be further simplified if the xy plane is a plane of symmetry or if the z axis is an axis of symmetry since the product of inertia terms will be zero. In this case, the angular momentum about G becomes

$$\mathbf{H}_G = I_z\omega_z\mathbf{k}. \qquad (11\text{-}23\text{c})$$

Note that Eq. (11-23c) applies to bodies which have rotation about an axis parallel to the z axis or plane motion parallel to the xy plane provided the z axis is an axis of symmetry or the xy plane is a plane of symmetry.

When a body rotates about a fixed (Z) axis, ω_x and ω_y are both zero. If point O is selected as the intersection of the Z axis and the plane of motion, the angular momentum of the body with respect to O becomes, from Eq. (11-21a) and (11-22a),

$$\mathbf{H}_O = \mathbf{R} \times m\mathbf{v}_G + \mathbf{H}_G$$
$$= \mathbf{R} \times m(\omega_z\mathbf{k} \times \mathbf{R}) + \omega_z(-I_{xz}\mathbf{i} - I_{yz}\mathbf{j} + I_z\mathbf{k})$$
$$= \omega_z[-I_{xz}\mathbf{i} - I_{yz}\mathbf{j} + (mR^2 + I_z)\mathbf{k}]$$
$$= \omega_z(-I_{xz}\mathbf{i} - I_{yz}\mathbf{j} + I_Z\mathbf{k}), \qquad (11\text{-}23\text{d})$$

where I_{xz} and I_{yz} refer to axes through G and I_Z is the moment of inertia with respect to the fixed axis of rotation. Note that $\mathbf{R}$ and $\mathbf{k}$ are mutually perpendicular so that $\mathbf{R} \times (\mathbf{k} \times \mathbf{R}) = R^2\mathbf{k}$.

If the body is symmetrical with respect to the plane of motion or the z axis, the product of inertia terms are zero, and the value of $\mathbf{H}_O$ reduces to

$$\mathbf{H}_O = (mR^2 + I_z)\omega_z\mathbf{k}$$

or

$$\mathbf{H}_{AR} = I_{AR}\omega\mathbf{k} \qquad (11\text{-}23\text{e})$$

where the subscript AR refers to the axis of rotation.

The angular momentum of a particle is defined [see Eq. (11-9)] as the moment of its linear momentum. The angular momentum of a system of particles, rigid or nonrigid, with respect to any point is the sum of the moments of the linear momenta of all the particles of the system with respect to that same point as shown by Eq. (11-19). The angular momentum of a rigid body, however, is equal to the moment of the linear momentum of the body only when certain conditions of symmetry are met, as shown in the next paragraph.

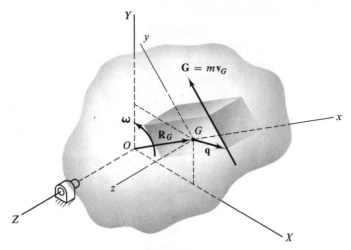

Figure 11-26

The rigid body of Fig. 11-26 is rotating about the Z axis through O parallel to the moving z axis. The xyz axes are fixed in the body and oriented as shown. The linear momentum of the body is

$$\mathbf{G} = m\mathbf{v}_G = m\omega R_G \mathbf{j},$$

and the position vector from the mass center to the intersection of $\mathbf{G}$ and the xz plane is

$$\mathbf{q} = x\mathbf{i} + z\mathbf{k}.$$

The moment of $\mathbf{G}$ with respect to the mass center is

$$\mathbf{H}'_G = \mathbf{q} \times \mathbf{G} = (x\mathbf{i} + z\mathbf{k}) \times (m\omega R_G \mathbf{j})$$

$$= xm\omega R_G \mathbf{k} - zm\omega R_G \mathbf{i}.$$

From Eq. (11-23b), the angular momentum about the mass center for this example is

$$\mathbf{H}_G = \omega(-I_{xz}\mathbf{i} - I_{yz}\mathbf{j} + I_z\mathbf{k}).$$

Since $\mathbf{H}'_G$ contains no coefficient of $\mathbf{j}$, these quantities can be equal only when I_{yz} is zero.

When the xy plane in Fig. 11-26 is a plane of symmetry, the expression for $\mathbf{H}_G$ reduces to $\omega I_z \mathbf{k}$. In this case, $\mathbf{H}'_G$ will be the same as $\mathbf{H}_G$ if G is in the xy plane (that is, if z is zero), and the value of x, the distance from the mass center to the action line of $\mathbf{G}$, is

Section 11-9
Angular momentum
of a rigid body or
system of particles

$$x = \frac{\omega I_z}{m\omega R_G} = \frac{mk_G^2}{mR_G} = \frac{k_G^2}{R_G},$$

where R_G is the distance from O to G. The distance from the axis of rotation to the linear momentum vector is

$$\frac{k_G^2}{R_G} + R_G,$$

which is identical to the value obtained in Art. 9-10 for the distance to the center of percussion. Consequently *the linear momentum vector acts through the center of percussion of a rotating rigid body that is symmetrical with respect to the plane of motion.*

An expression for the kinetic energy of a rigid body with plane motion was developed in Art. 10-6. By using Eq. (11-22) for the angular momentum of a rigid body with respect to its mass center, a concise expression for the kinetic energy of a rigid body with general motion can be developed.

As previously shown, the velocity of any particle P of the rigid body in Fig. 11-25 is

$$\mathbf{v}_P = \mathbf{v}_G + \boldsymbol{\omega} \times \mathbf{R},$$

and from Art. 10-3, the kinetic energy of the particle P is

$$dT = dm \frac{(\mathbf{v}_P)^2}{2} = \frac{dm}{2} \mathbf{v}_P \cdot \mathbf{v}_P$$

$$= \frac{dm}{2} (\mathbf{v}_G + \boldsymbol{\omega} \times \mathbf{r}) \cdot (\mathbf{v}_G + \boldsymbol{\omega} \times \mathbf{r})$$

$$= \frac{dm}{2} [v_G^2 + 2\mathbf{v}_G \cdot (\boldsymbol{\omega} \times \mathbf{r}) + (\boldsymbol{\omega} \times \mathbf{r}) \cdot (\boldsymbol{\omega} \times \mathbf{r})].$$

Since $\mathbf{v}_G$ and $\boldsymbol{\omega}$ do not vary over the body, the kinetic energy of the entire body can be written

$$T = \frac{v_G^2}{2} \int_m dm + \mathbf{v}_G \cdot \left(\boldsymbol{\omega} \times \int_m \mathbf{r}\, dm \right)$$

$$+ \frac{1}{2} \int_m [(\boldsymbol{\omega} \times \mathbf{r}) \cdot (\boldsymbol{\omega} \times \mathbf{r})]\, dm. \qquad (d)$$

The term $\int_m \mathbf{r}\, dm$ is zero, since $\mathbf{r}$ is measured from the mass center. The identity

$$(\mathbf{a} \times \mathbf{b}) \cdot \mathbf{c} = \mathbf{a} \cdot (\mathbf{b} \times \mathbf{c})$$

can be proved by writing the three vectors in terms of $\mathbf{i}$, $\mathbf{j}$, and $\mathbf{k}$

components and expanding (see any text on vector analysis). When **a** is replaced by **ω**, **b** by **r**, and **c** by **ω×r**, the identity shows that the last integral of Eq. (d) can be written as

$$\frac{1}{2}\boldsymbol{\omega} \cdot \int_m [\mathbf{r} \times (\boldsymbol{\omega} \times \mathbf{r})] \, dm = \frac{1}{2}\boldsymbol{\omega} \cdot \mathbf{H}_G$$

from Eq. (11-22). With these results, Eq. (d) reduces to

$$T = \tfrac{1}{2}mv_G^2 + \tfrac{1}{2}\boldsymbol{\omega} \cdot \mathbf{H}_G.$$

EXAMPLE 11-13

The homogeneous cylinder of mass m shown in Fig. 11-27 rolls without slipping on a horizontal plane with a counterclockwise angular velocity of ω rad per sec. Show that the angular momentum of the cylinder with respect to the instantaneous center, *IC*, is

$$\mathbf{H}_{IC} = I_{IC}\boldsymbol{\omega} = I_{IC}\omega\mathbf{k}.$$

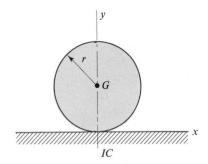

Figure 11-27

SOLUTION

Since the cylinder rolls without slipping,

$$\mathbf{v}_G = -r\omega\mathbf{i}.$$

From Eq. (11-23c),

$$\mathbf{H}_G = I_G\omega\mathbf{k},$$

and from Eq. (11-21a),

$$\mathbf{H}_{IC} = \mathbf{R} \times m\mathbf{v}_G + \mathbf{H}_G;$$

therefore

$$\mathbf{H}_{IC} = r\mathbf{j} \times m\mathbf{v}_G + \mathbf{H}_G$$

$$= r\mathbf{j} \times m(-r\omega\mathbf{i}) + I_G\omega\mathbf{k}$$

$$= mr^2\omega\mathbf{k} + I_G\,\omega\mathbf{k}$$

$$= (mr^2 + I_G)\omega\mathbf{k}$$

$$= I_{IC}\omega\mathbf{k}. \qquad\qquad \text{Ans.}$$

PROBLEMS

11-118 The homogeneous uniform 96.6-lb bar OA in Fig. P11-118 and the homogeneous 161-lb sphere are welded together to form a rigid body that is symmetrical with respect to the xy plane. The body has an angular velocity of 5 rad per sec counterclockwise about the z axis in the position shown.

(a) Determine the angular momentum with respect to the origin O.

(b) Locate the linear momentum vector of the body.

Figure P11-118

11-119 (a) If point A is at the top of the cylinder of Example 11-13, determine the angular momentum of the cylinder with respect to A. (b) Is the answer to part (a) equal to $I_A\omega$?

*11-120 The homogeneous bar OB in Fig. P11-120 has a mass of 10 kg and is attached by a pin to the homogeneous 30-kg cylinder C. The cylinder rolls without slipping on the circular track. The system is symmetrical with respect to the xy plane. Determine the angular momentum of the system with respect to point O when the bar has an angular velocity of 2 rad/s counterclockwise. The cylinder has a diameter of 1 m.

Figure P11-120

11-121 The 128.8-lb homogeneous bar AB in Fig. P11-121 moves on a horizontal xy plane. The bar has an angular velocity of 4 rad per sec clockwise, and the velocity of the mass center is shown on the figure. Determine
(a) The angular momentum of the bar with respect to point C.
(b) The kinetic energy of the bar.

Figure P11-121

*11-122 The radius of gyration of mass of the unbalanced 250-kg body in Fig. P11-122 is 300 mm with respect to an axis through the mass center G perpendicular to the plane of motion. The body rolls without slipping on the horizontal plane and in the position shown has an angular velocity of 2 rad/s clockwise.
(a) Determine the angular momentum of the body with respect to G.
(b) Locate the linear momentum vector with respect to G.

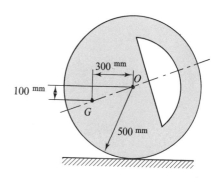

Figure P11-122

11-123 Two 1.61-lb weights are attached to two light rods as shown in Fig. P11-123. In the position shown, the angular velocity of rod OA is $8\mathbf{k}$ rad per sec, and the angular velocity of BC *relative* to OA is $40\mathbf{j}$ rad per sec.
(a) Determine the moment of momentum of the two weights with respect to point O.
(b) Calculate the linear momentum of the system and, if possible, locate the point at which it pierces the yz plane.

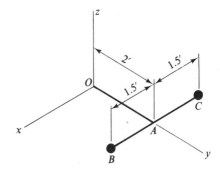

Figure P11-123

11-10 PRINCIPLE OF ANGULAR MOMENTUM

Equation (11-19) is an expression for the angular momentum of a system of particles with respect to any point. A useful relation can be developed for the time rate of change of the moment of momentum by differentiating Eq. (11-19) as follows:

$$\dot{\mathbf{H}}_{O'} = \sum_{i=1}^{n} [(\mathbf{R}_2 + \mathbf{r}_i) \times m_i(\ddot{\mathbf{R}} + \ddot{\mathbf{r}}_i)$$
$$+ (\dot{\mathbf{R}}_2 + \dot{\mathbf{r}}_i) \times m_i(\dot{\mathbf{R}} + \dot{\mathbf{r}}_i)]. \tag{a}$$

When O' is the fixed point O, $\mathbf{R}_2$ becomes $\mathbf{R}$, and

$$(\dot{\mathbf{R}} + \dot{\mathbf{r}}_i) \times (\dot{\mathbf{R}} + \dot{\mathbf{r}}_i) = 0.$$

Therefore Eq. (a) can be written

$$\dot{\mathbf{H}}_O = \sum_{i=1}^{n} [(\mathbf{R} + \mathbf{r}_i) \times m_i(\ddot{\mathbf{R}} + \ddot{\mathbf{r}}_i)]. \tag{b}$$

From Eq. (9-3),

$$\mathbf{F}_i = m_i \mathbf{a}_i = m_i(\ddot{\mathbf{R}} + \ddot{\mathbf{r}}_i),$$

and Eq. (b) reduces to

$$\dot{\mathbf{H}}_O = \sum_{i=1}^{n} [(\mathbf{R} + \mathbf{r}) \times \mathbf{F}_i] = \sum_{i=1}^{n} \mathbf{M}_{O_i} = \mathbf{M}_O, \tag{11-24}$$

where $\mathbf{M}_O$ is the sum of the moments with respect to O of the external forces acting on the system of particles. Equation (11-24) shows that *the time rate of change of the angular momentum of any system of particles, with respect to a point fixed in space, is equal to the moment of the external forces acting on the system about that point.* This relationship is known as the *principle of angular momentum for a fixed point.*

The principle of angular momentum can be derived for moments about the mass center by letting O' in Fig. 11-24 become G. In this case $\mathbf{R}_2$ is zero, and Eq. (a) becomes

$$\dot{\mathbf{H}}_G = \sum_{i=1}^{n} [\mathbf{r}_i \times m_i(\ddot{\mathbf{R}} + \ddot{\mathbf{r}}_i) + \dot{\mathbf{r}}_i \times m_i(\dot{\mathbf{R}} + \dot{\mathbf{r}}_i)]$$

$$= \sum_{i=1}^{n} [(\mathbf{r}_i \times \mathbf{F}_i) + (\dot{\mathbf{r}}_i m_i) \times \dot{\mathbf{R}} + 0] \tag{c}$$

since

$$\dot{\mathbf{r}}_i \times \dot{\mathbf{r}}_i = 0.$$

In Eq. (c),

$$\dot{\mathbf{r}}_i m_i = \frac{d}{dt}(\mathbf{r}_i m_i),$$

and the last term of Eq. (c) can be written

$$\frac{d}{dt}\left[\sum_{i=1}^{n}(\mathbf{r}_i m_i)\right] \times \dot{\mathbf{R}}$$

since the summation does not involve $\dot{\mathbf{R}}$ or d/dt. The vectors $\mathbf{r}_i$ are all measured from the mass center; therefore, the summation is zero, and Eq. (c) becomes

$$\dot{\mathbf{H}}_G = \sum_{i=1}^{n} (\mathbf{r}_i \times \mathbf{F}_i) = \sum_{i=1}^{n} \mathbf{M}_{G_i} = \mathbf{M}_G. \tag{11-25}$$

When the particles of the system constitute a rigid body Eq. (11-22a) and the various forms of Eq. (11-23) can be used to simplify the calculations.

Equations (11-24) and (11-25) will be utilized to develop closely related concepts of the next three articles that are useful for solving a variety of problems.

11-11 PRINCIPLE OF ANGULAR IMPULSE AND ANGULAR MOMENTUM

In Art. 11-10 the relation $\dot{\mathbf{H}} = \mathbf{M}$ was derived where the reference point was either a fixed point in space or the center of mass of a system of particles. The system of particles can be a rigid body or merely a group of particles. This relation is useful in the form developed in Art. 11-10, and it also leads to the principle of angular impulse and angular momentum. From Eq. (11-24),

$$\dot{\mathbf{H}}_O = \frac{d\mathbf{H}_O}{dt} = \mathbf{M}_O$$

or

$$d\mathbf{H}_O = \mathbf{M}_O \, dt$$

from which

$$\int_{\mathbf{H}_{Oi}}^{\mathbf{H}_{Of}} d\mathbf{H}_O = \int_{t_i}^{t_f} \mathbf{M}_O \, dt$$

and

$$\mathbf{H}_{Of} - \mathbf{H}_{Oi} = \int_{t_i}^{t_f} \mathbf{M}_O \, dt. \tag{11-26a}$$

The right-hand side of Eq. (11-26a) is defined as the angular impulse of the forces acting on a system of particles with respect to the fixed point O, and the left-hand side is the change in angular momentum of the system of particles with respect to the same point during the time interval from t_i to t_f.

By starting with Eq. (11-25), the same development that gave Eq. (11-26a) gives

$$\mathbf{H}_{Gf} - \mathbf{H}_{Gi} = \int_{t_i}^{t_f} \mathbf{M}_G \, dt, \tag{11-26b}$$

where G is the center of mass of the system of particles.

Equations (11-26a) and (11-26b) are valid for any system of particles. If the particles are so connected that they constitute a rigid body, Eq. (11-22a) is useful in determining the angular momentum of the body. When the body has plane motion parallel to the xy plane Eq. (11-23b) can be substituted into Eq. (11-26b) to give

$$\int_{t_i}^{t_f} \mathbf{M}_G \, dt = \mathbf{H}_{Gf} - \mathbf{H}_{Gi}$$

$$= (-I_{xz}\mathbf{i} - I_{yz}\mathbf{j} + I_z\mathbf{k})(\omega_{zf} - \omega_{zi}). \qquad (11\text{-}27a)$$

If the body is symmetrical with respect to the plane of motion, the product of inertia terms are zero, and the equation reduces to

$$\int_{t_i}^{t_f} \mathbf{M}_G \, dt = I_z\mathbf{k}(\omega_{zf} - \omega_{zi})$$

$$= I_G(\omega_f - \omega_i)\mathbf{k}, \qquad (11\text{-}27b)$$

where I_G is the moment of inertia of the body with respect to the axis through G perpendicular to the plane of motion.

When a body rotates about a fixed axis (say the Z axis) Eq. (11-26a) is valid if O is any point on the Z axis. If O is the intersection of the Z axis and the plane of motion, Eq. (11-23d) can be substituted into Eq. (11-26a) to give

$$\int_{t_i}^{t_f} \mathbf{M}_O \, dt = \mathbf{H}_{Of} - \mathbf{H}_{Oi}$$

$$= (-I_{xz}\mathbf{i} - I_{yz}\mathbf{j} + I_Z\mathbf{k})(\omega_{zf} - \omega_{zi}). \qquad (11\text{-}28a)$$

When the body is symmetrical with respect to the plane of motion or the z axis the angular momentum is given by Eq. (11-23e), and Eq. (11-26a) reduces to

$$\int_{t_i}^{t_f} \mathbf{M}_{AR} \, dt = I_{AR}(\omega_f - \omega_i)\mathbf{k}. \qquad (11\text{-}28b)$$

In addition to the angular impulse and momentum equations, the equations of linear impulse and momentum developed in Art. 11-2 provide additional information which may be needed to solve some problems. The general equation for linear impulse and momentum is

$$\int_{t_i}^{t_f} \mathbf{F} \, dt = \Delta\mathbf{G} = m(\mathbf{v}_{Gf} - \mathbf{v}_{Gi}),$$

which can be resolved into two or three component equations as needed.

When a symmetrical wheel rolls without slipping along a plane surface as in Fig. 11-27 a special angular impulse and momentum relationship can be developed using the instantaneous center of zero velocity. This point moves from point to point during a time interval and thus is neither a *fixed* point nor the mass center. The acceleration of the instantaneous center is perpendicular to the plane on which the wheel rolls and thus is toward the mass center of the symmetrical body.

Equation (9-20), from Chapter 9, is

$$\mathbf{M}_A = (\mathbf{r}_G \times \mathbf{a}_A)m + (\omega^2 I_{yz} - \alpha I_{xz})\mathbf{i}$$
$$- (\omega^2 I_{xz} + \alpha I_{yz})\mathbf{j} + \alpha I_z \mathbf{k},$$

and when point A is the instantaneous center, $\mathbf{r}_G$ is parallel to $\mathbf{a}_A$ and the first term on the right is zero. Since the wheel is assumed to be symmetrical with respect to the plane of motion, the product of inertia terms are zero. Therefore the equation reduces to

$$\mathbf{M}_{IC} = I_{IC}\boldsymbol{\alpha} = I_{IC}\frac{d\boldsymbol{\omega}}{dt}.$$

When this equation is integrated with respect to time it becomes

$$\int_{t_i}^{t_f}\mathbf{M}_{IC}\,dt = I_{IC}\int_{\omega_i}^{\omega_f}d\boldsymbol{\omega} = I_{IC}(\boldsymbol{\omega}_f - \boldsymbol{\omega}_i). \qquad (11\text{-}28c)$$

This equation indicates that the angular impulse of the external forces acting on the body, with respect to the (moving) instantaneous center, is equal to the change of the angular momentum of the body with respect to this axis. It should be emphasized, however, that this equation is valid *only* for a symmetrical wheel which rolls without slipping along a plane surface.

When the principle of angular impulse and angular momentum is used, it is usually better to draw separate free-body diagrams of each rigid body involved than to use combinations of bodies as is commonly done in using the principle of work and kinetic energy.

EXAMPLE 11-14

The sprinkler in Fig. 11-28(a) consists of a rotating horizontal tube AB with closed ends. The tube contains two orifices located on opposite sides of the tube, each at a distance r from the center of the tube. The volume rate of flow of water is Q ft^3 per sec, the velocity of efflux relative to the orifice is v fps perpendicular to the tube, and the tube has a constant angular velocity of ω rad per sec. Determine the frictional drag moment of the support on the tube.

SOLUTION

The principle of angular impulse and momentum will be applied to the tube and water for a differential time interval, dt. The mass of water flowing through the sprinkler in time dt is

$$dm = \rho Q\,dt,$$

where ρ is the density (mass per unit volume) of the water. The velocity leaving the orifice at

(a)

Figure 11-28(a)

695

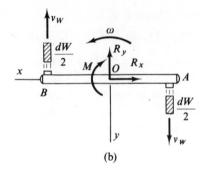

(b)

Figure 11-28(b)

A is

$$\mathbf{v}_W = \mathbf{v}_A + \mathbf{v}_{W/A} = -r\omega\mathbf{j} + v\mathbf{j} = (v - r\omega)\mathbf{j}.$$

The velocity of the water leaving the orifice at B has the same magnitude but is opposite in direction. The angular momentum of the tube and of the water in the tube is constant, and the angular momentum of the water entering the tube is zero (about the axis of rotation) so that the only *change* of angular momentum is that due to the momentum of the water leaving the tube. From the free-body diagram of the system in Fig. 11-28(b), the equation of angular impulse and momentum is

$$-M\mathbf{k}\, dt = d\mathbf{H} = -2\left(\frac{dm}{2}v_W\right)r\mathbf{k} = -Q\rho\, dt(v - r\omega)r\mathbf{k}$$

from which

$$M = Q\rho r(v - r\omega). \qquad\qquad \text{Ans.}$$

EXAMPLE 11-15

The body in Fig. 11-29(a) rotates about a horizontal axis at O and has a mass of 30 kg and a radius of gyration of mass with respect to

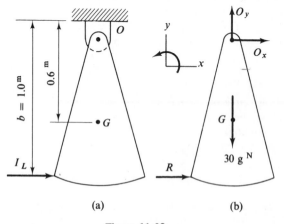

(a)   (b)

Figure 11-29

the axis of rotation of 0.80 m. The body has an angular velocity of 3.0 rad/s clockwise when an impulsive blow to the right is administered to the lower end of the body and changes the angular velocity of the body to 2.0 rad/s counterclockwise. Determine

(a) The linear impulse of the blow.
(b) The linear impulse of the horizontal component of the pin reaction at O during the time the blow was struck.

SOLUTION

(a) A free-body diagram of the body is shown in Fig. 11-29(b) where the force R is the magnitude of the impulsive blow. The body has rotation about a fixed axis and is assumed to be symmetrical with respect to the plane of motion; thus Eq. (11-28b) is valid. The angular impulse of the impulsive blow is the moment of the linear impulse; thus

$$\int Rb \, dt = I_{AR}(\omega_f - \omega_i) = mk_{AR}^2(\omega_f - \omega_i),$$

$$(1) \quad \int R \, dt = 30(0.8)^2[2 - (-3)],$$

$$\int \mathbf{R} \, dt = \underline{96.0 \text{ N} \cdot \text{s}} \rightarrow. \qquad \text{Ans.}$$

(b) The impulse of O_x can be obtained by applying either the principle of linear impulse and momentum or the principle of angular impulse and momentum about the mass center. When linear impulse and momentum is used the equation is

$$\left(\underset{+}{\curvearrowright}\right) \quad \Sigma \int F_x \, dt = m(v_{Gxf} - v_{Gxi}),$$

$$\int R \, dt + \int O_x \, dt = 30[0.6(2.0) - 0.6(-3.0)]$$

from which

$$\int O_x \, dt = 90.0 - 96.0 = -6.0,$$

$$\int \mathbf{O}_x \, dt = \underline{6.0 \text{ N} \cdot \text{s}} \leftarrow. \qquad \text{Ans.}$$

EXAMPLE 11-16

The wheel A of Fig. 11-30(a) weighs 1288 lb, rolls without slipping on the inclined plane, is symmetrical with respect to its geometrical axis, and has a radius of gyration of mass with respect to its geometrical axis of 2.0 ft. The homogeneous cylindrical drum B weighs 322 lb and is connected to A by an inextensible cable which wraps around both bodies. The couple $\mathbf{C}$ equals $(600 - 3t^2)\mathbf{k}$ ft-lb. When $t = 0$, $\boldsymbol{\omega}_B$ is $8\mathbf{k}$ rad per sec. Determine $\boldsymbol{\omega}_B$ when $t = 6$ sec.

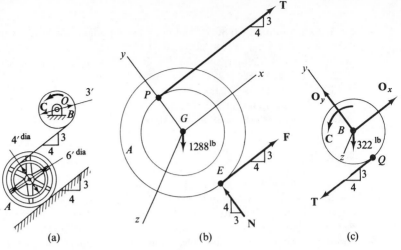

Figure 11-30

SOLUTION

Separate free-body diagrams of A and B are shown in Fig. 11-30(b) and 11-30(c). Body A has plane motion, and B has rotation about the z axis. The equations of linear and angular impulse and momentum for body A are

$$\mathbf{I}_{Ly} = \mathbf{G}_{yf} - \mathbf{G}_{yi} = m(\mathbf{v}_G)_{yf} - m(\mathbf{v}_G)_{yi} = 0,$$

and

$$\mathbf{I}_{Lx} = \mathbf{G}_{xf} - \mathbf{G}_{xi} = m(\mathbf{v}_G)_{xf} - m(\mathbf{v}_G)_{xi},$$ (a)

$$\int_{t_i}^{t_f} \mathbf{M}_G \, dt = \mathbf{H}_{Gf} - \mathbf{H}_{Gi} = \omega_{zf} I_z \mathbf{k} - \omega_{zi} I_z \mathbf{k}.$$

Note that $\mathbf{v}_{Gy}$ is zero and $\mathbf{H}_G$ is $I_z \omega_z \mathbf{k}$ due to symmetry—see Eq. (11-23c). The equations of linear and angular impulse and momentum for body B are

$$\mathbf{I}_{Lx} = m(\mathbf{v}_G)_{xf} - m(\mathbf{v}_G)_{xi} = 0,$$

and

$$\mathbf{I}_{Ly} = m(\mathbf{v}_G)_{yf} - m(\mathbf{v}_G)_{yi} = 0,$$ (b)

$$\int_{t_i}^{t_f} \mathbf{M}_O \, dt = \omega_{zf} I_z \mathbf{k} - \omega_{zi} I_z \mathbf{k}.$$

In Eq. (b), $\mathbf{v}_G$ is zero and $\mathbf{H}_O$ is equal to $\omega_z I_z \mathbf{k}$ from Eq. (11-23c).

There are five unknown forces ($\mathbf{T}$, $\mathbf{F}$, $\mathbf{N}$, $\mathbf{O}_x$, and $\mathbf{O}_y$) and five unknown velocities (ω_{Bf}, ω_{Ai}, ω_{Af}, $\mathbf{v}_{Gi}$, and $\mathbf{v}_{Gf}$). Since only six motion equations are available, four additional equations must be obtained from kinematics. The velocities of points P and Q are equal, since they are joined by an inextensible cable. Point E is the instantaneous center of body A. Consequently, the following kinematic equations can be written:

$$\mathbf{v}_P = \mathbf{v}_Q = -5\omega_A \mathbf{i} = 1.5\omega_B \mathbf{i},$$

where ω_A and ω_B are both assumed to be positive (counterclockwise). As a result,

$$\omega_A = -0.3\omega_B.$$

From the given data,

$$\omega_{Ai} = -0.3\omega_{Bi} = -0.3(8) = -2.4 \text{ rad per sec}$$

and

$$\omega_{Ai} = -2.4\mathbf{k} \text{ rad per sec (clockwise).} \tag{c}$$

Also

$$v_{Gi} = r\omega_{Ai} = 3(2.4) = 7.2 \text{ fps}$$

and

$$\mathbf{v}_{Gi} = 7.2\mathbf{i} \text{ fps.} \tag{d}$$

Similarly,

$$\omega_{Af} = -0.3\omega_{Bf}\mathbf{k} \text{ (clockwise if } \omega_{Bf} \text{ is positive),} \tag{e}$$

$$\mathbf{v}_{Gf} = 3(0.3)\omega_{Bf}\mathbf{i} = 0.9\omega_{Bf}\mathbf{i}. \tag{f}$$

The first of Eq. (a) and the first two of Eq. (b) involve $\mathbf{N}$, $\mathbf{O}_x$, and $\mathbf{O}_y$, which are not required. When Eq. (c), Eq. (d), Eq. (e), and Eq. (f) are substituted into the other three of Eq. (a) and Eq. (b) they become

$$\mathbf{I}_{Lx} = \mathbf{i}\int_0^6 F\,dt + \mathbf{i}\int_0^6 T\,dt - 0.6(1288)6\mathbf{i}$$

$$= m[\mathbf{v}_{Gf} - \mathbf{v}_{Gi}] = 40[0.9\omega_{Bf} - 7.2]\mathbf{i},$$

$$\int_0^6 \mathbf{M}_G\,dt = \mathbf{k}\int_0^6 (3F)\,dt - \mathbf{k}\int_0^6 (2T)\,dt$$

$$= I_G(\boldsymbol{\omega}_{Af} - \boldsymbol{\omega}_{Bf}) = 40(2)^2[-0.3\omega_{Bf} - (-2.4)]\mathbf{k},$$

and

$$\int_0^6 \mathbf{M}_O\,dt = \mathbf{k}\int_0^6 (600 - 3t^2)\,dt - \tfrac{3}{2}\mathbf{k}\int_0^6 \overset{\bullet}{T}\,dt$$

$$= I_z[\boldsymbol{\omega}_{Bf} - \boldsymbol{\omega}_{Bi}] = \tfrac{1}{2}(10)(\tfrac{3}{2})^2[\omega_{Bf} - 8]\mathbf{k}.$$

These three equations reduce to

$$\int_0^6 F\,dt + \int_0^6 T\,dt - 4637 = +36\omega_{Bf} - 288,$$

$$3\int_0^6 F\,dt - 2\int_0^6 T\,dt = -48\omega_{Bf} + 384,$$

and

$$3384 - 1.5\int_0^6 T\,dt = (11.25)\omega_{Bf} - 90.$$

The simultaneous solution of these three equations gives

$$\omega_{Bf} = -5.59$$

and

$$\boldsymbol{\omega}_{Bf} = \underline{-5.59\mathbf{k} \text{ rad per sec.}} \qquad \text{Ans.}$$

Since the mass center of A is at its geometric center, Eq. (11-28c) can be used where point O is at the instantaneous center. The equation would give the same result as the simultaneous solution of the last two equations of Eq. (a).

PROBLEMS

11-124 Equation (b) of Art. 11-9 is

$$\mathbf{H}_G = \sum m_i \mathbf{r}_i \times \dot{\mathbf{r}}_i.$$

Show by differentiating this expression that $\dot{\mathbf{H}}_G = \mathbf{M}_G$, where $\mathbf{M}_G$ is the moment about the mass center of the external forces acting on the system of particles.

11-125 (a) Water enters the horizontal pipe in Fig. P11-125 through a vertical inlet at O and discharges through two short nozzles at a distance L from the fixed axis at O. Determine the magnitude of the moment $\mathbf{M}$ which must be applied to the pipe to maintain a constant angular velocity $\boldsymbol{\omega}$ when the system is discharging Q ft^3 per sec of water. The density of water is ρ. The water leaves the nozzles with a velocity $\mathbf{v}_r$ relative to the nozzles.
(b) Determine the value of $\boldsymbol{\omega}$ if the moment M is zero.

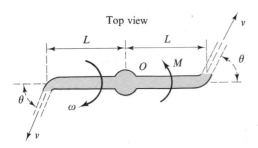

Figure P11-125

11-126 A flywheel is keyed to a shaft 4 in. in diameter, the combined weight being 644 lb. A constant turning moment of 3000 in-lb clockwise is exerted on the shaft causing the angular velocity to change from 60 to 420 rpm, both clockwise, in a time interval of 6 sec. The bearing (journal) friction on the shaft is 6 lb. Determine the radius of gyration of mass of the combined assembly with respect to the axis of rotation.

*11-127 The rotor of a turbine has a mass of 300 kg. When no other forces are acting, bearing friction will reduce the angular velocity from 150 rpm to zero in 2 minutes. The radius of gyration of mass with respect to the axis of rotation is 330 mm. Determine the frictional moment.

11-128 The armature of an electric motor weighs 640 lb, and its radius of gyration of mass with respect to the axis of rotation is 9 in. The radius of the shaft is 4 in. A period of 25 minutes is required for the bearing friction to bring the armature to rest from an angular velocity of 1800 rpm. Determine the coefficient of kinetic friction between the shaft and the bearings.

11-129 The 322-lb homogeneous disk in Fig. P11-129 has an angular velocity of 5 rad per sec counterclockwise when $t = 0$. The force P_v in pounds varies with time as indicated by the diagram. Determine the angular velocity of the disk when $t = 5$ sec.

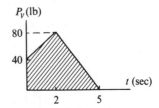

Figure P11-129

11-130 The force P_v in Problem 11-129 varies according to the equation $P_v = 3t^2$, where P_v is in pounds when t is in seconds. Determine the value or values of t when the angular velocity is again 5 rad per sec counterclockwise.

11-131 The 322-lb disk in Fig. P11-131 has a radius of gyration of mass of 2 ft with respect to a vertical axis through the mass center G and slides on a smooth *horizontal* plane. The force F_v in pounds, applied to a rope which wraps

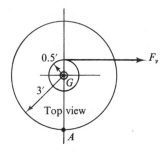

Figure P11-131

around the hub, varies according to the expression $F_v = 12t^2 + 20$, where t is in seconds. When $t = 0$, the velocity of G is 10 fps to the left, and the angular velocity of the disk is 5 rad per sec counterclockwise. The body is in the indicated position when $t = 5$ sec. Determine the velocity of point A at this instant.

*11-132 The force $\mathbf{F}_v$ in Fig. P11-132 is applied to a rope which wraps around a hub with a radius of 200 mm. The disk has a mass of 30 kg. The disk moves in a vertical plane and has a radius of gyration of mass of 0.5 m with respect to a horizontal axis through the mass center perpendicular to the plane of motion. The magnitude of the force $\mathbf{F}_v$ in newtons is given by the expression $F_v = 100 + 30t^2$, where t is in seconds. The system is released from rest when t is zero. When $t = 4$ s, determine

(a) The angular velocity of the disk.
(b) The velocity of point A on the rope.

Figure P11-132

11-133 The homogeneous rod AB in Fig. P11-133 weighs 96.6 lb and is at rest in the vertical position when an impulsive blow of 6 lb-sec is applied at the end B of the rod. Determine

(a) The angular velocity of the rod after the impulse has been applied.
(b) The change in the linear momentum of the rod.
(c) The time average force acting on the rod exerted by the horizontal component of the pin reaction at A if the impact time interval is 0.002 sec.

Figure P11-133

11-134 The 6-ft slender homogeneous 60-lb bar AB in Fig. P11-134 is suspended from a smooth pin at A and is at rest when the 10-lb ball C moving with a velocity of 20 fps to the left strikes the bar and rebounds to the right with 36% of its original kinetic energy. Determine

(a) The velocity of C after impact.
(b) The linear impulse of AB on C during the impact.
(c) The angular velocity of the bar after the impact.
(d) The linear impulse of the horizontal component of the pin reaction at A on the bar during the impact period.

Figure P11-134

*11-135 The wheel in Fig. P11-135 rolls without slipping and has a radius of gyration of mass of 75 mm with respect to a horizontal axis through the mass center G. The magnitude of the variable force P_v is given by the equation $P_v = 3t^2$, where P_v is in newtons when t is in seconds. The force P_v is applied to a rope which is wrapped around the hub. During the interval from $t = 2$ to $t = 5$ s the velocity of G changes from 200 mm/s to 1000 mm/s, both to the right. Determine the mass of the body.

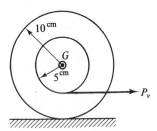

Figure P11-135

11-136 Body A in Fig. P11-136 weighs 16.1 lb and has a radius of gyration of mass with respect to its mass center of 0.20 ft. The variable force P_v is in pounds when t is in seconds. When $t = 0$ the velocity of the mass center is 2.0 fps downward. Determine the velocity of the mass center at $t = 5$ sec.

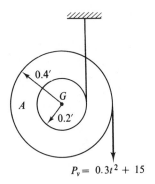

$$P_v = 0.3t^2 + 15$$

Figure P11-136

11-137 The solid homogeneous cylinder in Fig. P11-137 has a mass m and a radius r and rolls and slips on the horizontal plane. Initially the velocity of the center of the cylinder is v_0

down the plane, and the angular velocity of the cylinder is $4v_0/r$ counterclockwise. The coefficient of friction between the cylinder and plane is 0.1. Determine the time, in terms of v_0, that will elapse before the wheel will roll without slipping.

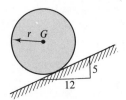

Figure P11-137

11-138 The homogeneous cylinder in Fig. P11-138 has a mass of 5 slugs and a radius of 1 ft. The variable couple $6t^2$ is in foot-pounds when t is in seconds. When $t = 0$, the angular velocity is zero. Determine
(a) The angular velocity when $t = 8$ sec.
(b) The value of t when the angular velocity is again zero.

Figure P11-138

11-139 The homogeneous 64.4-lb cylinder in Fig. P11-139 has an angular velocity of 20 rad

Figure P11-139

per sec clockwise when $t = 0$. Determine the time required to reduce the angular velocity to 10 rad per sec clockwise. The force $\mathbf{P}_v$ is in pounds when t is in seconds, and the coefficient of friction between the brake and cylinder is 0.25. Neglect the weight of the brake.

*11-140 The homogeneous cylinder A in Fig. P11-140 has a mass of 80 kg, and the mass of the hub can be neglected. The magnitude of the variable force P_v is in newtons when t is in seconds. The velocity of body B changes from 2 m/s downward when $t = 0$ to 4 m/s downward when $t = 5$ s. Determine the mass of B.

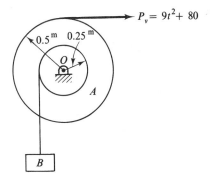

Figure P11-140

11-141 The symmetrical 64.4-lb drum A in Fig. P11-141 has a radius of gyration of mass of 6 in. with respect to its axis of rotation. The variable force $P = 12t^3 + 3t^2$ is in pounds when t is in seconds. The coefficient of friction between the 32.2-lb body B and the horizontal plane is 0.5. When $t = 0$, the angular velocity of A is 16 rad per sec counterclockwise. Determine the velocity of B when $t = 2$ sec.

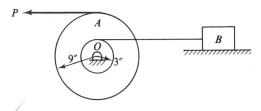

Figure P11-141

11-142 Wheel W in Fig. P11-142 weighs 644 lb, and body D weighs 322 lb. The radius of gyration of mass of W about its axis of rotation

is 2.0 ft. The coefficient of friction between W and the brake is 0.50. Neglect the weight of the brake. The moment of the couple C varies according to the equation $C = kt^2$, where C is in foot-pounds when t is in seconds. The velocity of D changes from 20 fps downward when $t = 0$ to 14 fps downward when $t = 5$ sec. Determine the constant k.

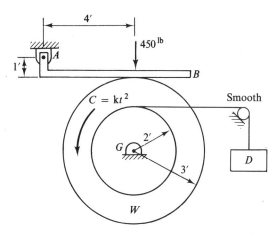

Figure P11-142

11-143 The 30-lb symmetrical body A in Fig. P11-143 has a radius of gyration of mass of 4 in. with respect to the horizontal axis through G. Body A rolls without slipping, and the moment of the variable couple $C = 0.3t^2$ is in inch-pounds when t is in seconds. When $t = 0$, the 4.0-lb body B has a velocity of 8 ips downward. Determine the angular velocity of A when $t = 10$ sec.

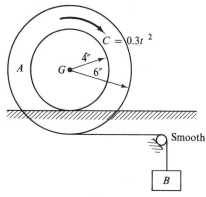

Figure P11-143

11-144 Body A in Fig. P11-144 weighs 30 lb, has a radius of gyration of mass with respect to a horizontal axis through its mass center of 3.0 in., and rolls without slipping. The magnitude of the moment of the variable couple is given by $C =$ $0.3t^2$, where C is in inch-pounds when t is in seconds. The angular velocity of body A changes from 20 rad per sec clockwise when $t = 0$ to 12 rad per sec counterclockwise when $t = 10$ sec. Determine the weight of body B.

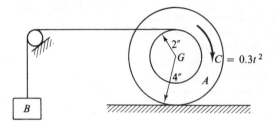

Figure P11-144

11-145 The symmetrical 644-lb body A in Fig. P11-145 rolls without slipping on the horizontal plane and has a radius of gyration of mass of 1.5 ft with respect to a horizontal axis through G. The homogeneous 128.8-lb disk B is acted upon by a variable couple $C_v = 3t^2$, where C_v is in foot-pounds when t is in seconds. The angular velocity of body A is 5 rad per sec clockwise when $t = 0$. Determine the angular velocity of body A 10 sec later.

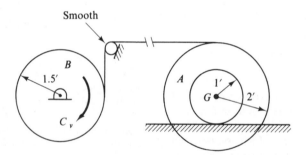

Figure P11-145

11-146 In Fig. P11-146, the homogeneous cylinder H weighs 322 lb and has a narrow slot cut into it for cord B, which has a negligible effect on its moment of inertia. Determine the time required for the center of the cylinder to reach a velocity of 9 fps to the right if it is at rest when $t = 0$. The variable force, of magnitude $P_v = 15t + 60$, is in pounds when t is in seconds. The coefficient of friction between H and the plane is 0.20.

Figure P11-146

11-147 Body A in Fig. P11-147 weighs 322 lb and has a radius of gyration of mass with respect to a horizontal axis through G of 2 ft. Body B weighs 96.6 lb and when $t = 0$ has a velocity of 10 fps downward. The moment of the variable couple is given by the expression $C_v = 60t^2$, where C_v is in foot-pounds when t is in seconds. Determine the velocity of B when $t = 10$ sec.

Figure P11-147

11-148 Body A in Fig. P11-148 weighs 60 lb and rolls without slipping along the horizontal plane. Body B is a homogeneous cylinder which weighs 10 lb. The moment of the variable torque C_v is given by the expression $C_v = 5.0 + 0.3t^2$, where C_v is in inch-pounds when t is in seconds. The angular velocity of A changes from 12 rad per sec counterclockwise when $t = 0$ to 30 rad per sec clockwise when $t = 5$ sec. Determine the radius of gyration of mass of A with respect to the horizontal axis through G.

Figure P11-148

11-149 The 322-lb body A in Fig. P11-149 has a radius of gyration of mass with respect to the horizontal axis through the mass center of 1.5 ft. The moment of the variable couple C_v is equal to $60t^2$ in foot-pounds when t is in seconds. The 64.4-lb body B has a velocity of 6 fps downward when $t = 0$. Determine the angular velocity of body A when $t = 5$ sec.

11-150 A flexible rope whose weight may be neglected has one end wrapped around the symmetrical pulley A and the other end around the

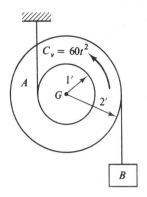

Figure P11-149

solid homogeneous cylinder B as shown in Fig. P11-150. The pulley has a radius of gyration of mass with respect to a horizontal axis through O of 9 in. and weighs 16.1 lb. The cylinder B weighs 24.15 lb. There is no initial slack in the rope, and the system starts from rest. Determine the time required for the mass center of B to attain a velocity of 50 fps downward.

Figure P11-150

*11-151 The wheel in Fig. P11-151(a) has a mass of 30 kg and a radius of gyration of mass with respect to a horizontal axis through G of 20 cm and rolls without slipping on the horizontal track. The force P_v varies in magnitude in the manner indicated in Fig. P11-151(b), and the magnitude of the clockwise torque T_v varies according to the equation $T_v = 3t^2$, where T_v is in newton·meters when t is in seconds. When $t = 0$, the velocity of G is 60 cm/s to the right. Determine the velocity of G when $t = 5$ s.

(a)

P_v (N)

(b)

Figure P11-151

*11-152 The homogeneous sphere in Fig. P11-152 has a mass of 7.0 kg. It is held above the horizontal plane and rotates with an angular velocity of 30 rad/s clockwise. The sphere is then lowered to the plane. The coefficient of friction between the sphere and plane is 0.1. Determine
(a) The period of time before the velocity of G becomes constant.

(b) The velocity of G after it becomes constant.
(c) The distance G has moved when the sphere begins to roll without slipping.

Figure P11-152

*11-153 The 150-kg homogeneous sphere A in Fig. P11-153 and the 40-kg body B are at rest when the 600-N force is applied to body B. Determine the velocity of the center of the sphere 0.5 s after the application of the force. Assume that the sphere does not slip on body B and that the friction between B and the horizontal plane is negligible.

Figure P11-153

11-12 CONSERVATION OF ANGULAR MOMENTUM

It was shown in Art. 11-10 that for any system of particles, rigid or nonrigid, the time rate of change of angular momentum, with respect to either a point fixed in space or the mass center of the system of particles, is equal to the resultant moment of the external forces acting on the system about that same point. When the moment of the external forces acting on a system of particles with respect to either a fixed point or the mass center is zero, the time rate of change of the angular momentum about the corresponding point is zero from Eq. (11-24) or (11-25), and the angular momentum is constant, or in other words, it is conserved. Therefore, *the angular momentum about a fixed point or the mass center is conserved or remains constant when the moment of the external forces about that point is zero.* In mathematical form, this statement is

$$\mathbf{H}_f - \mathbf{H}_i = 0. \tag{11-29}$$

If the resultant moment with respect to any fixed point or the mass center is relatively small and the time interval during which the moment acts is very small, the angular impulse can usually be neglected, and the angular momentum of the system of bodies can be assumed to be constant without introducing an appreciable error. The principle of conservation of angular momentum may be observed by means of slow-motion pictures of fancy dives. When a diver wishes to turn over in the air several times, he doubles up, holding his knees close to his chest. His moment of inertia about an axis through his mass center is small, and his angular velocity is relatively large. Just before the diver goes into the water, he straightens out, increasing his moment of inertia and decreasing his angular velocity so that he has nearly stopped turning when he enters the water.

An important practical use of the conservation of angular momentum occurs in stopping the spin of a space vehicle in flight. The so-called yo-yo despinner used in the Navy's Hugo III photographic-reconnaissance rocket is an example of a device designed to take advantage of the conservation of angular momentum.[7] With this device, the angular velocity of the spinning rocket can be reduced to zero or to any specific amount by transferring the angular momentum of the spinning mass to a pair of small connected masses as they are allowed to unwind from the main body.

When two bodies collide or react in any manner and the angular impulse on the system is zero with respect to some fixed point the angular momentum of the system with respect to the same point is conserved. It should be emphasized, however, that some of the kinetic energy of the system is usually lost during the impact or reaction.

In addition to conservation of angular momentum, it is often useful to write equations indicating that the linear momentum of a body or system of bodies is conserved in one or more directions. If the bodies are elastic and rebound after impact, the coefficient of restitution can be used to relate the relative velocity of departure of the points of contact to their relative velocity of approach (see Art. 11-4). It should be noted that the velocities of the points of contact are, in general, different from the velocities of the mass centers of the bodies. When eccentric impact is involved, the coefficient of restitution relates the *components* of the relative velocities along the line of impact.

EXAMPLE 11-17

The homogeneous 20-kg bar *AB* in Fig. 11-31(a) is falling with a velocity of 10 m/s downward and no angular velocity when the pin at *C* (extending from both sides of the bar) strikes the fixed bearings at *D*

---

[7] J. V. Fedor, "Theory and Design Curves for a Yo-Yo De-Spin Mechanism for Satellites," *Technical Note* D-708 (Washington, D.C.: National Aeronautics and Space Administration, 1961).

1.6 cm   0.4 cm

C

A $\quad$ B $\quad$ (a)

D

+

C_x

(b)

20 g $\quad$ C_y

$r_G = 0.6$ cm

Figure 11-31

with plastic impact (no rebound). Determine
(a) The angular velocity of the bar immediately after impact.
(b) The loss of kinetic energy of the bar during the impact period.
(c) The horizontal component of the reaction of the bearing on the bar immediately after the impact.

SOLUTION

(a) A free-body diagram of the bar during the impact period is shown in Fig. 11-31(b). Since the time of the impact is very small, the angular impulse of the weight with respect to the fixed bearing at C may be neglected during the impact period, and the angular momentum of the body about an axis through C is conserved; that is,

$$\mathbf{H}_{Ci} = \mathbf{H}_{Cf},$$

$$m v_{Gi} r_G = I_{AR} \omega_f,$$

$$20(10)(0.6) = [\tfrac{1}{12}(20)(2)^2 + 20(0.6)^2]\omega_f$$

from which

$$\omega_f = 8.64 \text{ rad/s}$$

and

$$\boldsymbol{\omega}_f = \underline{8.65 \text{ rad/s} \; }. \qquad \text{Ans.}$$

(b) The kinetic energy before impact is

$$T_i = \tfrac{1}{2} m v_{Gi}^2 = \tfrac{1}{2}(20)(10)^2 = 1000 \text{ N·m},$$

and the kinetic energy just after impact is

$$T_f = \tfrac{1}{2} I_{AR} \omega_f^2 = \tfrac{1}{2}[\tfrac{1}{12}(20)(2)^2 + 20(0.6)^2](8.65)^2$$

$$= 519 \text{ N·m}.$$

The loss of kinetic energy is

$$T_{loss} = T_i - T_f = 1000 - 519$$

$$= \underline{481 \text{ N·m}.} \qquad \text{Ans.}$$

(c) The horizontal component of the reaction can be determined by means of Newton's equation of motion. The horizontal component of the acceleration of the mass center is $r_G \omega^2$ to the right, and the equation of motion in the horizontal direction is

$$\sum F_x = m a_{Gx} = m r_G \omega^2,$$

$$\xrightarrow{+} C_x = 20(0.6)(8.65)^2 = 899 \text{ N},$$

and

$$\mathbf{C}_x = \underline{899 \text{ N} \rightarrow}. \qquad \text{Ans.}$$

EXAMPLE 11-18

Section 11-12
Conservation of
angular momentum

Bar AB in Fig. 11-32(a) weighs 25 lb and spins about its mass center on a smooth horizontal plane with an angular velocity of 5 rad per sec counterclockwise. Disk C weighs 5 lb and has a velocity of 50 fps as shown just before it strikes AB at point D. The coefficient of restitution between AB and C is 0.9. Determine the velocity of C just after impact and the corresponding angular velocity of AB.

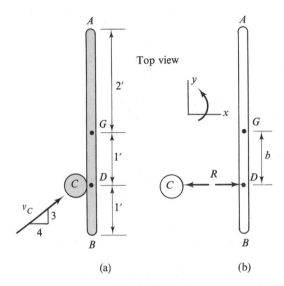

Figure 11-32

SOLUTION

Figure 11-32(b) is a free-body diagram of bar AB and disk C during the impact period, and it is seen that there is no impulse on either body in the y direction. Therefore, for C,

$$v_{Cyf} = v_{Cyi} = 0.6(50) = 30 \text{ fps}$$

and

$$\mathbf{v}_{Cyf} = 30 \text{ fps} \uparrow .$$

When the two bodies are considered as a system there are no external forces in the horizontal plane; therefore both the linear momentum and the angular momentum about any vertical axis must be conserved. All unknown linear and angular velocities are assumed to be in the positive directions as shown on the diagram. The conservation of linear momentum in the x direction gives

$$\sum mv_{xi} = \sum mv_{xf},$$

$$m_{AB}v_{Gxi} + m_C v_{Cxi} = m_{AB}v_{Gxf} + m_C v_{Cxf},$$

$$0 + \frac{5}{g}(40) = \frac{25}{g}v_{Gxf} + \frac{5}{g}v_{Cxf}$$

from which

$$40 = 5v_{Gxf} + v_{Cxf}. \qquad (a)$$

The angular momentum of the system about any vertical axis is conserved. When the vertical axis through G is selected, the angular momentum equation becomes

$$\sum \mathbf{H}_{Gi} = \sum \mathbf{H}_{Gf},$$

$$m_C v_{Cxi}(b) + \tfrac{1}{12}m_{AB}L^2\omega_i = m_C v_{Cxf}(b) + \tfrac{1}{12}m_{AB}L^2\omega_f,$$

$$\frac{5}{g}(40)(1) + \frac{1}{12}\frac{25}{g}(4)^2(5) = \frac{5}{g}v_{Cxf}(1) + \frac{1}{12}\frac{25}{g}(4)^2\omega_f$$

from which

$$220 = 3v_{Cxf} + 20\omega_f. \qquad (b)$$

The coefficient of restitution relates the relative velocities of approach and departure of the points of contact; that is,

$$e = -\frac{v_{Cxf} - v_{Dxf}}{v_{Cxi} - v_{Dxi}},$$

$$0.9 = -\frac{v_{Cxf} - v_{Dxf}}{40 - v_{Dxi}},$$

or

$$v_{Cxf} - v_{Dxf} = -0.9(40 - v_{Dxi}). \qquad (c)$$

Equations (a)–(c) contain five unknown quantities, v_{Gxf}, v_{Cxf}, ω_f, v_{Dxi}, and v_{Dxf}. The relative velocity equation for points D and G on body AB provides additional relationships for the initial and final velocities. The equation, using only the x components, is

$$v_{Dx} = v_{Gx} + v_{(D/G)x} = v_{Gx} + b\omega.$$

The initial velocity of D is

$$v_{Dxi} = v_{Gxi} + b\omega_i$$

$$= 0 + 1(5) = 5 \text{ fps}. \qquad (d)$$

The final velocity of D, in the x direction, is

$$v_{Dxf} = v_{Gxf} + 1\omega_f. \qquad (e)$$

Equation (d) can be substituted in Eq. (c) and the four equations (a), (b), (c), and (e) rearranged as follows:

$$5v_{Gxf} \quad + v_{Cxf} \qquad\qquad = 40 \qquad (a')$$

$$\omega_f + 0.15v_{Cxf} \qquad = 11 \qquad (b')$$

$$v_{Cxf} \quad - v_{Dxf} = -31.5 \qquad (c')$$

$$-v_{Gxf} - \omega_f \qquad\qquad + v_{Dxf} = 0 \qquad (e')$$

When Eq. (a') is multiplied by 0.2, the four equations can be added together to eliminate three of the four unknowns; thus

$$v_{Cxf}(0.2 + 0.15 + 1) = 8 + 11 - 31.5$$

from which

$$v_{Cxf} = -\frac{12.5}{1.35} = -9.26 \text{ fps}$$

and

$$\mathbf{v}_{Cxf} = 9.26 \text{ fps} \leftarrow.$$

The final velocity of C is

$$\mathbf{v}_{Cf} = v_{Cxf}\mathbf{i} + v_{Cyf}\mathbf{j}$$
$$= -9.26\mathbf{i} + 30.0\mathbf{j}$$
$$= 31.4 \text{ fps} \quad 72.8° \qquad \text{Ans.}$$

The value of v_{Cxf} can be substituted into Eq. (b') to give

$$\omega_f = 12.39 \text{ rad per sec,}$$

$$\boldsymbol{\omega}_f = \underline{12.39 \text{ rad per sec}} \quad \text{.} \qquad \text{Ans.}$$

PROBLEMS

*11-154 A slender homogeneous 2-m rod has a mass of 30 kg and rotates in a horizontal plane about a vertical axis which is 0.30 m from one end of the rod. The rod has an angular velocity of 4 rad/s clockwise when it is struck by a particle having a mass of 1 kg and a velocity of 120 m/s in the same direction as the velocity of the point of contact on the rod. The coefficient of restitution is zero. After impact the angular velocity of the rod is 8 rad/s. Determine the location of the point of impact relative to the axis of rotation.

11-155 The 3.0-ft bar in Fig. P11-155 has a mass of 0.10 slug and is suspended at rest in the vertical position. A ball with a mass of 0.01 slug is thrown horizontally with a velocity of 80 fps and strikes the lower end of the bar. After impact the ball rebounds with a velocity of 20 fps in the opposite direction. Determine
(a) The angular velocity of the bar after collision.
(b) The horizontal linear impulse imparted to the bar by the pivot during the impact period.
(c) The coefficient of restitution between the ball and bar.

11-156 The homogeneous bar in Fig. P11-156 weighs 64.4 lb and is free to rotate in a vertical plane about a smooth horizontal axis at O. The

Figure P11-155

Figure P11-156

bar is released from rest in the *OA* position. The bar has an angular velocity of 5 rad per sec counterclockwise when it strikes the 16.1-lb block *B*, which is at rest on the smooth horizontal plane. The coefficient of restitution between the bar and block is 0.5. Determine the velocity of the block immediately after impact.

*11-157 Bar *AB* in Fig. P11-157 has a mass of 30 kg and rotates about a vertical axis through the end *B*. The small body *C* has a mass of 5 kg and has a hole through it so that it can slide on the bar. A cord *D* keeps *C* from sliding when the bar rotates. When *C* is 1 m from *B* as shown the magnitude of the angular velocity of the bar is 50 rad/s. The cord is cut, and *C* slides out to the stop. Determine
(a) The final angular velocity of the bar.
(b) The loss of kinetic energy in the system.

Figure P11-157

11-158 The horizontal turntable in Fig. P11-158 weighs 1000 lb and is mounted on bearings whose friction may be neglected. The radius of gyration of the mass of the turntable with respect to its axis of rotation is 6.0 ft. A 200-lb man stands on the turntable at *A*. The turntable and the man are initially at rest. The man starts walking along the circular path on the platform with a velocity, relative to the platform, of 6.0 fps. Determine the magnitude of the absolute velocity of the man.

Figure P11-158

11-159 The slender homogeneous bar *AB* in Fig. P11-159 is rotating about its mass center

with an angular velocity ω_0 clockwise when it strikes the rigid stop at *D*. The coefficient of restitution for *AB* and *D* is 0.75. Determine the angular velocity of *AB* and the linear velocity of *G* just after impact.

Figure P11-159

11-160 Solve Problem 11-159 if the rod weighs 10 lb and is 5 ft long and if the angular velocity of *AB* is 20 rad per sec clockwise and the velocity of *G* is 30 fps to the right just before impact.

11-161 In Fig. P11-161, *A* is a 4.0-ft-diameter homogeneous disk which weighs 322 lb and is keyed to a vertical shaft rotating at 10 rad per sec as shown. The homogeneous cylinder *B* is 2 ft in diameter, weighs 322 lb, is not keyed to the shaft, and is not rotating. The mass of the shaft may be neglected. If the cylinder *B* is allowed to slide down the shaft until it comes in contact with *A*, determine
(a) The angular velocity of the two bodies after slipping has ceased.
(b) The frictional moments exerted on *B* by *A* if slipping occurs for 4 sec.
(c) The percentage loss of kinetic energy of the system.

Figure P11-161

11-162 The two homogeneous disks A and B in Fig. P11-162 are free to rotate on the small horizontal shaft. Disk A is 12 in. in diameter and weighs 40 lb; B is 18 in. in diameter and weighs 30 lb. The helical spring connecting the two disks has a torsional modulus of 20 ft-lb per rad; that is, a moment of 20 ft-lb must be applied to B to turn it 1 rad when A is held stationary. Disk A is held stationary, while B is turned two revolutions clockwise looking to the left. The two disks are then released simultaneously. Determine the angular velocity of each disk when the spring is unwound.

Figure P11-162

*11-163** A 10-kg homogeneous sphere S is fastened to the end of the 20-kg homogeneous rod OA as shown in Fig. P11-163. The 0.04-kg bullet B is fired with a horizontal velocity of 600 m/s. The bullet passes through the center of the sphere and emerges with a horizontal velocity of 100 m/s to the right. If the rod and sphere were stationary before impact, determine their angular velocity after the bullet passes completely through the sphere.

Figure P11-163

11-164 In Fig. P11-164 the smooth homogeneous 60-lb bar which is hanging vertically at rest is struck by the 6.0-lb body A with the velocity shown. If the coefficient of restitu-

tion is 0.8, determine the angular velocity of the bar immediately after impact.

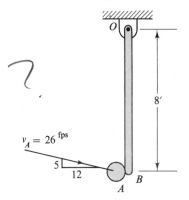

Figure P11-164

11-165 The homogeneous bar AB in Fig. P11-165 weighs 50 lb and rotates in a horizontal plane about a vertical axis at A. Body C weighs 10 lb and has a velocity of 20 fps as shown when it strikes the smooth bar. The bar has an angular velocity of 2 rad per sec counterclockwise just before impact. The coefficient of restitution is 0.4. Determine the velocity of body C immediately after impact.

Figure P11-165

*11-166** The homogeneous 2-m bar in Fig. P11-166 (p. 714) has a mass of 3 kg and rotates about a vertical axis through A. The ball D (considered as a particle) has a mass of 5 kg and can slide along the smooth bar. In the position shown the angular velocity of the bar is 4 rad/s, and the ball is held in place by a cord which passes through the smooth bearing. The ball is then pulled in with the cord and held 0.5 m from the axis of rotation. Determine
(a) The angular velocity in the new position.
(b) The tensions in the cord in the initial and final positions.
(c) The work done by the force P.

713

Figure P11-166

Figure P11-168

*11-167 When the 40-kg block *A* in Fig. P11-167 is in the position shown, 5 m of cord have been unwound from the drum. The drum is an 80-kg homogeneous cylinder and is initially at rest. Body *A* is dropped from rest from the indicated position. Determine the velocity of *A* just before the cord becomes taut and just after the impact when the speed of a point on the rim becomes equal to the speed of *A*.

Figure P11-167

11-168 Body *A* in Fig. P11-168 weighs 5.0 lb, *B* weighs 8.0 lb, and *C* is a solid homogeneous disk which weighs 20.0 lb. The two cords fastened to *A* and *B* are wrapped around *C* and are fastened to *C*. When *A* and *C* are released from rest there is 5 ft of slack in the cord between *B* and *C*. How far will *B* be lifted before coming to rest?

*11-169 The 5-kg homogeneous disk *D* in Fig. P11-169 rotates with a constant angular velocity of 10 rad/s counterclockwise. The 2-kg homogeneous rod *BC* rests on two smooth supports as shown. Rod *BC* is pushed to the left so that the small pin *A* on *D* strikes the end *B* of the rod. The coefficient of restitution for *A* and *B* is 0.80. Determine the angular velocity of each body just after the impact.

Figure P11-169

11-170 The homogeneous rectangular block *A* in Fig. P11-170 has a motion of translation with a velocity of 100 ips downward just before it strikes the *smooth* stop at *B*. The coefficient of restitution between *A* and *B* is 0.60. Determine

Figure P11-170

the angular velocity of A and the linear velocity of its mass center just after it strikes B.

11-171 Solve Problem 11-170 if block A has a small notch in it at B and there is no rebound after the impact.

11-172 The two identical slender bars in Fig. P11-172 spin on a smooth horizontal plane. Bar CD is pinned at D. Just before ends B and C collide the angular velocities are 30 rad per sec for AB and 5 rad per sec for CD, both coun-

terclockwise, and the linear velocity of G is zero. The coefficient of restitution for B and C is 0.80. Determine the angular velocity of each bar after impact.

*11-173 The three identical bars in Fig. P11-173 each have a mass m and are welded to form a rigid structure which is initially at rest. Body A also has a mass m and has an initial velocity of 10 m/s to the left. The coefficient of restitution is 0.80. Determine the velocity of A and the angular velocity of BCD just after impact.

Figure P11-172

Figure P11-173

11-13 SYMMETRICAL TOPS AND GYROSCOPES

In the entire study of dynamics, perhaps no motion is more interesting than that of a symmetrical top or gyroscope. Gyroscopes have long since passed the stage of being children's toys and now find application in such instruments as automatic pilots, stabilized gun platforms, and inertial guidance systems of aircraft, missiles, and submarines. Gyroscopic effects are noticeable in a variety of instances where a moving piece of machinery contains rotating elements, such as the propeller or jet engine rotor of an airplane, the wheels of motorcycles and bicycles, and spinning bullets.

The body of Fig. 11-33 is symmetrical with respect to the z axis and supported by a ball and socket joint, with negligible friction, at O. If the body is released from rest from the position

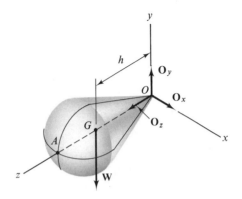

Figure 11-33

shown, it will rotate about the x axis, owing to the moment of the weight about O of

$$\mathbf{M}_O = Wh\mathbf{i}.$$

However, if the body is spinning about the z axis with a large angular velocity $\boldsymbol{\omega} = \omega\mathbf{k}$ when it is released, it will not rotate about the x axis but instead it will rotate or precess around the y axis. In this case, the x axis is called the *moment axis*, the y axis is called the *precession axis*, and the z axis is the *axis of spin*. *This tendency of a rapidly spinning body to turn about a second axis not parallel to the axis of spin, when acted on by a couple or torque about a third axis, is called gyroscopic action.*

Another important characteristic of a rapidly spinning body is its *rigidity in space* so effectively used for navigation. If the body in Fig. 11-33 is supported by frictionless supports called *gimbals* (see Fig. 11-34) which prevent any external moment from being applied to the body, the axis of spin will remain fixed in space. Both gyroscopic action and rigidity in space can be explained by means of Eq. (11-24),

$$\dot{\mathbf{H}}_O = \mathbf{M}_O,$$

where O is any point fixed in space.

Assume that a second ball and socket joint is placed at A in Fig. 11-33, and that the body is given a constant angular velocity $\boldsymbol{\omega}_z = \omega_z\mathbf{k}$. Since the body is not free to rotate about any axis other than the z axis, $\boldsymbol{\omega}_x$ and $\boldsymbol{\omega}_y$ are zero. Because of symmetry, both I_{xz} and I_{yz} are zero, and the mass center has zero velocity as it is on the axis of rotation. For this condition, Eq. (11-21a) becomes

$$\mathbf{H}_O = \omega_z I_z \mathbf{k}. \tag{a}$$

Gyroscopic action will take place when the support at A is gradually removed, allowing the external moment on the gyroscope with respect to O to become

$$\mathbf{M}_O = Wh\mathbf{i}.$$

Under these starting conditions, Eq. (11-24) becomes

$$\mathbf{M}_O = \dot{\mathbf{H}}_O$$

or

$$Wh\mathbf{i} = \frac{d}{dt}(\omega_z I_z \mathbf{k}) = \omega_z I_z \dot{\mathbf{k}} = \omega_z I_z (\boldsymbol{\omega} \times \mathbf{k}), \tag{b}$$

where $\boldsymbol{\omega}$ is the angular velocity of $\mathbf{k}$, that is, of the z axis. Equation (b) requires that $\boldsymbol{\omega} \times \mathbf{k}$ be in the direction of $\mathbf{i}$. Therefore,

$$\boldsymbol{\omega} = \omega_y\mathbf{j} \quad \text{and} \quad \boldsymbol{\omega} \times \mathbf{k} = \omega_y\mathbf{i}.$$

The value of ω_y can be obtained by substituting the preceding expression in Eq. (b) and is

$$\omega_y = \frac{Wh}{\omega_z I_z}. \tag{11-30}$$

Rotor

Rotor and inner gimbal (gyro)

Gyro and outer gimbal

A model gyroscope

Figure 11-34

Equation (11-30) shows that the initial motion is that motion defined as gyroscopic action. Once the motion starts, Eq. (a) is no longer correct for $\mathbf{H}_O$ because ω_y and v_G are no longer zero. The fact that the motion continues to be gyroscopic action will be demonstrated with a more general case.

When a gyroscope is supported by gimbals (see Fig. 11-34) the external moment about any point will be zero, and the principle of angular momentum gives

$$\dot{\mathbf{H}}_O = 0 \quad \text{or} \quad \mathbf{H}_O = \mathbf{C} = \text{a constant.}$$

The angular momentum will be constant only if the gyroscope continues to spin about the *fixed* z axis and thus it will remain rigid in space.

The two major features of the motion of gyroscopes have been explained for a simple case. A derivation for a more general motion can be developed as follows.

The body A in Fig. 11-35 is symmetrical about the z axis and mounted in a ball and socket joint with negligible friction so that it is free to rotate about any axis through the fixed point O. The XYZ coordinate system is fixed in space, and the xyz coordinate system

717

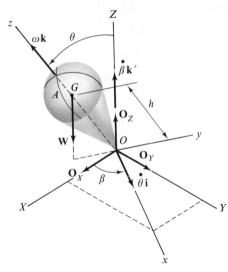

Figure 11-35

rotates about O in such a manner that the z axis is always the axis of symmetry of A and the x axis is always in the XY plane. The fact that the x axis is always in the XY plane in no way restricts or limits the orientation of the z axis. The x axis is retained in the XY plane for simplicity and convenience.

The relative angular velocity of body A with respect to the xyz coordinate system is $\boldsymbol{\omega} = \omega \mathbf{k}$, *and it is commonly called the rate of spin.* The angular velocity of the x axis with respect to the XYZ coordinate system is $\dot{\boldsymbol{\beta}} = \dot{\beta} \mathbf{k}'$ and is the *precessional angular velocity.* The angular velocity $\dot{\theta} \mathbf{i}$ (see Fig. 11-35) is called *nutation.* The general equations of motion of the gyroscope, in which $\omega \mathbf{k}$, $\dot{\beta} \mathbf{k}'$, and $\dot{\theta} \mathbf{i}$ are allowed to vary with time, have thus far defied exact solution even though attempts to solve them have been made for centuries. In almost all instrument applications of gyroscopes, nutation is undesirable, and much effort is expended to eliminate it.

In the following analysis, $\dot{\theta}$ is assumed to be zero and both ω and $\dot{\beta}$ are assumed to be constant. If $\dot{\theta}$ is zero, there will be no nutation, and θ will be constant. Consequently, the analysis is limited to steady-state precession.

The equation of motion for a gyroscope with the specified conditions is derived from the general relation $\dot{\mathbf{H}}_O = \mathbf{M}_O$. Body A is symmetrical with respect to the z axis; therefore,

$$I_{xz} = I_{yz} = 0 \quad \text{and} \quad I_{Gx} = I_{Gy} = I_G.$$

With these simplifications, Eq. (11-21a) for the angular momentum with respect to the fixed point O becomes

$$\mathbf{H}_O = \mathbf{R} \times (m\mathbf{v}_G) + \mathbf{H}_G$$
$$= (h\mathbf{k}) \times m[(\dot{\beta}\mathbf{k}') \times (h\mathbf{k})] + \omega_x I_G \mathbf{i} + \omega_y I_G \mathbf{j} + \omega_z I_z \mathbf{k}. \tag{c}$$

Note that the unit vectors $\mathbf{i}$, $\mathbf{j}$, and $\mathbf{k}$ are in the direction of the x, y, and z axes and that the $\mathbf{i}'$, $\mathbf{j}'$, and $\mathbf{k}'$ unit vectors are in the direction of the X, Y, and Z axes, respectively. The vector $\dot{\beta}\mathbf{k}'$ is perpendicular to the x axis because the x axis is in the XY plane; therefore, the Z axis is in the yz plane. The magnitudes of the components of $\dot{\beta}\mathbf{k}'$ in the y and z directions are $\dot{\beta} \sin \theta$ and $\dot{\beta} \cos \theta$, respectively, and the components of the absolute angular velocity of the body A in the x, y, and z directions are

$$\left.\begin{aligned}
\omega_x &= \dot{\theta} = 0, \\
\omega_y &= \dot{\beta} \sin \theta, \\
\omega_z &= \omega + \dot{\beta} \cos \theta.
\end{aligned}\right\} \tag{d}$$

and

718

When Eq. (c) is expanded, using these values, it becomes

$$\mathbf{H}_O = m\dot\beta h^2 \sin\theta\,\mathbf{j} + \dot\beta I_G \sin\theta\,\mathbf{j} + (\omega + \dot\beta \cos\theta)I_z\mathbf{k}. \qquad (e)$$

Since $\dot\theta$ is zero, the angular velocity of the rotating coordinate system is $\dot\beta\mathbf{k}'$; thus,

$$\dot{\mathbf{j}} = (\dot\beta\mathbf{k}')\times\mathbf{j}$$

and

$$\dot{\mathbf{k}} = (\dot\beta\mathbf{k}')\times\mathbf{k}.$$

All the quantities in Eq. (e) are constant with respect to time except $\mathbf{j}$ and $\mathbf{k}$, and the time derivative of Eq. (e) gives

$$\dot{\mathbf{H}}_O = (mh^2 + I_G)\dot\beta \sin\theta(\dot\beta\mathbf{k}'\times\mathbf{j}) + (\omega + \dot\beta\cos\theta)I_z(\dot\beta\mathbf{k}'\times\mathbf{k}).$$

Since $\mathbf{k}' = \sin\theta\mathbf{j} + \cos\theta\mathbf{k}$, the expression for $\dot{\mathbf{H}}_O$ becomes

$$\dot{\mathbf{H}}_O = (mh^2 + I_G)\dot\beta^2 \sin\theta[(\sin\theta\mathbf{j} + \cos\theta\mathbf{k})\times\mathbf{j}]$$

$$+ (\omega + \dot\beta\cos\theta)I_z\dot\beta[(\sin\theta\mathbf{j} + \cos\theta\mathbf{k})\times\mathbf{k}]$$

$$= [(I_z - mh^2 - I_G)\dot\beta^2 \sin\theta\cos\theta + \omega I_z\dot\beta\sin\theta]\mathbf{i}. \quad (11\text{-}31)$$

The moment of the external forces about the origin O is (see Fig. 11-35)

$$\mathbf{M}_O = Wh(\sin\theta)\mathbf{i},$$

and from the principle of angular momentum,

$$\mathbf{M}_O = \dot{\mathbf{H}}_O,$$

which becomes

$$Wh\sin\theta\mathbf{i} = [(I_z - mh^2 - I_G)\dot\beta^2 \cos\theta + \omega I_z\dot\beta](\sin\theta)\mathbf{i}$$

and

$$Wh = (I_z - I_y)\dot\beta^2 \cos\theta + \omega I_z\dot\beta. \qquad (11\text{-}32)$$

Note that $I_y = I_G + mh^2$.

Equation (11-32) can be used to determine the precessional angular velocity, $\dot\beta$, in terms of the magnitude of the spin angular velocity, ω, and the angle θ. Since the equation is a quadratic in $\dot\beta$ for values of θ other than 90°, there will, in general, be two distinct values of $\dot\beta$. By a more detailed analysis involving stability, it can be shown that, except in some unusual cases, the observed precessional angular velocity will be the smaller value obtained from Eq. (11-32).

For $\theta = 90°$, $\cos\theta$ is zero, and Eq. (11-32) reduces to Eq. (11-30); thus,

$$Wh = \omega I_z\dot\beta$$

or

$$\dot\beta = \frac{Wh}{\omega I_z}.$$

Equation (11-30) shows that as ω increases the precessional angular velocity decreases. In most instrument applications, it is desirable to have the precessional velocity as small as possible, which requires the use of a large rate of spin, ω. When ω is very large compared to $\dot{\beta}$, Eq. (11-30) approaches the exact solution of Eq. (11-32) for θ other than 90°.

Gyroscopic problems fall generally into two types: (1) The applied forces or moments are given, and either ω or $\dot{\beta}$ is required with the other known. (2) Both ω and $\dot{\beta}$ are given, and either the applied moment or the bearing reactions are required. In all cases, a free-body diagram is a necessary first step in the solution. When the moment is given the following additional steps are suggested:

1. Show the vector representation of $\mathbf{M}_O$ on the free-body diagram.
2. Calculate and show on the free-body diagram the y and z components of $\mathbf{H}_O$. One or both of these components will be in terms of an unknown angular velocity.
3. Determine the direction of rotation of $\dot{\beta}$ or ω by use of the fact that the gyroscope must precess in such a manner that the tip of the $\mathbf{H}_z$ or $\mathbf{H}$ vector moves in the direction of $\mathbf{M}$ when $\mathbf{H}_z$ or $\mathbf{H}$ is drawn with its tail at O.
4. Determine the magnitude of the unknown angular velocity from $\dot{\mathbf{H}}_O = \mathbf{M}_O$ or from Eq. (11-30) or (11-31).

When both angular velocities are given the following additional steps are suggested:

1. Determine $\mathbf{H}_O$ and show its z component on the free-body diagram. Use $\mathbf{H}_G$ when no point is fixed.
2. Determine the sense of $\mathbf{M}_O$ from the direction of precession and determine $\mathbf{M}_O$ from the equation $\dot{\mathbf{H}}_O = \mathbf{M}_O$.
3. If bearing reactions are required, use force-mass-acceleration relations as required.

The following examples illustrate these procedures.

EXAMPLE 11-19

A 6.00-in.-diameter thin homogeneous disk weighing 3.00 lb is mounted on a 9.00-in.-long rod of negligible weight as shown in Fig. 11-36(a). The disk is given a spin angular velocity of $-4800\mathbf{k}$ rpm, and the assembly is placed on the pedestal shown. Determine the steady-state precessional angular velocity of the rod.

SOLUTION

The free-body diagram of the disk and rod is shown in Fig. 11-36(b). The vector representation of $\mathbf{M}_O$ is shown on the free-body diagram in

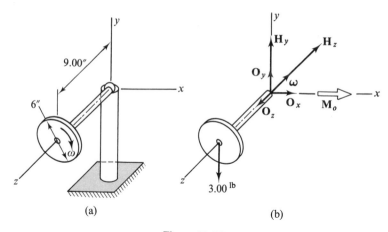

(a)   (b)

Figure 11-36

Fig. 11-36(b), and its value is

$$\mathbf{M}_O = 9(3.00)\mathbf{i} = 27.0\mathbf{i} \text{ in-lb.}$$

With no nutation, $\mathbf{H}_x = 0$, and $\mathbf{H}_y$ is constant in both magnitude and direction for steady-state precessional motion. The time derivative of $\mathbf{H}_y$ is zero, and thus $\mathbf{H}_y$ is not required for the solution. Therefore $\mathbf{H}_z$ is the only component of $\mathbf{H}_O$ required and is

$$\mathbf{H}_z = \omega_z I_z,$$

where

$$\omega_z = -4800\frac{2\pi}{60}\mathbf{k} = -503\mathbf{k} \text{ rad per sec}$$

and

$$I_z = \frac{1}{2}\frac{3.00}{386}(3)^2 = 0.0350 \text{ lb-sec}^2\text{-in.,}$$

thus giving

$$\mathbf{H}_z = -503(0.0350)\mathbf{k} = -17.58\mathbf{k} \text{ lb-sec-in.}$$

The vector representation of $\mathbf{H}_z$ is shown on the free-body diagram. From the free-body diagram, $\mathbf{M}_O$ will move the tip of $\mathbf{H}_z$ clockwise about the y axis looking down on the figure. Since $\dot{\mathbf{k}}$ is equal to $\boldsymbol{\omega}_y \times \mathbf{k}$, the rate of change of $\mathbf{H}$ is

$$\dot{\mathbf{H}}_z = \dot{\mathbf{H}}_O = -17.58\dot{\mathbf{k}} = -17.58(\boldsymbol{\omega}_y \times \mathbf{k})$$
$$= -17.58[(\omega_y\mathbf{j}) \times \mathbf{k}] = -17.58\omega_y\mathbf{i}.$$

From the principle of angular momentum,

$$\dot{\mathbf{H}}_O = \mathbf{M}_O,$$
$$-17.58\omega_y\mathbf{i} = 27.0\mathbf{i}$$

and

$$\omega_y = \dot{\beta} = \frac{-27.0}{17.58} = -1.536$$

or

$$\omega_y = -1.536\mathbf{j} \text{ rad per sec.} \qquad \text{Ans.}$$

EXAMPLE 11-20

The body of Example 11-19 is placed in the position shown in Fig. 11-37(a). All data are as given in Example 11-19. Determine the precessional angular velocity and compare it with the solution of Example 11-19.

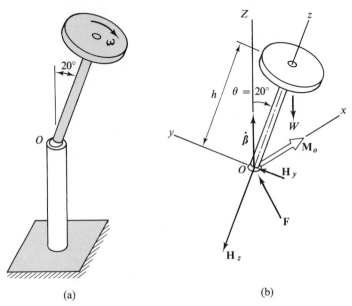

(a)   (b)

Figure 11-37

SOLUTION

Figure 11-37(b) shows a free-body diagram of the gyroscope. The force $\mathbf{F}$ is the reaction of the support at O. The vector representation of $\mathbf{M}_O$ is shown on the free-body diagram, and its value is

$$\mathbf{M}_O = Wh \sin\theta\mathbf{i}.$$

The angular velocities are [see Eq. (d)]

$$\omega_x = 0, \qquad \omega_y = \dot{\beta}\sin\theta\mathbf{j}, \qquad \omega_z = (\dot{\beta}\cos\theta - \omega)\mathbf{k},$$

where ω is the angular velocity of spin and $\dot{\beta}$ is assumed to be upward (in the positive Z direction). The angular momentum about O is

$$\mathbf{H}_O = \mathbf{H}_y + \mathbf{H}_z = I_y\dot{\beta}\sin\theta\mathbf{j} + I_z(\dot{\beta}\cos\theta - \omega)\mathbf{k}.$$

The only variables in this equation are $\mathbf{j}$ and $\mathbf{k}$, and their derivatives can be determined as follows:

$$\dot{\boldsymbol{\beta}} = \dot{\beta}\mathbf{k}' = \dot{\beta}(\sin\theta\mathbf{j} + \cos\theta\mathbf{k}),$$

$$\dot{\mathbf{j}} = \dot{\boldsymbol{\beta}} \times \mathbf{j} = -\dot{\beta}\cos\theta\mathbf{i},$$

$$\dot{\mathbf{k}} = \dot{\boldsymbol{\beta}} \times \mathbf{k} = \dot{\beta}\sin\theta\mathbf{i}.$$

The moment of the external forces about O is equal to the time rate of change of the angular momentum about O; therefore,

$$\mathbf{M}_o = \dot{\mathbf{H}}_o$$

or

$$Wh\sin\theta\mathbf{i} = I_y\dot{\beta}\sin\theta(-\dot{\beta}\cos\theta\mathbf{i}) + I_z(\dot{\beta}\cos\theta - \omega)(\dot{\beta}\sin\theta\mathbf{i})$$

from which

$$\dot{\beta}^2\cos\theta(I_z - I_y) - \dot{\beta}\omega I_z = Wh.$$

Numerical values for W and h are given in the problem statement. The value of ω is 503 rad per sec (see Example 11-19). The moments of inertia (see Chapter 6 or Appendix B) are

$$I_y = I_G + mh^2$$

$$= \frac{mR^2}{4} + mh^2$$

$$= \frac{3.00}{386}\left(\frac{3.00^2}{4} + 9.00^2\right) = 0.6470 \text{ lb-sec}^2\text{-in.}$$

and

$$I_z = 0.0350 \text{ lb-sec}^2\text{-in.} \qquad \text{(from Example 11-19).}$$

When the data are substituted into the quadratic equation for $\dot{\beta}$, it becomes

$$\dot{\beta}^2(\cos 20°)(0.0350 - 0.6470) - \dot{\beta}(503)(0.0350) = 3.00(9.00),$$

which reduces to

$$\dot{\beta}^2 + 30.57\dot{\beta} + 46.95 = 0.$$

The roots of this quadratic equation are

$$\dot{\beta} = -1.622 \quad \text{and} \quad -28.95.$$

The smaller value is the correct solution [see the comment following Eq. (11-32)], and the result is

$$\dot{\boldsymbol{\beta}} = \underline{-1.622\mathbf{k} \text{ rad per sec.}} \qquad \text{Ans.}$$

When this result is compared with the solution to Example 11-19, the error involved in assuming $\dot{\beta}$ to be independent of θ is

$$\text{error} = \frac{1.622 - 1.536}{1.622} = 5.30\%.$$

EXAMPLE 11-21

The rotating element of a jet engine weighs 193.2 lb and spins at 12,000$\mathbf{k}$ rpm. The radius of gyration of the mass with respect to the axis of rotation is 9.00 in., and the rotor is symmetrical about the axis AB in Fig. 11-38(a). Determine the bearing reactions on the rotor as the plane in which the engine is mounted makes a horizontal turn to the left with a radius of 4000 ft at a velocity of 480 mph.

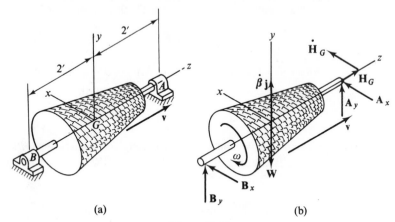

(a)   (b)

Figure 11-38

SOLUTION

Figure 11-38(b) is a free-body diagram of the rotor. The forward speed of the plane is

$$v = 480 \text{ mph} = 704 \text{ fps.}$$

Therefore, the precessional angular velocity is

$$\dot{\boldsymbol{\beta}} = \dot{\beta}\mathbf{j} = \frac{v}{R}\mathbf{j} = \frac{704}{4000}\mathbf{j} = 0.1760\mathbf{j} \text{ rad per sec,}$$

and the acceleration of the mass center is

$$\mathbf{a}_G = \frac{v^2}{R}\mathbf{i} = \left[\frac{(704)^2}{4000}\right]\mathbf{i} = 123.9\mathbf{i} \text{ fps}^2.$$

The rotor is symmetrical about the z axis; therefore, all products of inertia are zero. Also ω_x is zero from the statement of the problem. Therefore the component of the angular momentum along the z axis is

$$\mathbf{H}_z = \omega_z I_z \mathbf{k} = \frac{12,000(2\pi)}{60}\frac{193.2}{32.2}\left(\frac{9}{12}\right)^2\mathbf{k}$$

$$= 4241\mathbf{k} \text{ ft-lb-sec.}$$

Since $\mathbf{H}_y$ is constant in both magnitude and direction, its derivative will be zero. Consequently,

$$\dot{\mathbf{H}}_G = \dot{\mathbf{H}}_z = 4241\dot{\mathbf{k}} = 4241[(\dot{\beta}\mathbf{j})\times\mathbf{k}] = 4241(0.1760)\mathbf{i}$$

$$= 746\mathbf{i} \text{ ft-lb.}$$

From the free-body diagram in Fig. 11-38(b), the moment of the external forces about G is

$$\sum \mathbf{M}_G = (2\mathbf{k}) \times (A_x \mathbf{i} + A_y \mathbf{j}) + (-2\mathbf{k}) \times (B_x \mathbf{i} + B_y \mathbf{j}),$$

and this moment must be equal to $\dot{\mathbf{H}}_G$. Therefore,

$$746\mathbf{i} = (2A_x - 2B_x)\mathbf{j} + (2B_y - 2A_y)\mathbf{i}.$$

The component equations from this vector equation are

$$A_x = B_x$$

and

$$B_y - A_y = 373.$$

The force equation of motion $\sum \mathbf{F} = m\mathbf{a}_G$ gives

$$(A_x \mathbf{i} + A_y \mathbf{j}) + (B_x \mathbf{i} + B_y \mathbf{j}) - 193.2\mathbf{j} = \frac{193.2}{32.2} 123.9\mathbf{i},$$

or in component form,

$$A_x + B_x = 743.4,$$

and

$$A_y + B_y - 193.2 = 0.$$

The simultaneous solution of the four component equations gives

$$A_x = B_x = 371.7 \text{ lb}, \qquad A_y = -89.9 \text{ lb}, \qquad B_y = 283.1 \text{ lb}.$$

The bearing reactions are

$$\mathbf{A} = 371.7\mathbf{i} - 89.9\mathbf{j} = \underline{382(0.972\mathbf{i} - 0.235\mathbf{j}) \text{ lb}} \qquad \text{Ans.}$$

and

$$\mathbf{B} = 371.7\mathbf{i} + 283.1\mathbf{j} = \underline{467(0.796\mathbf{i} + 0.606\mathbf{j}) \text{ lb}.} \qquad \text{Ans.}$$

PROBLEMS

11-174 The homogeneous disk in Fig. P11-174 weighs 6.0 lb and rotates about bearings at the end of the shaft with an angular velocity of 3000 rpm in the direction shown. Body B weighs 10 lb, and the weight of the shaft is negligible. Determine the angular velocity of precession when the distance d is (a) 6 in., (b) 10 in.

***11-175** A homogeneous 20-cm-diameter disk has a mass of 1 kg and is placed on a pedestal so that the assembly precesses in a horizontal plane as shown in Fig. P11-175. If the precessional rate is $20\mathbf{i}$ rpm, determine the magnitude and direction of the spin of the disk about its own axle.

Figure P11-174

Figure P11-175

11-176 The armature of a motor is mounted on a rotating turntable as shown in Fig. P11-176. The 161-lb armature has a radius of gyration of mass of 0.5 ft with respect to its geometric axis. Assume that the armature is symmetrical with respect to the bearings, which are 2.0 ft between centers. Neglect bearing friction. The angular velocity of the armature is 1200 rpm in the direction indicated, and the angular velocity of the turntable is 4 rad per sec clockwise looking downward. Determine the components of the bearing reactions at A and B on the armature.

Figure P11-176

11-177 The propeller of a small airplane weighs 25.0 lb and has a radius of gyration of mass of 1.8 ft with respect to its axis of rotation. The plane has a speed of 120 mph when the propeller is rotating at 2000 rpm clockwise when viewed from the rear. The airplane is flying a level course when it makes a turn to the right on a 2000-ft radius. Will the gyroscopic effect of the propeller cause the plane to nose upward or downward? Determine the moment of the propeller on the air frame.

11-178 The rotor of the turbine in Fig. P11-178 may be approximated by a 16.1-lb homogeneous disk with a 6-in. radius. The turbine rotates clockwise at 12,000 rpm as viewed from the positive z axis. Determine the compo-

nents of the forces exerted by the bearings on axle AB if the angular velocity of the turbine housing is $\omega_x = 2$ rad per sec as shown.

11-179 The top in Fig. P11-179 weighs 0.80 lb. The principal radii of the gyration of mass for the origin are 0.60 in. for the axis of symmetry and 2.10 in. for any axis through the origin perpendicular to the axis of symmetry. The top spins about its geometrical axis with an angular velocity of 3600 rpm clockwise, looking down on the top. When the angle θ is 30°, determine the angular velocity of precession of the top about the Z axis.

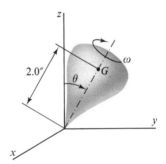

Figure P11-179

11-180 A homogeneous disk of radius r and weight w is traveling at a uniform speed, v, in a circular path of radius R on a horizontal plane. Due to gyroscopic effects, the disk leans at an angle θ, as shown in Fig. P11-180. Prove that the angle θ is given by

$$\tan \theta = \frac{3}{2g} \frac{v^2}{R}.$$

The wheel does not slip on the plane.

Figure P11-178

Figure P11-180

11-181 Two identical homogeneous cylinders with small flanges are mounted on a rigid axle as shown in Fig. P11-181. The wheels are set in bearings so each may rotate independently about the axle. The assembly is moving on a circular track on a horizontal plane as shown, and each wheel rolls without slipping. At a sufficiently large angular velocity, $\dot\beta$, one of the wheels will be on the verge of lifting off the track.

(a) Which wheel tends to lift off the track?

(b) Prove that the angular velocity for this condition is given by

$$\dot\beta^2 = \frac{2gL}{[3r(2R+L)]}$$

Figure P11-182

Figure P11-181

*11-182** The 20-cm-diameter homogeneous disk in Fig. P11-182 has a mass of 1.5 kg and is mounted on a 25-cm rod of negligible mass. The disk has a spin angular velocity of 5400**k** rpm. The rod makes an angle of 25° with the vertical axis through O. Determine the steady-state precessional angular velocity about the Z axis.

11-183 The rotor of a free gyroscope (one with two gimbals as in Fig. 11-34) is kept spinning at a high speed. The gyroscope is placed at the equator with its spin axis vertical (normal to the earth's surface) at noon. Determine the position of the spin axis relative to the normal to the earth's surface at 9:00 P.M.

11-184 The gyroscope in Problem 11-183 is placed at a point with a latitude of 30° north with its spin axis vertical at noon. Determine the angles the spin axis will make with the vertical and with the north-south meridian at 6:00 P.M.

11-185 The 3.00-lb disk in Fig. P11-185 is mounted on a massless shaft and collar which rotate with a constant angular velocity of 20**j** rad per sec. The disk rolls without slipping on the horizontal surface. The spring has a constant of 10 lb per in and is adjusted so that the shaft remains horizontal. Determine

(a) The normal force between the disk and the plane.

(b) The force in the spring.

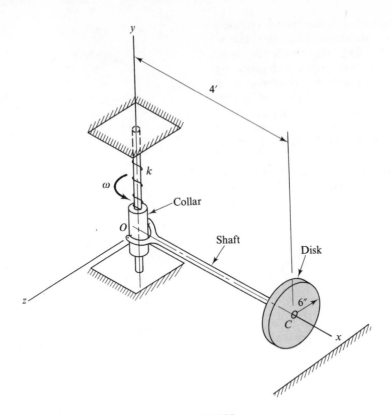

Figure P11-185

*11-186 The thin disk in Fig. P11-186 has a mass of 5.0 kg and is mounted on a rigid shaft with a negligible mass. The shaft is supported by a ball and socket at A and by a sleeve bearing at B. Points A and B are located on the turntable, which rotates about the vertical axis with an angular velocity ω_2 of 30 rpm as shown. The disk spins about the shaft with an angular velocity ω_1 of 600 rpm as indicated. Determine the forces on the shaft at A and B for the given position.

Figure P11-186

Mechanical Vibrations

12

12-1 INTRODUCTION

A *mechanical vibration* is an oscillatory, periodically repeated motion of a particle or body about a position of equilibrium. An engineer is frequently confronted with the problem of mechanical vibrations since they are encountered to some degree in almost all types of machinery and structures. Most vibrations are undesirable in machines and structures because they produce excessive stresses or repeated stresses; cause added wear; increase bearing loads; induce fatigue; create acute passenger discomfort in planes, ships, trains, buses, and automobiles; and absorb energy that could otherwise do useful work. The collapse of the Tacoma Narrows Bridge in 1940 is an example of structural failure due to excessive stresses produced by vibrations. The accuracy of precision instruments, tools, and machines may be impaired by excessive vibrations. Rotating machine parts require careful balancing in order to prevent damage from vibrations. When part of the propeller of an airplane is shot off or breaks off in flight the propeller is no longer symmetrical, and the vibrations from the engine may tear the engine from the plane unless it can be stopped in time. The vibrations produced in an automobile by the engine or by driving on rough roads set up repeated stresses in certain parts that can eventually lead to fatigue failure of the members.

Vibrations are sometimes used to produce desirable effects. For example, vibrators are used to compact concrete in the forms and to separate the grain from the chaff in threshing machines. Instruments which function properly on an airplane with a conventional engine may tend to stick when used in gliders or jet-powered planes because of the lack of vibration. In such instances, a vibrator is sometimes installed on the instrument panel.

(a) (b) (c)

Figure 12-1

When a particle or body that is supported by a system of springs, a shaft, a beam, or any other elastic system is disturbed from its position of equilibrium by the application and sudden removal of an additional force the particle or body will vibrate. Some common examples are (1) the vertical vibratory motion of body A in Fig. 12-1(a) when it is displaced vertically from its equilibrium position on a helical spring and then released; (2) the vertical vibratory motion of body B of Fig. 12-1(b) when it is displaced from its equilibrium position on a flexible springboard (cantilever beam) of negligible weight and released; and (3) the motion of the pendulum bob C in Fig. 12-1(c) as it swings in a vertical plane when supported by a cord of negligible weight.

Figure 12-2(a) shows a weight W suspended from a support by a spring and hanging at its equilibrium position. If the weight is pulled from its equilibrium position by a force **F** and released, it will, in the absence of any frictional forces, continue to oscillate about the equilibrium position indefinitely. Figure 12-2(b) shows a graph of the displacement, y, of the weight W from its equilibrium position as a function of time. One of the fundamental properties of mechanical vibrations is that the motion repeats itself at *definite intervals of time*. The *period, T,* of the oscillation is the *minimum amount of elapsed time before the motion starts to repeat itself.* The motion completed in one period is a *cycle.* The *frequency, f,* of the oscillation is the number of cycles which occur in a given unit of time; the customary unit is cycles per second, cps, or hertz, Hz. It should be noted that the frequency is the reciprocal of the period, or, in equation form,

$$f = \frac{1}{T}. \tag{12-1}$$

The *amplitude, A,* of the oscillation is the maximum displacement, either linear or angular, of the body from its equilibrium position.

Mechanical vibrations which are maintained by elastic, and

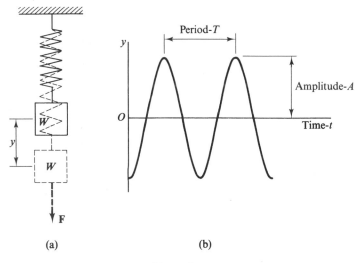

(a)   (b)

Figure 12-2

sometimes gravity, forces are called *free vibrations.* Once started, a free vibration, frequently called a *natural vibration,* continues to oscillate at its natural frequency. A *forced vibration* is produced and maintained by a periodic exciting force external to the system, and it takes place at the frequency of the exciting force. If the system is nearly frictionless and contains no dissipative elements, the amplitude of the forced oscillation will become very large when the frequency of the exciting force is close to the natural frequency of the system. Thus, although the frequency of a forced vibration is independent of the natural frequency of the elastic system, the resulting amplitude is affected by both frequencies. Vibrations are also classified as *damped* or *undamped.* When friction, air resistance, viscous damping (see Art. 12-5), and all other resisting forces are negligible the vibration is undamped. When any of these effects is appreciable the body has damped vibrations. In practice there is always a frictional damping force which will eventually stop a free vibration, even though it may be neglected for many purposes.

If the motion of a particle or body is constrained so that its position can be completely specified by one coordinate, it is said to have a *single degree of freedom.* If the system can vibrate in two directions or is composed of two particles that can each vibrate independently in one direction, it is said to have *two degrees of freedom* because two coordinates are required to specify the position of the system at any instant. For example, a system composed of a single particle, supported by a spring as indicated in Fig. 12-3(a), which vibrates in a vertical direction only, has one degree of freedom. A system composed of two particles which move vertically and are supported as indicated in Fig. 12-3(b) has two degrees of freedom

731

(a)　　　　　　　(b)　　　　　　　　　　(c)

Figure 12-3

because two coordinates are required to locate the particles at any instant. A system composed of one particle supported by four springs and constrained to move in a vertical plane as indicated in Fig. 12-3(c) has two degrees of freedom because two coordinates are required to locate the particle in the vertical plane at any instant. A single rigid body has, in general, six degrees of freedom, since it may have translation in three coordinate directions and it may also rotate about three coordinate axes. The following discussion is concerned only with systems which have a single degree of freedom or which vibrate in only one way at the instant being considered.

12-2 SIMPLE HARMONIC MOTION

A vibration has been defined as an oscillatory, periodically repeated motion. The simplest form of periodic motion is called *simple harmonic motion* and is defined as *any motion which has an acceleration which is proportional to the displacement from a fixed point on the path of the motion and which is directed toward the fixed point.* The motion of many vibratory systems having a single degree of freedom closely approximates simple harmonic motion, and a thorough knowledge of this concept is most helpful when analyzing such systems.

Expressions for velocity and acceleration of a point were obtained in Chapter 8 by differentiating a position function with respect to time. Consider line OB in Fig. 12-4(a), which has a length A and which rotates about O with a constant angular velocity ω. Point P is the projection of point B on the y axis, and the position of P at time t is

$$y = A \cos \omega t. \tag{a}$$

Note that each time the angle ωt increases by 2π rad, point P will have moved through one complete cycle, as shown in Fig. 12-4(b).

The velocity of P can be obtained by differentiating Eq. (a) with respect to time; that is,

$$v = \dot{y} = -A\omega \sin \omega t = A\omega \cos\left(\omega t + \frac{\pi}{2}\right). \tag{b}$$

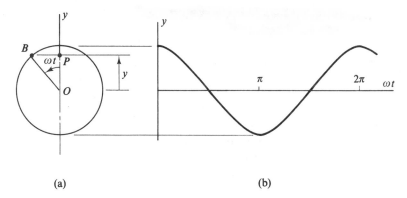

(a)   (b)

Figure 12-4

In a similar manner the acceleration of P is

$$a = \ddot{y} = -A\omega^2 \cos \omega t = A\omega^2 \cos(\omega t + \pi). \qquad (c)$$

Note that the only differences in the three equations are that $\dot{y}$ and $\ddot{y}$ are multiplied by ω and ω^2, respectively, and that their angles differ by $\pi/2$ and π from the variable ωt. The angles $\pi/2$ and π are known as *phase angles*. From Eq. (b) and (c) it is seen that the maximum values of the velocity and acceleration are $A\omega$ and $A\omega^2$, respectively. When Eq. (a) is substituted into Eq. (c) the acceleration becomes

$$\ddot{y} = -\omega^2 y \quad \text{or} \quad \ddot{y} + \omega^2 y = 0. \qquad (12\text{-}2)$$

This equation shows that the acceleration of P is proportional to its displacement, y, from the origin O, and the negative sign indicates that the acceleration is opposite the displacement; that is, the acceleration is directed toward O. By definition, therefore, point P executes simple harmonic motion. Any motion described by Eq. (12-2) is simple harmonic motion, and Eq. (12-2) is the *differential equation of motion* for simple harmonic motion.

The factor ω^2 in Eq. (12-2) is the square of the angular velocity of line OB. When line OB has rotated one revolution, 2π rad, point P has completed one cycle of motion. The time, T, required for this motion is the period of the vibratory motion of P, that is,

$$T = \frac{2\pi}{\omega},$$

and the frequency of the vibratory motion is

$$f = \frac{1}{T} = \frac{\omega}{2\pi}.$$

Since ω is closely related to the frequency of vibration, it is called the *circular frequency* of the motion.

12-3 FREE UNDAMPED VIBRATIONS OF A PARTICLE

In many systems whose mass and restoring forces have discrete values the small energy loss due to air resistance, internal friction of springs (hysteresis), or other friction forces can be neglected. Such systems are called *undamped* systems. It should be recognized that, although such systems do not actually exist, the errors introduced by neglecting these losses are small enough that the analysis often gives quite satisfactory engineering results. Experience has shown that the small damping forces which eventually stop a freely vibrating system have little effect on such quantities as frequency and period of vibration.

Figure 12-5(a) represents a particle B with a mass m which is suspended by means of a spring with a modulus k. In the equilibrium position the spring tension will be equal to the weight; that is,

$$ky_{st} = W = mg.$$

If the body is pulled downward a distance y and released, it will oscillate up and down. The free-body diagram in Fig. 12-5(c) shows the forces acting on the body when it is displaced y units downward. The magnitude of the force F is

$$F = k(y_{st} + y) = mg + ky,$$

and the equation of motion for B is

$$m\ddot{y} = \sum F = mg - F = mg - (mg + ky) = -ky$$

from which

$$\ddot{y} = -\frac{k}{m} y. \qquad (12\text{-}2a)$$

(a)   (b)   (c)

Figure 12-5

This equation shows that the acceleration of B is proportional to its displacement from the equilibrium position and is directed toward that position. Therefore the body has simple harmonic motion. When Eq. (12-2a) is compared with Eq. (12-2) the two expressions are identical if k/m is replaced by ω^2 or ω_n^2. Since f_n, the free undamped frequency of this system, is related to ω_n^2, the subscript n is used for all free undamped systems. The natural frequency of the vibration is

$$f_n = \frac{\omega_n}{2\pi} = \frac{1}{2\pi}\sqrt{\frac{k}{m}}.$$

Equation (12-2) or (12-2a) is a second-order linear differential equation with constant coefficients. Although there are many ways to solve this common differential equation, perhaps the simplest is to assume a solution and verify it by substitution into the differential equation. Note that the second derivative of the function must equal the function multiplied by a constant coefficient in order to satisfy the differential equation. Some exponential and trigonometric functions have this property, and it can be shown by substitution that the following trigonometric function will satisfy the differential equation:

$$y = B \cos \omega_n t + C \sin \omega_n t. \qquad (12\text{-}3a)$$

The quantities B and C are the constants of integration and can be determined from the initial starting conditions, and ω_n is the square root of k/m.

Equation (12-3a) can also be written as

$$y = B \cos \omega_n t + C \sin \omega_n t = A \cos(\omega_n t - \phi). \qquad (12\text{-}3b)$$

To show that this is possible Eq. (12-3b) can be expanded to give

$$B \cos \omega_n t + C \sin \omega_n t = A(\cos \omega_n t \cos \phi + \sin \omega_n t \sin \phi)$$

from which

$$(B - A \cos \phi)\cos \omega_n t + (C - A \sin \phi)\sin \omega_n t = 0.$$

If this equation is to be valid for any value of t, the coefficients in parentheses must both be zero. Therefore,

$$B = A \cos \phi \quad \text{and} \quad C = A \sin \phi$$

from which

and

$$\left. \begin{array}{c} A = (B^2 + C^2)^{1/2} \\[2mm] \tan \phi = \dfrac{C}{B}. \end{array} \right\} \qquad (12\text{-}4)$$

Since the maximum value of $\cos(\omega_n t - \phi)$ is 1, the amplitude of the oscillation is $A = (B^2 + C^2)^{1/2}$.

Equations for two other properties of the oscillation can be derived from Eq. (12-3b). The velocity of the particle is

$$\dot{y} = -\omega_n A \sin(\omega_n t - \phi),$$

and the magnitude of the maximum velocity of the particle is

$$\dot{y}_{max} = \omega_n A = \omega_n \sqrt{B^2 + C^2}. \tag{12-5a}$$

Similarly, the acceleration of the particle is

$$\ddot{y} = -\omega_n^2 A \cos(\omega_n t - \phi),$$

and the magnitude of the maximum acceleration is

$$\ddot{y}_{max} = \omega_n^2 A = \omega_n^2 \sqrt{B^2 + C^2}. \tag{12-5b}$$

Figure 12-6

When the amplitudes of oscillations are small, it is convenient to set up the equation of motion in terms of inches instead of feet. If the acceleration is in inches per second per second, the mass must be expressed in units of pound-second squared per inch instead of pound-second squared per foot (slugs). If the mass is calculated from the expression W/g, the value of g must be in inches per second per second, approximately 386 ips². When using SI units the value of g is 9.81 m/s² and all lengths are measured in meters.

In any problem involving free undamped vibrations, a free-body diagram of the vibrating particle should be drawn showing it displaced from its equilibrium position. The acceleration can be determined from the general motion equation in terms of the position of the particle. If the acceleration is of the form

$$a = \ddot{y} = -Ky,$$

where K is any positive constant and y is the position coordinate of the particle measured from its equilibrium position, the resulting motion will be simple harmonic motion. If the acceleration is not equal to $-Ky$, the motion will not be simple harmonic motion, although it may still be an oscillatory motion, and the solution must be obtained from the differential equation.

As an example, consider the oscillations of the simple pendulum in Fig. 12-6. The pendulum, consisting of a particle of weight W suspended by a cord of length L, swings in a vertical plane. The cord is inextensible, and its weight is negligible. The particle is moved out until the cord makes an angle θ_{max} with the vertical and is then released to oscillate in a vertical plane. A free-body diagram of the particle, when it is displaced any angle θ, is shown in Fig. 12-6, and the positive directions are indicated near the diagram. The tangential component of the acceleration can be determined from

$$\sum F_x = ma_x$$

or

$$-W \sin \theta = \frac{W}{g} a_x$$

from which

$$a_x = -g \sin \theta.$$

The tangential acceleration can be expressed in terms of the angular acceleration of the cord, that is, $a_x = L\ddot{\theta}$, and the differential equation becomes

$$L\ddot{\theta} = -g \sin \theta \quad \text{or} \quad \ddot{\theta} = -\frac{g}{L} \sin \theta.$$

When the maximum value of θ is small, $\sin \theta$ is approximately equal to θ, and the equation becomes

$$\ddot{\theta} = -\frac{g}{L} \theta,$$

which indicates that the pendulum has simple harmonic motion. If the angle θ is not small, the resulting motion will not be simple harmonic motion, although it will be an oscillatory motion.

EXAMPLE 12-1

The 7.72-lb body A in Fig. 12-7 moves along a smooth horizontal plane and is restrained by two identical springs each with a modulus of 4 lb-per in. The oscillation is started by moving body A 2 in. to the right of its equilibrium position and giving it a velocity of 30 ips to the left. Determine the period, frequency, amplitude, maximum velocity, and maximum acceleration of the motion. Both springs are in tension at all times.

SOLUTION

Figure 12-7(b) is a free-body diagram of body A displaced a positive distance, x, from the equilibrium position. At the equilibrium position, the tension in each spring must be the same, T_o, and as the mass is displaced to the right a distance x, the tensions in the left and right

Equilibrium position

(a)   (b)

Figure 12-7

springs become T_O+kx and T_O-kx, respectively. The differential equation of motion is obtained from Newton's equation of motion and the free-body diagram of Fig. 12-7(b) as follows:

$$\sum F_x = (T_O-4x)-(T_O+4x) = ma_x = \left(\frac{7.72}{386}\right)\ddot{x}$$

from which

$$\ddot{x}+400x = 0.$$

Note that k, x, and m are expressed in terms of pounds, seconds, and inches; therefore, $\ddot{x}$ will be in inches per second per second. The value of ω_n^2 is the coefficient of the displacement, x. Therefore $\omega_n = 20$ rad per sec, and the period is

$$T = \frac{2\pi}{\omega_n} = \frac{2\pi}{20} = 0.314 \text{ sec.} \qquad \text{Ans.}$$

The frequency is the reciprocal of the period; that is,

$$f = \frac{1}{T} = \frac{1}{0.314} = 3.18 \text{ cps} = 3.18 \text{ Hz.} \qquad \text{Ans.}$$

The solution to the differential equation of motion [see Eq. (12-3b)] is

$$x = B \cos(20t)+C \sin(20t) = A \cos(20t-\phi).$$

The constants B and C, or A and ϕ, can be obtained from the initial conditions. When $t = 0$, $x = +2.0$ and $\dot{x} = -30$. If the second form of the solution is used, the velocity is

$$\dot{x} = -20A \sin(20t-\phi). \qquad (a)$$

The initial conditions (when $t = 0$) give

$$x(0) = +2.0 = A \cos(-\phi) = A \cos \phi$$

and

$$\dot{x}(0) = -30 = -20A \sin(-\phi) = 20A \sin \phi.$$

These equations can be solved simultaneously to give

$$A = 2.50 \text{ in.,} \qquad \phi = \tan^{-1}(-0.750) = -0.644 \text{ rad.}$$

The angle ϕ is not needed for this solution, but its value is given to provide a check on the calculations. The amplitude of the motion is

$$A = 2.50 \text{ in.} \qquad \text{Ans.}$$

The maximum velocity and acceleration are

$$v_{max} = A\omega = 2.50(20) = 50 \text{ ips} \qquad \text{Ans.}$$

and

$$a_{max} = A\omega^2 = 2.50(20)^2 = 1000 \text{ ips}^2. \qquad \text{Ans.}$$

These values can be obtained from Eq. (a) by differentiation and substitution.

PROBLEMS

12-1 Two bodies with simple harmonic motions have equal maximum velocities, but their frequencies are in the ratio of 1 to 4. Determine the ratio of their amplitudes.

12-2 The position of a particle at any time t is given by the equation $x = 8 \cos \pi t/2$, where x is in inches and t is in seconds. Plot the displacement, velocity, and acceleration versus time curves for one complete cycle.

***12-3** The maximum velocity of a body with simple harmonic motion is 30 mm/s, and its frequency is 60 Hz. Determine the amplitude, maximum acceleration, and period of the motion.

***12-4** Vibration-measuring instruments indicate that a body is vibrating harmonically at a frequency of 480 Hz with a maximum acceleration of 500 cm/s². Determine the amplitude of the vibration and the maximum speed of the body.

12-5 A particle moves with simple harmonic motion, and the maximum magnitudes of its velocity and acceleration are 20 ips and 80 ips², respectively. Determine the magnitude of the acceleration of the point when the speed is 10 ips.

12-6 A particle moves with simple harmonic motion. The amplitude of the motion is 5 in., and the magnitude of the velocity is 12 ips when the point is displaced 3 in. from the position of maximum velocity. Determine the magnitude of the acceleration of the point for this position.

12-7 A particle has simple harmonic motion. When it is displaced 8 in. from the center of its path, the magnitudes of its velocity and acceleration are 30 ips and 200 ips², respectively. Determine
(a) The period of the motion.
(b) The maximum speed of the particle.

***12-8** The 6.0-kg body in Fig. P12-8 is displaced 0.20 m to the right of its equilibrium position and released from rest when $t = 0$. The modulus of the spring is 50 N/m. Determine the frequency of the resulting vibration and the solution of the differential equation.

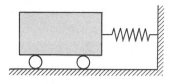

Figure P12-8

12-9 A weight of 25 lb is attached to a vertical spring with a modulus of 40 lb per ft. The weight is released when the spring is unstressed. Determine
(a) The distance the weight falls before coming to rest.
(b) The period of the motion.

12-10 The 48.3-lb block B is supported as indicated in Fig. P12-10 by a spring with a modulus of 8.0 lb per in. The velocity of B is 2.4 fps upward when B is 4.8 in. below its equilibrium position. Determine
(a) The amplitude of the free vibration of B.
(b) The maximum acceleration of B.

Figure P12-10

***12-11** The 8.0-kg ball in Fig. P12-11 swings in a vertical plane on a flexible wire with a length of 2.5 m. When $t = 0$, $\dot{\theta} = 0$ and $\theta = 0.1$ rad. Derive and solve the differential equation of motion of the body. Consider the body to be a particle.

Figure P12-11

12-12 Particle C in Fig. P12-12 weighs W lb and is supported by a 3-ft rod whose weight can be neglected. When the rod is displaced an angle θ of 5° and released from rest, C swings in a vertical plane as a simple pendulum. Determine the magnitude of the maximum velocity of C.

12-13 Block B of mass m is free to slide on the smooth rod CD as indicated in Fig. P12-13.

Figure P12-12

Figure P12-13

The modulus of the spring is k, and its unstretched length is b. The block is moved to the right a distance A and released from rest.
(a) Determine the acceleration of B as a function of the position coordinate x.
(b) Is the motion of B simple harmonic motion?
(c) If so, determine the period of the motion.

12-4 FREE UNDAMPED VIBRATIONS OF RIGID BODIES

Free vibrations of a particle were discussed in Art. 12-3 and the procedures developed there apply to rigid bodies that have translation without any rotation. When a rigid body is supported in such a manner that the body has rotational motion as it vibrates, the period and frequency of the resulting vibratory motion will depend on the moment of inertia of the mass of the body as well as on its mass and the forces that act upon the body.

For rectilinear motion, the criterion for simple harmonic motion is that the acceleration of the particle be of the form

$$a = \ddot{x} = -Kx.$$

An analogous equation for rotational motion in terms of the angular acceleration and position of the body was developed as

$$\alpha = \ddot{\theta} = -K\theta.$$

If the equations of motion for a body reduce to the preceding equation, the body has a *simple harmonic angular motion.* With angular harmonic motion, the angle θ is used in place of a distance x or y, and therefore the *amplitude of the motion is the maximum angular displacement of the body* from its position of equilibrium.

A *compound pendulum* is a rigid body of finite dimensions which oscillates about a fixed horizontal axis through the body. The period of vibration for small oscillations of the compound pendulum in Fig. 12-8(a) can be determined by obtaining an expression for the angular acceleration of the body in terms of its angular position. Figure 12-8(b) is a free-body diagram showing the body displaced through a positive angle θ from its equilibrium position. The moment equation of motion in scalar form is

$$\sum M_O = I_O \ddot{\theta},$$

which becomes

$$-Wr_G \sin \theta = I_O \ddot{\theta} \quad \text{or} \quad \ddot{\theta} = -\frac{Wr_G}{I_O} \sin \theta.$$

If the amplitude of the vibratory motion is small, the value of $\sin \theta$ is approximately equal to the value of θ in radians, and the moment equation becomes

$$\ddot{\theta} = -\frac{Wr_G}{I_O}\theta = -K\theta.$$

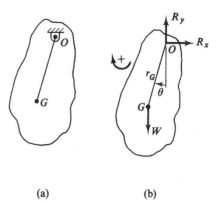

(a)   (b)

Figure 12-8

The natural circular frequency of the simple harmonic angular motion is

$$\omega_n = \sqrt{K} = \sqrt{\frac{Wr_G}{I_O}},$$

and the period of the vibration is

$$T = \frac{2\pi}{\omega_n} = 2\pi\sqrt{\frac{I_O}{Wr_G}}.$$

A *torsional pendulum* consists of a rigid body supported on a shaft as shown in Fig. 12-9. When the pendulum is twisted through a small angle θ and released, a simple harmonic angular motion results from the stresses developed in the shaft. Assume that a vibratory motion of body B in Fig. 12-9 is started by twisting the assembly through the small angle θ, as shown, and then releasing the pendulum. It is shown in texts on strength of materials that if the proportional limit[1] of the material in a *solid circular shaft* is not exceeded, the moment necessary to twist the shaft is proportional to the angle of twist and can be obtained from

$$T = \frac{JG}{L}\theta = \frac{\pi r^4 G}{2L}\theta = k\theta,$$

where

$$J = \frac{\pi r^4}{2} = \frac{\pi d^4}{32}$$

Figure 12-9

is the polar moment of inertia of the cross-sectional area of the shaft, G is the shearing modulus of elasticity of the material, L is the length

---

[1] The *proportional limit* is the highest unit stress at which strain (deformation) is proportional to the accompanying stress.

of the shaft, and θ is the angle, in radians, through which the shaft is twisted. The moment exerted by the shaft on the body B is equal in magnitude and opposite in sense to the moment, T, required to twist the shaft. The moment exerted by the shaft on B has a sense opposite to the sense of the angular displacement of the lower end of the shaft. The equation of motion for body B, in scalar form, is

$$\sum M_z = I_z \ddot{\theta}$$

or

$$-T = -k\theta = I_z \ddot{\theta}.$$

The angular acceleration is

$$\ddot{\theta} = -\frac{k}{I_z} \theta.$$

This last equation proves that body B has simple harmonic angular motion, and the natural circular frequency of the motion is

$$\omega_n = \sqrt{\frac{k}{I_z}} = \sqrt{\frac{\pi r^4 G}{2 L I_z}}.$$

The frequency and the period of the motion are

$$f = \frac{\omega_n}{2\pi} \quad \text{and} \quad T = \frac{2\pi}{\omega_n},$$

respectively.

The moment of inertia of a body can be determined experimentally by suspending it on a shaft as a torsional pendulum or by suspending it from some axis not through the mass center as a compound pendulum and measuring the period of vibration for oscillations of small amplitude.

The solution of problems in free vibrations of rigid bodies is further illustrated by the following examples.

EXAMPLE 12-2

The slender homogeneous rod BC in Fig. 12-10(a) weighs 8 lb, and the small 4.5-lb body E is welded to the rod. The spring has a modulus of 7 lb per in. Rod BC is horizontal when it is in equilibrium. The rod is displaced an angle of 8° clockwise and released from rest. Determine the natural circular frequency and the maximum angular velocity of the vibration.

SOLUTION

Figure 12-10(b) is a free-body diagram of the two bodies displaced an angle θ (positive) from the equilibrium position. The spring force is

$$T = T_o + k y_c,$$

where T_o is the tension in the equilibrium position and y_c is the displacement of the end of the rod. The bar is in equilibrium when

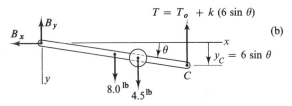

Figure 12-10

$\theta = 0$, and the moment equation about B gives

$$6T_o = 3(8.0) + 4(4.5)$$

from which

$$T_o = 7.0 \text{ lb.}$$

In the displaced position, the moment equation of motion, in scalar form, is

$$\sum M_B = I_B\ddot{\theta}$$

or

$$8.0(3 \cos \theta) + 4.5(4 \cos \theta) - (T_o + k6 \sin \theta)(6 \cos \theta)$$

$$= \left[\frac{1}{3}\frac{8.0}{386}(6)^2 + \frac{4.5}{386}(4)^2\right]\ddot{\theta}.$$

When the angle θ is small, $\cos \theta \approx 1$ and $\sin \theta \approx \theta$ and the equation reduces to

$$24 + 18 - (7 + 42\theta)6 = 0.435\ddot{\theta}$$

or

$$\ddot{\theta} = -579\theta.$$

This is the differential equation of a simple harmonic angular motion with a natural circular frequency of

$$\omega_n = \sqrt{579} = 24.1 \text{ rad per sec.} \qquad \text{Ans.}$$

The solution to the differential equation is

$$\theta = A \cos(\omega_n t - \phi).$$

In this example, $\theta = 8° = 8\pi/180$ rad when $t = 0$ and $\dot{\theta} = 0$ when $t = 0$. Therefore,

$$\frac{8\pi}{180} = A \cos(-\phi)$$

and

$$0 = \dot{\theta} = -A\omega_n \sin(-\phi).$$

From these equations, $\phi = 0$ and $A = 8\pi/180 = 0.1396$ rad. The solution to the equation is

$$\theta = 0.1396 \cos(24.1t)$$

from which

$$\dot{\theta} = -0.1396(24.1)\sin(24.1t) = -3.36 \sin(24.1t),$$

and the maximum angular velocity of the body is

$$\dot{\theta}_{max} = 3.36 \text{ rad per sec.} \qquad \text{Ans.}$$

Note the distinction between $\dot{\theta}_{max}$ and ω_n. These two quantities are sometimes confused since they have the same units.

EXAMPLE 12-3

The 64.4-lb body E in Fig. 12-11(a) is supported by an inextensible flexible belt which is wrapped around a flange on wheel C. The 96.6-lb wheel C is supported by a smooth bearing at O and is connected to a system of springs as indicated. The moduli of the springs are as indicated, and the mass of the springs and of bar AB is negligible. The radius of gyration of the mass of wheel C with respect to the axis of rotation through the mass center at O is 9.6 in. Body E is in the equilibrium position and has a velocity of 2 fps downward when $t = 0$. Determine the differential equation of motion of the system and the solution of the equation.

SOLUTION

Free-body diagrams of AB, C, and E are shown in Fig. 12-11(b). The resulting equations can be written in terms of y_D, y_E, θ_C, or some other

(a) (b)

Figure 12-11

variable. The variable θ will be used in this example, in which case $y_E = 0.75\theta$ and $y_D = 1.00\theta$. The angle θ is assumed to be positive when clockwise; therefore, downward is positive for y_E and upward is positive for y_D.

When the system is an equilibrium, $T = 64.4$ lb and $P = P_o = (9/12)64.4 = 48.3$ lb. When the system is given a positive displacement, the force P can be expressed as $P_o + \Delta P$, where ΔP is the change of P due to a change of θ. Let $\Delta_1 = \Delta_2$ be the stretch of springs with moduli k_1 and k_2 and Δ_3 be the stretch of the third spring. The deflections Δ_1 and Δ_2 are equal since AB is loaded symmetrically and since $k_1 = k_2$.

$$y_D = \Delta_1 + \Delta_3 = \frac{\Delta P/2}{k_1} + \frac{\Delta P}{k_3} = \frac{\Delta P}{2(54)} + \frac{\Delta P}{216} = \frac{\Delta P}{72}$$

or

$$\Delta P = 72y_D = 72\theta$$

and

$$P = P_o + \Delta P = 48.3 + 72\theta.$$

The equations of motion, in scalar form, are as follows: For body C,

$$\sum M_O = I_O\ddot{\theta}$$

or

$$0.75T - 1.00(48.3 + 72\theta) = \frac{96.6}{32.2}\left(\frac{9.6}{12}\right)^2\ddot{\theta} = 1.92\ddot{\theta},$$

and for body E,

$$\sum F_y = ma_E$$

or

$$64.4 - T = \frac{64.4}{32.2}a_E = 2.00(0.75\ddot{\theta}) = 1.50\ddot{\theta}.$$

The elimination of T in these equations gives

$$-72\theta = 3.045\ddot{\theta}$$

or

$$\ddot{\theta} = -\underline{23.6\theta}. \qquad\qquad \text{Ans.}$$

This differential equation of motion demonstrates that the system has simple harmonic motion with a natural circular frequency of

$$\omega_n = \sqrt{23.6} = 4.86 \text{ rad per sec.}$$

The solution of the differential equation can be written as

$$\theta = B \cos \omega_n t + C \sin \omega_n t.$$

From the initial condition $\theta = (\tfrac{4}{3})y_E = 0$ when $t = 0$, the value of B must be zero. The angular velocity, with $B = 0$, is

$$\dot{\theta} = C\omega_n \cos \omega_n t,$$

and since $\dot{\theta} = (\tfrac{4}{3})\dot{y}_E = (\tfrac{4}{3})2$ when $t = 0$,

$$\frac{8}{3} = C\omega_n$$

and

$$C = 0.548 \text{ rad.}$$

The solution of the differential equation of motion of the system is

$$\theta = \underline{0.548 \sin 4.86t}. \qquad \text{Ans.}$$

PROBLEMS

12-14 Body B in Fig. P12-14 weighs 20 lb, and the moduli of springs C and E are 5 and 15 lb per in., respectively. The speed of B while passing through its equilibrium position is 15 ips. Determine the amplitude and frequency of the vibration of B.

Figure P12-14

12-15 The springs in Problem 12-14 are arranged as shown in Fig. P12-15. All other data are unchanged. Determine the amplitude and frequency of the vibration of B.

Figure P12-15

12-16 When a small body which weighs W lb is supported by a spring with a modulus of k lb per in. as shown in Fig. P12-16(a), its natural frequency is found to be 2.0 Hz. When an additional spring with a modulus of 5.5 lb per in. is added, as shown in Fig. P12-16(b), the frequency is increased to 3.5 Hz. Determine the weight W and the modulus of the upper spring. Both springs are in tension during the entire vibration.

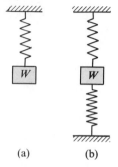

(a)   (b)

Figure P12-16

*12-17 The 3.0-kg mass C in Fig. P12-17 is supported by the rigid body ABC whose mass is negligible. The modulus of the spring at A is 3600 N/m. The arm BC is horizontal when the system is in equilibrium. Determine the period of the vibration of C when it is displaced slightly from its equilibrium position and released.

Figure P12-17

*12-18 Determine the distance BC in Problem 12-17 which will result in a frequency of vibration of 1.2 Hz. All other data are unchanged.

12-19 Bar AB in Fig. P12-19 is suspended from a smooth horizontal axis at O. When the spring S_1 at B is the only one acting, the frequency of the free vibrations is f_1 Hz. The modulus of S_1 is 15 lb per in. The natural frequency of the bar is to be doubled by adding a second spring S_2 at A. Both springs are in tension at all times, and the bar is in equilibrium in the horizontal position. Determine the modulus of S_2.

Figure P12-19

12-20 For the system shown in Fig. P12-20 derive and solve the differential equation of motion with the initial conditions $\theta = 0.1$ rad and $\dot\theta = 0.6$ rad per sec when $t = 0$. Neglect the weight of the rod.

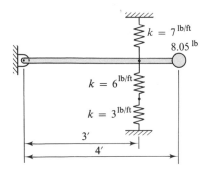

Figure P12-20

12-21 Determine the mass of the uniform bar in Problem 12-20 if the frequency of the free vibration is 0.50 Hz.

*12-22 The radius of gyration of the 8.0-kg body in Fig. P12-22 with respect to an axis through the center of gravity is 0.25 m. When $t = 0$, $\theta = \theta_0$ and $\dot\theta = 0$. Derive and solve the differential equation of motion for small values of θ and determine the period of the vibration.

Figure P12-22

12-23 The particle B in Fig. P12-23 is mounted on the rigid rod BC, whose weight may be neglected. The particle B weighs 0.50 lb, and the modulus of each of the springs is 10.0 lb per ft. The springs each have a tension of 2.00 lb when BC is vertical. B is moved 1 in. to the right and released from rest. Determine the frequency of the resulting vibration.

Figure P12-23

12-24 In Problem 12-23 determine the maximum weight of particle B for which member BC will vibrate. All other data are as given in Problem 12-23.

12-25 The solid homogeneous 20-lb sphere in Fig. P12-25 rolls without slipping when displaced from its equilibrium position. Initially each spring has a tension of 50 lb. The center of the sphere is moved 3.0 in. to the right and released from rest. Determine the resulting frequency and the maximum speed of the center of the sphere.

Figure P12-25

747

12-26 Two identical drums A and B rotate about fixed horizontal axes, as indicated in Fig. P12-26, with constant angular velocities of 60 rpm. The coefficient of kinetic friction between the 20-lb homogeneous plank C and the drums is 0.50. The plank is displaced 3.0 in. to the right from its position of equilibrium and released from rest.
(a) Show that the plank has simple harmonic motion.
(b) Determine the maximum speed of the plank during the resulting motion.

Figure P12-26

12-27 Solve Problem 12-26 if the drums are replaced by V-groove pulleys (see Fig. P12-27) and the plank by a cylindrical rod. The angle ϕ is 60°.

Figure P12-27

12-28 The mechanism in Problem 12-26 is inclined an angle β of 10° as shown in Fig. P12-28. The plank is released from rest in the position indicated. Determine the frequency and amplitude of the resulting motion.

12-29 The solid homogeneous 128.8-lb cylinder in Fig. P12-29 rolls without slipping. Neglect the weight of the hub for the cords which are

Figure P12-28

attached to the springs. The modulus of spring A is 8 lb per in., and the modulus of B is 2 lb per in. The springs are always in tension. When the cylinder is in equilibrium, a clockwise couple of 60 in-lb is gradually applied to the cylinder and then suddenly released. Determine the amplitude and period of the resulting free vibration.

Figure P12-29

***12-30** The 30-kg solid homogeneous cylinder in Fig. P12-30 rolls without slipping along the inclined plane. The wheel is released from rest when $t = 0$ and the spring tension is zero. Derive the differential equation of motion and its solution. Also determine the frequency of the vibration.

Figure P12-30

***12-31** The body in Fig. P12-31 is suspended from a smooth horizontal bearing. When the body is displaced slightly from the equilibrium position in Fig. P12-31(a) and released, it is found to oscillate 50 times in 68.0 s. The force R

(a)　　　　(b)

Figure P12-31

Figure P12-33

required to support the body in the position in Fig. P12-31(b) is 90 N. Determine

(a) The moment of inertia of the mass of the body with respect to its axis of rotation.

(b) The distance q from the axis of rotation to the center of percussion P.

12-32 A motor weighing 100 lb is mounted on two springs as indicated in Fig. P12-32. The springs are deflected 2.00 in. by the weight of the motor. The radius of gyration of the mass of the motor with respect to a horizontal axis through G perpendicular to AB is 4.00 in. Two 25-lb forces, one downward at A and the other upward at B, are applied and suddenly removed. Determine the period and amplitude of the resulting vibrations.

Figure P12-32

*12-33 The two bars in Fig. P12-33 each have a mass $m = 2$ kg and a length $L = 0.50$ m and are welded in the shape shown. The bars are held in equilibrium, with AB vertical, by the spring which has a modulus k. Show that the system will have simple harmonic motion with small angles only when k is more than $3mg/2L$, where g is the acceleration of gravity. Determine the frequency of the vibration when $k = 120$ N/m.

*12-34 Bar AB in Fig. P12-34 has a mass of 2 kg and CD has a mass of 4 kg. Each spring has

a modulus of 300 N/m and is stretched 0.10 m in the position shown. The member oscillates in the vertical plane, and the maximum velocity of D is observed to be 0.50 m/s. Determine the maximum tension in the springs.

Figure P12-34

*12-35 Solve Problem 12-34 if the body is inverted so that D is above C.

12-36 The solid homogeneous cylinder in Fig. P12-36 (p. 750) has a mass m and rolls along the curved surface without slipping. The cylinder is displaced slightly from its equilibrium position and is allowed to oscillate.

(a) Will the resulting motion be simple harmonic motion for small values of θ?

(b) If so, determine the frequency of the resulting motion.

Figure P12-36

12-37 Bars AB and CD in Fig. P12-37 are made of steel weighing 0.284 lb per in.^{3} The shearing modulus of elasticity of steel is 12,000,000 psi. Neglect the weight of CD and assume that bar AB is rigid. Bar AB has an angular velocity of 0.50 rad per sec when CD is not twisted. Determine
(a) The period of oscillation of AB.
(b) The maximum angular acceleration of AB.

Figure P12-37

*12-38** The solid homogeneous cylindrical disk in Fig. P12-38 has a mass of 1200 kg. The upper rod is steel with a shearing modulus of elasticity of $8.3(10^{10})$ N/m^2. The lower rod is aluminum with a shearing modulus of elasticity of $2.8(10^{10})$ N/m^2. Determine the frequency of the torsional vibration of the disk.

12-39 The torsional pendulum in Fig. P12-39 is to be used as an instrument for measuring increments of time. The pendulum is to have a period of 2.00 sec. The torsional stiffness of the wire is 4.10 in-lb per rad. The circular disk is 1.00 in. thick and is made of material having a density of 15.0 slugs per ft^3. Determine the diameter of the disk.

Figure P12-38

Figure P12-39

12-40 The torsion-bar suspension in Fig. P12-40 has been proposed as a mounting for an automobile wheel. The bar AB is assumed to be fixed at A and supported by a smooth bearing at C. Arm BD is fastened to the rod at B and carries the wheel bearing at D. The wheel weighs 75 lb and has a radius of gyration of mass of 12.0 in. about its geometrical axis. The arm BD is 10.0 in. long and is assumed to be rigid. The torsional modulus of AB is 20,000 in-lb per rad. Neglect the mass of AB and BD and assume that AB is twisted but does not bend. Determine the natural frequency of vibration of the wheel if
(a) The wheel is supported on its bearing at D with the brake released.
(b) The brake is locked so that the wheel must turn with the bearing and arm BD.

Figure P12-40

750

12-41 A weight W is suspended as shown in Fig. P12-41 on rigid rods with negligible weight. The system is free to rotate as indicated in the projected sketch. Determine the natural frequency of free vibrations of small oscillations.

Figure P12-41

12-42 A 30-in. slender rod is bent as shown in Fig. P12-42 and suspended by a pin at O. End B is pulled down 1.0 in. and released from rest. Determine the maximum speed of point A during the resulting motion and the frequency of the motion.

Figure P12-42

12-43 Rod AB and the small body C in Fig. P12-43 each have a mass m. Determine
(a) The distance x at which C should be located in order that the resulting frequency of small vibrations will be maximum.
(b) The frequency resulting from part (a).

Figure P12-43

12-5 FREE VIBRATIONS WITH VISCOUS DAMPING

A vibrating system contains damping if it possesses elements which remove energy from the system. There are several types of damping: *viscous damping*, which is experienced by bodies moving through fluids with moderate velocities; *coulomb damping*, which arises from the relative motion of dry surfaces; and *structural* or *solid damping*, sometimes called mechanical *hysteresis*, which is caused by the internal friction of the elastic material. Only viscous damping is considered here because viscous dampers or dashpots are often used to control or limit vibrations, and the sliding of well-lubricated surfaces can be closely approximated by assuming viscous damping. Viscous damping results in a direct straightforward solution and is often used as a good approximation for other types of damping.

The free vibrations of systems with viscous damping are transient oscillations that gradually decrease in amplitude and eventually die out altogether. Although this might seem to indicate that the study of free, damped motion would be of little practical interest, it provides a desirable background for the study of forced, damped vibrations.

Body A in Fig. 12-12(a) has a weight W, is supported by a spring of modulus k, and is acted on by a viscous damper. The damper is shown as a dashpot, and c is the *coefficient of viscous damping*. The damping force is proportional to the velocity and its magnitude is $c\dot{y}$. The damping force always opposes the direction of the velocity as shown in the free-body diagram of Fig. 12-12(b).

The positive direction for y, $\dot{y}$, and $\ddot{y}$ is downward, and the free-body diagram is drawn with y and $\dot{y}$ both assumed to be positive. When the body is in equilibrium, F_d is zero and T is equal to W. In the displaced position, the tension is

$$T = W + ky,$$

(a)   (b)

Figure 12-12

and the equation of motion in the y direction is

$$\sum F_y = W - (W + ky) - c\dot{y} = m\ddot{y}$$

or

$$m\ddot{y} + c\dot{y} + ky = 0. \qquad (12\text{-}6)$$

The solution of any linear, ordinary differential equation with constant coefficients can be obtained by assuming that

$$y = Be^{rt}$$

and determining values of B and r which satisfy the differential equation and the initial or boundary conditions. Substitution into the differential equation gives

$$Be^{rt}[mr^2 + cr + k] = 0.$$

The differential equation will be satisfied if

$$B = 0$$

or if

$$mr^2 + cr + k = 0.$$

The first equation is a trivial solution, and the second equation gives

$$r_{1,2} = \frac{-c \pm \sqrt{c^2 - 4mk}}{2m}. \qquad (12\text{-}7)$$

The general solution to the differential equation must have two arbitrary constants and can be written as

$$y = Be^{r_1 t} + Ce^{r_2 t}. \qquad (12\text{-}8)$$

The constants B and C can be determined from the initial conditions for the motion, and the constants r_1 and r_2 are obtained from Eq. (12-7). The behavior of the system depends on the quantity within the radical of Eq. (12-7). The quantity can be zero, positive, or negative and the radical will be zero, real, or imaginary, respectively.

The *critical damping coefficient* is defined as the value of c which will make the radical zero; therefore,

$$c_{cr} = 2\sqrt{km}.$$

The critical damping coefficient represents the minimum amount of damping for nonvibratory motion. The solution to Eq. (12-6) will have three distinct forms, depending on whether c is greater than, equal to, or less than c_{cr}.

Case I: $c > c_{cr}$. In this case, the radical in Eq. (12-7) is real, and both roots are real and negative. The motion is nonvibratory or *aperiodic*, and such a system will return slowly to its equilibrium position with not more than one (usually none) reversal of motion.

Case II: $c = c_{cr}$. In this case, in which the damping is critical, the two roots in Eq. (12-7) are identical and are equal to

$$r = -\frac{c_{cr}}{2m} = -\frac{2\sqrt{km}}{2m} = -\sqrt{\frac{k}{m}} = -\omega_n.$$

In this case, Eq. (12-8) will have only one arbitrary constant, and therefore it is not the general solution. The general solution, when $c = c_{cr}$, is

$$y = (B + Ct)e^{rt} = (B + Ct)e^{-\omega_n t}.$$

This statement can be verified by substitution into Eq. (12-6). The motion given by this equation is also aperiodic, and it is of special importance because it represents the dividing point between aperiodic motion and damped oscillatory motion.

Critical damping is the minimum damping for which a system will return to rest without oscillation. Furthermore, the system will return to rest in the *least possible time* for any specified initial conditions.

Case III: $c < c_{cr}$. In this case, the system is underdamped, the radical in Eq. (12-7) is imaginary, and the roots are

$$r_{1,2} = -\frac{c}{2m} \pm i\sqrt{\frac{k}{m} - \left(\frac{c}{2m}\right)^2} = \alpha \pm i\omega_d,$$

where $i = \sqrt{-1}$, $\alpha = -c/(2m)$, and

$$\omega_d = \sqrt{\frac{k}{m} - \left(\frac{c}{2m}\right)^2}$$

is called the damped natural frequency.

When these values of r_1 and r_2 are substituted into Eq. (12-8) it becomes

$$y = B_1 e^{(\alpha + i\omega_d)t} + C_1 e^{(\alpha - i\omega_d)t} = e^{\alpha t}(B_1 e^{i\omega_d t} + C_1 e^{-i\omega_d t}).$$

By means of a series expansion, it can be shown that

$$e^{\pm i\omega_d t} = \cos \omega_d t \pm i \sin \omega_d t,$$

and after some algebraic reduction, the solution to the differential equation becomes

$$y = e^{\alpha t}(B \cos \omega_d t + C \sin \omega_d t) \qquad (12\text{-}9a)$$

$$= Ae^{\alpha t} \cos(\omega_d t - \phi), \qquad (12\text{-}9b)$$

where B and C or A and ϕ are constants which are determined from the initial conditions. This motion, called *time-periodic*, is oscillatory in nature, but the amplitude, $Ae^{\alpha t}$, varies with time. Since the constant $\alpha = -c/(2m)$ is negative, the amplitude will always decrease with time.

Damped free vibration ($c < c_{cr}$) does not repeat itself as does free undamped motion; therefore, it does not have a period or frequency as defined for free undamped vibrations. It is customary, however, to call the time interval between pairs of successive "zeroes or peaks" of the motion the *damped period*. When Eq. (12-9) and (12-3) are compared the damped period and frequency are seen to be

$$T_d = \frac{2\pi}{\omega_d} \quad \text{and} \quad f_d = \frac{\omega_d}{2\pi}.$$

The value of ω_d is

$$\omega_d = \sqrt{\frac{k}{m} - \left(\frac{c}{2m}\right)^2},$$

and it is apparent that ω_d will always be less than $\omega_n = \sqrt{k/m}$ for any positive value of c (less than c_{cr}).

Figure 12-13 shows the response of the system in Fig. 12-12 for

Figure 12-13

Figure 12-14

the three cases. The motion is started with the same positive velocity and with zero displacement for all three curves.

The amount of damping in a system is indicated by the rate of decay of the oscillation. The rate can be indicated by a quantity called the *logarithmic decrement*, δ, which is defined as the logarithm of the ratio of any two successive positive (or negative) amplitudes. Figure 12-14 represents the response of a system with a relatively small amount of damping. The amplitude at time t_1 is

$$y_n = Ae^{-(c/2m)t_1},$$

and the next amplitude is

$$y_{n+1} = Ae^{-(c/2m)(t_1+T_d)}.$$

The ratio of the two amplitudes is

$$\frac{y_n}{y_{n+1}} = \frac{Ae^{-(c/2m)t_1}}{Ae^{-(c/2m)(t_1+T_d)}} = e^{cT_d/(2m)},$$

and the logarithmic decrement is

$$\delta = \ln e^{cT_d/(2m)} = \frac{cT_d}{2m}. \tag{12-10}$$

The logarithmic decrement may be expressed in various forms since

$$T_d = \frac{2\pi}{\omega_d} = \frac{2\pi}{\sqrt{(k/m)-(c/2m)^2}}.$$

The ratio $\zeta = c/c_{cr}$ is called the *damping factor* and is convenient in expressing many of the preceding equations. When $\sqrt{k/m}$ is replaced by ω_n, $2\sqrt{km}$ by c_{cr}, and c by ζc_{cr}, the following results can be derived:

$$\omega_d = \omega_n\sqrt{1-\zeta^2},$$

$$T_d = \frac{2\pi}{\omega_n\sqrt{1-\zeta^2}},$$

and

$$\delta = \frac{2\pi\zeta}{\sqrt{1-\zeta^2}}.$$

It is suggested that these expressions be verified by substitution.

EXAMPLE 12-4

The 9.66-lb body E in Fig. 12-15(a) is fastened to the rod DF whose weight can be neglected. The spring has a modulus of 9.00 lb per ft, and the coefficient of the dashpot is 2.40 lb-sec per ft. The system is in equilibrium when DF is horizontal. The rod is displaced 0.10 rad clockwise and released from rest when $t = 0$. Determine

(a) The equation of motion of the rod.
(b) The frequency of the motion.
(c) The logarithmic decrement of the motion.

SOLUTION

(a) A free-body diagram of the rod, displaced in the positive direction, is shown in Fig. 12-15(b). When the body is in equilibrium, θ and $\dot{\theta}$ are both zero, and $T = T_0 = (\frac{4}{5})9.66$. The moment equation for the body with the displacement and velocity indicated in Fig. 12-15(b) is

$$4(9.66) - 4F_d - 5T = I_0\ddot{\theta}$$

or

$$4(9.66) - 4(2.40)(4\dot{\theta}) - 5\left[\left(\frac{4}{5}\right)9.66 + 9(5\theta)\right] = \frac{9.66}{32.2}(4.00)^2\ddot{\theta}.$$

This equation can be reduced to

$$4.8\ddot{\theta} + 38.4\dot{\theta} + 225\theta = 0$$

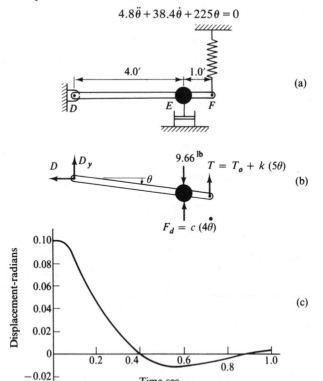

Figure 12-15

or

$$\ddot{\theta} + 8.00\dot{\theta} + 46.9\theta = 0.$$

The general solution can be obtained by substituting $\theta = De^{rt}$ in this equation to give

$$De^{rt}(r^2 + 8r + 46.9) = 0 \qquad (a)$$

from which

$$r_{1,2} = -4.00 \pm i5.56 = \alpha \pm i\omega_d.$$

The equation of motion is

$$\theta = e^{-4t}(B \cos 5.56t + C \sin 5.56t).$$

When $t = 0$, $\theta = 0.1$; therefore,

$$B = 0.100 \text{ rad.}$$

The angular velocity of the rod is

$$\dot{\theta} = e^{-4t}(-4B \cos 5.56t - 4C \sin 5.56t - 5.56B \sin 5.56t$$
$$+ 5.56C \cos 5.56t)$$

and when $t = 0$, $\dot{\theta} = 0$; therefore,

$$0 = -4B + 5.56C$$

and

$$C = \frac{4(0.100)}{5.56} = 0.0720 \text{ rad.}$$

The solution is

$$\theta = e^{-4.00t}(0.100 \cos 5.56t + 0.0720 \sin 5.56t). \qquad \text{Ans.}$$

The graph of this solution is shown in Fig. 12-15(c). For critical damping, the coefficient of r in Eq. (a) should be $\sqrt{4(46.9)} = 13.70$. The ratio of 8 to 13.70 is 0.584, which is the damping factor. With this amount of damping, the vibrations are damped out very quickly.

(b) The circular frequency of the motion is

$$\omega_d = 5.56 \text{ rad per sec,}$$

and the frequency is

$$f_d = \frac{\omega_d}{2\pi} = \frac{5.56}{6.28} = 0.884 \text{ Hz.} \qquad \text{Ans.}$$

(c) The logarithmic decrement is

$$\delta = \ln \frac{\theta_0 e^{-4t_1}}{\theta_0 e^{-4(t_1 + T_d)}} = 4T_d = \frac{4}{f_d} = \frac{4}{0.884} = 4.52. \qquad \text{Ans.}$$

In this expression, θ_0 is the maximum value of the quantity in parentheses in the answer to part (a) and occurs at $t = t_1$. Since the values of θ_0 and t_1 both drop out in this evaluation, it is not necessary to calculate either t_1 or θ_0.

PROBLEMS

*12-44 Body A in Fig. P12-44 has a mass of 2.0 kg, and the spring has a modulus of 50 N/m. The system is critically damped by the dashpot. Body A is moved 5.0 cm to the right of the equilibrium position and released with an initial velocity of 50 cm/s to the left. Determine
(a) The value of c_{cr}.
(b) The position and velocity of A when $t = 0.20$ s.
(c) The position and velocity of A when $t = 1.00$ s.

Figure P12-44

*12-45 Solve Problem 12-44, parts (b) and (c), if the damping coefficient is $24 \, \text{N} \cdot \text{s/m}$.

12-46 Body B in Fig. P12-46 has a mass of 2.0 slugs and the mass of the T-shaped bar can be neglected. The spring modulus is 80 lb per ft, and the damping coefficient of the dashpot is 16 lb-sec per ft. The system is in equilibrium when AB is horizontal. When $b = c = 2$ ft and the system is disturbed from its position of equilibrium, determine
(a) The type of motion which will exist.
(b) The frequency of oscillation (if any exists).
(c) The damping factor.

Figure P12-46

*12-47 Body A in Fig. P12-47 has a mass of 5 kg, the spring has a modulus of 300 N/m, and the coefficient of the viscous damper is

$180 \, \text{N} \cdot \text{s/m}$. The mass of the T-shaped member is negligible, and the system is in equilibrium when OA is horizontal. Determine the distance y which will result in critical damping.

Figure P12-47

*12-48 The damper coefficient in Problem 12-47 is $20 \, \text{N} \cdot \text{s/m}$, and the distance y is 30 cm. All other data are the same as in Problem 12-47. Body A is pulled 4.0 cm downward from the equilibrium position and released from rest. Determine
(a) The logarithmic decrement of the resulting motion.
(b) The damping factor ζ.

*12-49 Determine the velocity of body A in Problem 12-48 the first time OA is horizontal after being released.

12-50 The body in Fig. P12-50 weighs 96.6 lb, the spring stiffness is 39 lb per ft, and the dashpot coefficient is 12 lb-sec per ft. Determine the critical damping coefficient and the logarithmic decrement.

Figure P12-50

12-51 The rigid T-bar of negligible weight in Fig. P12-51 rotates in a vertical plane about a horizontal axis at O. The bar is caused to vibrate by displacing the weight and releasing it from rest. Determine the damped frequency of vibration and the ratio of the positive amplitudes of the third and fourth cycles. What is the ratio of the amplitudes of the first and third cycles?

Figure P12-51

12-52 The body in Fig. P12-52 weighs 24.15 lb and is supported by three springs and three dashpots as shown. The spring moduli are $k_1 = k_2 = 10$ lb per ft and $k_3 = 7$ lb per ft. The dashpot coefficients are $c_1 = c_2 = 0.05$ lb-sec per ft and $c_3 = 0.08$ lb-sec per ft. The body is displaced 4 in. downward from equilibrium and released from rest. Determine the number of oscillations which will occur before the amplitude of the vibration is reduced to 20% of its original value.

Figure P12-52

12-53 The homogeneous bar in Fig. P12-53 has a mass of 0.8 lb-sec² per ft and is in equilibrium when it is horizontal. The spring modulus is 3.0 lb per ft, and the dashpot coefficient is 1.5 lb-sec per ft. The rod is given an angular displacement of 0.06 rad and an angular velocity of 1.4 rad per sec, both clockwise, when $t = 0$. Solve the differential equation of motion for θ as a function of time and determine the angular position and velocity of the bar when $t = 1.0$ sec.

12-54 Set up the differential equation of motion for the inverted pendulum of Fig. P12-54.

Figure P12-53

The mass of the L-shaped rod can be neglected. Derive an expression for the critical damping coefficient.

Figure P12-54

12-55 The disk in Fig. P12-55 has a moment of inertia of 0.300 lb-sec²-in. about the z axis and is suspended from a wire as shown. The frequency of torsional vibrations in air (assume zero damping) is 0.50 Hz. When the disk is immersed in a fluid and caused to oscillate by turning the disk 1.00 rad and releasing it, the amplitude of the oscillation is reduced to 0.20 rad after five complete cycles.

Figure P12-55

(a) Determine the damping coefficient of the disk and fluid.

(b) What would be the value of c if it were assumed that the frequency did not change due to damping?

12-56 Body m in Fig. P12-56 has a mass of 0.025 lb-sec^2 per in., and the weight of the bar can be neglected. During 20 cycles of free vibration of the system, the amplitude of the motion of D decreases from 3.0 to 1.0 in. in 10 sec. Determine

(a) The damping coefficient c.

(b) The spring stiffness k.

(c) The critical damping coefficient c_{cr}.

(d) The logarithmic decrement of the system.

Figure P12-56

12-6 ROTATING VECTORS

Rotating vectors, as indicated in Art. 12-2, can be used to provide a graphical interpretation of some of the properties of simple harmonic motion.

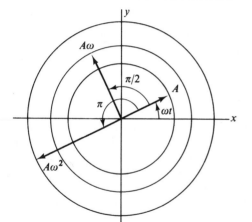

Figure 12-16

A simple harmonic motion can be described by the equation

$$y = A \sin \omega t. \qquad (a)$$

If a vector of length A rotates about a fixed point with a constant angular velocity ω as shown in Fig. 12-16, its projection on the y axis is given by Eq. (a). This figure assumes that t was zero when the vector was horizontal.

Differentiation of Eq. (a) with respect to time gives

$$\dot{y} = A\omega \cos \omega t = A\omega \sin\left(\omega t + \frac{\pi}{2}\right)$$

and

$$\ddot{y} = -A\omega^2 \sin \omega t = A\omega^2 \sin(\omega t + \pi).$$

These expressions are similar to Eq. (a) except that the amplitude is multiplied by ω for $\dot{y}$ and by ω^2 for $\ddot{y}$ and the angle ωt is increased by $\pi/2$ for $\dot{y}$ and by π for $\ddot{y}$. Rotating vectors to represent $\dot{y}$ and $\ddot{y}$ are included in Fig. 12-16. The angles $\pi/2$ and π are called *phase angles*. All three vectors rotate with the same angular velocity ω, and the velocity vector is said to *lead* the displacement vector by a phase angle of $\pi/2$ rad. Note that the velocity attains its maximum value one-fourth of a cycle *before* the displacement becomes maximum. Similarly, the acceleration leads the velocity by a phase angle of $\pi/2$ rad and the displacement by π rad.

Figure 12-17 shows the displacement, velocity, and acceleration

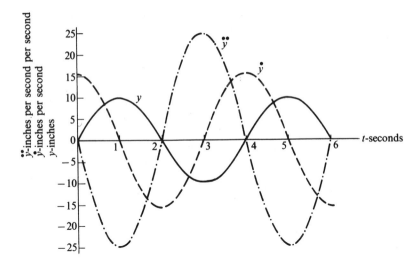

Figure 12-17

of the harmonic motion just discussed when plotted as functions of time. In this figure, A is 10 in. and ω is $\pi/2$ rad per sec. Note that y is maximum at $t = 1$ sec, while $\dot{y}$ is maximum at $t = 0$, and the acceleration passed through its maximum value 1 sec earlier.

Two harmonic motions with the same frequency but with different phase angles can be combined by vector addition to give the resultant motion. Consider two motions given by

$$y_1 = A_1 \sin(\omega t + \phi_1), \qquad y_2 = A_2 \sin(\omega t + \phi_2).$$

The sum of these motions is

$$
\begin{aligned}
y_1 + y_2 &= A_1 \sin(\omega t + \phi_1) + A_2 \sin(\omega t + \phi_2) \\
&= A_1(\sin \omega t \cos \phi_1 + \cos \omega t \sin \phi_1) \\
&\quad + A_2(\sin \omega t \cos \phi_2 + \cos \omega t \sin \phi_2) \\
&= (A_1 \cos \phi_1 + A_2 \cos \phi_2)\sin \omega t + (A_1 \sin \phi_1 + A_2 \sin \phi_2)\cos \omega t.
\end{aligned}
$$

The last expression can also be written as

$$y_1 + y_2 = A \sin(\omega t + \phi),$$

where

$$A = [(A_1 \cos \phi_1 + A_2 \cos \phi_2)^2 + (A_1 \sin \phi_1 + A_2 \sin \phi_2)^2]^{1/2} \qquad \text{(b)}$$

and

$$\tan \phi = \frac{A_1 \sin \phi_1 + A_2 \sin \phi_2}{A_1 \cos \phi_1 + A_2 \cos \phi_2}. \qquad \text{(c)}$$

Figure 12-18 shows the rotating vectors $\mathbf{A}_1$ and $\mathbf{A}_2$ together with their resultant $\mathbf{A}$. The sum of the projections of $\mathbf{A}_1$ and $\mathbf{A}_2$ on line OC is $A_1 \cos \phi_1 + A_2 \cos \phi_2$, and the sum of their projections perpendicular

761

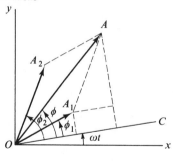

Figure 12-18

to OC is $A_1 \sin \phi_1 + A_2 \sin \phi_2$. Thus the magnitude of vector **A** is

$$A = [(A_1 \cos \phi_1 + A_2 \cos \phi_2)^2 + (A_1 \sin \phi_1 + A_2 \sin \phi_2)^2]^{1/2},$$

which is the same as Eq. (b). From a similar analysis, the angle ϕ in Fig. 12-18 is given by Eq. (c). Therefore, any number of harmonic motions with the same frequency and with various amplitudes and phase angles can be added or subtracted either algebraically or by vector addition of the rotating vectors which represent the motions.

12-7 FORCED VIBRATIONS

When a force which varies periodically is applied to a body mounted on springs or other elastic supports a forced vibration of the body will result. A forced vibration of a body can also be produced by giving the body supporting the vibrating system a periodic motion. For example, if a weight is hung from a spring and the upper end of the spring is moved vertically with a periodic motion, forced vibrations of the weight will be produced. In the last example, the variable force is transmitted through the spring to the weight.

A body subjected to a periodic force and arbitrary initial conditions will have a combination of free and forced vibrations at the start of the motion. In all practical cases, however, the damping forces will eliminate the free vibrations (often called *transients*), and the remaining motion is called the *steady-state* vibration.

The period and frequency of free vibrations depend on the mass of the body, on the stiffness of the elastic support, and on the damping coefficient. The amplitude of free vibrations depends on the starting conditions and in general on the circular frequency. The frequency of steady-state forced vibrations depends on the frequency of the applied load but not on the characteristics of the vibrating body. The amplitude of the steady-state forced vibrations depends on the magnitude and frequency of the applied load and on the frequency of the free vibrations but not on the starting conditions.

Any periodically varying applied force will result in forced vibrations. A common type of variable force is one which can be expressed as a sine or cosine function of time.

Figure 12-19 represents a body of mass m, supported by a spring whose modulus is k and a dashpot whose coefficient is c. The body is acted on by the force $F = F_O \sin \omega t$, where F_O is the

Figure 12-19

amplitude of the variable force and ω is its circular frequency. A free-body diagram of the body displaced a distance y from the equilibrium position is shown in Fig. 12-19(b). The spring force is equal to $W + ky$, where W is the weight of the body and is also the spring force when the body is in equilibrium and the force F is zero. The differential equation of motion, in scalar form, is

$$\sum F_y = m\ddot{y},$$

$$W - (W + ky) - c\dot{y} + F_O \sin \omega t = m\ddot{y}$$

or

$$m\ddot{y} + c\dot{y} + ky = F_O \sin \omega t. \qquad (12\text{-}11)$$

The general solution to Eq. (12-11) is made up of two parts: the complementary function when the right-hand side is zero (see Art. 12-5) and the particular solution which satisfies the equation as written. The complementary function decreases exponentially with time and contains two constants of integration. It is often called the *transient* term. The particular solution is the *steady-state* solution and is usually the one of primary interest. The transient solution has already been discussed in Art. 12-5 and will not be considered further.

The steady-state solution can be obtained by assuming y to be of the form

$$y = A \sin(\omega t + \phi) = B \sin \omega t + C \cos \omega t, \qquad (12\text{-}12)$$

substituting it into Eq. (12-11), and solving for either A and ϕ or B and C. A convenient graphical representation of Eq. (12-11) can be obtained by means of rotating vectors. If the steady-state solution is assumed to be $y = A \sin(\omega t + \phi)$, the various terms of Eq. (12-11) are

$$m\ddot{y} = -Am\omega^2 \sin(\omega t + \phi) = Am\omega^2 \sin(\omega t + \phi + \pi),$$

$$c\dot{y} = Ac\omega \cos(\omega t + \phi) = Ac\omega \sin\left(\omega t + \phi + \frac{\pi}{2}\right),$$

and

$$ky = Ak \sin(\omega t + \phi).$$

Notice that the quantity $m\ddot{y}$ leads $\mathbf{y}$ by π rad and $c\dot{y}$ leads $\mathbf{y}$ by $\pi/2$ rad. The vector sum of these three rotating vectors must be equal to $F_O \sin \omega t$ from Eq. (12-11). Figure 12-20(a) shows three vectors with magnitudes $Am\omega^2$, $Ac\omega$, and Ak, which, when added vectorially, must equal F_O. The vector $\mathbf{A}$ is in the direction of ky (opposite $m\ddot{y}$) and lags the force $\mathbf{F}$ by the angle ϕ. In this case, ϕ will be negative.

Since the amplitude of the resulting motion occurs in three of the rotating vectors, it is often convenient to divide all amplitudes by A to get Fig. 12-20(b). Note that if $k = m\omega^2$, $(\omega = \omega_n = \sqrt{k/m})$, the applied force will lead the displacement by $\pi/2$ rad. This case is called *resonance*. If c is zero (no damping), the vector polygon becomes a straight line, and ϕ is either zero or π rad, depending on whether k is more or less than $m\omega^2$. When there is no damping and $k = m\omega^2$, the

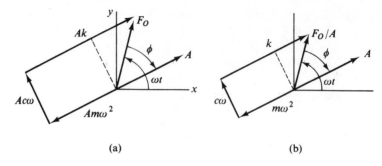

(a)   (b)

Figure 12-20

steady-state amplitude becomes infinite. The rate at which the amplitude will increase can be determined by obtaining the particular solution to Eq. (12-11) subject to the conditions $c = 0$ and $k = m\omega^2$. The solution will be of the form

$$y = A_0 t \cos \omega t.$$

This solution can be demonstrated by direct substitution. The buildup of the amplitude may be quite rapid, and unless stops to prevent excessive motion are provided, damage to the equipment will result.

Solutions for the amplitude and phase angle, satisfying Eq. (12-11), can be determined from Fig. 12-20. Thus,

$$\left(\frac{F_0}{A}\right)^2 = (k - m\omega^2)^2 + (c\omega)^2$$

or

$$A = \frac{F_0}{\sqrt{(k - m\omega^2)^2 + (c\omega)^2}}.$$ (a)

Similarly,

$$\tan \phi = \frac{c\omega}{k - m\omega^2}.$$ (b)

When other than rectilinear motion is involved, a differential equation similar to Eq. (12-11) must be derived, and the same analysis can be used to obtain the linear or angular quantities used to derive the equation. When the numerator and denominator of Eq. (a) are both divided by k and when c is replaced by $\zeta c_{cr} = \zeta 2\sqrt{mk}$ and $\sqrt{k/m}$ is replaced by ω_n Eq. (a) becomes

$$A = \frac{F_0/k}{[(1 - \omega^2/\omega_n^2)^2 + (2\zeta\omega/\omega_n)^2]^{1/2}}.$$ (a')

The ratio of A to F_0/k is called the *magnification factor*. It is the ratio by which the quantity F_0/k, the amount the spring would stretch due to a static load F_0, must be multiplied to give A. The curves in Fig. 12-21 show the variation of the magnification factor with ω/ω_n for various values of ζ.

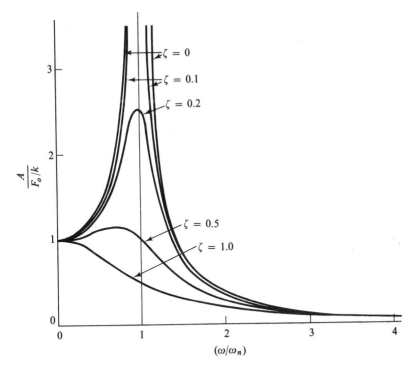

Figure 12-21

A differential equation of motion similar to Eq. (12-11) will result from the rotation of an unbalanced mass or from the harmonic oscillation of a body supporting the spring-mass system as shown in the following examples.

EXAMPLE 12-5

The motor in Fig. 12-22(a) is mounted on two springs, each with a modulus of $k/2 = 60$ lb per in. The dashpot has a coefficient of 0.80 lb-sec per in. The motor, including the unbalanced mass B, weighs 38.6 lb, and the unbalanced body B weighs 1.0 lb and is located 3.0 in. from the center of the shaft.
(a) The motor rotates at 300 rpm. Determine the amplitude and phase angle (relative to the position of B) of the resulting motion.
(b) Determine the maximum and minimum force on the motor exerted by the springs, by the damper, and by the springs and damper together when the motor rotates at 300 rpm.
(c) Determine the resonant speed and the resulting amplitude of the motion.

SOLUTION

(a) Figure 12-22(b) is a free-body diagram of the motor without the weight B, and Fig. 12-22(c) is a free-body diagram of B. The variable y gives the position of the center of the motor shaft measured from the

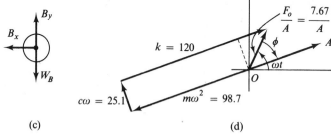

Figure 12-22

equilibrium position, and y_B is the vertical coordinate of B. From the figure,

$$y_B = y + e \sin \omega t;$$

therefore,

$$\ddot{y}_B = \ddot{y} - e\omega^2 \sin \omega t.$$

The equations of motion in the y direction for the two bodies, in scalar form, are

$$(m - m_B)\ddot{y} = (W - W_B) + B_y - (W + ky) - c\dot{y}$$

and

$$m_B(\ddot{y}_B) = m_B(\ddot{y} - e\omega^2 \sin \omega t) = W_B - B_y.$$

When these two equations are added to eliminate B_y, the result is

$$m\ddot{y} - m_B e\omega^2 \sin \omega t = -ky - c\dot{y}$$

or

$$m\ddot{y} + c\dot{y} + ky = m_B e\omega^2 \sin \omega t = F_O \sin \omega t,$$

where

$$F_O = m_B e\omega^2.$$

The equation is the same as Eq. (12-11) and can be solved by means of a vector diagram similar to Fig. 12-20(b). From the data,

$$\omega = \frac{300}{60}(2\pi) = 31.4 \text{ rad per sec,}$$

$$m\omega^2 = \frac{38.6}{386}(31.4)^2 = 98.7 \text{ lb per in.,}$$

$$c\omega = 0.8(31.4) = 25.1 \text{ lb per in.,}$$

and

$$F_O = \frac{1.0}{386}(3.0)(31.4)^2 = 7.67 \text{ lb.}$$

These values are shown on Fig. 12-22(d), and from the right triangle,

$$\left(\frac{7.67}{A}\right)^2 = (120 - 98.7)^2 + (25.1)^2,$$

which gives

$$A = \underline{0.233 \text{ in.}} \qquad \text{Ans.}$$

Also from the diagram,

$$\tan\phi = \frac{25.1}{21.3}$$

from which

$$\phi = 49.7° \text{ (y lags behind } y_B). \qquad \text{Ans.}$$

(b) The magnitude of the force exerted by the springs is

$$F_{sp} = W + ky = 38.6 + 120y,$$

and the maximum and minimum values are

$$(F_{sp})_{max} = 38.6 + 120(0.233) = \underline{66.56 \text{ lb,}} \qquad \text{Ans.}$$

$$(F_{sp})_{min} = 38.6 + 120(-0.233) = \underline{10.64 \text{ lb.}} \qquad \text{Ans.}$$

The magnitude of the force exerted by the damper is

$$F_d = c\dot{y} = 0.8\dot{y},$$

and the maximum (and minimum) value of F_d is

$$(F_d)_{max} = 0.8\dot{y}_{max} = 0.8A\omega = 0.8(0.233)(31.4)$$

$$= \underline{5.85 \text{ lb.}} \qquad \text{Ans.}$$

The variable part of the spring force and the damper force are 90° out of phase, as shown by the vector diagram; therefore, the amplitude of the resultant variable force is

$$F_v = [(27.96)^2 + 5.85^2]^{1/2} = 28.6 \text{ lb,}$$

and the maximum and minimum resultant force of the spring and damper on the motor is

$$F_{max,min} = 38.6 \pm 28.6 = \underline{67.2 \text{ lb}} \quad \text{and} \quad \underline{10.0 \text{ lb.}} \qquad \text{Ans.}$$

(c) The resonant speed occurs when $m\omega^2 = k$ or when

$$\omega = \omega_n = \sqrt{\frac{k}{m}} = \sqrt{\frac{120}{0.1}} = 34.64 \text{ rad per sec}$$

$$= \frac{34.64}{2\pi}(60) = \underline{331 \text{ rpm}}. \qquad \text{Ans.}$$

Since

$$F_0 = m_B e\omega^2,$$

the amplitude at resonance is

$$A = \frac{F_0}{c\omega} = \frac{m_B e\omega}{c} = \frac{1.0(3.0)(34.6)}{(386)(0.8)} = \underline{0.337 \text{ in.}} \qquad \text{Ans.}$$

EXAMPLE 12-6

Body D in Fig. 12-23(a) weighs 19.3 lb and is supported by a spring with a modulus of 60 lb per in. Body B at the upper end of the spring is given a vertical harmonic motion by the crank, which has an angular velocity of 40 rad per sec as shown. The length of the crank is 0.50 in.
(a) Determine the amplitude and phase angle of the motion of D when the damping coefficient of the damper is 0.60 lb-sec per in. and when the damper is disconnected.
(b) Determine the range of values of ω (if any) which will limit the amplitude of the motion of D to 0.80 in. when $c = 0$.

SOLUTION

(a) A free-body diagram of body D is shown in Fig. 12-23(b). The downward direction is positive. Note that the tension in the spring is equal to the weight of D plus the spring modulus multiplied by the

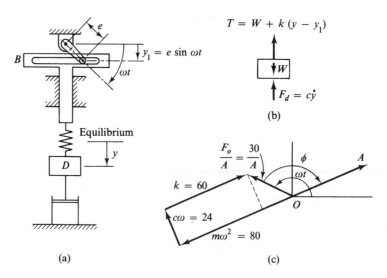

(a)

(b)

(c)

Figure 12-23

amount the spring is stretched. The differential equation of motion in the y direction is

$$m\ddot{y} = W - c\dot{y} - [W + k(y - y_1)]$$

or

$$m\ddot{y} + c\dot{y} + ky = ky_1 = ke \sin \omega t.$$

This is the same as Eq. (12-11) if ke is set equal to F_0.

The steady-state solution is assumed to be of the form

$$y = A \sin(\omega t + \phi),$$

and the vector diagram shown in Fig. 12-23(c) can be drawn when numerical values are substituted for m, c, k, e, and ω. The numerical values for the diagram are

$$m\omega^2 = \frac{19.3}{386}(40)^2 = 80 \text{ lb per in.,}$$

$$c\omega = 0.60(40) = 24 \text{ lb per in.,}$$

$$k = 60 \text{ lb per in.,}$$

and

$$F_0 = ek = 0.50(60) = 30 \text{ lb.}$$

From the diagram,

$$\left(\frac{30}{A}\right)^2 = (24)^2 + (80 - 60)^2,$$

which gives

$$A = \underline{0.960 \text{ in.}} \qquad \text{Ans.}$$

The phase angle is

$$\phi = \tan^{-1}\frac{24}{-20} = \underline{129.8°}. \qquad \text{Ans.}$$

This angle indicates that D will reach its lowest position 129.8° after B is at its lowest position. This angle corresponds to a time interval of

$$t = \frac{\phi}{\omega} = \frac{129.8\pi}{180(40)} = 0.0566 \text{ sec.}$$

When c is equal to zero, the polygon in Fig. 12-23(c) is a straight line, and since k is less than $m\omega^2$, y will lag y_1 by 180°. The amplitude of y is obtained from

$$\frac{30}{A} = 80 - 60 = 20$$

or

$$A = \underline{1.500 \text{ in.}} \qquad \text{Ans.}$$

(b) When there is no damping, the value of A will be less than 0.80 in. if the difference between $m\omega^2$ and k is greater than $F_0/A = 30/0.80 = 37.5$. In this case, two possible solutions exist, one with y in phase with y_1 and $k > m\omega^2$, and a second solution with y out of phase with y_1 and $k < m\omega^2$. The first solution will not exist for small amplitudes of A. In this example,

$$60 - 0.05\omega_1^2 = 37.5 \quad \text{and} \quad \omega_1 = 21.2 \text{ rad per sec}$$

or

$$0.05\omega_2^2 - 60 = 37.5 \quad \text{and} \quad \omega_2 = 44.2 \text{ rad per sec.}$$

Thus the amplitude of the motion of D will be less than 0.80 in. if ω is less than 21.2 rad per sec or greater than 44.2 rad per sec.

PROBLEMS

12-57 Body W in Fig. P12-57 weighs 8.05 lb, the modulus k_1 is 20 lb per ft, and k_2 is 16 lb per ft. The displacement of E is given by the expression $y_E = 4 \cos 2t$ where y_E is in feet and t is in seconds. Determine the amplitude of the motion of W.

Figure P12-57

12-58 Body W in Fig. P12-58 weighs 60 lb, and the spring moduli are 5 lb per in. for S_1 and 2 lb per in. for S_2. The force F in pounds is given by the expression $F = 2 \cos 2t$, where t is in seconds.
(a) Derive the steady-state solution for the motion of W.
(b) Determine the maximum velocity of W.

Figure P12-58

12-59 A 50-lb ball is supported as shown in Fig. P12-59. The mass of the bar can be neglected, and the spring modulus is 2 lb per in. Roller E has a harmonic motion given by the expression $x_E = 0.5 \cos 7t$, where x_E is in inches and t is in seconds. Obtain the steady-state solution for the motion of B.

Figure P12-59

12-60 Determine the components of the reaction at O in Problem 12-59 when
(a) Arm OB is horizontal.
(b) Body B is in its lowest position.

12-61 Block E in Fig. P12-61 has a harmonic motion given by the equation $y_E = 0.500 \sin 5t$(y_E is in feet). The modulus of S_1 is 8.0 lb per ft and the modulus of S_2 is 15.0 lb per ft. The weight of the bars supporting the 32.2-lb weight can be neglected. Determine the steady-state solution for the vibrating system. The springs can act in either tension or compression.

Figure P12-61

12-62 An instrument panel which weighs 20 lb is to be mounted on a machine which vibrates with a frequency of 1200 rpm and an amplitude of 0.25 in. It is necessary to reduce the amplitude of the vibration of the instruments to not more than 0.10 in., and it is proposed to use a spring mounting to accomplish this reduction.

Determine

(a) The range or ranges of spring stiffness which will be satisfactory.

(b) The amplitude of the *variable* part of the force transmitted by the spring when the amplitude of motion of the panel is 0.10 in.

12-63 The curves in Fig. 12-21 indicate the amplitude of motion of a body with forced vibration if the force causing the vibration has a constant amplitude. With an unbalanced rotor as shown in Fig. 12-22, the amplitude of the force producing the vibrations is not constant. Draw a curve, similar to Fig. 12-21, showing the variation of the amplitude of the motion of the body with the speed of the motor when $c = 0$.

12-64 A small variable-speed motor weighing 18.0 lb is mounted on an elastic beam as indicated in Fig. P12-64. The motor rotates with a 2.0-lb eccentric weight 2.0 in. from the center of the shaft. When the motor is not running, the motor and the eccentric weight cause the beam to deflect 0.50 in. Determine

(a) The speed of the system at resonance.

(b) The amplitude of the forced vibrations when the motor is running at 300 rpm.

Figure P12-64

12-65 Would it be possible to reduce the amplitude of the forced vibration of the motor in part (b), Problem 12-64, by fastening an additional weight to the motor? If so, what weight should be added to reduce the amplitude of the vibration to 0.500 in.?

*12-66 Block B in Fig. P12-66 is connected to body A by a viscous dashpot. The mass of B is 2.0 kg, and the coefficient of the dashpot is

Figure P12-66

40 N · s/m. Body A has simple harmonic motion given by the equation $x_1 = e \sin \omega t$, where e is 50 mm and ω is 20 rad/s. Determine the amplitude of the steady-state motion of B and the phase angle by which B leads or lags A.

12-67 Body A in Fig. P12-67 weighs 5.0 lb, and the viscous damper has a coefficient of 0.10 lb-sec per in. The magnitude of the force F is given by the equation $F = F_0 \sin \omega t$, where F_0 is 2.0 lb and ω is 10 rad per sec. Determine the amplitude of the motion of A and the maximum force in member B.

Figure P12-67

12-68 A block which weighs 60 lb is suspended by a spring with a modulus of 48 lb per in. and is subjected to a shaking force $F = 2 \sin 30t$ (F is in pounds and t is in seconds). The block is connected to the floor through a dashpot with a coefficient of 4.0 lb-sec per in. Determine the amplitude of the force transmitted through the dashpot.

12-69 Block A in Fig. P12-69 has simple harmonic motion with an amplitude e and a circular frequency ω. The frequency of the motion is less than the resonant frequency. Body B has a mass m, and the moduli of the spring and damper are k and c, respectively. Determine the amplitude of the motion of B and indicate whether it leads or lags the motion of A.

Figure P12-69

12-70 A body weighing 30 lb is mounted on four springs (in parallel) each having a modulus of 2.0 lb per in. The body is acted on by a force with an amplitude of 3.0 lb and a circular frequency of 10 rad per sec. This force produces

excessive vibrations, and three possible remedies are suggested:
(a) Add a weight of 15 lb to the body.
(b) Add a fifth spring with the same modulus.
(c) Place a dashpot with a damping coefficient of 0.90 lb-sec per in. between the body and foundation.
Determine the effect of each of these proposed remedies, applied separately, on the amplitude of the resulting vibrations and on the variable portion of the force transmitted to the foundation.

12-71 Body A in Fig. P12-71 weighs 193 lb, the damping coefficient c is 5.0 lb-sec per in., and the spring modulus k is 40 lb per in. Body B has harmonic motion due to the uniform rotation of the crank. The crank arm e is 2.0 in., and it rotates at 10 rad per sec.
(a) Determine the amplitude of the motion of A and its phase relationship with respect to the motion of B.
(b) (1) Would a small increase in k increase or decrease the amplitude of the motion of A?
(2) Would a small increase in c increase or decrease the amplitude of the motion of A?

Figure P12-71

*12-72** The 10-kg body in Fig. P12-72 is mounted on springs with a modulus of 7000 N/m. The force $F = 40 \sin 30t$ (F in newtons and t in seconds) acts on the body as shown. A dashpot is to be placed between the body and foundation to limit the amplitude of the motion to 1.20 cm. Determine
(a) The coefficient of the dashpot.
(b) The amplitude of the variable portion of the force transmitted to the foundation.

12-73 Solve part (b), Example 12-6, for $A_{max} = 0.80$ in. and $c = 0.60$ lb-sec per in.

12-74 Solve part (b), Example 12-6, for $A_{max} = 0.30$ in. and $c = 0$.

Figure P12-72

12-75 The system in Fig. P12-75 is adjusted to be in equilibrium when AB is horizontal and x_E is zero. Body B weighs 50 lb, the spring modulus is 6.0 lb per in., and the damping coefficient is 1.5 lb-sec per in. The position of block E (in inches) is given by the expression $x_E = 5.0 \sin 5t$. Determine the amplitude of the motion of B and its maximum velocity.

Figure P12-75

*12-76** The 30-kg body A of Fig. P12-76 is connected to two springs, each with a modulus of 350 N/m, and a dashpot with a coefficient of 120 N·s/m. The force P (in newtons) acting on A is given by the equation $P = 20.0 \cos 3t$, where t is in seconds. Obtain the steady-state solution for the motion of A and determine the maximum force transmitted to the wall by the springs and dashpot.

Figure P12-76

12-77 The 386-lb air compressor of Fig. P12-77 is supported by four spring-dashpot mounts, each having a spring modulus of 250 lb per in. and a damping coefficient of 5.0 lb-sec per in. The piston of the compressor weighs 3.86 lb, has a 4.00-in. stroke, and moves in a vertical cylinder. The crankshaft rotates at 40 rad per sec. Assume that the motion of the piston *with respect to the cylinder* can be approximated by

$$y_{P/C} = A \sin \omega t.$$

Neglect the weights of the crank and connecting rod. Determine the amplitude of the vertical vibrations of the compressor.

12-78 The two disks in Fig. P12-78 are coupled by two springs, each of modulus k, and a damper, coefficient c. The springs are always under tension, and the cable does not slip on either wheel. Each disk has a moment of inertia with respect to its axis of rotation of I. A variable torque causes disk A to have simple harmonic rotation with an amplitude of θ_0 and a circular frequency of ω.

(a) Determine the value of ω for resonance.

(b) Determine the range of values of ω, if any, for which an increase in c will result in a decrease in the amplitude of disk B.

Figure P12-77

Figure P12-78

Appendix A

International System of Units

The International System of Units, commonly called SI units, was formally adopted in 1960 by the International Organization for Standardization (ISO). The ISO is composed of representatives of all the major industrial countries of the world. SI is a modification of the metric system of units and is under continuous study for necessary refinement and additions.

The seven fundamental or base units defined in this system are length, mass, time, thermodynamic temperature, electric current, luminous intensity, and amount of substance. Mass, length, time, and sometimes temperature are the quantities commonly encountered in engineering mechanics. Since mass instead of force is a base unit or quantity, SI is an absolute system of units in contrast to the conventional gravitational system used by engineers in the United States in which force is a fundamental quantity. Quantities involving units other than those selected as base units are formed as multiples of base units and sometimes given special names. Table A-1 lists the base units and SI symbols together with selected derived units.

TABLE A-1
SI UNITS AND SYMBOLS

	Physical Quantity	*Name of Unit*	*SI Symbol*
Base unit	Length	meter	m
	Mass	kilogram	kg
	Time	second	s
Derived unit	Area	square meter	m^2
	Volume	cubic meter	m^3
	Force	newton	$N, kg \cdot m/s^2$
	Moment of a force	newton $\cdot$ meter	$N \cdot m$
	Work	joule, newton $\cdot$ meter	$J, N \cdot m$
	Power	watt, joule per second	$W, J/s, N \cdot m/s$
	Linear velocity	meter per second	m/s
	Linear acceleration	meter per second squared	m/s^2
	Density, mass per unit volume	kilogram per cubic meter	kg/m^3
	Frequency	hertz, cycles per second	Hz

Prefixes, as listed in Table A-2, are used to describe quantities which are much larger or smaller than the base units or derived unit in order to use numbers which are more easily comprehended. Thus a length of 120 000 meters would be stated as 120 km, and a length of 25 500 000 meters would be 25.5 Mm. It is recommended that prefixes be selected which will result in numerical values between 0.1 and 1000. Use of prefixes hecto, deca, deci, and centi is discouraged unless the recommended prefix results in inconvenient numbers. For example, the volume of a cube which is 35 mm on a side would be either 42 875 mm^3 or 0.000 042 875 m^3. In such a situation the volume

TABLE A-2
SI PREFIXES

Multiplication Factor	*Prefix*	
	Name	*Symbol*
10^{12}	tera	T
10^9	giga	G
10^6	mega	M
10^3	kilo	k
10^2	hecto	h
10	deka	da
10^{-1}	deci	d
10^{-2}	centi	c
10^{-3}	milli	m
10^{-6}	micro	μ
10^{-9}	nano	n
10^{-12}	pico	p
10^{-15}	femto	f
10^{-18}	atto	a

should be expressed as 42.9 cm³. Errors in calculations involving prefixes can be avoided more easily if all quantities are expressed in SI units by replacing prefixes by powers of 10, e.g., 25.5 Mm = $25.5(10^6)$m = $2.55(10^7)$m.

The use of SI units is becoming more commonplace in the United States, but a complete changeover from English units will probably not take place for many years. During the transition period engineers and scientists will need to be familiar with both systems and will frequently find it necessary to convert quantities from one system to the other. Table A-3 contains selected factors useful in engineering mechanics.

TABLE A-3
CONVERSION FACTORS

	English to SI	*SI to English*
Length	1 in. = 25.40 mm	1 mm = 0.03937 in.
	1 ft = 0.3048 m	1 m = 39.37 in.
		= 3.281 ft
	1 mile = 1.609 km	1 km = 0.6214 mile
Mass	1 slug (1 lb-sec²/ft) = 14.59 kg	1 kg = 0.06853 slugs
		= 0.06853 lb-sec²/ft
Area	1 in.² = 6.452 cm²	1 cm² = 0.1550 in.²
	1 ft² = 0.09290 m²	1 m² = 10.76 ft²
Volume	1 in.³ = 16.39 cm³	1 cm³ = 0.06102 in.³
	1 ft³ = 0.02832 m³	1 m³ = 35.31 ft³
Force	1 lb = 4.448 N	1 N = 0.2248 lb
	= 4.448 kg · m/s²	
Moment of force	1 in-lb = 0.1130 N · m	1 N · m = 0.7376 ft-lb
	1 ft-lb = 1.356 N · m	= 8.851 in-lb
Work or energy	1 ft-lb = 1.356 N · m	1 J = 0.7376 ft-lb
	= 1.356 J	
Power	1 ft-lb/sec = 1.356 W	1 W = 0.7376 ft-lb/sec
	1 hp = 745.7 W	1 kW = 1.341 hp
Velocity	1 in/sec = 25.40 mm/s	1 m/s = 3.281 ft/sec
	1 ft/sec = 0.3048 m/s	1 km/hr = 0.6214 mph
	1 mile/hr = 0.4470 m/s	= 0.9113 ft/sec
	= 1.609 km/hr	
Acceleration	1 in/sec² = 25.40 mm/s²	1 m/s² = 3.281 ft/sec²
	1 ft/sec² = 0.3048 m/s²	
Density	1 slug/ft³ = 515.3 kg/m³	1 Mg/m³ = 1.941 slug/ft³
	1 lb-sec/ft⁴ = 515.3 kg/m³	
Spring modulus	1 lb/in. = 175.1 N/m	1 N/m = 0.06852 lb/ft
	1 lb/ft = 14.59 N/m	1 kN/m = 5.710 lb/in.
Second moment of mass	1 slug-ft² = 1.355 kg · m²	1 kg · m² = 0.7378 slug-ft²
	1 lb-sec²-ft = 1.355 kg · m²	
Load intensity	1 lb/in.² = 6.895 kN/m²	1 kN/m² = 0.1450 lb/in.²
	1 lb/ft² = 47.88 N/m²	1 kN/m² = 20.89 lb/ft²

Moments of Inertia of Masses

B-1 DEFINITIONS

In the analysis of the motion of rigid bodies, the moment of inertia (and sometimes the product of inertia) of the mass of the body with respect to an axis (or a pair of planes) is frequently required.

The moment of inertia or second moment of an element of mass dm (see Fig. B-1) *with respect to any axis or line is defined as the product of the mass of the element and the square of the distance from the axis to the element.* Thus the moment of inertia of the element of mass with respect to the x axis is

$$dI_x = r^2 \, dm = (y^2 + z^2) \, dm.$$

---

*This Appendix is PART B of Chapter 6 in *Engineering Mechanics*, 2nd Vector Edition; Vol. I: Statics.

Similar expressions can be written for the y and z axes. The quantity $y^2\,dm$ is sometimes referred to as the *second moment* of the mass of the element with respect to the xz plane. Similarly, $z^2\,dm$ is called the second moment of the mass of the element with respect to the xy plane. These quantities are useful primarily as part of the expression for dI_x. Since they do not arise individually in physical equations, they are seldom computed separately.

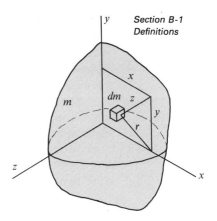

Figure B-1

The product of inertia of an element of mass dm (as in Fig. B-1) with respect to a pair of orthogonal coordinate planes is defined as the product of the mass of the element and the coordinate distances from the planes to the element. For example, the product of inertia of the element with respect to the xz and yz planes is

$$dI_{xy} = xy\,dm.$$

The sum of the moments (or products) of inertia of all elements of mass of a body with respect to any axis (or pair of planes) is defined as the moment (or product) of inertia of the body with respect to the indicated reference. The expressions for the axes of Fig. B-1 are

$$I_x = \int (y^2 + z^2)\,dm, \qquad I_y = \int (z^2 + x^2)\,dm, \qquad I_z = \int (x^2 + y^2)\,dm$$

and
$$(B\text{-}1)$$

$$I_{xy} = \int xy\,dm, \qquad I_{yz} = \int yz\,dm, \qquad I_{zx} = \int zx\,dm.$$

The equation for I_x may also be written as

$$I_x = \int y^2\,dm + \int z^2\,dm = I_{xz\text{ plane}} + I_{xy\text{ plane}}, \qquad\qquad (a)$$

and similar relationships exist for I_y and I_z. It is important to note the difference between the moment of inertia of mass with respect to the xy plane, $I_{xy\text{ plane}}$, and the product of inertia I_{xy}. Equation (a) can be stated as follows: *The moment of inertia of the mass of a body with respect to any axis is equal to the sum of its moments of inertia with respect to two perpendicular planes which intersect along the axis.* The moments of inertia with respect to planes are useful only in that they sometimes simplify the determination of the moments of inertia of bodies with respect to the line of intersection of two planes.

Moments and products of inertia of mass have the dimensions of mass multiplied by length squared, mL^2. When force, length, and time are selected as the fundamental quantities, mass is FT^2L^{-1} (see Art. 1-8), and the second moment of mass has dimensions of $(FT^2L^{-1})L^2$ or

v

FT^2L. Common units for moment of inertia are lb-sec^2-ft, kg·m^2, or other combinations having the same dimensions. If the mass of the body, W/g, is expressed in slugs (lb-sec^2/ft), it is usually desirable to express the moment of inertia in slug-ft^2 instead of slug-in.^{2} to avoid a combination of feet and inches.

The moment of inertia of mass is always a positive quantity, since it is the sum of the products of elements of mass, which are inherently positive, and distances squared. The product of inertia of a body may be positive, negative, or zero, since the two coordinates have independent signs, and the product will be positive for coordinates with the same sign and negative when they have opposite signs.

The product of inertia of the mass of a body with respect to two perpendicular planes is zero if either of the planes is a plane of symmetry. This is true because the elements will always occur in pairs on opposite sides of the plane of symmetry one of which will have a positive and one a negative product of inertia. Any plane of symmetry will, of course, pass through the mass center of the body, and the perpendicular plane used in computing the product of inertia may or may not pass through the mass center. It should be noted, however, that planes through the mass center are not necessarily planes of symmetry and that the product of inertia of the body with respect to such planes will not necessarily be zero.

B-2 PARALLEL-AXIS AND PARALLEL-PLANE THEOREMS FOR MASSES

The moment of inertia of the mass of a body with respect to any axis can be expressed in terms of the moment of inertia of mass of the body with respect to a parallel axis through the mass center as indicated in the following development. In Fig. B-2, the x', y', and z' axes are parallel to the x, y, and z axes and pass through the mass center, G, of the body. The moment of inertia of the element of mass, dm, with respect to the z axis is

$$dI_z = (x^2 + y^2)\, dm$$
$$= [(x_G + x')^2 + (y_G + y')^2]\, dm$$
$$= [(x_G^2 + y_G^2) + (x'^2 + y'^2) + 2x_Gx' + 2y_Gy']\, dm$$

and

$$I_z = (x_G^2 + y_G^2)\int dm + \int (x'^2 + y'^2)\, dm + 2x_G\int x'\, dm + 2y_G\int y'\, dm.$$

From Fig. B-2, $(x_G^2 + y_G^2)$ is equal to r_G^2. Also, $\int (x'^2 + y'^2)\, dm$ is the moment of inertia of the mass of the body with respect to the z' (centroidal) axis, and the two expressions $\int x'\, dm$ and $\int y'\, dm$ are both equal to zero, since they represent the first moment of the mass with

Figure B-2

respect to the centroidal planes parallel to the yz and xz planes, respectively. Thus the equation can be written

$$I_z = I_{z'} + mr_G^2 = I_G + mr_G^2 \qquad \text{(B-2)}$$

where I_G is the moment of inertia of the mass of the body with respect to an axis through the mass center parallel to the z axis. In general, *the moment of inertia of the mass of a body with respect to any axis is equal to the sum of the moment of inertia of the mass with respect to a parallel axis through the mass center and the product of the mass and the square of the distance between the two axes.*

A similar theorem can be proved for the moment of inertia of the mass of a body with respect to a *plane.* The theorem states that the moment of inertia of the mass of a body with respect to any plane is equal to the sum of the moment of inertia of the mass with respect to a parallel plane through the mass center of the body and the product of the mass of the body and the square of the distance between the two planes.

A similar relationship can be developed for products of inertia. The product of inertia of the element of mass in Fig. B-2 with respect to the xz and yz coordinate planes is

$$dI_{xy} = xy \, dm = (x_G + x')(y_G + y') \, dm$$
$$= (x_G y_G + x_G y' + x' y_G + x' y') \, dm.$$

The product of inertia of the body is

$$I_{xy} = x_G y_G \int dm + x_G \int y' \, dm + y_G \int x' \, dm + \int x' y' \, dm$$
$$= x_G y_G m + I_{xy_G} \qquad \text{(B-3)}$$

since $\int y'\,dm = \int x'\,dm = 0$. Equation (B-3) indicates that *the product of inertia of the mass of a body with respect to any two perpendicular planes is equal to the sum of the product of inertia of the mass with respect to the two parallel planes through the mass center of the body and the product of the mass and the two coordinates of the mass center from the given planes.*

PROBLEMS

B-1 The moment of inertia of mass of the 322-lb homogeneous sphere in Fig. PB-1 with respect to the x axis is 134 slug-ft². Determine its moment of inertia with respect to the a axis.

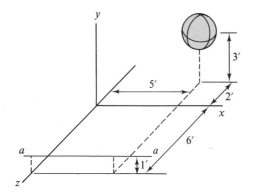

Figure PB-1

B-2 Determine the moment of inertia of mass of the sphere of Problem B-1 with respect to
(a) The y axis.
(b) The z axis.

B-3 For the sphere of Problem B-1, determine the following products of inertia of mass:
(a) I_{xy}, (b) I_{xz}, (c) I_{yz}.

B-4 Determine the second moment of mass of the sphere of Problem B-1 with respect to
(a) The xy plane.
(b) The xz plane.

B-5 The homogeneous tetrahedron or pyramid in Fig. PB-5 weighs 5.80 lb and has a product of inertia of mass I_{xy} (with respect to the

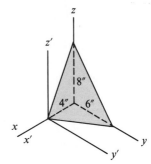

Figure PB-5

yz and xz planes) of 0.0180 lb-sec²-in. Determine the product of inertia $I_{x'y'}$ (with respect to the $y'z'$ and $x'z'$ planes).

B-6 Figure PB-6 represents a homogeneous right circular cone. Which, if any, of the quantities I_{xy}, I_{yz}, and I_{zx} are zero? Determine in terms of the mass m and the dimensions a and h any of the quantities $I_{x'y'}$, $I_{y'z'}$, and $I_{z'x'}$ which can be calculated without integration.

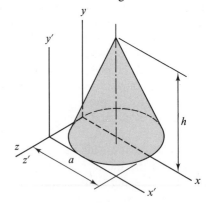

Figure PB-6

B-3 MOMENTS AND PRODUCTS OF INERTIA BY INTEGRATION

In determining moments and products of inertia of mass by integration, the element can be selected for single, double, or triple integration. The choice of element depends on whether the moment or product of inertia is to be obtained and on the orientation of the reference axis or planes. The element should be selected in such a manner that either (1) all parts of the element are the same distance from the reference axis (or from the reference planes for products of inertia), or (2) the moment or product of inertia of the element with respect to the reference axis or planes is known or can be determined. When triple integration is used, the element always satisfies the first requirement, but this condition is not necessarily satisfied by elements requiring single or double integration.

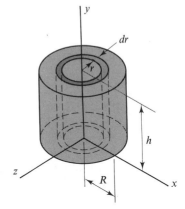

It will be helpful if the element selected is shown on a sketch of the body and dimensioned completely.

EXAMPLE B-1

Figure B-3

Determine the moment of inertia of the mass of a homogeneous right circular cylinder with respect to the geometrical axis. Express the result in terms of the mass of the cylinder.

SOLUTION

A cylinder of radius R and height h is shown in Fig. B-3. A thin cylindrical element, all parts of which are the same distance from the geometrical axis, is selected, and its dimensions are shown in the figure. The mass of the element can be expressed as the product of the volume and the density (mass per unit volume):

$$dm = \rho \, dV = \rho 2\pi rh \, dr,$$

where ρ is the density of the material.

The mass of the cylinder is

$$m = \int_0^R \rho 2\pi rh \, dr = \rho \pi R^2 h.$$

The moment of inertia of the mass of the element with respect to the y axis is

$$dI_y = r^2 \, dm = \rho 2\pi r^3 h \, dr$$

and

$$I_y = \int_0^R \rho 2\pi r^3 h \, dr = \frac{\rho \pi R^4 h}{2}.$$

The moment of inertia of the mass is thus expressed in terms of the density and dimensions of the body. The moment of inertia can be expressed in terms of the mass by multiplying by m and dividing by its equivalent in terms of the density and dimensions of the body as follows:

$$I_y = \frac{\rho\pi R^4 h}{2}\frac{m}{\rho\pi R^2 h} = \frac{1}{2}mR^2. \qquad \text{Ans.}$$

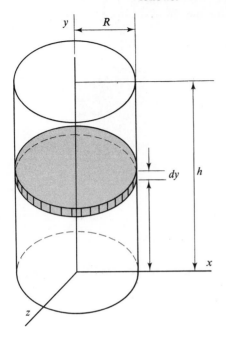

Figure B-4

EXAMPLE B-2

Determine the moment of inertia of the mass of a homogeneous right circular cylinder with respect to
(a) A diameter in the base of the cylinder.
(b) A diameter through the mass center of the cylinder.

SOLUTION

(a) A cylinder of radius R and height h is shown in Fig. B-4. The x axis is a diameter in the base. Unless double or triple integration is used, an element having all points the same distance from the x axis would be quite involved. However, the moment of inertia with respect to the x axis is equal to the sum of the moments of inertia with respect to the xy and xz planes; that is.

$$I_x = I_{xy\ plane} + I_{xz\ plane}.$$

The value of $I_{xy\ plane}$ is equal to $I_{yz\ plane}$ from symmetry, and

$$I_y = I_{xy\ plane} + I_{yz\ plane} = 2I_{xy\ plane} = \frac{1}{2}mR^2 \qquad \text{(from Example B-1)}.$$

Therefore

$$I_{xy\ plane} = \frac{1}{4}mR^2.$$

All points of the element in the shape of a flat disk as shown are the same distance from the xz plane. The mass of the element is

$$dm = \rho\,dV = \rho\pi R^2\,dy,$$

and its moment of inertia with respect to the xz plane is

$$dI_{xz\ plane} = y^2\,dm = \rho\pi R^2 y^2\,dy.$$

The moment of inertia of the mass of the cylinder with respect to the xz plane is

$$I_{xz\ plane} = \int_0^h \rho\pi R^2 y^2\,dy = \rho\pi R^2\frac{y^3}{3}\Big]_0^h = \frac{\rho\pi R^2 h^3}{3},$$

which can be expressed in terms of the mass as

$$I_{xz \text{ plane}} = \frac{\rho \pi R^2 h^3}{3} \frac{m}{\rho \pi R^2 h} = \frac{1}{3} mh^2.$$

The moment of inertia about the x axis is

$$I_x = I_{xy \text{ plane}} + I_{xz \text{ plane}} = \frac{mR^2}{4} + \frac{mh^2}{3} = \frac{m}{12}(3R^2 + 4h^2). \quad \text{Ans.}$$

(b) The moment of inertia of the cylinder with respect to an axis through the mass center parallel to the x axis is

$$I_G = I_x - m\left(\frac{h}{2}\right)^2 = \frac{m}{12}(3R^2 + 4h^2 - 3h^2) = \frac{m(3R^2 + h^2)}{12}. \quad \text{Ans.}$$

EXAMPLE B-3

Determine the moment of inertia of the mass of a solid homogeneous sphere with respect to any centroidal axis.

SOLUTION

Figure B-5 represents a sphere with a radius R. The y axis is selected as the centroidal axis for this example. The element shown is a thin cylindrical disk. The moment of inertia of the element of mass with respect to the y axis is (see Example B-1) $dmr^2/2$. The mass of the element is

$$dm = \rho \pi x^2 \, dy = \rho \pi (R^2 - y^2) \, dy$$

Figure B-5

since the equation of the circle in the xy plane is $x^2 + y^2 = R^2$. The mass of the sphere can be obtained by integration or by multiplying the volume of the sphere by the density of the material. In either case,

$$m = \frac{4\rho \pi R^3}{3}.$$

The moment of inertia of the mass of the cylindrical element is

$$dI_y = \tfrac{1}{2}(dm)x^2 = \tfrac{1}{2}(\rho \pi x^2 \, dy)x^2 = \tfrac{1}{2}\rho \pi (R^4 - 2R^2 y^2 + y^4) \, dy,$$

and the total moment of inertia is

$$I_y = \int_{-R}^{R} \frac{1}{2} \rho \pi (R^4 - 2R^2 y^2 + y^4) \, dy = \frac{8\rho \pi R^5}{15}.$$

The moment of inertia can be expressed in terms of the mass of the sphere as follows:

$$I_y = \frac{8\rho \pi R^5}{15} \frac{m}{(4\rho \pi R^3/3)} = \frac{2mR^2}{5}. \qquad \text{Ans.}$$

EXAMPLE B-4

A solid of revolution is formed by revolving the area bounded by the curve $y^2 = ax$, the line $x = a$ and the x axis around the x axis (see Fig. B-6). The density of the material in the solid is proportional to the distance from the x axis.
(a) Determine the moment of inertia of the mass of the body with respect to the x axis. Express the result in terms of the mass of the body.
(b) The body weighs 16 lb and a is 4 in. Determine the moment of inertia of the mass of the body with respect to the x axis.

SOLUTION

(a) The hollow cylindrical element of mass in Fig. B-6 is selected because the density of the element is constant and all parts of the element are the same distance from the moment axis. The mass of the element is

$$dm = \rho \, dV = \rho 2\pi y(a - x) \, dy,$$

and since

$$\rho = Ky \quad \text{and} \quad x = \frac{y^2}{a},$$

$$dm = K2\pi y^2 \left(a - \frac{y^2}{a}\right) dy = K2\pi \left(ay^2 - \frac{y^4}{a}\right) dy.$$

The mass of the body is

$$m = K2\pi \int_0^a \left(ay^2 - \frac{y^4}{a}\right) dy = K2\pi \left[\frac{ay^3}{3} - \frac{y^5}{5a}\right]_0^a = \frac{4}{15}\pi Ka^4.$$

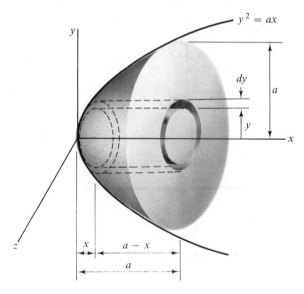

Figure B-6

The moment of inertia of the mass of the element with respect to the x axis is

$$dI_x = y^2 \, dm = K2\pi\left(ay^4 - \frac{y^6}{a}\right) dy,$$

and the moment of inertia of the body is

$$I_x = K2\pi\int_0^a \left(ay^4 - \frac{y^6}{a}\right) dy = K2\pi\left[\frac{ay^5}{5} - \frac{y^7}{7a}\right]_0^a$$

$$= K2\pi\left(\frac{2a^6}{35}\right) = \frac{K4\pi a^6}{35}\frac{m}{\frac{4}{15}\pi Ka^4} = \frac{3}{7}ma^2. \qquad \text{Ans.}$$

(b) The mass of the body is equal to its weight divided by the acceleration of gravity; that is,

$$m = \frac{W}{g} = \frac{16}{32.2} = 0.497 \frac{\text{lb-sec}^2}{\text{ft}} = 0.497 \text{ slug.}$$

The length a should have the same units of length as the acceleration g; thus $a = 4$ in. $= \frac{1}{3}$ ft, and the moment of inertia becomes

$$I_x = \frac{3}{7}(0.497)\left(\frac{1}{3}\right)^2 = 0.0236 \text{ lb-sec}^2\text{-ft} = 0.0236 \text{ slug-ft}^2. \qquad \text{Ans.}$$

If g and a were both expressed in inches, the result would be

$$I_x = \frac{3}{7}\frac{16}{12(32.2)}(4)^2 = 0.284 \text{ lb-sec}^2\text{-in.} \qquad \text{Ans.}$$

EXAMPLE B-5

Determine the product of inertia of the mass of the homogeneous tetrahedron or pyramid in Fig. B-7 with respect to the xz and yz coordinate planes.

SOLUTION

An element parallel to one of the reference planes (the xz plane) is selected, since the plane of the element will be a plane of symmetry for the element. The product of inertia of the mass of the element with respect to a set of planes through the mass center of the element (parallel to the specified planes) is zero since one of the planes is a plane of symmetry. The mass of the element is

$$dm = \rho \, dV = \frac{\rho x_1 z_1}{2} \, dy.$$

The expressions $x_1 = (a/b)(b - y)$ and $z_1 = (c/b)(b - y)$ can be obtained from similar triangles and the mass of the element becomes

$$dm = \frac{\rho ac}{2b^2}(b - y)^2 \, dy.$$

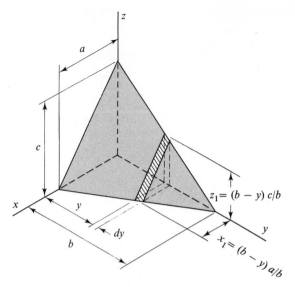

Figure B-7

The mass of the body is

$$m = \frac{\rho ac}{2b^2} \int_0^b (b-y)^2 \, dy$$

$$= -\frac{\rho ac}{2b^2} \frac{(b-y)^3}{3} \Big]_0^b = \frac{\rho abc}{6}.$$

The product of inertia of the mass of the element is

$$dI_{xy} = dI_{x_G y_G} + x_G y_G \, dm$$

$$= 0 + y \frac{x_1}{3} \, dm$$

$$= y \frac{a}{3b} (b-y) \frac{\rho ac}{2b^2} (b-y)^2 \, dy$$

$$= \frac{\rho a^2 c}{6b^3} (b^3 y - 3b^2 y^2 + 3by^3 - y^4) \, dy$$

and the product of inertia of the mass of the body is

$$I_{xy} = \frac{\rho a^2 c}{6b^3} \int_0^b (b^3 y - 3b^2 y^2 + 3by^3 - y^4) \, dy$$

$$= \frac{\rho a^2 c}{6b^3} \left[\frac{b^3 y^2}{2} - b^2 y^3 + \frac{3by^4}{4} - \frac{y^5}{5} \right]_0^b$$

$$= \left(\frac{\rho a^2 b^2 c}{120} \right) \left(\frac{m}{\rho abc/6} \right) = \underline{\frac{mab}{20}}. \qquad \text{Ans.}$$

EXAMPLE B-6

A homogeneous solid of revolution is formed by revolving the shaded area in Fig. B-8(a) about the a axis. The equation of the curve is $y^2 = b(b-x)$, where $b = 0.300$ m, and the density of the material is 800 kg/m³. Determine the moment of inertia of the mass of the body with respect to the z axis.

SOLUTION

The differential element of mass shown in Fig. B-8(b) is a thin cylindrical disk with the a axis as its geometrical axis. The c axis through the mass center of the differential disk is parallel to the z axis. The mass of the element is the product of the density, ρ, and the volume, dV; hence

$$dm = \rho \, dV = \rho\pi(b-x)^2 \, dy = \frac{\rho\pi y^4}{b^2} \, dy.$$

From Example B-2, the moment of inertia of the mass of the disk with respect to the c axis is

$$dI_c = \frac{dm(3R^2 + h^2)}{12}$$

$$= \frac{dm(b-x)^2}{4} + \frac{dm(dy)^2}{12}.$$

Since the last term is a differential of third order, it can be neglected when it is added to the first-order differential. Therefore

$$dI_c = \frac{\rho\pi(b-x)^4 \, dy}{4} = \frac{\rho\pi y^8 \, dy}{4b^4}.$$

From Eq. (B-2), the moment of inertia of the differential element with respect to the z axis becomes

$$dI_z = dI_c + dm(b^2 + y^2)$$

$$= \frac{\rho\pi y^8}{4b^4} \, dy + \frac{\rho\pi y^4}{b^2} (b^2 + y^2) \, dy.$$

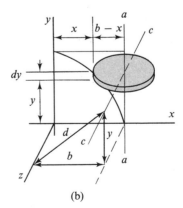

(a)   (b)

Figure B-8

The moment of inertia of the mass of the entire body can be determined by integration as follows:

$$I_z = \frac{\rho\pi}{4b^4}\int_0^b y^8 \, dy + \frac{\rho\pi}{b^2}\int_0^b (b^2y^4 + y^6) \, dy$$

$$= \rho\pi b^5 \left(\frac{1}{36} + \frac{1}{5} + \frac{1}{7}\right) = 800(\pi)(0.300)^5(0.3706)$$

$$= \underline{2.26 \text{ kg} \cdot \text{m}^2}. \qquad \text{Ans.}$$

PROBLEMS

B-7 In Example B-2 it was shown that the moment of inertia of mass of a solid homogeneous right circular cylinder with respect to an axis through the mass center and perpendicular to the axis of the cylinder is $m(3R^2+h^2)/12$. For a slender rod, an approximate value for this moment of inertia is $mh^2/12$. Determine the minimum ratio of the height, h, to the diameter, D, of the cylinder if the error when using the approximate formula is not to exceed 1%.

B-8 Determine the moment of inertia of the mass, m, of the homogeneous slender rod in Fig. PB-8 with respect to the x axis. Assume that the rod lies in the xy plane.

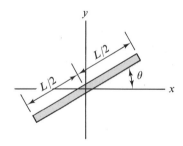

Figure PB-8

B-9 Determine the product of inertia of mass, I_{xy}, of the rod in Problem B-8.

B-10 The density of a slender 6-ft rod of constant cross section varies directly as the square root of the distance from the left end. The rod weighs 128.8 lb. Determine the moment of inertia of mass with respect to an axis (or plane) perpendicular to the rod at the left end.

B-11 The cross section of a homogeneous hollow cylinder of mass m is shown in Fig. PB-11.

Determine by integration the moment of inertia of mass with respect to the axis of the cylinder.

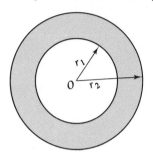

Figure PB-11

*B-12 The shaded area in Fig. PB-12 is revolved about the x axis to form a homogeneous cone with a density of 2.0 Mg/m³. Determine the moment of inertia of the mass with respect to
(a) The x axis.
(b) The y axis.

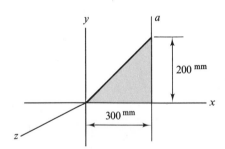

Figure PB-12

*B-13 The shaded area in Fig. PB-12 is revolved about the a axis to form a homogeneous solid with a density of 1.2 Mg/m³. Determine the moment of inertia of mass with respect to
(a) The a axis.
(b) The x axis.

*B-14 The mass generated in Problem B-12 has a variable density. Determine the moment of inertia of mass, in terms of the mass, with respect to
(a) The x axis when the density varies directly as the distance from the x axis.
(b) The y axis when the density varies directly as the distance from the yz plane.

*B-15 The mass generated in Problem B-13 has a variable density. Determine the moment of inertia of mass, in terms of the mass, with respect to
(a) The a axis when the density varies directly as the distance from the a axis.
(b) The x axis when the density varies directly as the distance from the xz plane.

B-16 The solid homogeneous pyramid in Fig. PB-16 weighs $150\,\text{lb/ft}^3$.
(a) Determine, by integration, the product of inertia I_{xy} (with respect to the yz and xz planes).
(b) Check the solution of part (a) by using the result of Example B-5 and the parallel-axis theorem.

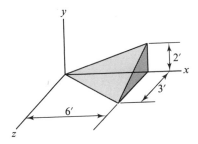

Figure PB-16

B-17 Determine I_y for the pyramid in Problem B-16.

B-18 A homogeneous solid is formed by revolving the shaded area in Fig. PB-18 about the y axis. The material of the body weighs $48.3\,\text{lb/ft}^3$. Determine the moment of inertia of mass of the solid with respect to the x axis. The equation of the curve is $x^2 = y$.

B-19 Solve Problem B-18 if the density of the material varies as the square of its distance from the xz plane. Express the moment of inertia in terms of the mass of the body.

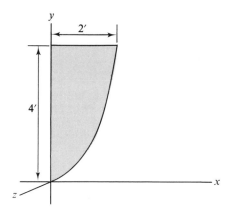

Figure PB-18

B-20 The shaded area in Fig. PB-20 is revolved about the a axis generating a homogeneous solid weighing $64.4\,\text{lb/ft}^3$. Determine the moment of inertia of mass of the solid with respect to the a axis. The equation of the curve is $y^2 = 4x$.

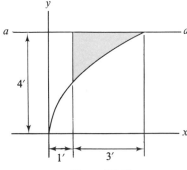

Figure PB-20

B-21 Determine the product of inertia, I_{xy}, for the mass of Problem B-20.

B-22 Determine the moment of inertia of mass with respect to the y axis of the homogeneous body formed by revolving the shaded area in Fig. PB-22 about the y axis. The equation of the curve is $xy = 4b^2$.

B-23 The density of the mass of Problem B-22 varies directly as the distance from the axis of revolution. Determine the moment of inertia with respect to the y axis.

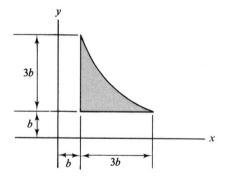

Figure PB-22

B-24 One element of an eccentric cone with a circular base lies along the z axis as shown in Fig. PB-24. The material in the cone is homogeneous. Determine the moment of inertia of the mass of the body with respect to the z axis.

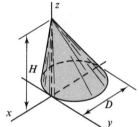

Figure PB-24

B-25 The triangular area in Fig. PB-25 is revolved about the center line to generate a homogeneous body. Determine the moment of

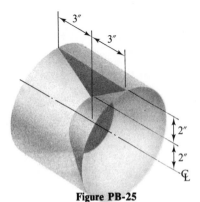

Figure PB-25

inertia of the mass of the body with respect to the axis of rotation. The body is made of aluminum weighing 0.128 lb/in.³

B-26 Figure PB-26 represents a homogeneous right triangular prism. Determine the products of inertia I_{xy} and I_{xz} of the mass of the body in terms of the mass and dimensions.

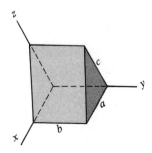

Figure PB-26

B-27 The homogeneous body in Fig. PB-27 is one-fourth of a circular cylinder. Determine the

Figure PB-27

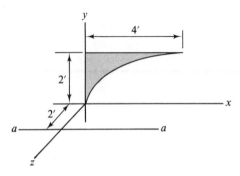

Figure PB-28

products of inertia I_{xy} and I_{yz} of the mass of the body.

B-28 The shaded area in Fig. PB-28 is revolved about the y axis forming a solid whose density is proportional to the square of the distance from the xz plane. Determine the moment of inertia of the mass with respect to the a axis, which lies in the xz plane. The equation of the curve is $y^2 = x$.

B-4 RADIUS OF GYRATION OF MASS

It was pointed out in Art. B-1 that the dimensions of the moment of inertia of mass are mass multiplied by length squared. *The radius of gyration of the mass of a body with respect to an axis is defined as the length which must be squared and multiplied by the mass of the body to give the moment of inertia of the body with respect to the axis.*

This definition is expressed mathematically in the following equation:

$$I_x = k_x^2 m.$$

Figure B-9·

The radius of gyration does not indicate the distance from the given axis to a fixed point in the body (such as the mass center). The parallel-axis theorem can be used to demonstrate that the radius of gyration of the mass of a body with respect to any axis is greater than the distance from the axis to the mass center of the body. From Fig. B-9

$$I_x = I_G + md^2,$$

$$k_x^2 m = k_G^2 m + md^2,$$

$$k_x^2 = k_G^2 + d^2,$$

and

$$k_x = (k_G^2 + d^2)^{1/2},$$

which indicates that k_x is greater than d. The radius of gyration of the mass of a body with respect to an axis is the distance from the axis at which the entire mass of the body can be imagined to be concentrated and still have the same moment of inertia as the distributed mass.

The radii of gyration of the mass of a body with respect to two parallel lines 1 ft apart do not in general differ by 1 ft. In fact, there is no particularly useful physical interpretation or meaning of a radius of gyration. It is merely a convenient means of expressing the moment of inertia of the mass of a body in terms of its mass and a length.

PROBLEMS

B-29 Determine the radius of gyration of mass of a homogeneous slender rod of length L with respect to an axis
(a) Perpendicular to the axis of the rod through the mass center.
(b) Perpendicular to the axis of the rod at one end.
(c) Parallel to the rod at a distance d from the axis of the rod. Assume that d is large compared to the radius of the rod.

B-30 A homogeneous solid of revolution is generated by revolving the shaded area in Fig. PB-30 about the y axis. Determine the radius of gyration of the mass of the body with respect to the y axis. The equation of the curve is $x^2y = 4$.

B-31 The shaded area in Fig. PB-31 is revolved about the x axis forming a solid the density of which varies directly as the square of the distance from the yz plane. Determine the radius of gyration of mass with respect to the x axis.

Figure PB-30

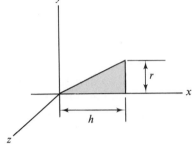

Figure PB-31

B-5 MOMENTS AND PRODUCTS OF INERTIA OF COMPOSITE MASSES

When the moment of inertia of a body is to be determined, it is frequently convenient to resolve the body into a number of simple shapes, such as cylinders, spheres, and rods. The parallel-axis theorem can be used to determine the moment of inertia of each part if the moment of inertia of each part with respect to a parallel axis through the mass center is known. The moment of inertia of the composite body is the algebraic sum of the values for each part, all calculated for the same axis. A similar statement applies to products of inertia. A table listing moments and products of inertia of several bodies is included on p. xxii.

EXAMPLE B-7

The flywheel in Fig. B-10 is made of steel weighing 490 lb/ft³. The spokes have elliptic cross sections. Determine the moment of inertia of the mass of the wheel with respect to the axis of the wheel.

SOLUTION

The rim and hub can be considered solid cylinders with smaller cylinders removed, and the spokes can be assumed to be slender rods. The density of the material is

$$\rho = \frac{490}{32.2} = 15.22 \text{ slugs/ft}^3.$$

The moment of inertia of the mass of the rim with respect to the

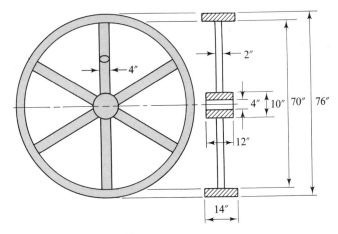

Figure B-10

axis is

$$I_O = \tfrac{1}{2}m_{\text{solid}}(r_{\text{solid}})^2 - \tfrac{1}{2}m_{\text{hole}}(r_{\text{hole}})^2$$

$$= \frac{1}{2}\left[\pi\left(\frac{38}{12}\right)^2\frac{14}{12}(15.22)\right]\left(\frac{38}{12}\right)^2 - \frac{1}{2}\left[\pi\left(\frac{35}{12}\right)^2\frac{14}{12}(15.22)\right]\left(\frac{35}{12}\right)^2$$

$$= 2804 - 2018 = 786 \text{ slug-ft}^2.$$

The moment of inertia of the mass of the hub is

$$I_O = \frac{1}{2}\left[\pi\left(\frac{5}{12}\right)^2\left(\frac{12}{12}\right)15.22\right]\left(\frac{5}{12}\right)^2 - \frac{1}{2}\left[\pi\left(\frac{2}{12}\right)^2\left(\frac{12}{12}\right)15.22\right]\left(\frac{2}{12}\right)^2$$

$$= 0.721 - 0.0185 = 0.703 \text{ slug-ft}^2.$$

The area of the elliptical cross section of the spokes is $\pi ab/4$, where a and b are the major and minor axes. The moment of inertia of the spokes is

$$I_O = 6\left(\frac{ml^2}{12} + md^2\right)$$

$$= 6\left\{\frac{1}{12}\left[\frac{\pi}{4}\left(\frac{2}{12}\right)\left(\frac{4}{12}\right)\left(\frac{30}{12}\right)15.22\right]\left(\frac{30}{12}\right)^2\right.$$

$$\left. + \left[\frac{\pi}{4}\left(\frac{2}{12}\right)\left(\frac{4}{12}\right)\left(\frac{30}{12}\right)(15.22)\right]\left(\frac{20}{12}\right)^2\right\}$$

$$= 6[0.865 + 4.61] = 32.8 \text{ slug-ft}^2.$$

The total moment of inertia of the mass of the flywheel is

$$I_O = 786 + 0.7 + 32.8 = \underline{820 \text{ slug-ft}^2}. \qquad \text{Ans.}$$

Notice that the moment of inertia of the hub has a negligible effect on the result and that the spokes contribute only about 4 per cent of the total moment of inertia. It is thus apparent that the shape of the hub and spokes is relatively unimportant in calculating the moment of inertia of the mass of the flywheel.

Body	Moment of inertia	Product of inertia
Cylinder	$I_x = I_z = \dfrac{m}{12}(3R^2 + h^2)$ $I_y = mR^2/2$	$I_{xy} = I_{yz} = I_{zx} = 0$
Sphere	$I_x = I_y = I_z = 2mR^2/5$	$I_{xy} = I_{yz} = I_{zx} = 0$
Rectangular prism	$I_x = m(b^2 + c^2)/12$ $I_y = m(c^2 + a^2)/12$ $I_z = m(a^2 + b^2)/12$	$I_{xy} = I_{yz} = I_{zx} = 0$
Triangular wedge	$I_x = m(2b^2 + 3c^2)/36$ $I_y = m(2a^2 + 3c^2)/36$ $I_z = m(a^2 + b^2)/18$	$I_{xy} = -mab/36$ $I_{yz} = 0$ $I_{zx} = 0$
Cone	$I_x = I_z = 3m(h^2 + 4R^2)/80$ $I_y = 3mR^2/10$	$I_{zy} = I_{yz} = I_{zx} = 0$

PROBLEMS

B-32 The homogeneous semicircular cylinder of height h and radius R shown in Fig. PB-32 has a mass m. Determine its moment of inertia of mass with respect to the x axis.

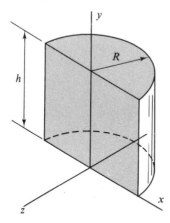

Figure PB-32

B-33 Determine the moment of inertia of the composite mass shown in Fig. PB-33 with respect to the line ab. The disks A and B weigh 48.3 lb each and are parallel to the yz plane, and the uniform slender rods C and D weigh 322 and 161 lb, respectively.

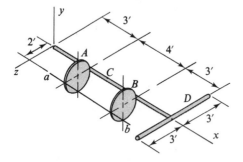

Figure PB-33

B-34 Body A of Fig. PB-34 weighs 322 lb, has its mass center at G, and has a radius of gyration of mass with respect to an axis through O perpendicular to the plane of the figure of 2.5 ft. The homogeneous bar BC weighs 96.6 lb, and the thin cylindrical shell D weighs 128.8 lb.

Determine the radius of gyration of the mass of the composite body with respect to an axis through E perpendicular to the plane of the figure.

Figure PB-34

B-35 In Fig. PB-35, the slender homogeneous rods A and B and the thin homogeneous half-ring C (mass center at G) are welded together to form a rigid frame which rotates about the horizontal axis at O. The weights of A, B, and C are 64.4, 32.2, and 64.4 lb, respectively. Determine the moment of inertia of the mass of the assembly with respect to the axis of roatation.

Figure PB-35

*B-36 A solid homogeneous cylinder with a density of 800 kg/m³ has an off-center 300-mm-diameter hole bored in it as shown in Fig. PB-36. Determine the moment of inertia of the body with respect to the a axis.

B-37 Determine the moment of inertia of the mass of the 322-lb homogeneous slender rod AB and the 966-lb nonhomogeneous cylinder C in Fig. PB-37 with respect to an axis through A perpendicular to the plane of the figure. The radius of gyration of the mass of body C with respect to an axis through O perpendicular to the plane of the figure is 3 ft, and the mass center of the cylinder is at G.

Figure PB-36

Figure PB-37

Figure PB-38

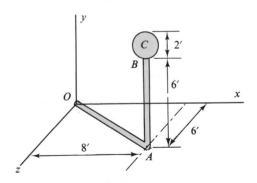

Figure PB-39

B-38 The rigid assembly in Fig. PB-38 is composed of a homogeneous 161-lb sphere *A*, homogeneous slender rods *B* and *C* weighing 16.1 lb per linear ft, and a homogeneous 128.8-lb cylinder *D*. The yz plane is a plane of symmetry. Determine
(a) The moment of inertia of the mass with respect to the z axis.
(b) The product of inertia I_{yz} for the assembly.

B-39 The homogeneous sphere *C* in Fig. PB-39 weighs 322 lb. The homogeneous bar *OA* lies in the xz plane and weighs 193.2 lb. The homogeneous bar *AB* is parallel to the y axis

and weighs 161 lb. Determine
(a) The radius of gyration of mass of the composite body with respect to the y axis.
(b) The product of inertia I_{xy} of the mass of the body.

***B-40** The mass of the homogeneous hollow cylinder *A* in Fig. PB-40 is 4.5 Mg. Bodies *B* and *C* are homogeneous slender bars with masses of 2.0 and 1.5 Mg, respectively. Determine the radius of gyration of the body with respect to the *a* axis.

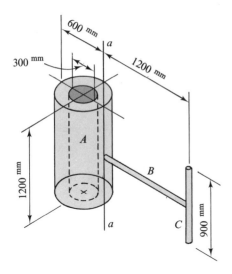

Figure PB-40

B-41 Body A in Fig. PB-41 is a sphere, B is a slender rod, and C is a hollow cylinder. The three bodies are homogeneous and weigh 161, 96.6, and 644 lb, respectively. Determine the moment of inertia of mass of the composite body with respect to the *a* axis.

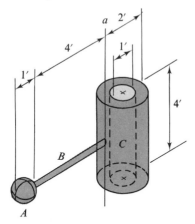

Figure PB-41

B-42 The 644-lb body A in Fig. PB-42 is a homogeneous sphere; bar BC and the cylinder D are homogeneous and weigh 128.8 and 966 lb, respectively. The xy plane is a plane of symmetry. Determine
(a) The radius of gyration of mass with respect to the y axis.
(b) The product of inertia of mass, I_{xy}.

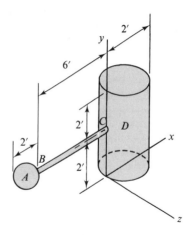

Figure PB-42

B-43 The frustum of the right circular cone in Fig. PB-43 weighs 64.4 lb per ft³. Determine the moment of inertia of mass with respect to the axis of symmetry.

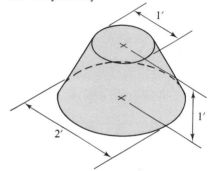

Figure PB-43

B-44 Determine the radius of gyration of mass, with respect to the x axis, of the homogeneous body in Fig. PB-44.

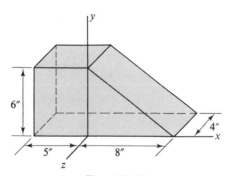

Figure PB-44

B-45 The homogeneous thin cylindrical shell C, weighing 32.2 lb, is connected by a slender homogeneous bar weighing 32.2 lb to the nonhomogeneous body A in Fig. PB-45. Body A weighs 96.6 lb (mass center at G) and has a radius of gyration of mass with respect to an axis through O perpendicular to the plane of the figure of 9 in. Determine the radius of gyration of the mass of the system with respect to an axis perpendicular to the plane of the figure through the mass center, B, of the rod.

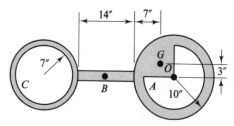

Figure PB-45

B-46 The assembly in Fig. PB-46 is composed of three homogeneous bodies, a 1932-lb cylinder, a 64.4-lb slender rod, and a 322-lb sphere. The xy plane is the plane of symmetry. Determine
(a) The radius of gyration of mass with respect to the x axis.
(b) The product of inertia of mass, I_{xy}.

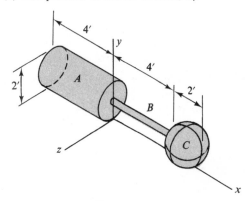

Figure PB-46

*B-47 The two balls and rods of a governor shown in Fig. PB-47 are made of brass with a density of 8.6 Mg/m³. The rods have diameters 10 mm, and the balls have diameters of 80 mm. Determine the moment of inertia, with respect to the vertical axis, of the mass of the balls and

Figure PB-47

rods when the angle θ is 20°. How great an error is introduced if the mass of the rods is neglected?

B-48 All bodies in Fig. PB-48 are homogeneous, and the system is symmetrical with respect to the xy plane. Body A is a sphere with a mass of 5 slugs, C is a disk having a mass of 14 slugs, and bars DE and BO have masses of 3 slugs and 2 slugs, respectively. Determine
(a) The radius of gyration of mass with respect to the z axis.
(b) The product of inertia of mass I_{xy}.

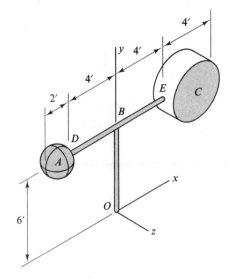

Figure PB-48

Answers to Problems

CHAPTER 8

8-2 (a) 11.0 fps ←, (b) 24.0 fps² →,
(c) 16.0 ft →

8-4 (a) 0, 4 sec, (b) 40 rad ↘, (c) 68 rad

8-6 (a) 9 ft →, (b) 15.83 ft
(c) $\mathbf{a}_{0\,sec} = 8$ fps² ←, $\mathbf{a}_{1.6\,sec} = 8$ fps² →

8-8 99.9 cm

8-10 (a) 7.0 rad per sec² ↘, (b) 6.0 rad ↘

8-12 4.0 m right of origin

8-14 $\mathbf{x} = 1.746\mathbf{i}$ in., $\mathbf{v} = 4.87\mathbf{i}$ ips, $\mathbf{a} = 5.44\mathbf{i}$ ips²

8-16 39.0 rad

8-18 (a) 0.244 ft →, (b) 0.1189 fps ←
(c) 0.1161 fps² →

8-20 0.613 sec

8-22 $\mathbf{x} = 5.06\mathbf{i}$ ft, $\mathbf{v} = 6.75\mathbf{i}$ fps,
$\mathbf{a} = 6.75\mathbf{i}$ fps²

8-24 $x = x_0 e^{v_0^2/2k}$, $\mathbf{a} = -(k/x_0)e^{-v_0^2/2k}\mathbf{i}$

8-26 $\mathbf{x} = 52.1\mathbf{i}$ ft, $\mathbf{v} = 3.70\mathbf{i}$ fps, $\mathbf{a} = -0.686\mathbf{i}$ fps²

8-28 $x = [(2kt + v_0^2)^{3/2} - v_0^3]/3k$, $k \rightarrow L^2T^{-3}$

8-30 $\omega = 0.1875$ rad/s ↙, $\alpha = 0.0406$ rad/s ↘

8-32 $\omega = 1.732$ rad per sec ↙, $\alpha = 2.89$ rad
per sec² ↙

8-34 (a) $v = 2620$ fps, $a = 0$,
(b) $v = 0$ $a = 3.49(10^4)$ fps²

8-38 $v_i^2 + v_f^2 = 2aQ$

8-40 52.0 cm/s² ←

8-42 $\Delta t = 2.42$ sec, $\mathbf{v}_f = 44.5$ ips ↑

8-44 $\mathbf{v}_i = 20$ fps $\leftarrow$, $\mathbf{v}_f = 30$ fps $\rightarrow$,
$\mathbf{a} = 10$ fps$^2 \rightarrow$

8-46 (a) $\mathbf{a} = 18.89$ fps$^2 \leftarrow$, $\mathbf{v}_f = 13.33$ fps $\leftarrow$
(b) $\mathbf{a} = 47.8$ fps$^2 \leftarrow$, $\mathbf{v}_f = 186.7$ fps $\leftarrow$

8-48 (a) 327 ips$^2 \downarrow$, (b) 6.69 in.

8-50 (a) 9 fps$^2 \rightarrow$, (b) 72 ft, (c) 12 sec

8-52 12 sec

8-54 15.15 sec

8-56 61.8 sec

8-58 4 sec

8-60 1.677 m 26.60°

8-62 24.0 ips$^2 \rightarrow$

8-64 16.67 mm/s$^2 \rightarrow$

8-66 $v_A s_A + v_B (s_B + \sqrt{s_A^2 + s_B^2})$

8-68 (a) 0, (b) 6.00 ips $\uparrow$, (c) 5.00 ips$^2 \uparrow$

8-70 (a) 10.00 ips $\uparrow$, (b) 18.82 ips $\uparrow$

8-72 $v_A(2s_A + s_B) + v_B(s_A + 2s_B)$

8-74 7.47 ips $\rightarrow$

8-76 40 mm/s $\leftarrow$

8-78 $\mathbf{v}_A = 7.75$ ips $\downarrow$, $\mathbf{v}_B = 7.75$ ips $\uparrow$,
$\mathbf{v}_C = 11.62$ ips $\uparrow$

8-80 62.5 ft

8-82 (a) 4.00 sec, (b) 96.0 ft

8-84 (a) B wins by 198 ft, (b) 660 ft,
(c) 2.40 miles

8-86 (a) 16.87 fps $\downarrow$, (b) 11.21 ft
(c) 18.40 ft

8-88 (a) 107.7 ft 21.8°, (b) 4.71 fps 45°

8-90 $v = 91.4$ mph at 5500 ft

8-92 30.8° east of north

8-94 133.3 ft

8-96 (a) $(6\mathbf{i} - 3\mathbf{j})$ m, (b) $(8\mathbf{i} - \mathbf{j})$ m/s
(c) $(4\mathbf{i} + 1.5\mathbf{j})$ m/s^2

8-98 (a) $(-\mathbf{i} + \mathbf{j} + 0.368\mathbf{k})$ in.
(b) $(2\mathbf{i} + 3\mathbf{j} - 0.368\mathbf{k})$ ips
(c) $(2\mathbf{i} + 6\mathbf{j} + 0.368\mathbf{k})$ ips^2

8-100 $(0.1338\mathbf{i} + 0.233\mathbf{j})$ m/s^2

8-102 (a) $\left(\dfrac{x-b}{b}\right)^2 + \left(\dfrac{y}{c}\right)^2 = 1$
(b) $\mathbf{r} = (6\mathbf{i} - 3\mathbf{j})$ in., $\mathbf{v} = 9.42\mathbf{i}$ ips

8-104 $3.60\mathbf{i}$ ips^2

8-106 $\mathbf{v}_A = b[(\cos\theta - \sin\theta)\mathbf{i} + \cos\theta\mathbf{j}]\dot\theta$
$\mathbf{a}_A = b\{[\ddot\theta(\cos\theta - \sin\theta)$
$\quad - \dot\theta^2(\cos\theta + \sin\theta)]\mathbf{i}$
$\quad + (\ddot\theta\cos\theta - \dot\theta^2\sin\theta)\mathbf{j}\}$

8-108 25.0 fps 7, 24

8-110 $-27.7\mathbf{i}$ ips^2

8-114 14.14%

8-116 40.2°

8-118 Home run (Cleared fence 1.94 ft)

8-120 34.2° or 69.8°

8-122 170.0 fps 31.2°

8-124 68.2 fps 61.0°

8-126 20.5 fps

8-128 73.1 fps 50.9°

8-130 $\mathbf{v} = 3.00$ fps $\rightarrow$,

$\mathbf{a} = 2.69$ fps^2 42.0°

8-132 35.8 m/s^2 26.6°

8-134 12.00 fps $\downarrow$

8-136 22.0 m/s^2 74.9°

8-138 (a) $(19.08\mathbf{i} + 15.90\mathbf{j})$ ips^2
(b) $(4.92\mathbf{i} - 5.90\mathbf{j})$ ips^2,
(c) 127.0 in.

8-140 825 ft

8-142 $\mathbf{a}_t = (37.5\mathbf{i} + 36.0\mathbf{j})$ fps²,
$\mathbf{a}_n = (-37.5\mathbf{i} + 39.0\mathbf{j})$ fps²

8-144 (a) $\mathbf{a}_t = (3.33\mathbf{i} + 5.00\mathbf{j} - 0.612\mathbf{k})$ ips²,
$\mathbf{a}_n = (-1.329\mathbf{i} + 1.006\mathbf{j} + 0.980\mathbf{k})$ ips²
(b) $\mathbf{r}_c = (-5.67\mathbf{i} + 4.53\mathbf{j} + 3.81\mathbf{k})$ in.

8-146 $\mathbf{a}_R = -4330\mathbf{e}_r$ ips², $\mathbf{a}_T = 2000\mathbf{e}_\theta$ ips²

8-148 166.6 cm/s² ∠ 19.23°

8-150 (a) 147.1 ips², (b) 56.0 ips²

8-152 (a) 15.78 ips ∠ 17.66°

(b) 114.8 ips² ∠ 60.0°

8-154 $\theta = 120°$, $\mathbf{v} = -6.00\mathbf{i}$ ips

8-156 $\mathbf{v} = 11.18$ ips ∠ 71.6° ,

$\mathbf{a} = 103.1$ ips² ∠ 31.0°

8-158 $\mathbf{v} = (-1.25\mathbf{e}_r + 23.6\mathbf{e}_\theta - 2.5\mathbf{k})$ ips,
$\mathbf{a} = (-74.0\mathbf{e}_r - 7.85\mathbf{e}_\theta)$ ips²

8-160 (a) $-c_2\mathbf{e}_r + (bc_1 - c_1 c_2 t)\mathbf{e}_\theta$
$$-\frac{2hc_2}{b^2}(b - c_2 t)\mathbf{k}$$
(b) $\mathbf{v}_0 = (-0.5\mathbf{e}_r + 2\pi\mathbf{e}_\theta - 2\mathbf{k})$ ips
$\mathbf{v}_{10} = (-0.5\mathbf{e}_r + \pi\mathbf{e}_\theta - \mathbf{k})$ ips

8-162 6.24 fps ←

8-164 $v = -r\sin\theta\dot{\theta}$,
$a = -r(\sin\theta\ddot{\theta} + \cos\theta\dot{\theta}^2)$

8-166 (a) 131.1 ft, (b) 15.90 ft

8-168 $\mathbf{a} = (2\mathbf{i} - 4.5\mathbf{j})$ cm/s²

8-170 $s = 3\tan(3t + \pi/4)$

8-172 (a) $\omega_{CE} = 6.0$ rad/s , $\omega_{CD} = 15$ rad/s

(b) 110.4 ips ∠ 42.8°

8-174 0.028 rad per sec

8-176 106.4 ips →

8-178 2.89 cm/s ∠ 60°

8-180 $\mathbf{v}_B = 14$ ips ←, $\mathbf{v}_C = 18$ ips ←,

$\mathbf{v}_D = 6$ ips →, $\mathbf{v}_E = 13.42$ ips ∠ 63.4°

8-182 13.33 ips →

8-184 (a) 0.640 rad per sec

(b) 8.00 ips ∠ 16.26°

(c) 11.20 ips ∠ 36.9°

8-186 16.79 ips ∠ 35.9°

8-188 $\omega_{BC} = 0$, $\omega_{CD} = 3.00$ rad/sec

8-190 (a) 23.1 ips ∠ 30°

(b) 21.4 ips ∠ 21.05°

8-192 (a) 48.8 ips ∠(5,12) , (b) 6.25 rad/sec

8-194 $\mathbf{v}_A = 25$ ips ∠(3,4) , $\mathbf{v}_B = 40.3$ ips ∠(7,4),

$\mathbf{v}_D = 15$ ips →

8-196 $\mathbf{v}_P = 10.17$ ips ←,

$\mathbf{v}_Q = 17.07$ ips ∠ 35.9°

8-198 $\mathbf{v}_A = 1.00$ m/s ←,

$\mathbf{v}_B = 3.81$ m/s ∠ 23.2°

8-200 (a) 25 rpm . (b) 18.51 cm/s ∠ 45°

8-202 (a) 6.67 rpm , (b) 10.47 cm/s ↓

8-204 $\mathbf{v}_A = 90$ ips ∠(4,3) , $\mathbf{v}_B = 54$ ips ↑

8-206 56.0 cm/s ←

8-208 24.0 cm/s ←

8-210 $\omega_{AB} = 0$, $\mathbf{v}_Q = 132.0$ cm/s →

8-212 104.8 cm/s ∠ 13.24°

8-214 $\mathbf{v}_B = 33.0$ ips ←, $\mathbf{v}_F = 14.71$ ips →

8-216 $\mathbf{v}_C = (2v_A/9)(3\mathbf{i} - \mathbf{j})$,
$\boldsymbol{\omega} = (v_A/27)(2\mathbf{i} - 3\mathbf{j} - 3\mathbf{k})$

8-218 $(2.50\mathbf{i} - 0.833\mathbf{j} - 0.833\mathbf{k})$ rad per sec

8-220 (a) $(-8\mathbf{i} + 4\mathbf{j})$ rad per sec
(b) $-88\mathbf{k}$ ips

8-222 $\omega_1(-\mathbf{i} + \mathbf{j} + \mathbf{k})/3$ rad per sec

8-224 $\mathbf{v}_A = 52.9$ ips $\overline{\diagdown}\,40.9°$

$\mathbf{a}_A = 381$ ips^2 $\diagup\!\!\!\!\diagdown\,41.3°$

8-226 $\mathbf{v}_0 = r\omega \rightarrow$, $\mathbf{a}_0 = \vec{r\alpha} + r^2\overset{\uparrow}{\omega^2}/R.$

8-228 48.3 cm/s^2 $\overline{\diagdown\!\!\!\searrow}\,24.4°$

8-230 170.9 ips^2 $\diagup\,35.0°$

8-232 30.0 rad per sec^2 ⤵

8-234 $\mathbf{v}_C = 0$, $\mathbf{a}_C = 50.0$ ips^2 $\downarrow$

8-236 $\omega = 6.00$ rad per sec ⤵,
$\boldsymbol{\alpha} = 11.43$ rad per sec^2⤵

8-238 68.4 ips^2 $\downarrow$

8-240 (a) 4.00 sec, (b) 22.1 ips^2 $\overline{\diagdown}\,26.6°$

(c) 33.0 ips^2 $\overline{\diagdown}\,32.8°$

8-242 $\boldsymbol{\alpha}_{AB} = 18.33$ rad per sec^2⤵,
$\boldsymbol{\alpha}_{BC} = 4.67$ rad per sec^2 ⤵

8-244 $\mathbf{v}_A = 13.71$ cm/s $\leftarrow$,
$\mathbf{a}_A = 20.4$ cm/s^2 $\leftarrow$

8-246 $(\frac{4}{33})(27\mathbf{i} + 80\mathbf{j} - \mathbf{k})$ rad per sec

8-248 (a) $\frac{4}{3}\omega_2^2\mathbf{k}$, (b) $\frac{4}{75}\omega_2^2 r(-24\mathbf{i} - 25\mathbf{j})$

8-250 $-\frac{5}{27}r\omega_1^2(\mathbf{j} + 2\mathbf{k})$

8-252 (a) $(25.7\mathbf{i} - 58\mathbf{j} + 20\mathbf{k})$ rad/s^2
(b) $(42\mathbf{i} + 67\mathbf{j})$ cm/s^2

8-254 $\boldsymbol{\omega}_{AB} = 3.52$ rad/s ⤵, $\boldsymbol{\alpha}_{AB} = 9.50$ rad/s^2 ⤵

8-256 $\omega = 0.769$ rad/s ⤵, $\alpha = 0.917$ rad/s^2 ⤵

8-258 $\boldsymbol{\alpha}_A = 170$ rad/s^2 ⤵, $\mathbf{a}_{P/A} = 850$ *ips*^{2} $\downarrow$

8-260 $\mathbf{v}_P = (-32\mathbf{i} + 15\mathbf{j})$ fps,
$\mathbf{a}_P = (-20\mathbf{i} - 134\mathbf{j})$ fps^2

8-262 (a) 0.896 rad/sec^2 ⤵

(b) 108.9 ips^2 $\diagdown\,64.2°$

8-264 16.55 rad per sec^2 ⤵

8-266 $\mathbf{v} = 24$ cm/s $\rightarrow$, $\mathbf{a} = 14$ cm/s^2 $\rightarrow$

8-268 (a) 12.15 ips $\leftarrow$, (b) 4.91 rad per sec ⤵
(c) 176.5 ips^2 $\leftarrow$

8-270 $\mathbf{v}_C = (7.5\mathbf{i} + 18.0\mathbf{j} - 25.2\mathbf{k})$ fps
$\mathbf{a}_C = (-57.2\mathbf{i} + 11.25\mathbf{j} - 18.00\mathbf{k})$ fps^2

8-272 $\mathbf{v}_P = (-22\mathbf{i} + 4\mathbf{j} + 6\mathbf{k})$ fps,
$\mathbf{a}_P = (-4\mathbf{i} - 40\mathbf{j})$ fps^2

8-274 (a) $(-8\mathbf{i} + 5\mathbf{j} + 3\mathbf{k})$ rad per sec^2
(b) $(20\mathbf{i} + 4\mathbf{j} - 20\mathbf{k})$ ips
(c) $(-46\mathbf{i} + 55\mathbf{j} - 25\mathbf{k})$ ips^2

8-276 (a) $(8\mathbf{i} + 3\mathbf{j} + 6\mathbf{k})$ rad per sec^2
(b) $(-2\mathbf{i} + 30\mathbf{j} + 9\mathbf{k})$ ips^2

8-278 $\mathbf{v} = b\omega_0(\mathbf{i} - 2\mathbf{j})$,
$\mathbf{a} = (b\omega_0^2/2)(3\mathbf{i} + 4\mathbf{j})$

8-280 $\mathbf{v}_P = (2b\omega_0/9)(2\mathbf{i} + \mathbf{j} - 2\mathbf{k})$,
$\mathbf{a}_P = (4b\omega_0^2/27)(-\mathbf{i} + \mathbf{j} - 2\mathbf{k})$

8-282 $\mathbf{v}_C = 50$ ips $\diagup\!|\,{}^3_4$

$\mathbf{v}_P = 36.1$ ips $\diagup\!|\,{}^3_2$

8-284 116.5 ips^2 $\diagup\,65.8°$

8-286 $\mathbf{v}_C = (-12\mathbf{i} - 16\mathbf{j} - 12\mathbf{k})$ ips
$\mathbf{a}_C = (-60\mathbf{i} - 80\mathbf{j} + 76\mathbf{k})$ ips^2

8-288 $\mathbf{v}_C = 30.0$ ips $\leftarrow$, $\mathbf{a}_C = 651$ ips^2 $\overline{\diagup\!\!\!\leftarrow}\,7.94°$

8-290 3.75 rad per sec^2 ⤵

8-292 (a) 33.8 ips $\rightarrow$, (b) 55.6 ips^2 $\rightarrow$

8-294 $\boldsymbol{\omega}_{AB} = 1.479$ rad/s ⤵,
$\boldsymbol{\alpha}_{AB} = 25.0$ rad/s^2 ⤵

8-296 $\mathbf{v}_A = 214$ ips $\overline{\diagdown}\,30°$,

$\mathbf{a}_A = 38,700$ ips^2 $\diagup\,49.1°$

8-298 $\boldsymbol{\omega}_C = (10\mathbf{i} + 5\mathbf{j} + 3\mathbf{k})$ rad per sec,
$\boldsymbol{\alpha}_C = (-15\mathbf{i} + 30\mathbf{j} - 50\mathbf{k})$ rad per sec^2

CHAPTER 9

9-2   (a) $m = 5.28$ slugs, $W = 134.3$ lb
      (b) $24{,}600$ fps $= 16{,}760$ mph

9-4   (a) 26.8 fps^2 ,

      (b) 5.37 fps^2 , (c) $a = 0$

9-6   (a) 3.19 fps ↗, (b) 0.658 fps ↙

9-8   (a) $\tan^{-1}(-\mu)$
      (b) $g[\mu + (P/W)(1 + \mu^2)^{1/2}]$

9-10 (a) $T_{AB} = 139.8$ N, $T_{BC} = 103.0$ N
      (b) 0.490 m/s^2 ←

9-12 (a) 3.22 fps^2 ↓, (b) 13.50 lb, (c) 6.00 lb

9-14 24.5 lb

9-16 1.44 lb

9-18 11.59 fps^2

9-20 $T = 4.64$ lb, $\alpha = 2.01$ rad/s^2 ↻

9-22 (a) $T_{AB} = m(9\omega^2 - 24.15)$,
      (b) 1.638 rad per sec
      (c) 1.465 rad per sec

9-24 0.259

9-26 3.29 rad per sec

9-28 $\Delta t = 0.509$ sec, $\Delta s_B = 0.833$ ft

9-30 $a_A = 8.00$ fps^2 →, $a_B = 4.00$ fps^2 →,
      $a_C = 6.00$ fps^2 →

9-32 3.27 fps^2

9-34 (a) $\omega = 11.35$ rad/sec, $T = 2.33$ lb
      (b) 7.59 rad/sec

9-36 (a) 2.548 lb, (b) 14.77 fps^2 →
      (c) 24.0 fps^2 ↓

9-38 $a_B = 3.63$ m/s^2 ↓, $T = 12.37$ N

9-40 353 N

9-42 0.932 lb

9-44 24.4 lb ←

9-46 4.72 rad per sec

9-48 $a_A = 4.20$ m/s^2 ↑, $T_A = 140.1$ N,
      $T_B = 224$ N

9-50 Body has translation. a_P is parallel to
      line HG

9-52 (a) 27.9 fps^2 ,
      $R_A = R_B = 2.50$ lb

9-54 (a) $T_B = 30$ lb, $T_C = 60$ lb
      (b) 6.44 fps^2 ↑

9-56 (a) 70.0 lb →, (b) 21.5 fps^2 →

9-58 $T_A = 3.09$ lb, $T_B = 20.1$ lb,

      $T_C = 51.0$ lb, $a = 30.8$ fps^2

9-60 $a = 3.22$ fps^2 →, $R_L = 420$ lb ↑,
      $R_R = 480$ lb ↑

9-62 $a = 6.44$ fps^2 , $\theta = 10.30°$

9-64 $B_x = 3.00$ lb ←, $B_y = 5.02$ lb ↑

9-66 $a = 1.200\, g$. Not resonable since tires
      would slip unless $\mu > 1.2$

9-68 $A_x = A_z = 0$, $B_x = ma$,
      $B_y = m(4g - 3a)/8$, $C_y = m(4g + 3a)/24$,
      $D_y = m(4g + 3a)/12$

9-70 112.5 lb ←

9-72 $C_x = 174.2$ lb ←, $D_x = 18.94$ lb →
      $D_y = 231$ lb ↑

9-74 0.577

9-76 (a) 4.57 rad per sec ↻

      (b) $N = 53.8$ lb

      $F = 8.46$ lb

9-78 11.47 rad per sec^2 ↻

9-80 187.5 rad per sec^2 ↻

9-82 0.677 ft

9-84 $P = 121.2$ N ←, $C_x = 82.8$ N ←
      $C_y = 304$ N ↑

9-86 0.0344 in.

9-88   164.4 N

9-90   3.28 rad per sec ↻

9-92   0.496 m/s² ↑

9-94   2.52 ft

9-96   (a) 1.502 rad/s² ↻, (b) 286 lb   ∠ 81.8°

9-98 T_A = 5 mg/8, T_B = 3 mg/8

9-100   21.16 lb ↖ 70.9°

9-102   23.9 rad/sec²

9-104   (a) 59.4 rad/s²
   (b) C_x = 3.40 lb ↓, C_y = 2.37 lb ↙,
   D_x = 1.687 lb ↓, D_y = 0.088 lb ↙

9-106 T = 18.55 lb,
   R_D = 7.45(0.987$\mathbf{i}$ + 0.161$\mathbf{k}$) lb

9-108   (a) 4.00 fps² ←, (b) 0.373

9-110   (a) 37.2 rad/s² ↻, (b) 12.19 fps ↓

9-112   3.01 rad per sec² ↻

9-114 $\mathbf{a}_B$ = 6.71 fps² →, $\mathbf{a}_C$ = 25.5 fps² ↓

9-116   (a) α = 2.40 rad/s² ↻,

   $\mathbf{a}_G$ = 0.720 m/s² (5, 12)

   (b) α = 10.72 rad/s² ↻,

   $\mathbf{a}_G$ = 0.684 m/s² (5, 12)

9-118   107.5 kg

9-120   (a). 60.0 lb →, (b) 1360 ft-lb ↻

9-122   (a) 427 N ←, (b) 1.885 m/s² →
   (c) 2.26 rad/s² ↻

9-124   (a) 96.6 rad per sec² ↻, (b) 64.4 fps² ↑

9-126 $\mathbf{a}_G$ = 12.08 fps² ↘ 30°,

   α = 8.05 rad/s² ↻

9-128   (a) 372 ft-lb ↻, (b) 4.32 fps² →

9-130   28.3 ft ↘ (4, 3)

9-132   40.0 rad per sec ↻

9-134 $\mathbf{a}_A$ = 4.32 fps² ↗ (3, 4),

   $\boldsymbol{\alpha}_B$ = 10 rad/s² ↻

9-136   (a) 10.59 rad/s² ↻, (b) 1.918 lb

9-138   8.49 rad/s² ↻

9-140   0.1697 rad/s² ↻

9-142   67.1 lb ∠ 72.6°

9-144   0.275 ft

9-146   (a) 10.00 lb →, (b) 15.85 fps →

9-148   164.4 N

9-150   (a) 372 ft-lb ↻, (b) 4.32 fps² →

9-152 P = 40 lb ←, C = 50.7 ft-lb ↻

9-154   2.40 m/s² ←

9-156 C_x = 0.552 lb →, C_y = 34.7 lb ↑,
   D_x = 0.868 lb →, D_y = 33.4 lb ↑

9-158   6.58 fps² →

9-160   (a) 2.87 rad/s² ↻

   (b) 382 lb ↙ 40.4°

9-162   12.15 fps² ↗ (3, 4)

9-164 A_x = 3590 lb →, A_y = 1498 lb ↑

9-166   (a) 4.42 rad/s ↻, (b) 78.2 lb ↘ 14.59°

9-168   (a) $\mathbf{a}_G$ = 0, T = 8.00 lb
   (b) $\mathbf{a}_G$ = 2.58 fps² →, T = 4.00 lb

CHAPTER 10

10-2   14.30 fps to the right

10-4   0.738 m to the left of original starting position

10-6   2160 lb/ft

10-8   3 fps

10-10 6.89 m/s down the slot

10-12 (a) $v = \sqrt{36.3x - 4x^2}$
(b) 8.08 ft

10-14 2.68 m/s to the right

10-16 (a) 38.6 in-lb
(b) 0.879 in.
(c) 11.58 lb

10-18 35.1 lb/ft

10-20 204 lb/ft

10-22 88.5 lb/ft

10-24 8.27 fps to the right

10-26 29.5 ips down the slot

10-28 (a) $v = \sqrt{2gR(1 - \cos\theta)}$
(b) 48.2°

10-30 17.94 fps

10-32 (a) 12.40 fps to the left
(b) 8.87 lb

10-34 16.05 fps to the left

10-38 (a) $v = \sqrt{2gR(1 - \cos\theta)}$
(b) 48.2°

10-40 (a) 16 ft-lb
(b) 16 ft-lb
(c) 16 ft-lb

10-42 only (b) $V = -\left(\dfrac{x^2}{2} + \dfrac{y^2}{2} + \dfrac{z^2}{2}\right) + C$

10-44 yes, 5.39 rad/sec clockwise

10-46 89.0 lb/ft

10-48 1.485 ft downward

10-50 66.8 lb upward

10-52 281 N, 4 to the left and 3 downward

10-54 32.5 lb

10-56 126.7 lb, 4 to the left and 3 downward

10-58 5.42 rad/sec clockwise

10-60 174.8 lb/ft

10-62 counterclockwise, 11.31 rad/sec clockwise

10-64 691 N/m

10-66 1.644 slugs

10-68 3.49 rad/sec clockwise

10-70 30.7 lb/ft

10-72 11.23 lb/ft

10-74 110.4 N/m

10-76 4.99 fps to the right

10-78 6.37 m/s downward

10-80 11.94 N · m

10-82 76.2%

10-84 (a) 36.4 hp
(b) 40.9 hp

10-86 (a) 1.894 hp
(b) 78.9%

10-88 (a) 77.8 kg
(b) 2290 N · m/sec

10-90 83.1%

10-92 33.6°

10-94 75.5°

10-96 (a) $T = 28.3$ lb along cord
$N = 47.4$ lb upward
$F = 7.56$ lb to the left
(b) $T = 58.7$ lb upward
$N = 5.67$ lb upward
$F = 0$

10-98 (a) 39.2 N
(b) 40.6 m/s downward

10-100 41.8°

CHAPTER 11

11-2 (a) 140 N · s up the plane
(b) 398 N · s up the plane

11-4 34.4 fps to the right

11-6 (a) 18 lb to the right
(b) $s = t^4/4$
(c) 729 ft-lb
(d) 36 lb to the right

11-8   83.2 N · s, 74 to the right and 38 downward

11-10   2.74 sec

11-12   29.8 fps to the right

11-14 $(22.8\mathbf{i} - 17.6\mathbf{j})$ m/s

11-16   0.545

11-18   (a) 1.605 sec
(b) 7.25 fps to the right

11-20   28.2 fps up the plane

11-22   20.7 fps to the left

11-24   (a) No, $\mathbf{v}_{min} = 17.12$ fps to the right
(b) 30 fps to the right

11-26 $\mathbf{v}_A = 17.92$ fps right
$\mathbf{v}_B = 14.33$ fps right
$\mathbf{v}_C = 10.75$ fps left

11-28   136.1 lb, 60.9 right, 121.8 downward

11-30   79.4 fps

11-32   (a) 909 lb right
(b) 82.7 hp
(c) 131.2 ft-lb/lb

11-34   59.8°

11-36   (a) 6.67 fps right
(b) 99.5 ft-lb

11-38 $v = \dfrac{M+m}{m}[2gL(1-\cos\theta)]^{1/2}$ m/s

11-40   0.1391 ft

11-42   (a) 26 fps left
(b) 27,960 lb/ft

11-44   2.30 ft

11-46 $\left(\dfrac{gr}{21}\right)^{1/2}$

11-48 $(29.7\mathbf{i} - 4.28\mathbf{k})$ m/s

11-50   (a) $\mathbf{v}_A = 4$ m/s right, $\mathbf{v}_B = 22.0$ m/s right
(b) 12800 N right

11-52   0.789

11-54   6.34 ft

11-56   16.39 fps, 6.6 right and 15 upward

11-58 $\mathbf{v}_A = 2$ fps left
$\mathbf{v}_B = 13.42$ fps, 6 right and 12 upward

11-60 $\mathbf{v}_A = 13.58$ fps, 3.2 left and 13.2 upward
$\mathbf{v}_B = 8.53$ fps, 6.6 right and 5.4 upward

11-62   17.55 m/s, 17.28 right and 3.04 upward

11-66 $\mathbf{v}_A = 15.13$ m/s, 7.44 right and 13.17 upward
$\mathbf{v}_B = 17.56$ m/s right

11-68 $\dfrac{m_A m_B(v_A^2 - v_B^2)(1-e^2)}{(m_A + m_B)(m_A v_A^2 + m_B v_B^2)}$

11-70   (a) $\mathbf{H}_O = 13mL^2\dot{\theta}\mathbf{k}$,
(b) $\ddot{\theta} = -5g\sin\theta/13L$

11-72   (a) 6 fps ↓

(b) 2 fps ⟋ 18.43°

(c) (5.66 m) lb-sec ↘ 70.5°

11-74   15 lb/ft

11-76   9 ft

11-78   (a) 0.327 ft
(b) 8.97 ft

11-80   (a) 19.46 Mm, (b) 17.72 Mm/hr

11-82   (a) 15,800 mph, (b) 12,920 miles

11-84   (a) 0.450, (b) 32.2 Mm/hr

11-86   (a) 14.40 Mm, (b) 3.64 hr

11-88 $r = r_0 \sec\dfrac{\sqrt{h^2-k}}{h}\theta$

11-90   5.32 fps²

11-92   (a) 3.67 hr, (b) 7070 miles

11-94   6.00 kN

11-96   (a) 3610 fps, (b) 14,580 fps

11-98   813 ft

11-100   14.65 fps ↑

11-102   16.01 fps ↓

11-104   31.6 ft

11-106 $25.4[\ln(5+y)-0.1y-1.708]^{1/2}$

11-108 31.7 fps ↓

11-110 12.92 lb ↓

11-112 37.6 lb/ft³

11-114 $F = 139.8$ lb, $R = 242$ lb

11-116 (a) 10.00 fps ←, (b) 0.500

11-118 (a) 354**k** ft-lb-sec
(b) 3.22 ft below O

11-120 555**k** kg · m²/s

11-122 (a) 45.0 kg · m²/s ⤾

(b)

11-126 1.408 ft

11-128 0.00659

11-130 7.07 sec

11-132 (a) 27.7 rad/s ⤿, (b) 0.97 m/s ↑

11-134 (a) 12.0 fps →
(b) 9.94 lb-sec →
(c) 1.333 rad per sec ⤾
(d) 2.48 lb-sec →

11-136 5.00 fps ↑

11-138 (a) 35.2 rad per sec ⤿, (b) 8.97 sec

11-140 39.7 kg

11-142 13.92 ft-lb per sec²

11-144 2.84 lb

11-146 4.44 sec

11-148 1.810 in.

11-150 1.886 sec

11-152 (a) 0.874 s, (b) 857 m/s →
(c) 0.374 m

11-154 0.874 m

11-156 25.7 fps →

11-158 5.00 fps

11-160 $\omega = 20.3$ rad per sec ⤿,
$\mathbf{v}_G = 11.95$ fps ←

11-162 $\omega_A = 113.0$ rad per sec ,

$\omega_B = 67.0$ rad per sec

11-164 1.246 rad per sec ⤿

11-166 (a) 14.75 rad/s, (b) $P_i = 120.0$ N,
$P_f = 544$ N, (c) 295 N · m

11-168 5.43 ft

11-170 $\omega = 23.4$ rad per sec ⤿,
$\mathbf{v}_G = 33.7$ fps ↓

11-172 $\omega_{AB} = 0.857$ rad per sec ⤸
$\omega_{CD} = 10.43$ rad per sec ⤸

11-174 (a) 0.491 rad per sec

(b) 0.1638 rad per sec

11-176 $\mathbf{A}_x = 120$ lb ↖, $\mathbf{A}_y = 234$ lb ↓
$\mathbf{B}_x = 120$ lb ↖, $\mathbf{B}_y = 395$ lb ↑

11-178 $\mathbf{A}_x = 87.9\mathbf{i}$ lb, $\mathbf{A}_y = 0$,
$\mathbf{B}_x = -71.8\mathbf{i}$ lb, $\mathbf{B}_y = 0$

11-182 $0.882\mathbf{k}'$ rad/s

11-184 75.5° from vertical
64.3° from north

11-186 $\mathbf{R}_A = 10.06(0.965\mathbf{i}' + 0.261\mathbf{j}')$ N,
$\mathbf{R}_B = 53.6(-0.500\mathbf{i}' + 0.866\mathbf{j}')$ N

CHAPTER 12

12-4 $A = 0.550\ \mu$m, $v_{max} = 1.658$ mm/s

12-6 27.0 ips²

12-8 $f = 0.459$ Hz, $x = 0.2\cos 2.89t$

12-10 (a) 0.500 ft, (b) 32.0 fps²

12-12 0.857 fps

12-14 Amp = 0.763 in., $f = 3.13$ Hz

12-16 $W = 6.52$ lb, $k = 2.67$ lb/in.

12-18 0.459 m

12-20 $\theta = 0.1333\sin 4.5t + 0.1000\cos 4.5t$

12-22 $\theta = \theta_0\cos 3.04t$, $T = 2.07$ s

12-24 6.00 lb

12-26 (b) 12.04 ips

12-28 $f = 0.633$ Hz, Amp $= 4.23$ in.

12-30 $\theta_{\text{wheel}} = -0.219 \cos 5.16t$ rad
$f = 0.822$ Hz

12-32 $T = 0.201$ sec, $\theta_m = 0.1111$ rad $= 6.37°$

12-34 35.8 N

12-36 (a) Yes, (b) 0.466 Hz

12-38 4.02 Hz

12-40 (a) 5.11 Hz, (b) 3.27 Hz

12-42 $(v_A)_{\text{max}} = 2.63$ ips, $f = 0.374$ Hz

12-44 (a) 20 N $\cdot$ s/m
(b) $x = 0$, $\dot{x} = 92.0$ mm/s $\leftarrow$
(c) $x = 1.348$ mm $\leftarrow$,
$\dot{x} = 5.05$ mm/s $\rightarrow$

12-46 (a) Underdamped,
(b) 0.477 Hz, (c) 0.316

12-48 (a) 0.818, (b) 0.1291

12-50 $c_{\text{cr}} = 21.6$ lb-sec per ft, $\delta = 4.19$

12-52 13 cycles

12-54 $2[mf^2(2e^2k - fmg)]^{1/2}/d^2$

12-56 (a) 0.0220 lb-sec/in., (b) 63.2 lb/in.
(c) 2.51 lb-sec/in., (d) 0.0549

12-58 (a) $x = 2.48 \cos 2t$, (b) 4.96 ips

12-60 (a) $\mathbf{O}_x = 24.2$ lb $\leftarrow$, $\mathbf{O}_y = 50.0$ lb $\uparrow$
(b) $\mathbf{O}_x = 29.6$ lb $\leftarrow$, $\mathbf{O}_y = 57.7$ lb $\uparrow$

12-62 (a) $0 < k < 234$ lb/in., (b) 81.8 lb

12-64 (a) 265 rpm
(b) 0.918 in. (out of phase)

12-66 Amp $= 35.4$ mm, B lags A $135°$

12-68 1.588 lb

12-70 (c) Most improvement,
(b) Least improvement

12-72 (a) 88.9 N $\cdot$ s/m (b) 88.9 N

12-74 $\omega > 56.6$ rad per sec

12-76 $x = 0.0357 \cos(3t - 0.697)$,
$R_{\text{max}} = 28.1$ N

12-78 (a) $r\sqrt{2k/I}$, (b) $\omega < \sqrt{2}\,\omega_n$

B-1 684 slug-ft^2

B-3 $I_{xy} = 150$ slug-ft^2,
$I_{xz} = -100$ slug-ft^2,
$I_{yz} = -60.0$ slug-ft^2

B-5 -0.0721 lb-sec^2-in.

B-7 $h/D = 8.62$ (min)

B-9 $(mL^2 \sin 2\theta)/24$

B-11 $\frac{1}{2}m(r_2^2 + r_1^2)$

B-13 (a) 0.611 kg $\cdot$ m^2
(b) 0.396 kg $\cdot$ m^2

B-15 (a) 0.036 m
(b) 0.017 m

B-17 629 slug-ft^2

B-19 11.47 m

B-21 140.7 slug-ft^2

B-23 5.60 mb^2

B-25 0.490 lb-sec^2-in.

B-27 $I_{xy} = mR^2/2\pi$, $I_{yz} = 2mRL/3\pi$

B-29 (a) 0.289L
(b) 0.577L
(c) d

B-31 0.598r

B-33 79.5 slug-ft^2

B-35 88.5 slug-ft^2

B-37 2180 slug-ft^2

B-39 (a) 9.01 ft
(b) 680 slug-ft^2

B-41 150.2 slug-ft^2

B-43 1.217 slug-ft^2

B-45 14.60 in.

B-47 $I_y = 0.1775$ kg $\cdot$ m^2, Error $= 4.05\%$

Index